Yu. V. Riznichenko

Problems
of Seismology

Selected Papers

With 129 Figures and 23 Tables

Translated from the Russian by
M. Volochkovich, G. Leib, O. Odintseva

Mir Publishers Moscow

Springer-Verlag Berlin Heidelberg GmbH

Yurii Vladimirovich Riznichenko †

ISBN 978-3-662-09448-8   ISBN 978-3-662-09446-4 (eBook)
DOI 10.1007/978-3-662-09446-4

Title of the Original Russian edition:
Yu. V. Riznichenko: Izbrannye trudy – Problemy Seismologii

32/3140-543210 - Printed on acid-free paper

Preface

This volume is a selection of the most significant papers Yu.V. Riznichenko wrote during his 30-year career and concerns the development of fundamental problems of seismology. The book is in five parts.

The first is devoted to source seismicity. It starts with a chapter on the "elementary unit" of seismicity, viz. the earthquake source. The size of an earthquake is in itself a significant problem in seismology and is discussed here. The main energetic parameters of the source are considered, namely, seismic energy, magnitude, seismic moment, geometric size of the main fault, displacements along the fault, and stress drop. The main results of comparison, mutual adjustment and correlations between these parameters are presented.

The following chapters are a logical continuation of the first: from an individual earthquake source to a set of earthquakes in space and time, viz. a seismic regime. The chapters convey the now classical principles of seismic regime parametrization and mapping techniques for long-term average seismic activity based on basic principles.

How to determine the maximum possible earthquake is a central question. Both the theoretical (or gnostic) and practical i.e. seismic zoning, aspects are considered. Methods of determining and mapping the maximum possible earthquakes are proposed, in particular, by correlating them with seismic activity and, in general, from seismological and other geophysical, geological, geomorphological and geodetical data sets.

A number of problems concerning the seismic regime are complemented by a theoretical study of time and space variation of a region's seismicity. From this research an energetic model of a seismic regime has been developed that differs greatly from existing models because of its deterministic, continuous and physically justified character.

The last chapter in the first section is devoted to the seismic magnitudes of underground nuclear explosions. Yu.V. Riznichenko was a member of the International Commission of Experts at the Geneva Disarmament Conference. His contributions were concerned with discrimination between the seismic effects of explosions and those of natural earthquakes.

The scientific ideas in the second and third sections are naturally based on the work of the first.

The second discusses the theoretical principles for quantitative assessment of seismic hazard. Practical applications of scientific results specially interested Riznichenko. He introduced a new quantitative parameter of seismic hazard, namely, seismic "shakeability". It fulfils a modern economic requirement for rational engineering in earthquake zones as part of a building code.

Seismic shakeability is directly associated with seismic risk. It is the long-term average frequency of shocks with a given intensity at a given site. The intensity can be measured either on a macroseismic scale or from appropriate physical or engineering characteristics of the Earth's surface oscillations or by the response of a construction given spectral or time-spectral assumptions. Maps of seismic shakeability for intensities of 7 to 9 points on the Soviet scale were compiled for all seismically active regions of the USSR and were used to prepare the 1978 official map of seismic zoning for the USSR.

In the second section, the energetic basis of the traditional macroseismic formulas derived by Kovesligeth, or Blake-Shebalin is shown using a simple correlations and logical postulates. It is shown that this approach can be used to study the anisotropy or structural inhomogeneity of a medium.

The third section is on the seismic flow of rock masses. Riznichenko created the basis of this new trend in seismology. He proposed the concept of seismic flow, meaning tectonic deformation (tectonic flow of rock mass) of the Earth's crust and upper mantle over large time-space domains, resulting from all the earthquakes that occur there. In his analysis of seismic and general seismotectonic flows Riznichenko used a hydrodynamic approach. The theoretical principles for the Earth are given which make it possible to estimate the deformation rates of spatial volumes of the Earth's crust and upper mantle during the seismic flow of rock masses. Estimates are given for the Caucasus, Baikal rift zone, Vrancea area and southern Central Asia. Riznichenko's method is now applied to recent deformations in the crust and upper mantle in most seismically active regions in the USSR.

The fourth section of the book concerns laboratory modelling of seismic phenomena and the study of rock pressure in boreholes by ultrasound. At present investigations by this method have been widely applied and current techniques are improvements on those in the mid-50s. Therefore only the basic principles of the method are presented here. Various ways have been proposed for seismic wave field modelling to study the properties of continuous media. The geoacoustic method is considered in a study aimed at forecasting individual earthquakes in terms of time, location and energy.

The fifth section consists of papers on structural seismology. This was the field in which Riznichenko started his career as a seismologist and he obtained important theoretical results in it over several years. His work on geometric seismology is now fundamental.

The last section explores the potential of the main geophysical methods in the study of the Earth's interior. Riznichenko compares the resolving capacity of each method and explains why seismological methods deserve priority in the solution of one of the main problems of geophysics.

A study of the general properties of travel-time curves for seismic waves, the inverse spatial problem for travel-times of head waves, and the stability of its solution is discussed.

The book is completed by two papers on the theory of seismic instruments.

The book is of interest to seismologists, seismic prospecters, geophysicists, tectonic specialists in earthquake engineering and mining.

Contents

Contents

Yurii Vladimirovich Riznichenko.
A Brief Scientific Biography

On January 1, 1981 Yurii Vladimirovich Riznichenko passed away. A world-renowned Soviet seismologist, he made major contributions to various branches of geophysics. He was an associate member of the Academy of Sciences of the USSR, professor, doctor of physical and mathematical sciences, director of the Laboratory of Earthquake Modelling of the Institute of Earth Physics of the Academy, and the chief editor of the journal *Izvestiya Academii Nauk SSSR, Seriya Physika Zemli*. It will not be easy for Soviet geophysics to fill this gap.

Yurii Vladimirovich Riznichenko was born in 1911 in Kiev, into a geologist's family. In 1935, upon graduation from the Kiev Mining and Geology Institute, he joined the Red Army. In 1938, he took a position at the Institute of Theoretical Geophysics (presently the Institute of Earth Physics of the Academy of Sciences of the USSR), wrote his thesis under the guidance of Academician G.A. Gamburtsev, and was awarded the degrees of candidate of science and doctor of science. At various times he was the head of the laboratory for modelling of seismic events, the section of seismic exploration, the section of earthquake physics, and the laboratory for earthquake modelling. In 1958 he was elected an associate member of the Academy of Sciences of the USSR.

The scope of Dr. Riznichenko's scientific interests was wide, and he made a major impact on the development of seismology. His activities embraced the physical principles of seismic prospecting, interpretation of seismic prospecting data, simulation of seismic wave processes, seismo-acoustic methods for the study of rock pressure, seismic regime and earthquake physics, quantitative methods of study of seismic activity and seismic hazards, and the seismic flow of rock masses. He wrote several brilliant papers with a total of about 200 publications, many of which were translated into other languages.

Dr. Riznichenko's work on seismic prospecting (over 50 publications during 20 years) make him one of the founders of geophysical prospecting in the USSR. The method of time fields, which he developed during the war while on strenuous field expeditions prospecting for oil, constituted the core of his doctoral dissertation, which was completed in the same year as his candidate's thesis (1943). This method has become the main tool for interpreting systems of travel-time functions in refracted wave studies. In the late 1940s, the school of G.A. Gamburtsev, with the active participation of Dr. Riznichenko, developed the correlation method for refracted waves (CMRW), followed by development under Riznichenko's guidance of the method of seismic modelling which was a unique tool for studying the dynamic characteristics of waves in real media.

Besides his applied work Riznichenko was concerned with the theoretical aspects of seismic prospecting. In an early series of articles published in the 1940s, he examined the properties of seismic waves in inhomogeneous stratified media. He presented a wide-ranging analysis of the basic seismic characteristics of a medium, viz. the velocity of wave propagation. This question had an obvious practical importance, and Riznichenko would return to it again during the 1950s and 1960s.

In the forties and fifties, Riznichenko regularly took part in seismic prospecting expeditions. Experiments on how to record and study seismic waves of different types were staged under his guidance. In the Arctic, he investigated the physical properties of the permafrost layer, determining the conditions for applying seismic methods. In East Apsheron and Bashkiria, the possibilities of the correlation method of refracted waves were analyzed in a study of oil fields. In Central Asia, he pioneered the development of a method of high-frequency seismic prospecting for mineral deposits, and later took part in geoseismic sounding studies of the Caspian Sea and in the Far East seas. During these years, he worked in close collaboration with Grigorii Aleksandrovich Gamburtsev.

Work on seismic wave modelling was initiated and directed by Dr. Riznichenko at the Institute of Theoretical Geophysics in 1944-1946, where he proposed and substantiated the ultrasound method of modelling. The first impulse-ultrasound seismoscope was created at his laboratory, and modelling was begun on two-dimensional solid models, which later helped on a great number of important seismologic and seismic prospecting problems. Currently, seismic modelling is the main tool for studying wave fields in complex media that defy mathematical analysis.

Simultaneously, during the 1950s, the seismoacoustic method for studying the stressed state of a rock mass developed under the guidance of Dr. Riznichenko as a response to the need to prevent abrupt blow-outs of coal and gas and mine explosions. The method was suggested by Dr. Riznichenko, who also provided general scientific guidance, visited the mines where experiments were conducted, and attended crucial experiments. Dr. Riznichenko treated mine explosions as a model for earthquakes, so that developments of the stressed state studies for mining led to studies in earthquake prediction, which later produced the method of seismic sounding of focal zones.

The seismoacoustic method was adopted by the mining industry and is now commonly used in field explorations. It is also employed to study the stressed state of a rock mass in hydroengineering construction.

The method of acoustic logging of bore holes was introduced using Riznichenko's ideas around the same time. It yielded a greater degree of detail in seismic observations in bore holes. It is presently an important routine step in mineral prospecting.

After the death of Academician Gamburtsev in 1955, Dr. Riznichenko, replaced him as the Chief of the Department of Earthquake Physics of the Academy's Institute of Earth Physics, and he become involved in studies of source seismology. Riznichenko made an enormous contribution to the concepts and methods of seismic investigations. He spent a great amount of time in fruitful collaboration with a newly organized group of the Institute's Comprehensive Seismologic Expedition, taking part in the development of methods that later became quite common, such as the method of time fields and vertical travel times in hypocenter locations. This work was based on the experience of seismic prospecting and the energy classification of earthquakes. In the late 1950s, Dr. Riznichenko commenced work on the concept of seismic regime stability while developing the method of quantitative description of seismicity. His efforts led to initial results from studies of minor earthquakes in the Gharm experimental field being brought together in *Methods of Detailed Studies of Seismicity*. This was published in 1960 and still retains its value.

Work on the quantitative description of seismicity led Dr. Riznichenko to a major theme of his activities, namely the development of quantitative methods for estimating seismic hazard. Dr. Riznichenko introduced and defined the basic concepts of the theory of seismic activity, maximum potential earthquakes, and seismic shakeability, and suggested the essentially new idea of seismicity mapping based on a regular correlation of the number of major and minor earthquakes. His first publication on the subject appeared in 1958, followed by papers printed in 1962 and 1965. The first maps of seismic shakeability in isolines of average periods of repetition of earthquakes of different intensity, calculated by computer, were published in 1967. This was one year before a similar methodology (calculation of "seismic risk" was started in the United States, and five years before the first results were published there.

Subsequently, Dr. Riznichenko developed a consistent approach for the prediction of long-term seismic hazard based on the analysis of main parameters of a seismic regime and determination of the maximum possible earthquake by comprehensive (seismological, geologic, geodetic, and geophysical) data.

In order to map seismic hazards and evaluate seismic shakeability for zones of high seismic risk in the USSR, Dr. Riznichenko organized in 1969 and became the head of a commission of the Interagency Council on Seismology and Earthquake Engineering of the Academy of Sciences of the USSR. He regularly visited seismically active regions, explaining his methods and extending guidance to specialists and graduate students in academies of sciences in the Union republics and in other organizations in Moldavia, the Ukraine, the Caucasus, Central Asia, Siberia, and the Far East. He brought up a generation of seismologists imbued with the concept of quantitative seismic studies. One result of these efforts was the publication in 1979 of the book *Seismic Shakeability over the USSR*,

and an *Atlas of Seismic Shakeability*. The results of these studies were
used to compile a new map of the seismic zones of the USSR. The method
of seismic zoning was based on the probability of earthquakes of different
shaking intensity and was presented by Dr. Riznichenko in a review on
seismic zoning of the USSR published in 1980.

These studies continued with the introduction into the definition of
seismic hazard, new parameters reflecting the spectral composition of
seismic oscillations. This is particularly important for the design and construction of industrial and housing complexes. Dr. Riznichenko, working
with the State Committee for Construction, gained an understanding of
the needs of structural engineers and, together with his students, developed a theory of spectral-temporal seismic shakeability.

In the early 1970s, Riznichenko began to develop his theory of seismic flow of the rock masses, which was a logical continuation of quantitative methods of description of seismicity, although oriented towards a
better physical analysis of the geotectonic process. He derived expressions for rates of deformation of the seismic flow of rock masses, initially
for the vertical component and later for the general case. In the past few
years, due to the efforts of seismologists in Moldavia, the Caucasus and
Tadjikistan, this laborious and subtle work yielded its first results. The
involved geometry of macrodeformations in large tectonic zones caused
by earthquakes began to be understood, making it possible to correlate
them with geodetic observations.

Along with quantitative methods of seismic description, Dr. Riznichenko developed a physical concept of seismic processes. In 1968, the
first semiqualitative variant of this scheme was published. It was the energy model of seismic regimes, which Dr. Riznichenko subsequently developed with his physics students. The results of this analysis of
spatial-temporal behaviour of seismicity now confirm Dr. Riznichenko's
ideas.

In many papers, especially his later ones, Dr. Riznichenko was concerned with the seismic energy balance in the Earth's interior.

Consistent methodology was a characteristic trait of Dr. Riznichenko's scientific work. The result and practical applications were extremely important. He always strove to formulate his results in a form
applicable to practice, always preferring "engineering" solutions to pure
theory. He was fond of writing popular science and believed that the activities of a scientist should ultimately be oriented towards practice, aimed
at society's and mankind's benefit.

Dr. Riznichenko was deeply involved in education. In 1945-1950, he
lectured at the Moscow Geological Prospecting Institute and in 1950-1957
at Moscow University. He was adviser to scores of graduate students who
wrote their candidate's and doctoral theses under his guidance. His
reviews of dissertations were profound and constructive. Dr. Riznichenko
was a conscientious teacher who gave his students not only knowledge
but also the skills of thorough and demanding scientific research so

characteristic of himself. The style of his own writing combined precision with imagination and pithiness with lucidity.

Dr. Riznichenko took part in international activities. He attended general assemblies of the International Union of Geodesy and Geophysics, congresses and conferences, delivered lectures and maintained regular correspondence with colleagues in the German Democratic Republic, Czechoslovakia, Canada, France, Britain, the United States, and other countries. He took part in the Geneva Conference on the nuclear test ban, and later in a project on seismic risk assessment in India, Japan, Turkey, and Rumania. He was the first Soviet scientist to be elected, in 1957, vice-president of the International Association of Seismology and Physics of Earth's Interior. His travels abroad promoted the prestige of the Soviet science and the strengthening of international contacts with scientists of other countries.

Riznichenko's non-scientific activities were also wide and diversified. He was a participant in the Pugwash Conference of Scientists for Peace, a member of a section of the Committee on Lenin and State Prizes of the USSR, and for several years a member of the Bureau of the Section of Geology, Geophysics and Geochemistry of the Presidium of the Academy of Sciences of the USSR.

From 1962 until his death, Dr. Riznichenko was the chief editor of the journal Izvestiya Academii Nauk SSSR, Seriya Physika Zemli, contributing a great deal of energy to this work. His attitude to papers submitted for publication was attentive and objective and personal relations with the author were irrelevant, although he was gentle and friendly in his communications to them. He was fond of saying that every author tends to believe that his paper is the most important. He devoted much attention to the quality of published material, and wrote long editorial notes or discussed the work with authors. Dr. Riznichenko was himself a prolific author and wrote regularly for the journal. His manuscripts, while fully prepared for the print were always submitted to the routine reviewing and be frequently put his reviewer's suggestions to use.

Dr. Riznichenko's outstanding activities received official recognition. He was twice awarded the Order of the Red Banner and a number of medals. Yurii Vladimirovich Riznichenko was a principled, broad-minded person, indefatigable worker, dedicated to the cause of knowledge, a life-loving and vigorous person, invariably well-disposed and considerate with people. His ardent love for science was contagious. In difficult circumstances, people who had the privilege of working with him always received a helping hand.

I

Source Seismology

As long as human society has existed, people have regarded earthquakes as a hazard. In recent decades, it has turned out that besides their adverse impact on human life, they provide us with an endless source from which we can gain knowledge about the globe, namely its structure and internal motions. This knowledge can be used for the benefit of human society. Let us consider the general ideas and evolution of this development to appreciate it.

A study of the structure of the Earth is first geoscience, and one in which the seismic method gained the first major results. Seismology is now a major geophysical method in this field. It resulted in the development of *structural* seismology, to which seismic prospecting is related. Physically, a relatively short-period elastic seismic wave penetrates the Earth's interior more easily than electromagnetic waves as it is more transparent to them. They carry more information than electromagnetic waves, for which the Earth's interior is quite opaque, and more information than quasistable electric, magnetic, gravity or heat fields, whose resolution capacity is lower than rapidly changing wave fields.

The second field of study—the Earth's motion—is the domain of source seismology. Potentially, it too has great possibilities, but it began being developed later than structural seismology nor has it yet matured. This field of knowledge is now undergoing a rapid development, even exponential growth. This process is consistent with general tendencies in present-day world geophysics.

Now, source seismology and the physics of earthquakes face problems typical of every developing branch of the natural sciences. Besides the provision of observational data with the necessary quantity and quality—the permanent problem — there are others, namely establishing the typical, mass, and average fundamental ratios between the main properties of each phenomenon and their parameters; developing qualitative assumptions about their physics; formalizing these assumptions, that is transfering them into quantitative assertations, so as to make theoretical, mathematical models, which are the basis for any properly developed physical theory; to provide for a gradual transfer from an empirical to theoretical and physical methodology of the study. Note that besides the establishment and explanation of typical reference average functions, one should also study deviations from them. It is the latter that helps us penetrate more deeply into the essence of any natural object, infinitely complex in its origin, which can be just poorly depicted by any theoretical model.

The study of deviations helps to determine the degree of approximation of the simulated model to the reality, to find its weak points and ways to improve them, probably by sophisticating the model, by introducing elements which can add to the understanding of those factors, the effects of which were first ignored.

In the past, when methods of earthquake prediction were not developed, estimation and spatial distribution of long-term average seismicity and seismic danger— seismic zoning—were the only preventive measures from earthquakes that people knew. This aspect of the problem will not lose its acuity in the near future, even when forecasting of individual earthquakes in time, will be a reality. Buildings and structures are erected for many years, sometimes centuries. They should be able to resist individual natural events irrespective of when they may occur.

We hope that our readers will not only find something new and educational, but also something which will cause doubts, make them think harder, stimulate research in geodynamics, using the methods of source seismology and the physics of earthquakes, a field which is now truly begun. We are sure that seismology will lead both to the solution of new problems in geodynamics and in more traditional and simple problems of the Earth's structure.

1. Earthquake Evaluation

1.1. Earthquake Characteristics

An earthquake's value characterizes its focus. The effect of the source impact on the Earth's surface is characterized by its intensity. We must assess an earthquake's source value from observations of its effect on the Earth's surface. Therefore, it is difficult to refer to the value of an earthquake without mentioning its intensity. However, in our study we shall emphasize the source.

An earthquake's source value may be assessed by the following parameters: seismic energy; magnitude; seismic moment; geometrical dimensions of a source, namely its length, width, height, volume of the main shock or its area, and the fore- and aftershocks; the size of the main fault, namely the length, width, displacement along the main fault; the same characteristics of the accompanying faults; the total earthquake energy released in the source; the total, released and effective stresses and strains in the source area; the rupture velocity etc. This list of source parameters affecting its value could be augmented by parameters which determine the orientation of the elements, since the latter affect the details of intensity distribution over the Earth's surface.

The spectral-temporal and statistical features of seismic oscillations which emanate from the source are directly related to the above characteristics of the source. These features also determine the main seismic parameter of the source, that is the seismic energy. Spectra seem to close the logical circle of the definition of the source properties which characterize its value.

To complete our description of the problem, we mention two more aspects. The first is simpler. It is the location of the hypocenter (determination of an epicenter and focal depth), and also the time origin of the source. The second is more complex, viz., the physics of the source, in particular its mechanics, and the set of sources that in fact form both an "individual" complex source in a large earthquake and in the seismic regime as a whole. In turn, the seismic region is related to the seismic flow of the rock masses and is a part of their rupturing or continuous tectonic flow, and the entire geodynamics of the area.

The range of problems concerning earthquake value is unlimited. We shall select some elementary ones which have acquired a more or less definite structure. Let us recall the past, look at the present and think of the future.

1.1.1. Energy

Seismic energy. Seismic energy E is the energy of elastic seismic waves radiated by the source. B. Golitsyn first introduced this notion (1915). Essentially his approach started from the energy density of the surface seismic waves per unit of the front length as determined at the point of observation. This value is integrated along the entire front, with its centre at the focus being considered a "point" source of seismic radiation with circular symmetry. As a result and given wave attenuation in the medium, we have the energy of the source emitted as waves of a given type. This idea can easily be generalized to bulk waves propagating from the focus when it is considered a source of seismic radiation with spherical symmetry.

This elementary scheme is complicated by the necessity to consider first the inhomogeneity of the medium through which the waves propagate, and also the attendant circumstances, such as changes in the direction and amplitudes of oscillations at various points on the Earth's surface, etc. Besides, the earthquake source, in fact, has no circular or spherical symmetry— a deep blast under the Earth's surface could be a closer concept. The asymmetry of the source radiation is partly discovered by determining the energy at many stations surrounding the epicentre, and by averaging the results.

Golitsyn's approach is now the basic one when determining the energy of a source at middle distance away, i.e. at large distance compared with its geometrical dimensions, and small distance compared to those where the energy is strongly distorted by the Earth's structural inhomogeneity, which is difficult to explain. In fact, this method is applied in practice without these restrictions, since it is difficult to observe them.

Earthquake energy E_0. Another local approach to the problem of earthquake source determination is to treat it as a fracture and dislocation in a stressed elastic medium. Reid started this approach with his theory of "elastic rebound". There we consider the energy of an earthquake to be the difference between the potential energies of the medium before and after the fracturing and dislocation. Part of this energy is spent on non-elastic processes in the source, namely the destruction of material, friction along the rupture, plastic deformations, physicochemical transformations, and also to the work of mass transportation in the gravity field. More of the energy is spent on forming seismic waves; it is transferred as seismic energy of the source. Although the seismogenic potential energy of the medium is mainly elastic, it is also gravitational. The medium from which the potential energy comes is the Earth as a whole. However, the further away the fault, the less energy comes from it (the St. Venant principle in the theory of elasticity). Therefore, in practice, the reservoir of earthquake energy is only the local volume around the rupture, dimensions of the volume being of the same order as the length of the rupture: 98% of the released energy comes from an area whose diameter is five times greater than that of the rupture [C. Archambeau]. Given the "confidential" proportion of the energy, it is possible to determine the approximate dimension of the source as of the effective supply of the earthquake energy, in particular the seismic energy.

Thus, Golitsyn's wave seismic energy of the source is only one part of the Reid released potential energy spent both to form seismic waves and for non-elastic processes in the source. How great is a fraction? This question was debated many times, arising from data on the observation of earthquakes and seismodislocations on the Earth's surface, from the study of rock blasts in mines, and also from laboratory experiments on the destruction of rock samples. For shallow crustal earthquakes, when there is no considerable remelting of the material near the rupture, the "earthquake efficiency" seems to be only some decimals of a percent of the total earthquake energy. When partial melting occurs along the rupture in the crust or, in particular, under the crust where the melting could be higher the "earthquake efficiency" is not much greater, thus reaching the total potential energy of the source; its value depending directly upon the value of an earthquake [Riznichenko, Wyss]. What is the share of seismic energy in the total earthquake energy in each individual case? This is one of the most complicated and important theoretical problems of seismology.

It is much easier and secure to measure the seismic energy of the source than to determine the released potential energy, geometrical dimensions of the source, or displacements along it and so on. If we admit conventionally that an earthquake's value should be characterized by a single figure only, we would not find a better physical value than seismic energy. However, this scalar characteristic alone does not seem to suit many of the demands of theory and practice.

1.1.2. Magnitude

Definition. At present, it is a conventional index of the seismic energy of a source. It expresses the earthquake source as value in the form of the intensity of seismic oscillations at a standard distance r_0 from the source. Ch. Richter invented the term magnitude and B. Gutenberg developed the notion. The magnitude is usually defined by the amplitude A or the amplitude-to-period ratio of the oscillations (A/T) in a definite wave $r_0 = 100$ km from the source, the oscillations being record by a frequency-selective standard seismograph. When oscillations are recorded at different distances r from the source, as is usual, the magnitude is reduced to the standard distance r_0 by "calibration functions" in the form of graphs or tables showing the average empirical dependence of the magnitude for crustal earthquakes as a function of distance r, and, in the case of subcrustal events, as a function of source depth h. The set of definitions and functions which allow the magnitude to be estimated from observations is called a "magnitude scale". We speak about a particular magnitude scale in relation to the wave type considered, the oscillation component, the value in question (A or A/T), the type of instrument, or some other factors such as region or even author. There are several dozen such scales and new ones continue to appear. Sometimes, they are mutually inconsistent and getting scales of different magnitudes to agree has become a major international problem. Gutenberg allocated a physical sense to magnitude [Gutenberg et al] when he related it to the source's energy. This is possible by correlating the magnitude with the energy defined by the Golitsyn method at middle distances from the source.

Rautian's energy scale $K = \log E$ (E in Joules) [T. Rautian] is widely applied in the USSR, and is a definite form for the magnitude correlation with the energy. This scale is based on the same elements as any other magnitude, and like a magnitude scale it starts from the energy density ε at the observation site, or from the amplitude A, or from the sum of amplitudes $A_P + A_S$, or from the A/T ratio. An empirical calibration function of distance r is plotted. The Rautian scale takes $r_0 = 10$ km as the standard radius for the reference sphere. Standard seismometers are named VEGIK and SK (Short- and middle-period Kirnos). The magnitude is transformed into seismic source energy E by integrating the energy density ε over the surface of the reference sphere or over the surface of the source, the radius of which is a direct function of the size of earthquake.

The same logic is also followed in the system of macroseismic descriptive intensities. Here the seismic scale intensities I are taken to be functions of earthquake source value K or M, the source depth h, and the epi- or hypocentral distance r. The system can be adopted to the requirements of seismic zoning and earthquake resistant design and construction. The intensities are expressed by macroseismic field equations derived by Shebalin (Sen.) [Shebalin]. The well-known macroseismic formulae by Köveslighethy, Blake, et al [Medvedev, Sponheuer] came, respectively, before and after his correlation between the macroseismic and instrumental intensities as functions of the distance r (like magnitude calibrating functions) was studied by Djibladze [Djibladze].

Magnitude has been the major index of earthquake source value, for some decades, since 1935 despite all its defects, starting from the pioneer research of Richter. Seismic energy seems to have taken second place. So why has magnitude been more successful than energy?

Advantages. The success of magnitude arises primarily because two different problems can be distinguished, namely, the determination of earthquake's source value and the determination of the Earth's properties along the wave path from the source to the observation point. A knowledge of the source-to-point properties of the medium is substituted during magnitude estimation by empirical intensity/distance relations without any interpretation of the latter; such an interpretation can be made separately if necessary. When the seismic energy is to be determined, a knowledge of the structure and properties of the source-to-point medium or, at least a definite model, is absolutely necessary. When the model is not perfect, source energies determined from observations at different distances, and in case of horizontal inhomogeneity even from different azimuths, will agree poorly between themselves and also not be consistent between themselves or near-field measurements [Riznichenko et al]. This problem is partly considered in [Riznichenko].

A general strategy of the way out of the embarassing situation is to combine both approaches, the energetic one and magnitude one, as it was realized in [Riznichenko, Rautian]. The magnitude approach should be applied to rather long source distances, which enables us to reduce the observed intensities to those eventually observed at relatively short distances, for instance, at the standard distance r_0. The direct or reduced near-field observations can then be used for an energy calculation using the Golitsyn's scheme. Data on structure and the properties near the source can be determined by additional applications of the seismological and gravimetric methods, DSS etc.

It should be noted that the restrictions on the distance to the source remain those on the application of Golitsyn's method, being for small earthquakes (about $K < 10$) several kilometers, for intermediate ones ($10 < K < 15$) several tens or hundreds of kilometers, and for large ones ($K > 15$) several hundreds or thousands of kilometers. In the latter case we are in the range of teleseismic distances, where the calculations are problematic. This has caused difficulties when making seismic energy calculations for large earthquakes.

Defects. There are two main defects to the magnitude method. The first is its tenuous, inferential character without a functional relation to a distinct objective physical parameter of the earthquake source which does not depend on the measurement techniques. The second is its unidimensional scalar character.

Seismic energy E, seismic moment M_0 and various other scalar characteristics of the earthquake source have deprived this defect in principle. But the second defect is inherent in all scalar physical values.

Since the magnitude is conditional because it is determined according to a scale we cannot ask the obvious question, that is which kind of magnitude is the best earthquake source value. We cannot ask the question unless we agree, in advance, the distinct meanings of the terms, seismic energy, destructive economic effect etc. While it is important to have an inherent convergence of the results obtained by the particular scale (e.g. M_{LH} or m_{pv}), this does not answer the principal question. We can find magnitude with a better inherent convergence but it also has a weaker, more scattered correlation with the value which we want to consider, intuitively or consciously, as the earthquake value, e.g., the seismic energy. A magnitude based on a scale can behave stably, showing little variation with changes in uncontrolled factors. However, it may also show only slight variation with the factors we know to be decisive in estimating an earthquake's source value.

Seismologists in many countries, beginning with Gutenberg and Richter have expended much effort on adjusting the different magnitude scales and in establishing a unified integral scale, to which all other scales may be referred. This activity continues, but with only mediocre success. The reasons are clear. The failure arises because each magnitude is dependent on many factors, producing an immense number of combinations. We can select only one kind of magnitude as basic and this, as mentioned above, cannot be solved within the framework of the magnitude idea in principle.

A possible approach is to correlate each kind of magnitude with an exact physical source parameter that can represent the size or value of the earthquake, for example, the seismic energy measured for a fixed range of oscillation frequencies. The closeness of the correlation for magnitude could be used as the criterion of its quality. Such a comparison could also solve the complicated problem of how to correlate different magnitude scales. This would be done by finding a magnitude-independent technique for determining the seismic source energy. But then, what need would there be for magnitude?

Now to consider the second defect of magnitude, its scalar nature. An earthquake is an intricate train of events related both to the source and to its manifestations at the Earth's surface. These must be presented in the form of simplified models, necessary to understand the phenomenon better, to present it in an appreciable form, and to compare it with observations. Moreover, the consistency with observation is essential so that the models can be used for other problems of theory and practice. Earthquake models are more complex, the greater the amount of differentiated data that must be obtained, the more complex the earthquake models.

Many of the facts so far accumulated, both scientific and practical, demonstrate that one scalar value is insufficient either to treat the effects within the seismic source or to discuss its impact at the Earth's surface. The scalar approach leads to many contradictions, many solvable by a spectral approach.

The disadvantage of the scalar approach is inherent in techniques involving magnitude, seismic source energy, shaking intensity, energy density at the site, maximum accelerations or velocities etc.

Before studying movements within the source and at the site (i.e. shaking intensity), we consider another important source parameter. This is the seismic moment M_0 which also has, in its simplest form, a scalar nature.

1.1.3. Seismic Moment

Seismic moment M_0 is the long-period ($T \to \infty$) property of the dislocation caused by the rupture in the medium that remains after the earthquake, viz. $M_0 = \mu S D$, where μ is the shear modulus, S is the rupture surface, and D is the finite displacement along the rupture (averaged over the surface).

The idea of an earthquake source, like that of a dislocation, arises from field observations of ruptures exposed at the Earth's surface. Reid started to analyse this aspect in 1910 and later Kasahara generalized the idea in 1957. The length L of the rupture and the displacement

D along it can be determined directly from geological and geodetic data. If, from geological arguments, we assume a depth for the rupture then it is possible to determine its width W, then its area is $S = LW$ and the geometrical moment is SD. The modern theory of seismic moment was developed due to the efforts of many scientists, namely Eshelby, Stekethy, Knopoff, Chinnery, and Maruyama. For some time the long-period static strain field near the dislocation in the elastic medium was considered as the geometrical moment. In 1966, Aki first effectively applied the seismic moment M_0 to the processing of seismological observations by developing the theory of moment calculations from the long-period part of the spectrum of the surface waves. He also associated seismic determination with field data. Later, seismological determination of the moment and individual dimensions of the rupture: L, W, S, displacement D and stress drop $\Delta\sigma$ at the rupture were made from the body wave spectra, while Brune determined them from shear waves. Wyss and Hanks determined them from longitudinal ones. The seismic moment has now been determined for many earthquakes and a correlation established between the moment M_0 and magnitude M [Ryall, et al].

The data presented by Aki for world events with magnitude $M = 5$ to 8.9 (e.g., the events of Sanriku, 1933; Alaska, 1964; Tokachi-Oki, 1968; Kuril Is., 1963; Niigata, 1964; Spain, 1964; Parkfield, 1966), those presented by Wyss and Brune [Ryall] for earthquakes with $M > 2$ in the western United States, California, and along the San-Andress fault, and those presented by Ryall, et al. for micro-events in Nevada with $M = 0$ to 2, were used to obtain a rough average correlation function for M and M_0 (dyne·cm = erg = 10^{-7} J):

$$\log M_0 \pm 0.6 = 15.4 + 1.6M = 11.8 + 0.9K \qquad (1.1)$$

The confidence interval of ± 0.6 (Fig.1) contains approximately 70% of the observed points $(M, \log M_0)$. It is interesting that such a log-linear function covers such a large interval of measured magnitudes, viz. from $M = 0$ to 9; thus, it covers over 15 orders of seismic energy E.

Randal and Kostrov (in more complete form) generalized the theory of seismic moment and determination of the spatial location of the rupture and displacement along it by introducing a seismic moment tensor

$$M_{0ik} = M_0(b_k n_i + b_i n_k) \qquad (1.2)$$

where b_{ijk} and n_{ijk} are column vectors in the direction of the residual displacement and normal to the rupture surface. This is a deviation in the tensor: $M_{0kk} = 0$. The scalar seismic moment is the principal value of the seismic moment tensor M_{0ik}.

At present, the main way to determine the seismic moment M_0 (or M_{0ik}) is seismological—from the spectra of surface (Rayleigh and Love) or body (P, S) waves [Aki]. It can be also determined from strong motion accelerograms [M. Trifunac]. The moment M_0 is proportional to spectral density A_0 in the low frequency range (as $f \rightarrow 0$) of the amplitude displacement spectrum. It can also be derived from the spectra of the record taken at one station after corrections are made for the frequency characteristics of the instruments, the source radiation pattern (determined by observations from a number of stations and by determining the geometrical parameters of the source) given wave divergence and absorption between the source and observation point [Riznichenko et al]. The moment tensor M_{0ik} with its principal value is determined from observations of the low-frequency $f \rightarrow 0$ parts of the spectra at a number of stations [Kostrov]. In theory, we can certainly determine the seismic moment and other source parameters related to residual displacements from other observa-

Fig. 1. Correlation between seismic moment M_0 and earthquake energetic values
K, M.

Instrumental determination of M_0: *1*—large
global earthquakes (after Aki); *2*—average values
for individual regions; *3*—earthquakes in the
western USA (after Wyss and Brune);
4—microearthquakes in Nevada (after Ryall et
al); *5*—M_0 determinations from field data. *Heavy
line* is the averaging according to Eq. (2.1);
dashed lines are the limits of the 68% confidence
interval; the curve is Aki's model; *thin lines*
represent the stress drop $\Delta\sigma = 0.1$ and 100 bar.
Numbers on the graph denote individual earthquakes: *1-14* after Aki, *15* after Shteinberg et al.;
1—Sanrikum, Japan, 1933; *2*—Alaska, 1964;
3—Tokachi-Oki, Japan, 1968; *4*—Iturup, Kuril Islands, 1963; *5*—Rat Island, 1965; *6*—Niigata,
Japan, 1964; *7*—Spain, 1964; *8*—Parkfield,
California, 1966; *9*—Thrakia, Greece, 1966;
10—Illinois, USA, 1968; *11*—Aleutian Islands.;
12, *13*—Northern Atlantic; *14*—Azore;
15—Daghestan, Caucasus, 1970.

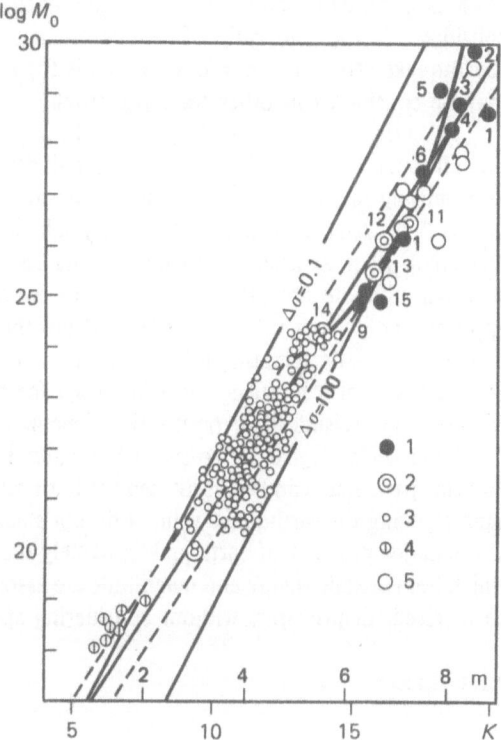

tions. These allow them to be detected, that is to record zero frequency of oscillations
$f \rightarrow 0 (T \rightarrow \infty)$, by repeated geodetic measurements, observations by strain seismometers or
strainmeters, and tiltmeters.

Merits and demerits of M_0. The main merit of the seismic moment M_0, compared with
magnitude M in its various modifications, is that it has a clearly formulated physical value,
also true of seismic energy. In some aspects it is even better than seismic energy since it
preserves its meaning in the near field zone, for which seismic energy cannot be defined
clearly.

However, the seismic moment has the same basic defect as the magnitude and seismic
energy have. It does not contain a frequency dependence. This defect remains even when
a tensor seismic moment is introduced. Since the seismic moment is only related to the residual
displacement, it has no direct relation to the seismic energy in the frequency range where
$f \gg 0$. These frequencies are responsible for Earth tremors, building destruction and the effects on the human senses. Thus, "slow" earthquake or quick creep may generate a remarkable
moment M_0 with a negligibly small magnitude M and energy E of the shocks in the seismological, engineering frequency range.

It is interesting that energy is the dimension of seismic moment. It is the energy necessary
to shift the sides of the rupture at the force due to friction stress, equal to shear modulus.
This theoretical friction is much greater than the actual one, the difference being approximately the same as between the physical and technical strength. Therefore, for the main rupture

$M_0 \gg E$ in all cases. However, E may approach M_0 due to high-frequency oscillations not considered in M_0. The ratio of seismic moment to seismic energy is closely related to the "earthquake efficiency" $\eta = E/E_0$ which is the analog of technical efficiency. This problem may be considered in other ways [Kostrov].

Recently, seismic moment has been of interest to seismologists both in theory and practice, although its application has not been completely understood. It seems, however, extremely promising in solving problems on the seismic flow of rock, though, not yet used widely in this field. A general quantitative approach to the problem was formulated in 1965 [Riznichenko], when an attempt was made using assumptions about the seismic energy and conventional ultimate strength. This was on the eve of a new era of seismic moment [Aki]. The application of "moment" to the calculation of the total flow of rock masses, sets of ruptures, and earthquakes originating in one rupture plane, was a success [Brune]. We hope such an approach will also be successful when used for finding the spatial distribution of a set of sources when seismic flow occurs [Riznichenko].

Some seismologists are so enthusiastic about the concept of moment M_0 that they hope that this parameter can be substituted for both magnitude and seismic energy as a parameter characterizing the earthquake value. I do not share this hope as M_0 has a weaker connection to the major property of earthquakes, namely the "shaking" effect at high frequencies, with which both seismic energy and magnitude are associated. I believe an earthquake value cannot be assessed, in principle, without considering spectra.

1.1.4. Spectra

The spectral analysis of the oscillations $S(\omega)$ during earthquakes has, alongside the study of the form of oscillations in time $f(t)$ (their wave-forms), been a traditional branch of general and engineering seismology. We shall consider the basic types of spectral presentations.

Types of spectral presentation. The predominant periods visible in the seismograms [Riznichenko] were usually used to characterize the frequency of seismic oscillations. Spectra were plotted as the distribution of visible periods. The computerization of science means we only need such primitive, manual techniques for rough estimates.

At present, two major spectral presentations have been developed in seismology, both general and engineering. The first is the Fourier spectra, and the second is response spectra. In both cases we consider the response at the output of the simplest oscillation system to the input impulse. However, in the case of Fourier spectra, the system has zero damping, whereas the amplitude at the output at each frequency is recorded only after the impulse at the input is over. In the case of response spectra, a system has several fixed values of damping (usually they are small, precritical, and zero), and the amplitude at the output is recorded when it is at its maximum; this may not necessarily be at the end of the impulse at the input.

No unambiguous direct correlation exists between amplitude Fourier spectra or any response spectra. It goes through the impulse form at the input $f(t)$ or, what is equivalent, through its complex Fourier spectrum (amplitude and phase). However, for the common forms of seismic oscillations $f(t)$, the amplitudes of the Fourier spectrum A_F usually differ slightly from the amplitudes of the corresponding response spectrum at zero damping ($A_F \leqslant A_0$); this is clear from many comparisons [Riznichenko et al]. Therefore, it is possible to approximate one spectrum the other. Applying this procedure should help seismologists and engineers to come into closer contact, which is important for both parties.

Engineers prefer to use response spectra since they are primarily interested in the maximum oscillation amplitudes of a system simulating a construction. Seismologists prefer to describe ground motion using Fourier spectra (this is especially true in western countries), as the latter are commonly accepted in science and in many fields of engineering. In the USSR, seismologists seem to use response spectra as widely as Fourier spectra, although they show more interest in Fourier spectra. In the USSR, the response spectra are popular in general seismology because of the widespread use of analogue instruments and spectral studies using frequency-selective seismic stations (FSSS) popularized by Zapol'sky [Vinogradov et al]. These instruments yield response spectra more easily than Fourier spectra, which can only be determined by manually plotting response envelope oscillations; "FSSS spectra" are response spectra. The FSSS technique partially compensates for the insufficient attention paid in Soviet seismological practice to magnetic records or play-back records of seismic oscillations and their computer processing, and also for inadequate digitizing of ordinary photo-seismograms and their processing by computer.

FSSS seismograms can be easily handled in a manual spectral-temporal analysis (moving window analysis). Such an analysis may be carried out by computer or by a digital or analog filtration of broad-band seismograms. Whatever the method, we have either a response envelope spectrum or a moving window spectrum [Trifunac].

The strategy of spectrum generalization. Given the numerous factors affecting the oscillation spectra of earthquakes, it is best to select a small number of "major" ones and to determine their average functions of the spectra, and to ascribe the combined effect of the other factors to random scattering analogous to scattering caused by measurement errors. Factors which have been considered important include the value of the earthquake (or blast), viz. its magnitude, seismic energy, or seismic moment; the focal depth viz. crustal, or subcrustal; the distance between the source (hypocenter) and the point of observation (which will include in some indirect form the structure and properties of the medium along the wave paths); and geological and local soil conditions in the recording zone. The mean spectrum system can be constructed as a global, regional or local one. It can be constructed for the waves bearing the most of the energy (shear waves, as a rule), or individually for each major wave type (shear, longitudinal, or surface), or lastly, for each wave separately.

Two main approaches have been established to assess experimentally how spectra depend on the major factors. The first or differential method is to study how the effect of one factor determines all the rest. The same procedure is repeated for each factor in turn [Riznichenko et al]. Then, an empirical "theoretical" model may be simulated to account for the effect of all factors simultaneously. The second or integral approach is to set up initially a general flexible theoretical model which accounts for the combined effect of all the major factors and which contains a set of constants. The values of this set of these parameters is determined so as to optimize the results between the estimates from the model with the observed data [Riznichenko].

Both approaches are predominantly empirical. In addition, there is, as in other fields of research, a theoretical approach which may be applied to the spectra of seismic oscillations; in this case we have the dynamic elasticity theory and the continual theory of dislocations; the direct design of the shape and spectra of the oscillations arising from the dimensions and orientation of the rupture and displacements along it, the rate of rupture, etc. [Balakina et al]. Some theoretical approaches may be used when compiling and using an empirical model, e.g. an account of the rupture orientation and displacement if any [K. Khattri] and an account of the effect of local conditions by calculating the frequency characteristics of the sequence of superficial layers [Plotnikova et al]. Some parameters of the dislocation or other theoretical model may be fixed in a general or numerical form and in the basis of an empirical theoretical model. Then, the general approach will become a "semi-empirical" one and this seems to be the most efficient.

The first experimental approach, the differential one, is typical of much previous work. It has become almost traditional and is still used. The second, integral, is comparatively new in its application to seismic spectra. It seems to be more promising. In fact, to obtain each result by the first approach it is necessary to take a comparatively small sample of the observed data. When the data is widely scattered it considerably reduces the stability of the results. There are more serious defects in the individual approaches. The second approach is free from them. The technical advantages of the first approach are indubitable, viz. the simplicity of the individual estimates. This was important when data was manually processed. Now, in the age of computers this advantage becomes unimportant, and we can turn to the second approach.

The potential of the spectral or, to be exact, the spectral-temporal approach in seismology is considerable both in theory and practice. In theory, the principal advantage comes when determining the value of an earthquake at the source from seismological data: i.e. seismic energy, moment, size of rupture and displacement along it, rate of rupturing, stress drop; the fundamental papers on this topic and on the theory of cleave rupturing in general are [Aki et al]. In practice, spectral analysis is gaining acceptance in seismic engineering [S. Medvedev, Trifunac] and is dominant technique when determining the degree of seismic danger with physical and engineering indices of oscillation intensity and probability, i.e. when calculating the spectral-temporal shakeability [Riznichenko].

It seems reasonable, therefore, to construct a fundamental spectral-temporal system of seismic data, that is a global system similar to the travel-time system of Jeffreys and Bullen or the magnitude system of Richter and Gutenberg. The system would combine the spectral-temporal characteristics of seismic oscillations at various distances from the source with the spectral-temporal characteristics of oscillations in the source as the origin of the oscillations. The compilation of such a system has already been initiated [Riznichenko]. This system should include magnitude, seismic energy, and seismic moment as special cases which can be derived from it.

1.1.5. Source Parameters

We shall only consider those parameters of the source we listed which are directly related to the earthquake value. Three (E, M, M_0) we discussed above. Now we shall describe the others, those parameters which determine the geometric dimensions of the source and stress drop.

The geometric dimensions of the source are the mean radius R or diameter $2R$, length L, width W (vertical extension in case of a vertical rupture plane), area of the rupture surface S and displacement D along it. These characteristics have not all been completely formalized, nor indeed has the concept "source". However, we should remember that the order of magnitude, whereas the precise nature of the notion "source" depends to a great extent on the character of the data and the hypotheses used to process it.

Many papers have discussed various estimates of the geometric dimensions of the source. The basic initial data and many ways of discussing them are the first direct observations of tectonic ruptures exposed at the Earth's surface due to large crustal earthquakes [Frorensov et al]. The second set of data arises from observations of the size of the aftershock area, conventionally identified with the source area of the main shock [Riznichenko et al]; a comparison with the surface ruptures verifies the validity of this approach. The third set of data arises from assumptions about the ultimate strength of the material in the source area at the values, say of the seismic energy, strength of the material, etc. considering also the data from the first and second groups [M. Gzovskii]. The fourth set of data is based on geodetic measurements of ground motion and tilts due to earthquakes, and their comparison with the theoretical statistical dislocation model [M. Chinnery et al]. The fifth set of data comes from the frequency characteristics of oscillations during earthquakes: the dominant periods [Riznichenko et al] and oscillation spectra [Balakina et al].

Applications of the various methods, those mainly concentrating on methods one, two and three, are reviewed in [Riznichenko et al], while method five in its modern spectral form is reviewed in [Chinnery].

At present, spectral methods are the primary technique. Field geological and geodetic methods also remain important as spectral estimates must be correlated with the data. Methods based on ultimate strengths are now outmoded in this science. Now, the strength characteristics of the medium in the focal zone can be found from the seismological spectral data itself. Strength characteristics such as stress drop on the rupture surface $\Delta\sigma$ is determined simultaneously with the rupture dimensions, for instance, its mean radius R and displacement D. Table 1 gives rough estimates of K, M, M_0, R, L, W and D, which may be regarded as averages obtained from many references on many events mainly by spectral methods.

Table 1. The Comparison of Some Energetic Characteristics of Earthquakes with Source Size

M	3	4	5	6	7	8	9
K	9	11	13	15	17	18	20
log M_0 (M_0 erg)	20	22	24	25	27	28	30
R, km	0,5	1	3	10	20	50	150
L, km	1	5	10	50	100	500	1000
W, km	0.5	1	2	5	10	20	50
D, m	0.001	0.01	0.05	0.2	1	5	50

Stress drop. The mean shear stress released at the rupture in the source, or in other words the "stress drop" $\Delta\sigma$ can be determined from moment M_0 of the earthquake and the geometry of the rupture surface. With a circular rupture of radius R, the stress drop can be determined [Brune, Hanks.] on the basis of T. Mikumo, Molnar formulae in [Ehelby, Haskell] by using the following simple ratio

$$\Delta\sigma = 7/16 \ (M_0/R^3) \tag{1.3}$$

whereas in the case of a rectangular rupturing pattern, *LW*, it becomes ratio [Chinnery]

$$\Delta\sigma = \frac{(2M_0/3\pi)(4/L^2 + 3/W^2)}{(L^2 + W^2)^{1/2}} \tag{1.4}$$

For equal rupture areas $\pi r^2 = LW$, the $\Delta\sigma$ calculated from the first formula is approximately twice that the second one [Mikumo], that is within the usual accuracy for determining the value.

The stress drop for crustal sources determined by spectral analysis was from several units to several tens of bars. On average, $\Delta\sigma$ increases with focal depth *h*. A review by Mikumo (Fig. 2), based on personal definitions and derived from work by Brune and Allen for crustal sources and Berckhemer and Jakob, Fukao and others for intermediate and deep undercrustal sources, shows that the mean ratio for earthquakes with $M \geqslant 5.8$ (Mikumo did not risk writing out the formula because of the wide scatter of data):

$$\log \Delta\sigma \pm 0.3 = 1.68 + 0.0022h \tag{1.5}$$

The corresponding averages are approximately:

h, km	0	100	200	500	600
$\Delta\sigma$, bar	50	80	150	600	1000

The width of the confidence band corresponds in approximately 70% of cases: it includes 19 of the 27 points given by Mikumo in his review. Note that the error $0.3 = \log 2$ is double $\Delta\sigma$. This uncertainty is of the same order as that of an indefined knowledge of the rupturing surface geometry.

An increase in the stress drop $\Delta\sigma$ with depth *h* at the same magnitude *M* and moment M_0 of earthquakes might be caused by smaller area of rupture, smaller displacement and greater friction, under conditions of greater hydrostatic pressure at depth [Mikumo].

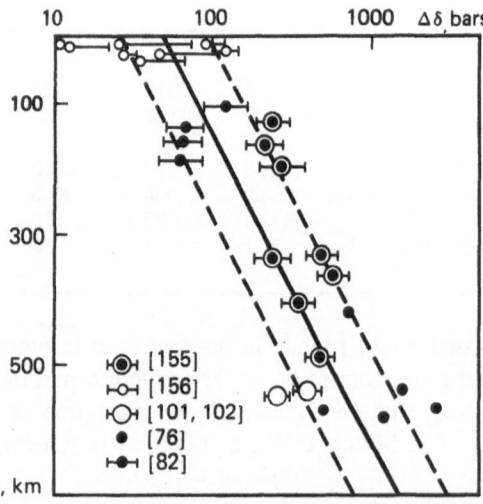

Fig. 2. Correlation of the stress drop $\Delta\sigma$ with the focal depth, (after Mikumo and other authors). The averaging function is $\log \Delta\sigma = 0.3 = 1.7 + 0.2\,h$. References are indicated together with the confidence intervals for $\log \Delta\sigma$

Models of source mechanism solutions. The above estimates of the source parameters were obtained by different methods, but in most cases they are correlated with the Brune spectral model. We should avoid discussion the basis and first mention simple functions deduced from it, which are used for observational data processing (for instance [See Rayll, Brune, Allen]).

The displacement spectrum in S waves obtained at a distance r from the source and corrected for attenuation etc. [Ben-Menachem et al] is plotted in log Ω, log f coordinates as two straight lines (where Ω is the spectral density in cm·s, $f = 1/T$ is the oscillation frequency in Hz, T is the period in s). The first line is horizontal within $0 < f < f_0$, that is with an angle f^0, and the second line slopes down for $f \geqslant f_0$, with an angle f^{-2} at higher frequencies. The frequency f_0 where these lines cross is the corner frequency. The zero ordinate $\Omega_0 = \Omega|_{f \to 0}$ of a spectral curve, i.e. the altitude of the horizontal segment, determines the seismic moment M_0, The corner frequency f_0 describes the geometrical dimensions of source R. The formula used in practice for seismic moment M_0 is [W. Thatcher]:

$$M_0 = 4\pi \varrho r V_S^3 \Omega_0 / R_{\theta\varphi} \qquad (1.6)$$

Here, $R_{\theta\varphi}$ is a function of the direction of the source radiation. If the directions of the rupturing plane and displacement along it are unknown, the mean value of $R_{\theta\varphi} = 1/0.6$ is assumed. In particular, if we assume a medium density $\varrho = 2.7$ g/cm^3, shear wave velocity $V_S = 3.2$ km/s, $R_{\theta\varphi} = 1/0.6$, and consider doubling of the displacement in the wave at the free surface, and a correction $\sqrt{2}$ for the usage of one component instead of a total vector, we get the following for a distance $r = 100$ km

$$M_0 = 1.31 \times 10^{25} \, \Omega \qquad (1.7)$$

here M_0 is in dyne·cm = erg; and Ω is in cm·s.

The radius R of a circular rupturing plane is defined by Brune as

$$R = 2.34 \, V_S / 2\pi f_0 \qquad (1.8)$$

The stress drop $\Delta\sigma$ can easily be determined from (1.3).

The Brune model seems rather diagrammatic. It ignores significant factors such as the finite velocity of crack propagation, the rupturing velocity (in [Brune] it is assumed to be infinite), and the complex, saw-tooth character of the rupture, which enriches the spectrum of the oscillations of general displacement by high frequency components. Numerous other models have been proposed, both before and after that of Brune [Balakina et al]. They have common features with Brune's model and notable differences. Thus, the ratios we gave earlier between the spectral values Ω, f_0 (etc.) and the source parameters M_0, R, $\Delta\sigma$ are not the only ones possible. Some differences between the models and results are discussed in [Hanks]. However, the temporal function of the process in the source postulated by Brune seems fairly satisfactory; it is close to that obtained by solving problems numerically on stress relaxation in the case of a crack [B. Burridge, M. Hanson et al].

Cleave ruptures propagating at a finite rate was considered in [Ida et al]. One of the first papers on source rupture propagating in one direction was written by Ben-Menachem. He found an orientation function for the surface wave radiation. This result was used to find, from observed data, the direction of the rupture, its length and rupture velocity [Ben-Menachem]. A calculation was performed for body waves in a theoretical paper [Ben-Menachem]. Knopoff and Gilbert were the first to study two-dimensional seismic rupture propagation. They were followed by others [Moskvina et al]. From [Khattri et al] comes evidence that the downward rupture velocity is less than that along the rupture; thus, in many papers, these velocities are assumed to be different for different directions of one and the same rupture. A good review of the papers on radiation orientation, as a function of the character of rupture propagation is given in [Khattri].

The role of the high frequency radiation due to the irregular saw-tooth character of the rupture is considered in [Aki, Haskell]. It was discovered that the intensity of the high frequency component depends upon the rupture width W, if it is considerably less than the length $W \ll L$ [J. Savage]. High frequency oscillations do not contribute to seismic moment M_0, although they carry most of its energy in K or M. They should not be excluded when considering seismic danger or engineering seismology.

We can see that the Brune model does not include all the important problems of modern seismology. This seems to be true of any theoretical model. So, clearly that besides a theoretical model, a reliable empirical spectral-temporal model for all earthquakes must be compiled. They should reflect in a systematic, general and average form, the typical features of events based on observed data, rather than speculative suppositions.

This review shows that the spectral-temporal characteristics of seismic oscillations, with a clear physical meaning and which allow the determination of significant source parameters, in particular its energy, can substitute for the conventional expression of earthquake value at the source by the term "magnitude".

To make it real and acceptable on a large scale, a great deal of observational data must be processed. That is only possible if systematic usage of modern techniques using playback recording and computer processing. So far, the main method of the data processing remains manual, thus "magnitude"—our old friend—will continue despite its many defects.

We should be prepared to exchange magnitude for spectral-temporal and energetic indices. While the former exists, the main technical problem remains the same, that is how to adjust various magnitude scales to conform with each other, and seismic energy, seismic moment and the other physical parameters of the source. We are concerned, nevertheless, that this problem can only be solved roughly, to a first approximation.

1.2. Source Dimensions of Crustal Earthquakes and Seismic Moment*

The topic of earthquake value was considered in Sec. 1.1 and significant source parameters such as length L along the strike, width W, rupture area S and mean radius R, displacement D along the rupture, seismic moment M_0, and stress drop $\Delta\sigma$ were discussed. Rough numerical estimates of these parameters were also given.

However, rough estimates are not sufficient when solving quantitative seismic problems, e.g. the determination of the maximum possible earthquakes K_{max} from seismological, geological and geophysical, geodetic and other data [Riznichenko], the calculation of the parameters of seismic flow of rock masses [Kostrov et al] compared with neotectonic and recent movements, and the development of a general physical theory of seismology compatible with the facts. We should, therefore, have at our disposal more definite, realistic averages over the range of variation. A reliable agreed system of average and typical values is necessary to

---

* See [Riznichenko].

discover and analyse possible regional, local and individual deviations of the observed source parameters from our standard.

This section is aimed to satisfy these requirements, at least for the present and foreseeable future. As new data are collected they should be generalized and the standards reassessed.

1.2.1. Initial Values and Comparison Techniques

There are now many papers by various workers who have postulated and discussed correlations between K and M on one side, and the other source parameters—L, W, R, D, M_0, $\Delta\sigma$ on the other. These papers can be divided roughly into two groups. The first group is devoted to the establishment of empirical correlations in a form determined by the best approximation of the processed data [Chinnery, Press]. The second group considers general ratios attributed to a proposed model of the source. They are compared with observed data, and numerical ratio parameters are derived to verify the model. Despite the attractiveness of the second approach, that is its physical orientation, the first approach seems likey to give an unbiassed reflection of reality at a given stage. However, it is not free of the subjectivity of some concepts and models. Since the data for correlation are not usually obtained from direct measurements (such field determinations of the rupture length or displacement D), but indirectly (by zones of aftershocks, seismic wave spectra etc.), certain concepts (on deep-sited faults, possible ratios between the length and width of the fault) and models (for instance, the Brune source model of shear when using body wave spectra are implicit).

The basic method for obtaining the average dependence of L, W etc. upon K or M is to construct the correlation fields of observed points using suitable coordinates and to draw averaging curves. In log L, K coordinates, for example, these curves are usually straight lines, and the problem of averaging is linearized so as to determine the "fit" of the coordinate system. It is also easy to see how good the correlation is. This may be achieved in different ways. Firstly, by the simplest graphical analysis techniques to obtain confidence bands at a given level of probability (examples will be given later). Secondly, by some more complex statistical technique, namely least squares, correlation and regression analysis, in particular orthogonal regression or maximum likelihood.

These procedures can be applied to individual pairs of compared values e.g. (L, K), (W, K), if L, W, etc. can be considered mutually independent (as is sometimes done). However, they are not independent even at a superficial level. We should try to get internally consistent results for the entire set of values.

Source parameters are related to each other firstly through the seismic moment $M_0 = \mu SD$, one of the principal independently determined values. For the same rupture area S, the length L and width W are dependent upon its form. If it is a rectangle, $S = LW$, if it is an ellipse, $S = \pi(L/2)(W/2)$. Considering that the width of a real rupture at its middle is probably greater than that at its ends, we shall analyse the "elliptical" variant, where:

$$M_0 = \mu\pi R^2 D = (\mu\pi/4)LWD, \tag{1.9}$$

from this expression we can find a ratio between R, L and W as

$$(2R)^2 = LW \tag{1.10}$$

We look first at (1.9) and (1.10). Which of the values in these equations may reasonably be assumed to be initial ones determined from the correlational fields, and which are to be calculated from (1.9) and (1.10)? Only then can the results of these calculations be compared to other correlation fields to verify the absence of pronounced contradictions over the set of observations. A better approach is to formulate a general empirical or semi-empirical mathematical model which embraces the set of ratios. Then to find numerical solutions to its parameters that, in a specific sense optimize the agreement between the model and the set of observations. Examples of this approach can be found from experience of other seismo-logical problems, e.g. the construction of the average system of spectra [Riznichenko], the determination of maximum possible earthquakes from complex data [Riznichenko]. However, we shall avoid this method at this stage.

Our attempt is to treat the topic without prejudice, as far as the concepts and models are concerned. Thus, it seems natural to chose as initial values those which can be determined directly from the observed data, such as magnitude M and its energetic analog K, and the seismic moment M_0. We can also try to derive parameters from such values as L and D, which are accessible, at least in some cases, to direct measurement. We can then consider in greater detail some of the individual values in (1.9) and (1.10).

1.2.2. Review of the Values

We express all these values as a function of energy value K or magnitude M of an earthquake. The (K, M)-correlation is taken in its usual Gutenberg-Richter-Rautian form

$$K = 4 + 1.8M \tag{1.11}$$

Thus, the estimated seismic energy at a reference sphere with radius 10 km $E = 10^K$ Joules does not acquire more physical meaning than it deserves. The value of larger earth-quakes, starting approximately from $K > 15, M > 6$, sensed by the world network of stations, is estimated primarily in magnitudes M or magnitudes m, which are correlated with K. For such events, we shall calculate K from M by formula (1.11). Their actual seismic energy, with respect to frequency range and other conditions, may differ greatly from $E = 10^K$ J. For small earthquakes, sensed within the USSR by regional networks of stations, the value of K is determined directly. This usually correlates rather well with M, in a corresponding sphere, and gives a more consistent estimate of the energy. However, we shall not procede in greater detail about the comparison of magnitude and energy. At present, this topic requires a fun-damental reconsideration with a spectral basis [Riznichenko, et al].

The seismic moment M_0, because of its importance in modern seismology, can be com-pared with a basic concept as seismic energy E, and its derivatives, the energy K and magni-tude M of earthquakes. It follows from the definition of $M_0 = \mu SD$ that the physical meaning of M_0 is the potential work or the possible energy which could be spent to overcome friction forces at a rupture surface to move the sides an average (square) distance D if the surface density of the forces equals to elastic shear modulus μ, that is close to the theoretical strength of a defectless crystalline material. The effective macroscopic, physical or "geophysical" strength of a nonideal, heterogeneous rock material in an earthquake source is considerably less and depends on many factors.

The major geometric dimensions of a source as a dislocation are closely related to M_0. M_0 may, in principle, be determined from non-seismic data such as field geological descrip-tions and geodetic measurements which has sometimes been done. However, most M_0 values, which are known for many earthquakes, and in particular for small deep-sited events, are taken from the same seismograms from which K and M are determined. A comparison of M_0, values obtained by a pure seismological method, with other estimates—geological, geodetic, etc.—was made for many events. In particular, one paper [Hanks, Wyss] presents a convincing example of such comparison. It is shown that M_0, R and $\Delta\sigma$ determined from the known Brune model [Brune], which is utilized for most definitions of this type, agrees well with the estimates of L and D from geological and geodetic data. As for proper seismological determinations of M_0, using a good modern observational technique and data processing, they can as reliable as those possible for K and M estimates.

The various aspects of a rupture process at the source, depend upon many factors. These include: elastic, nonelastic, strength and other material properties, structural inhomogeneity in the rock mass, stress and strain conditions, type of displacement along the rupture or set of ruptures, etc.. The indices of seismic energy at the source, namely K and M on one side, and residual deformation M_0 on the other, do not have mutually unambiguous functional relations. However, it is possible to establish an average correlation betweeen them, and as seismology is developing, this correlation, despite local discrepancies, becomes more and more stable, on average.

Bearing this in mind, the seismic moment M_0 together with magnitude M and energy value K, and potentially the seismic energy E, determined from seismic oscillation spectra, should be considered as one of the basic initial source parameters. The relations between M_0 and K and M can be detected from the correlational field.

The rigidity μ of the medium, also a term in (1.9), does not depend upon K and M and is not determined from the correlational field. The μs are known for various rocks from laboratory tests of rock samples (for instance [B. Belikov et al]). Under real conditions, μ depends on seismic velocities and the densities, which are also correlated with velocities. Since sources of crustal earthquakes are predominantly sited in the upper part of the consolidated crust with granite velocities (viz., $V_p = 6$ km/s, $V_S = 3.5$ km/s), $\mu = 3.0 \times 10^{11}$ dyne/cm^2 seems suitable. Values close to μ are typical of many granites and some dense metamorphic rocks. To account for local horizontal and vertical variations, other values should be considered.

The length L of the source and the length of the rupture are some of the few indices of earthquake value that are accessible to direct measurements *in situ*—usually along the strike. Another index is D-displacement. However, this is only possible in rare cases of shallow and large crustal sources when the major tectonic rupture reaches the Earth's surface. Even then, we cannot always be certain that a major rupture does not continue beyond the margins under the surface. More often, the information on the length L of the rupture in the source and its width W is gained from indirect data, by knowing *the mean radius R* of the source as the index of area S of the rupture surface $S = \pi R^2$, which in its turn is in the expression for seismic moment M_0 (1.9). Values of R and M_0 are derived from seismic oscillations spectra.

The value L (respectively W) is sometimes treated as the length (width) of the "source" considered as a space, and a 3-D body, where the main rupture, its components and accompanying ruptures with considerable plastic deformations (in their general sense) occur during the earthquake. Such a 3.D body is identified from the area of the aftershocks associated with the event. It is clear that parameters such as rupture and body of the source with nonelastic deformations and aftershock area cannot be identified simply by their parameters. In fact, seismic energy E is not radiated from the rupture surface. It is radiated only from the body of the source, where nonelastic deformations prevail, but also from the surrounding quasi-elastic area. The same is true for the residual released strain forming seismic moment M_0. Aftershock area is only delineated gradually a long time after the main rupture has passed accompanied by the earthquake characteristics of K, M, E, M_0, etc. The fact that aftershocks do occur in an area where the source originated may only serve as indirect evidence for processes, originating their in it, and the values of their generalized parameters. It seems reasonable to attribute aftershocks and the development of seismic activity following the main shock to a diffusive wave propagating slowly from the source area after its generation [Riznichenko et al]. Nevertheless, the various phenomena are elements of a single process, "physics" or "geophysics" should consider them jointly. So far, it does not seem possible to arrange the parameters of these phenomena into an logical elegant system without further detailed analysis. Nevertheless, seismologists have learned from the repeated observations of individual cases that estimates of the geometrical dimensions of these objects approximately coincide, and this experience should not be ignored.

Despite this L, often with R, as defined from various standpoints and by various methods for the source, may be expressed through K and M from correlational fields, and they may be used as the initial data when estimating other source parameters.

The width W of the source or rupture poorly reflects the role of the initial parameter for further estimations. In fact, if the width W is the downward width of the source's rupturing surface, it is difficult to judge its lower edge. This can be done using L and M_0 at known μ, L and D. It is often believed that large earthquakes with a near vertical rupture surface, are limited from below by the depth of the upper asthenosphere boundary, viz., by the layer of lower velocity and viscosity. The higher yield limit of the material within this layer seems to prevent the accumulation of elastic strain there and so there are no conditions for ruptures to exist there. Over the long-term during earthquake development these suppositions are true. For the short-term process of the source rupture, that is during the earthquake proper, the situation changes. Displacements of lithospheric blocks when they rupture may cause severe short-term elastic-plastic deformations in the underlying asthenosphere, the deformations concentrating in a thin plate which extends the lithospheric rupture down somewhat to the asthenosphere. It is of no importance whether these deformations cause rupture or provide for a concentrated flow. Given the rather generalized macroscopic consideration, which is carried out when M_0 is determined, continuous displacement in the plate body cannot be distinguished from rupture along the plane. It is only necessary that the plate thickness be less than the characteristic length of wave λ. For the oscillation period in shear waves for large earthquakes $T = 10$ to 20 s or more and the velocity V_S reaches 3.5 km/s, while λ lies between 35 and 70 km. This distance is comparable with the observed width of the zone of aftershocks.

Other interpretations of the source, e.g. as a 3-D nonelastic area or as the aftershock area, are fraught with the same difficulties when determining its width W, the difficulties originating from determination of the length L. However, another problem appears in the case of W addition to those typical of L. There is no clear correlation between the width of the aftershock area in the plane and the rupture width, in particular where it is steeply inclined. Because of these logical problems in determining the value, we should not utilize it as the primary parameter for estimates. It is better determined as a function of other values from (1.9), (1.10).

The mean radius R of the source as with M_0 is determined directly from spectral observations, and so may be regarded as an initial value. Essentially, the mean diameter $2R$ of the rupture surface for small earthquakes is not usually distinguished from its length L or width W. For large earthquakes where the linear dimensions of the source lie close to the crustal thickness, assuming a source length L greater than its width W, the diameter $2R$ of the source is naturally assumed to be different from L and W.

Despite a first impression that R is less sensible and more distant from the primary information compared with L, it is, in fact, more "initial" than L in case of instrumental spectral observations. Taking this into consideration, we put R together with M_0 as a primary value when determining other source parameters.

Displacement D along the rupture is a principal focus parameter and is sometimes accessible to direct observation (field examination of a rupture). For the majority of earthquakes, where the focus does not come to the surface, D can be derived from M_0 and S or R from spectral observations. It is reasonable to establish its relation with K and M from the correlational fields and to utilize it as one of the initial parameters for further estimates.

1.2.3. M_0, L and D as Average Functions of K and M as Derived from Correlational Fields

The correlation between *seismic moment* and the earthquake values K, M has been considered by many scientists. Figure 1 illustrates the correlational field M_0, M or K for earthquakes of various sizes taken from various publications.

Most of the points in Fig. 1 were obtained from the instrumental spectral definitions of M_0. For some large earthquakes, M_0 was determined from field geological data (large circles in Fig. 1). It is clear that the general spacing of the points yielded by instrumental and field data are consistent.

The average dependence between M_0 and M is given in [Thatcher, Hanks] for Southern California as $\log M_0 = 16.0 + 1.5\,M$. Our averaging over the set for the world (Fig. 1) is expressed in (1.1). Our world function (1.1) is the average for the California data, which is no worse than that given in [Thatcher et al]. In [Thatcher et al], the deviation of points for Southern California from the straight line (1.1) is $\sigma_{\log} M_0 = \pm 0.4$, that is less than for the world as a whole.

It can be seen from Fig. 1, that the observations range from $K = 6$ to 20, and for $M = 1$ to 8.9. The figure also contains the averaging curve suggested by Aki in his paper where it is drawn to support his theoretical model of the focus ω^2. It is clear that the Aki model is probably in better agreement with the observations, but only for the segment $M = 5$ to 8, or the range $M = 5$ to 8.9 in [Aki]. Beyond this range (both to the left and right) our average (1.1) seems to be more effective for a wide range of earthquake values K and M.

In addition to the data on M_0, for general information Fig. 1 also shows the *stress drop* $\Delta\sigma$ in the focus; the straight lines are for $\Delta\sigma = 0.1$ and 100 bar. These lines were drawn using the Brune model of the focus, and estimates of Thatcher and Hanks. It seems as if $\Delta\sigma$ has the tendency to increase with increasing K and M. However, this point is disputable. The spread of $\Delta\sigma$ obtained from observations is great. There are cases of large values even for small earthquakes. Some general ideas supported by diagrammatic theoretical estimates [Riznichenko] lead to the conclusion that in large earthquakes K_{max}, seismic flow of rock masses is characterized by reduced viscosity: rocks, broken by frequent ruptures, are more pliable and the general stress σ is less. We may conclude that although large earthquakes cause large relative stress drops $\Delta\sigma/\sigma$ in greater volumes, the absolute value of $\Delta\sigma$ may not increase.

We can add that stress drops of the source estimated from observations, namely $\Delta\sigma = 0.1\text{-}100$ bars $= 10^5\text{-}10^8$ dyne/cm^2 (1 bar $= 1$ kg/cm$^2 = 10^6$ dyne/cm^2) are smaller by a factor of $10^3\text{-}10^6$ than the theoretical strength (which is about $\mu = 3 \cdot 10^{11}$ dyne/cm$^2 = 3 \cdot 10^5$

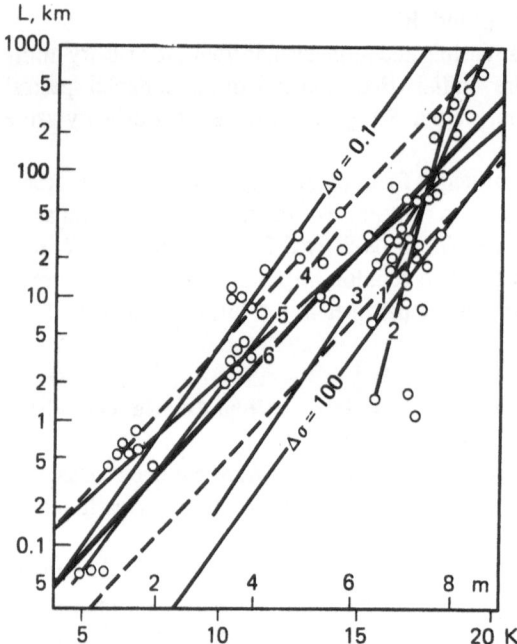

Fig. 3. Correlation of the source length L with the earthquake energetic values K, M.

Straight lines are averages: *1*—after Tocher; *2*—Iida; *3*—Press; *4*—Wyss and Brune; author's averages: *5*—without and *6*—with consideration of other source parameters according to Eqs. (2.12) and (2.13), respectively. *Dashed lines* are limits of the 68% confidence interval by averaging *6*; *thin straight lines* represent the stress drop $\Delta\sigma$ = 0.1 and 100 bar

bars), but at its upper observed limit $\Delta\sigma$ approaching the technical, engineering strength of solid rocks. This is a temporary resistance equal to that determined from short-term sample tests under room conditions. For example, the technical compression strength of granite is 1000 to 2000 kg/cm^2, and its shear strength is 50 to 100 kg/cm^2 (bars).

To establish a correlation between *the focus length* (or rupture at the focus) and the earthquake value K or M, we shall use the summary by Ryall et al from the data of Tocher, Iida, Press, and Wyss and Brune for $M > 3$, and Ryall et al for $M < 3$. Figure 3 shows the correlational field from [Ryall]. The straight lines *1* to *4* show the averaging variants for observations as proposed by various authors [Ryall, Iida, Press, Wyss, Brune]. We can see that these averages are remarkably different. Ryall and other authors did not risk developing their own general average, so we have done so.

Figure 3 shows two versions of averaging. The first is shown by straight line *5*; it is a simple visual averaging of the points in the given field. Its equation is

$$\log L_{km} \pm 0.5 = -1.7 + 0.21\,K = -0.88 + 0.37\,M \tag{1.12}$$

The second is given by straight line *6*; its equation (1.13) is corrected by M_0 and D for an assumption of ratio W/L

$$\log L_{km} \pm 0.5 = -2.27 + 0.24\,K = -1.29 + 0.44\,M \tag{1.13}$$

Identical transformations of equations (1.12) and (1.13) into

$$\log L_{km} \pm 0.5 = 1.38 + 0.21(K-15) = 1.35 - 0.37(M - 6) \tag{1.12a}$$

$$\log L_{km} \pm 0.5 = 1.39 + 0.24(K - 15) = 1.25 - 0.44(M - -6) \tag{1.13a}$$

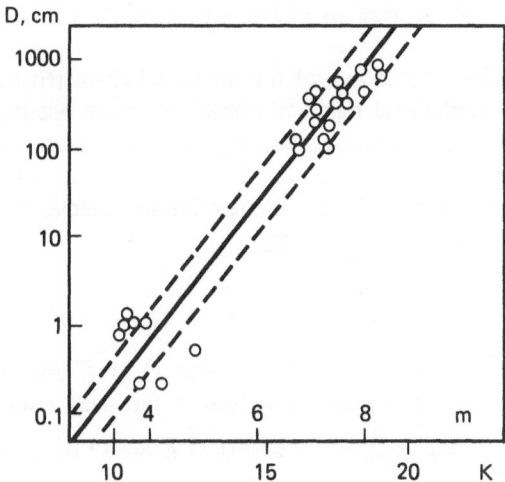

Fig. 4. Correlation between the displacement D along the rupture, and earthquake energetic values K, M (after Chinnery).

Straight line is the average by Chinnery according to (Eq. 2.14); *dashed lines* are the limits of the 68% confidence interval

show that, over a significant range of earthquake values near $K = 15$ and $M = 6$, the averages only differ slightly although the difference is larger for very small earthquakes with $K = 0$ or $M = 0$.

In Fig. 3, the scatter of the correlational field points near the plotted averaging lines does not seem to be completely random, as it should be statistically. These are initial data. Our averages (1.12) and (1.13) for $K = 4$ to 13, $M = 0$ to 5 are close to the average line *4* made by Brune and Wyss, and for $K = 15$ to 18, $M = 6$ to 8 if it is close, line *3* constructed by Press. All the averaging lines, in particular ours, and except for number *4* by Brune and Wyss, met within a small area near $K = 17.5$, $M = 7.5$; viz log $L_{km} = 1.8$. A 68%-confidence band for both of our averaging versions (1.12) and (1.13) includes line *4* of Brune and Wyss in its entirety. However, if line *4* is compared to lines *5* and *6*, it reflects the observed data on large earthquakes $K > 15$, $M > 6$ less well. Lines *1* to *3* do not agree with the observed data for small events $K < 11$, $M < 4$.

We shall stress that we prefer equation (1.13) to describe the relation between L and K or M as it is consistent with the behaviour of other source parameters, too. It was obtained from the observations for $K = 5$ to 19, $M = 0.5$ to 8.5.

The correlation between *displacement D* along the rupture and magnitude M was analyzed by Chinnery, who obtained the $M = 1.32$, log $D_{cm} + 4.27$. His correlational field in our coordinates is shown in Fig. 4. Its general character is better than for L in Fig. 3; the point scattering seems to be close to random. The width of our 68%-confidence band is $\sigma_{\log D} D = \pm 0.4$. Thus, the average ratio between D and K and M may be written as

$$\log D_{cm} \pm 0.4 = -4.9 + 0.42\, K = -3.2 + 0.76\, M \tag{1.14}$$

which is verified by the observations for $K = 9$ to 19 and $M = 3$ to 8.5.

1.2.4. Estimations of Mutually Consistent Focus Parameters

Given our review of the individual characteristics of focus parameters and analysis of the correlational fields, it seems best, when obtaining consistent values of the parameters, to choose reference ratios (1.1), (1.12) and (1.14) for the three parameters M_0, L and D determined from the correlational fields. As for R and W, they should be calculated using the functions M_0, L and D from the conforming equations (1.9) and (1.10) by formulae

$$R = \sqrt{M_0/\mu \pi\, D} \tag{1.15}$$

$$W = (2R)^2/L \tag{1.16}$$

If we express M_0, L and D in K and M by (1.1), (1.12) and (1:14) and assume $\mu = 3 \cdot 10^{11}$ dyne/cm^2, we obtain for mean radius R of the source

$$\log R_{km} = -2.6 + 0.23\, K = -1.67 + 0.42\, M \tag{1.17}$$

and

$$\log W_{km} = -2.91 + 0.26\, K = -1.86 + 0.47\, M \tag{1.18}$$

for its mean width W.

We used this scheme for numerical calculations but the results did not satisfy us. Consider the values for the length L and width W of the source at internal points and at the ends of the range of earthquake values K and M:

K	M	L, km	W, km	L/W
5	0.6	0.21	0.025	8.4
15	6.1	24	10	2.4
20	8.9	260	210	1.2

Note the ratios L/W (length of the source to its width). In the middle of the range, which is best supported by observations, at $K = 15$ and $M = 6.1$ the ratio $L/W = 2.4$ does not surprise us. But for small events with $K = 5$ and $M = 0.6$ the ratio is unnaturally large: $L/W = 8.4$, whereas for large events with $K = 20$, $M = 8.9$, it is too small, $L/W = 1.2$. In fact, we would expect the opposite to be true.

Small earthquake foci are sited wholly within the quasi-homogeneous crustal layers, mainly in a granite layer. Under quasi-homogeneous conditions, the shape of the focus tends to be isometric with $L/W = 1$. Large earthquakes foci capture more heterogeneous layers of the crust and upper mantle as the rupture surface increases $S = \pi L W/4$. When the seismic process tends to concentrate within the granite layer, the focus shape especially with a steeply inclined rupture surface, should be more elongated along the layer in a horizontal direction. The ratio L/W should increase. Less favourable for rupture is the asthenosphere which lies 100 to 200 km under continents and 50 to 100 km under oceans. Although its upper boundary depth, as was mentioned earlier, should not be the absolute limit of the rupture's lower edge (that is the value of W for vertical rupturing), the existence of the asthenosphere at such depths should make W smaller than L for rather large events. There is also direct evidence for this: the form of the aftershock area, which is to some extent related to the focus shape, for small and intermediate events is usually close to isometrical, whereas with large earthquakes it is much longer.

Thus, earlier results should be corrected: small foci should be rounded, large ones be elongated. This should be done without changing (for the homogeneous μ = const approximation) the rupture area S as associated with M_0, and therefore its mean radius R. This correction was realized in the following way.

Only two parameters—M_0 and D—were left as initial ones to be determined from the correlational fields. R is derived using them from formula (1.15). Thus, we have established a ratio between L and W in R, which is expressed in (1.16). Then, an additional condition is introduced: we know the average behaviour of L/W for varying K, that is we have a function $L/W = f(K)$. At this stage, we shall stick to a linear relation and assume it specifically to be

$$L/W = 1 + 0.1\,K \text{ for } K \geqslant 0 \tag{1.19}$$

Estimate (1.19) results in $L/W = 1$ at $K = 0$, $L/W = 3$ at $K = 20$, and $L/W = 2$ at $K = 10$. These values of L/W at the margins and an internal point in the K range seem acceptable. The choice of $L/W = f(K)$ (or M) as a conforming factor is important because although the absolute values of L and W vary widely, by tens of orders of magnitude, and the averaging straight lines cannot be accurate due to data scatter, the ends of the lines at the range margins may diverge or converge greatly for even slight variations in their tilt. This makes L/W fairly unstable at the range margins. A control of L/W within the range, including margins, is the only effective solution to the problem.

These theories formalize the solution of the problem. Numerical calculations should follow with the verification of the solution's conformity to the correlational fields not directly used in the calculations.

Certainly, (1.19) is only an approximate, semi-intuitive expression of the probable average relation between L/W and K, and thus M. In principle, L/W may and should be studied using seismological, geological-geophysical and geodetic information, as well as using theoretical and laboratory experiments. References to papers on this topic can be found in [Riznichenko]. They are the studies of wave fields in relation to source dimensions and rupturing velocities in various directions; geodetic measurements of vertical and horizontal residual displacements over shallow foci. The observations were compared with the dynamic and static models of the focus as a dislocation. Macroseismic near-field data also may be considered. After sets of such material are obtained, they should be statistically processed. However, this is a problem for the future.

At present, estimates of the source parameters M_0, R, L, W and D performed in accordance with our corrected technique including L/W (1.19) are given in Tables 2 and 3. To ease its use in practice, Table 2 presents integer values of earthquakes for K, whereas Table 3 gives them for M. The ratio between the energy value K of an earthquake and its magnitude M is assumed in its common form (1.11).

The agreement between this present solution and the correlational field for L was verified and the results were reported earlier when Fig. 3 was considered. Our second average correlation between L and K, M, corrected by L/W, differs insignificantly from the uncorrected one. It is worth mentioning that the averages (Fig. 3) proposed by Press, Tocher, and Iida have the same defect, but it is more distinct, that is the L/W values for small earthquakes are more unrealistic. The Wyss and Brune average is free from this defect. However, as can be seen from Fig. 3, it is less satisfactory for large earthquakes with $K > 15$ and $M > 6$.

Table 2. Parameters of Earthquake Core Sources

K	M	M_0, dyne $\cdot$ cm	L, km	W, km	R, km	D, cm
5	0.6	$0.19 \cdot 10^{17}$	0.089	0.059	0.036	0.0018
6	1.1	$0.15 \cdot 10^{18}$	0.16	0.098	0.062	0.0040
7	1.7	$0.12 \cdot 10^{19}$	0.28	0.16	0.11	0.011
8	2.2	$0.90 \cdot 10^{19}$	0.49	0.27	0.18	0.028
9	2.8	$0.70 \cdot 10^{20}$	0.87	0.45	0.31	0.073
10	3.3	$0.54 \cdot 10^{21}$	1.5	0.76	0.54	0.19
11	3.9	$0.42 \cdot 10^{22}$	2.7	1.3	0.92	0.51
12	4.4	$0.32 \cdot 10^{23}$	5.7	2.1	1.6	1.3
13	5.0	$0.25 \cdot 10^{24}$	8.3	3.6	2.7	3.5
14	5.6	$0.19 \cdot 10^{25}$	14	6.0	4.6	9.2
15	6.1	$0.15 \cdot 10^{26}$	25	10	7.9	24
16	6.7	$0.12 \cdot 10^{27}$	44	17	14	64
17	7.2	$0.90 \cdot 10^{27}$	75	28	23	170
18	7.8	$0.70 \cdot 10^{28}$	130	48	40	440
19	8.4	$0.54 \cdot 10^{29}$	230	80	69	1200
20	8.9	$0.42 \cdot 10^{30}$	410	140	120	3100

That our solution is considered with the observed correlational field for a source width W was verified by Chinnery, where such a field is fixed in the interval $M = 3$ to 8.5. Over most of the interval, $M = 3$ to 7, our solution $W(K, M)$ agrees well with it. However, in the interval $M = 7$ to 8.5, our values log W are greater than of Chinnery by $\Delta \log W = 0.3$-0.5. This discrepancy goes beyond the 68%-confidence band. However, the maximum width of the focus $W \leqslant 20$ km, shown in [Chinnery], seems to be underestimated by 280 km. Then, $L/W = 14$; it is even greater if it agrees with R, an improbable result. Perhaps Chinnery assumed $W < 20$ km for the largest earthquakes of the World, because he did not believe *a priori* that a near vertical rupture can extend considerably beyond the crustal layer. This

Table 3. Parameters of Earthquake Core Sources

M	K	M_0, dyne $\cdot$ cm	L, km	W, km	R, km	D, cm
1	5.8	$0.10 \cdot 10^{18}$	0.14	0.090	0.056	0.0033
2	7.6	$0.40 \cdot 10^{19}$	0.40	0.22	0.15	0.019
3	9.4	$0.16 \cdot 10^{21}$	1.10	0.56	0.39	0.11
4	11.2	$0.63 \cdot 10^{22}$	3.0	1.4	1.0	0.62
5	13.0	$0.25 \cdot 10^{24}$	8.3	3.6	2.7	3.50
6	14.8	$0.10 \cdot 10^{26}$	23	9.2	7.3	20
7	16.6	$0.40 \cdot 10^{27}$	62	24	19	120
8	18.4	$0.16 \cdot 10^{29}$	170	60	51	660
9	20.2	$0.63 \cdot 10^{30}$	470	150	130	3800

is approximately 20 km thick in the transition zones from continent to ocean, where such earthquakes do occur as a rule. Under this circumstance, the verification of our solution with the correlational field for W from Chinnery seems quite satisfactory.

Thus, we have made sure that our solution agrees with all the correlational fields attributed to it.

One detail should be stressed. Condition (1.19) breaks the log-linear relation between L and W and K, M. However, the violation is negligible, and within the range $0 \leqslant K \leqslant 20$, $-2 \leqslant M \leqslant 9$ we can use the following approximate average log-linear functions for L and W:

$$\log L_{km} = -2.266 + 0.244K = -1.289 + 0.440 M \qquad (1.20)$$

$$\log W_{km} = -2.340 + 0.223K = -1.448 + 0.401 M \qquad (1.21)$$

Equation (1.20) was derived when discussing Fig. 16.

The final way we used to obtain consistent source parameters is not the only one possible. Instead of M_0 and D as initial values we could have chosen M_0 and R. R is primary for instrumental spectral determinations, and D is secondary. However, R is not accessible for direct control by field observation, whereas D is sometimes accessible for large earthquakes. The advantages of M_0 and R are as debatable as the choice of M_0 and D. The agreement between our results and the data on R is achieved, in fact, by making them conform with L (Fig. 3); as for small and intermediate earthquakes, for which R is really primary compared to D, the data on L are practically the data on $2R$ obtained from spectral processing.

No doubt other formulations of the problem are possible. One which seems very attractive is the simultaneous determination of an optimal set of parameters of the general empirical or semi-empirical mathematical model. We considered this earlier for the problem of average spectra [Riznichenko et al.] and in the definition of K_{max} by complex data [Riznichenko]. However, its usage for average focus parameters lies in the future.

Thus, we have considered how to value the average functions of focus parameters for crustal earthquakes, namely seismic moment M_0, mean radius R, length L, width W and displacement D along the rupture, consisted with energy K and magnitude M. It can be shown that independently determining each of these parameters as functions of K and M from correlational fields given considerable scattering of the observed points and inaccuracies in the plotting of the averaging curves (straight lines) can cause great divergences in the relative values of parameters such as L/W. This can be avoided by including the procedure for adjusting the results of the terms L/W and R themselves into the expressions for them as functions of K and M.

The result of these estimates may serve as a standard for M_0, R, L, W and D as average functions of K, and M for the case of crustal foci. The series of numerical values may be used for theoretical studies of the focus, for numerical problems on focus ensembles, such as the estimation of parameters of seismic flow of rock masses. They can also be used for comparing various regional, local, and individual focus parameters as functions of K, and M with standard during investigation of local conditions and causes of observed variations.

2. Quantitative Presentation of Seismicity

2.1. Seismic Regime[*]

The seismic regime of some area is the set of earthquakes within the area in space and time. The earthquakes in this set may be connected. In some cases their connections are pronounced, for instance, between a powerful earthquake and its aftershocks; in others the connections are not distinct, at first glance, so the aim of research is to reveal them, if they do exist, and to study their features and regularities.

Earthquake sets have been studied in certain regions for a long time particularly in Japan and California. Gamburtsev outlined the problem distinctly, since he saw in it the key to a methodology for predicting severe earthquakes. Some of his ideas were presented in October, 1953 to a session of the Seismological Council (Moscow, USSR).

Gamburtsev, like many others, believed that earthquakes might occur in the weakened zones separating stronger areas and crustal blocks. He called such weakened zones, or active deep faults, "seismic sutures". He defined certain relations between earthquakes and the corresponding migration of severe earthquake foci within the system of seismic sutures. The lifetime of a suture is assumed to be rather long. Thus the hypothesis is supported that, there is an average stability in a seismic regime for a system of seismic sutures, over the entire suture system and long time interval (several hundred years).

Gamburtsev proposed that severe earthquakes correspond to the breaking of "larger seals between crustal blocks. The character of seismic regime, as well as of slow motions, prior to the break of strong seals should differ from that of the weak seals".

Gamburtsev did not restrict himself to these discussions. To study seismic regime of the regions in detail so as to develop methods of earthquake prediction, he organized in 1953, the larger Tadzhik integrated seismological expedition (TCSE) on the basis of Garm seismological expedition [Gamburtsev]. The centre was organized at the Institute of Earth Physics, Acad. Sci. USSR and the Seismological Institute of the Tadjik Acad. Sci. It was headed first by Gamburtsev and then by Nersessov together with a large seismological team.

2.1.1. Principal Characteristics of a Seismic Regime[**]

The principal and simplest geometrical (or kinematic) focus parameters of an individual earthquake are the coordinates x, y, z of the hypocenter and the time t_0 the earthquake originates.

The main dynamic characteristic of an earthquake is its seismic energy E, that is the energy of the seismic waves radiated from the focus area. The set of five values (x, y, z, t, E) gives the simplest quantitative physical characteristic of an individual earthquake with respect to its focus. Each i-th earthquake may thus be described by a five-dimensional number. In a five-dimension space Π_s, each earthquake is a point with coordinates $(x, y, z, t, E)_i$.

---

[*] From [Riznichenko].

[**] From [Riznichenko]. Co-authors: V. I. Bune, I. L. Nersessov.

Other parameters may be also used to describe each earthquake. For instance, values defining focus dynamics, its spectral aspect, etc. may be used.

In some cases, any of the five principal values may be excluded from consideration, e.g. energy E, or focal depth z. However, to make the situation specific, we shall deal with a space $\Pi = \Pi_s$ as defined.

A seismic regime in Π space will be the set of all the points for the individual earthquakes. These points will be spaced discretely with empty space between them.

The study of a seismic regime concerns how these points are distributed in space and what governs this distribution. This analysis may be carried out either by considering the discrete points $(x, y, z, t, E)_i$ directly or by constructing secondary functions with continuous arguments for x, y, z, t, E, viz. the coordinates of Π space that yield the general characteristics of the $(x, y, z, t, E)_i$ distribution. The second way seems more convenient.

Earthquake density. The most complete description of a seismic regime that reflects the concept of a regime as a set of points in Π_s space, is the earthquake density N_* in the space. This is the distribution density of the earthquake "points" in a physical space x, y, z in time t with energy E. The density $N_* = n/\Delta\Pi$, where $\Delta\Pi$ is a volume element of Π space, and n is the number of events that fall within this volume element. Calculated densities N_* are ascribed to the centers of these elements. Further, it is assumed that N_* changes gradually as a function of Π space, taking the estimated value at the center of the $\Delta\Pi$ element. Hence, the picture for some values of N_* is smoothed. Other techniques for calculating the points and smoothing exist: the use of partly overlapping or moving volume elements $\Delta\Pi$ with distinct boundaries (in one-dimension case this is the moving interval technique); and the technique of moving volume elements $\Delta\Pi$ with fuzzy boundaries. The calculation technique for earthquake density N_* should be chosen according to the conditions of the geophysical problem relevant when analysing the given events. In particular, when determining the dimensions and shape of the elements $\Delta\Pi$ in physical space x, y, z, we should consider the geological and geophysical setting in the study region: the strike and dimensions of the tectonic structures of large fold systems, deep fault zones, etc. and the shape and position of earthquake sources, etc.

Earthquake recurrence. The earthquake density N_* in Π_s is directly associated with the earthquake recurrency N in an interval of energy in an area x, y, z. If Δt is a time element, this relation will be: $N = \int_\Omega N_*\ dx\ dy\ dz\ dE$, where the area of integration Ω extends to the given area of space x, y, z and the given interval of earthquake energy E. The recurrence N defined by this expression is an average for the time interval $\Delta t = 1$.

Let ω denote the volume of the entire area Ω in space x, y, z. Then, $N = \bar{N}_*\omega$, where $\bar{N}_*$ is the average density N_* in this volume. In case of a fixed volume of physical space x, y, z and interval of energy E, the average density $\bar{N}_*$ of earthquakes and their recurrencies N will only differ by a constant factor $\omega = $ const with dimension L^3E.

For a fixed area of space x, y, z and various "classes" ΔE, each energy class $E = E_i$ corresponds to its recurrency value $N = N_i$, so that $N = N(E)$. Here E_i is an energy average within an interval ΔE (it is usually E plotted along E axis in a logarithmic scale at the midpoint of the interval). In fact, the earthquake recurrence as a function of energy was the subject of an analysis by Gutenberg and Richter, Kavasumi, Bune.

Instead of recurrence N, it might seem more convenient to consider the density N_* of the recurrence normalized to the volume in which sources are sited; it equals to the density of earthquake number distribution in Π_s calculated for an element $\Delta\Pi$. The element is determined by the elements Δx, Δy, Δz, Δt and the energy class E. This will be called in short "recurrence" and sometimes "normalized recurrence", if necessary.

When changing from non-normalized recurrencies to normalized ones N^*, we shall not change the slope γ of the recurrence curve by earthquake energy, plotted in log-log coordinates. However, we shall be able to compare the seismic activity of the areas determined by absolute value of numbers N^*.

Recurrence curve and its parameters A and γ. When plotting the recurrence curve, we attribute the number of earthquakes to the interval of the energy logarithm, ascribing this value N_k to the geometric center of interval E_k. For the intermediate energies E'_k, the ordinate N'_k is related to an energy interval of the same width, but with its center at point E'_k.

It should be stressed that we referred above logarithmic equal intervals of energy log $E = $ const. As a rule, in seismology, $\Delta \log E = 1$ and N_k is the number of earthquakes where energy falls within the interval $\log E_k = \pm 0.5$. The choice of some other constant will change the level of the curve $N(E)$. At the same time, any other way of selecting the energy interval (for instance, $\Delta E = $ const) and the classification of earthquakes by another dynamic parameter, such as magnitude M or amplitude A, will cause variations in the value of γ.

Experience in the field has shown that within this range of energy variations, the angle coefficient $\gamma = - \Delta \log N^*/\Delta \log E$ is constant.

The lower limit on the energy E for which linear interpolation is possible almost coincides with the minimum energy of the "representative" earthquakes considered in the existing observation system. The upper limit for E is more difficult to determine because severe events are scarce. Sometimes it is determined from the bend in the observed curve $N^*(E)$ in the given area. More often, however, we have to break the straight-line $N^*(E)$ at the maximum energy of the event observed in the given or neighboring regions under similar geological and geophysical conditions. Given those restrictions, each curve $N^*(E)$ may be defined by two values, namely the recurrency $N_k^*(0)$ attributed to some fixed class $K = K^{(0)}$ of earthquake energy chosen preferably within the known straight-line area $N^*(E)$, and parameter γ.

If the slopes γ of the $N^*(E)$ curve for various parts of a space x, y, z are similar, (as was found in most regions for direct field observations), it is sufficient to compare corresponding $A = N_k^*(0)$, related to a fixed class $K = K^{(0)}$ of earthquake energy, and to compare the average seismic activity of the regions. The number $A = N_k^*(0)$, which, under these conditions, is a general index of seismic activity or the seismic activity of the area, is called the "activity".

It is convenient in practice to determine the activity A from a representative earthquake energy class $K^{(0)}$ which has numerous stations that are not independent of the distribution of the stations in a network. The value A may be determined from the recurrence of earthquakes both from class $K^{(0)}$ and also considering other classes using the averaging line of the recurrence curve in its straight-line segment.

Conventionally, we may compare the seismic activity A of regions when the slopes γ of the curves $N^*(E)$ are different. However, when these regions are characterized by both parameters—A and γ—as ratio, their activity will depend upon the choice of fixed class.

In general, the seismic activity A should be attributed to a volume of physical space Δx, Δy, Δz, so that we can consider volume seismic activity. However, where the dependence of A upon the depth can be ignored, all earthquake foci can be conventionally condensed to the horizontal surface x, y. Then, activity A can be related to an area of this surface. In such cases, we can conventionally speak about the surface of seismic activity.

Under conditions of a relatively dense network of highly sensitive stations and observed relatively high seismic activity, it is useful to use average recurrency $N_k^*(0)$ of earthquakes as a unit of seismic activity. This equals an earthquake of class 7 ($K^{(0)} = 7$) with energy $E = 10^{7 \pm 0.5}$ J per year in volume $10 \times 10 \times 10$ km^3 for the volume seismic activity, or over the area $S = 10 \times 10 = 100$ km^2 for the surface activity. We shall denote this unit A_7 (the index corresponds to $K = 7$). Since most foci in these regions are sited within the depth (z) interval down to 10 km, numbers $A = A_7$ for volume and surface seismic activity almost coincide. There is obviously no need to specify each time what type of activity is meant: volume or surface, but it is necessary to distinguish them when earthquake foci with large depth intervals occur and their depth distribution cannot be ignored.

In regions where the network of stations is sparse and equipped with the same instruments, class 7 earthquake energy cannot be representative any longer. To define seismic activity it is more convenient to choose any other unit based on the common principle that A_7 is only related to some higher class, which is more representative in this situation. At the same time, it seems reasonable to enlarge the volume or area of consideration. Denoting indices (1) and (2) values of "old" and "new" activity units, respectively, we can write the following ratio between these units:

$$A^{(2)}/A^{(1)} = (S_1/S_2) \cdot 10^{\gamma(k^{(2)} - k^{(1)})} \tag{2.1}$$

This formula may naturally be used to work out new units of activity for more detailed observations.

For less detailed observations, we suggest the unit A_{10} of seismic activity, which is an average earthquake recurrence equal to one earthquake of class 10 of energy $E = 10^{10 \pm 0.5}$ Joule per year in volume 10^4 km^3 for volume seismic activity and over an area $S = 1000$ km^2 for surface activity. By substituting into (2.1) the assumed values of A_7 and A_{10} and supposing that $\gamma = 0.43$, we have $A^{(2)}/A^{(1)} = 1.95$, that is A_{10} is approximately twice as large as A_7. Respectively, activity in the same region at $\gamma = 0.43$ will be half as much in case of A_{10} than for A_7.

The value of $A = A(x, y, z)$ may be used to compile maps (in the plane xy) or cross sections (in the xz plane) of the seismic activity, and to plot curves (in the xt plane) of seismic activity changes in time t and space along line x. In all these cases, the average value as A obtained by integration within the volumes adjacent to the points of planes xy, xz or xt are ascribed to these points. To compile such maps and other similar curves, we may also use data on earthquakes of other classes, not only of the class which is formally used for A definition.

Density of seismic energy of earthquake sources. In order to characterize a seismic regime other than with recurrence N^* of earthquakes or related values (slope γ of recurrence curve and seismic activity A), we may use the density E^* of the seismic energy of earthquake foci in Π_s space with coordinates x, y, z, t, E. The density E^* will be function E^* $(x, y, z, t,$

E) determined as $E^* = \Sigma E/\Delta\Pi = nE/\Delta\Pi = \bar{E}N^*$, where ΣE is the sum of the seismic energies of all the earthquake foci in the volume element $\Delta\Pi$ of space Π_s; E is the average energy of any individual earthquake corresponding to this volume. Value E^* has dimension $L^{-3}\,T^{-1}$.

The value E^* is associated with the volume density W^* of the power of the summary seismic energy flow in the earthquake sources in physical space

$$W^* = \int_{E_1}^{E_2} E^* dE, \tag{2.2}$$

where E_1 and E_2 are boundaries of the range within which the earthquake energy varies and within which the power is determined. In the most interesting case, $E_1 = 0$ and $E_2 \to \infty$. Physically, it is clear that W^* should have finite value in this case as well.

If we delineate a definite volume V in space x, y, z with volume v, the power W of an earthquake focus for this volume will be

$$W = \int_v W^* dx\, dy\, dz = \bar{W}^* v, \tag{2.3}$$

where $\bar{W}^*$ is the average density of power for the volume of physical space for the given range of energies E.

If the earthquake foci are conventionally condensed at the horizontal surface x, y, then by analogy with the surface seismic activity, we can refer to the surface density of the power of earthquake sources. This value can be written with an expression similar to (2.3). In fact, this value, determined at the infinite upper limit ($E_1 = 0$, $E_2 = \infty$) of integration, is achieved practically by summation of the energies of all observed earthquakes. It was proposed in [M. Bath] for a quantitative estimate of seismic regions and for the corresponding mapping. The logarithm of this value, sometimes called the seismicity coefficient [Bune], may be used for the same purpose. Note that in order to map seismicity it was proposed to sum the square roots of earthquake energy [Bath, Ritsema] similar to papers by Benioff or to use magnitudes M as Gutenberg and Richter did. However, both methods seem to have a less distinct physical sense.

The energies E^* and W^* and others have a convenient property in that they can be directly summed or integrated within any range of energy variations, whereas values connected with the recurrence N^* and others can only be directly summed over a rather narrow energy interval. It should be kept in mind, that in order to use this property of energy values, we need to know the recurrency curve $N(E)$ in full, including probable areas of function bending as $E \to 0$ and $E \to \infty$, and not only the area of straight-line segment $\gamma = $ const. The fact is that if $\gamma = $ const, the integral of energy

$$W^* = \int_0^\infty E^* dE \tag{2.4}$$

within infinite limits, diverges at any numerical values of γ.

In fact, let us assume that the power density of earthquake foci of a fixed energy class K is a given finite value $W_0^* = W^* \Big|_{E_\alpha}^{E_\beta}$ where E_α and E_β are boundaries of this class. We shall consider, as is usual, that the classes correspond to energy orders, that is the average energy E of individual events of two neighboring classes differ by 10 times. First, we take $\gamma = 1$. In this case number N of earthquakes decreases by 10 times when energy class E increases by one. The summed energy of all the events in neighboring classes is obviously the same and equals $NE = cW_0^*$ (here, c = const is a conforming dimensional factor). The same will be true for any class of this sequence. If earthquake classes form an unbounded sequence in both directions of the given class K, the summed energy of all classes of events increases unlimitedly in both directions (both, towards stronger and weaker earthquakes), and, therefore, the integral (2.4) diverges. Now, let us assume that $\gamma > 1$. This will correspond to the increase in the number of weak events, thus amplifying the divergence of the energy sums in this direction. The assumption that $\gamma < 1$ will cause an amplification of energy divergence but towards severe earthquakes. Thus, energy integral (2.4) diverges at any values of γ = const.

Practically, the value γ is in the linear segment of the recurrency curve, and is usually 0.4-0.5 (i.e. $\gamma < 1$), when the energy integral diverges towards stronger events. This can explain the situation when, in the analysed range of energies, seismicity estimated from the summed energetic indices is usually very unstable: each severe event changes the seismicity estimate in the region. On the contrary, seismicity estimates, derived from the values related to recurrence (seismic activity A), are usually much more stable. This clearly is very useful in practical terms, in particular for long-term seismic zoning.

Physically, it is clear that unlimited power density W^* of the summed seismic energy of earthquake foci cannot exist. Therefore, the assumption of unlimited linear extension of the recurrence curve $N(E)$, that is the regularity γ = const for $\gamma < 1$ towards severe events, and γ = const for $\gamma > 1$ towards weak earthquakes, does not seem consistent.

We shall not consider the behavior of the recurrence function $N(E)$ for weak events. We shall note, however, that at normally observed values of γ, the summed effect of all weak earthquakes upon the total energy balance of earthquakes is not great. As for severe events, their role in this balance is dominant and, therefore, the behavior of the recurrence function $N(E)$ in this area is of principal concern.

For world catastrophies this problem is solved by the Gutenberg and Richter curves. At present, however, it is not clear how stable the geometry (slope) of the curve $N^*(E)$ is, especially in areas of great energy in various regions. When the real behavior of the function $N^*(E)$ in very energetic regions under various conditions is clear, and the function $\gamma = \gamma(E)$ is determined with sufficient reliability, the integral energy characteristics W^* of the seismic regime will probably have wide practical applications.

The power density W^* of an earthquake focus is related to the recurrence N^* by the following expression:

$$W^* = \int_{-\infty}^{\infty} EN^* dK, \qquad (2.5)$$

where $N^* = dn/dK$, $E = 10^k$, n is the number of earthquakes per element of Π_s space. Since the function $N^*(E)$ is known—it is expressed by the recurrence curve—formula (2.5), it allows us to estimate the average value W^* with greater reliability than with usual estimates of "energy flow" of earthquake foci. These are achieved by a direct summation of the energy E_i of totality Q of observed earthquakes over the given area S for the given time t:

$$W^* = \frac{1}{St} \sum_{i=1}^{Q} E_i$$

Note that a similar approach with a recurrence law may be applied to the estimate of the "conventional strain" ε, introduced by Benioff. For average value ε, by analogy with (2.5), one can write:

$$\varepsilon = \int\limits_{-\infty}^{\infty} \sqrt{E} N^* dk$$

This formula may be used to obtain more realible average values of ε instead of the simple summation:

$$\varepsilon = \frac{1}{St} \sum_{i=1}^{Q} \sqrt{E_i}$$

The use of energy values to present seismic regime dynamics in time, in particular as function of conventional strain ε, is of special concern.

Dispersion measure R of earthquake recurrence. After analyzing seismic regime by distribution of points x, y, z, t, E in Π_s or analyzing the functions and parameters N^*, A, γ etc. characterizing this distribution, we can distinguish two aspects of seismic regime, namely systematic and random.

The study of the systematic aspect concerns averages obtained by complete averaging within limited or, by coordinates, formally unlimited areas. So, when compiling earthquake recurrence curves $N^*(E)$, a complete averaging of observed values N^* for the entire time t of observations is usually performed. By analogy, maps of seismic activity $A(x, y)$, etc. are compiled.

The random aspect mainly concerns the speed of N^* (and other indices of seismic regime), determined in small volumes of Π_s space about the corresponding average values $\overline{N}^*$ related to large volumes; the relations (regressive or correlative) between parameters, characterizing the seismic regime, etc. These problems are treated by mathematical statistics and probability.

The main problem in this field is the spread of the observed recurrence N^* of the earthquakes with energy E and time t, the problem of seismic regime fluctuations. It is related to the accuracy and reliability of estimates made for the long-term average characteristics of seismic regime. The data for which are collected, of course, during the relatively short period of time of actual observations. It is also important to establish these characteristics with the desired accuracy during the observations.

Fluctuations in the seismic regime may be considered in the first approximation, using the concept of the stability of an average regime over time. The next approximation may be the supposition that the average parameters of the regime are linear in time or periodic in the long term, etc. However, in most cases, the lack of reliable actual data, means we must stick to the first approximation and to use the simplest methods to analyze the distribution of random values. In this case, the spread of N^* (or A, etc.), determined for a series of relatively short observation periods, may be estimated by the standard (mean square) deviation σ_{N^*} of N^* from the long-term averages of N^*. These can be regarded as close to the "real" ones characterizing, a presumably constant, average seismic regime. Instead of σ_{N^*}, we may use another parameter obtained from statistics and the theory of errors. However, this does not change the essence of the problem.

However, the σ_{N*} depends on the duration of observation, the size of analysed areas and various other conditions. In this sense they cannot be used as natural characteristics of the spread of the earthquake process proper.

The value $R = \sigma_N/\sqrt{N}$ does not have this defect. N is the recurrence itself, or the recurrence density, etc. and σ_N is the corresponding standard deviation. So, if N is a monthly average of earthquake number in the given energy class, derived, say, from annual observations, σ_N is the standard error for a single determination of the monthly average. The value R characterizes the scatter in the spread of earthquake recurrence. On average, the spread R turns out to be close to one in several areas of the study region of TCSE, irrespective of the duration of the observations, the dimensions of the areas, the level of seismic activity, or the level of earthquake energy for energies in the range $E_{max}/E_{min} > 10^{10}$. The equation $R = 1$ indicates the absence of any relation between the moment of earthquake origin for which the average recurrence is calculated. However, in some cases, that is within individual areas and in some particular time intervals, for instance during the aftershocks resulting from a severe earthquake, the spread shows a distinct deviation from the average. It may be caused also by other violations in the seismic regime, its deviation from the "normal" long-term average and, respectively, with the deviation of the parameters A and γ from their long-term averages. Large R corresponds to an increase in the instability of the area's seismic regime in the range of E, t, etc. The value R, like A and γ, is a major parameter of a seismic regime. It characterizes its "random" aspect, whereas A and γ characterize its "systematic" aspect.

2.1.2. Analysis of Seismic Regime Parameters

We shall consider the principal ways of processing and presenting material on a seismic regime by organizing data accumulated in an expedition. These methods are used to study the dynamics of a seismic regime over time, or in space and time, the distribution of earthquake recurrence by energy released, the fluctuation of a seismic regime, and the correlations between the major parameters of a seismic regime.

Temporal variations of seismic regime. The simplest method of reflecting the temporal variations of a seismic regime is to show the distribution pattern of discrete earthquake points in a five-dimensional space Π_s. However, it is not possible to project the complete pattern onto a plane, which has only two dimensions. We must, therefore, exclude some of the dimensions, retaining only the most significant for a specific purpose, including, naturally, time t. Then, for the coordinates excluded from our consideration, earthquake points in adjacent areas of Π_s space are summed by coordinates.

1. Graphs of temporal variations of seismic regime. If a definite volume of x, y, z space or an area of xy surface with a corresponding extension down to a depth z is established, a graph of the seismic regime's variation in E, t coordinates may be plotted. This graph shows all the earthquakes, whith foci within the volume. The scale of t is assumed to be linear as long as there is no need to specify any particular points in time. If a particular point such as the start of a severe earthquake, exists, it is convenient to study the fore- and aftershocks using an irregular scale, whose "scaling" increases as it approaches the specified point. This scale may be logarithmic, for instance, although this is not very suitable as a logarithmic scale has no reference point and it must be chosen arbitrarily. The scale may be of the form

$\sqrt[m]{t}$ for $m > 1$. The energy scale should be non-linear (for instance, logarithmic or $\sqrt[m]{E}$ type) so as not to obliterate weak earthquakes, whith is much smaller energy than severe events on the same graph.

2. Space-time graphs of seismic regimes in the x, t plane. Such graphs are useful when earthquake fice are elongated in one direction (for instance, along the axis of a ridge). In this case the x spatial coordinate corresponds to this direction. Earthquake foci falling within a belt on an epicentral map, with the x axis in the middle, are dropped along the perpendicular to line x. Within the x, y, z space, this belt corresponds to a volume with depth z. If these fice are sited within a relatively narrow depth range, their differentiation with depth may be ignored. The earthquakes are indicated on the graph in the x, t plane by closed circles, the centers of which have corresponding coordinates along the x (space) and t (time) axes, whereas the energy is expressed by the radius of the circle.

When the number of earthquakes is large, it is convenient to plot the points using the data by the decade or half-month, etc. but not the individual earthquake time. In this case, the earthquake points in an x, t graph are spaced along the lines $t = $ const at regular intervals Δt. It follows from experience that a reasonable choice of interval Δt will not distort the overall picture of the graph, but will save time in compilation.

Since the energetic classification of earthquakes is exponential $E = 10^k$ Joules ($K = 1$, 2, 3 ...), the scale of the circle radii for the energy should also be naturally exponential.

However, we cannot assume radii proportional to $\sqrt{E}$ or $\sqrt[3]{E}$, which would be desirable from geometrical considerations, because they would increase too rapidly with increasing E. Therefore, the radii scale must only be illustrative. This does not allow automatic photometry of the graph to average it with the definition of numerical indices.

It does not seem possible to show on a plane graph the temporal variations of a seismic regime in a two-dimensional area x, y, or to indicate simultaneously earthquakes with various energies. A series of graphs is necessary to do this. It is possible to use a series of maps of epicentres or seismic activity compiled for a number of time periods.

Attempts have also been made to plot $E = E(t)$ graphs with smoothed values. These graphs were found to be unclear. Smoothed graphs with space-time variations of the seismic x, t plane, but based on the same principle as maps of seismic activity are more useful.

3. Benioff conventional strain accumulation graphs. When Benioff tried to establish a correlation between the energy E_i of individual earthquakes in a region and the strain ε_i originating as the result of these earthquakes, he assumed $\varepsilon_i \sim \sqrt{E_i}$. To obtain the summed strain in a given area resulting from a sequence of earthquakes, he wrote the sum $\varepsilon_\Sigma = \Sigma C_i \varepsilon_i$, where C_i are the coefficients depending on the character of the focus and the elastic properties of the medium. However, in practice C_i are usually ignored, in favour of the simpler formula:

$$\varepsilon_\Sigma = \Sigma \sqrt{E_i}. \tag{2.6}$$

This formula was used to determine the ordinates of a stepped graph of a conventional "strain" accumulation originating as seismic energy was released during a series of earthquakes; the time t of the earthquakes was plotted along the abscissa.

This compilation is based on the assumption that each earthquake in the sequence creates strain of the same type and accordingly causes relative motion in the same direction in the medium. Only in this case can the strains be summed arithmetically as in (2.6).

This assumption seems to be reasonable for a sequence of severe events associated with motion along a large fault, but for numerous weak earthquakes it is not justifiable. Moreover, when studying a seismic regime, it is the weak earthquakes that are the main concern. Individual shocks may be associated with different directions of motion in the earthquake foci. It is only their combined effect, accumulated over time, that acquires more regular character of a plastic flow in some direction.

It was therefore assumed that over short time intervals Δt_k—almost every ten days— the strain ε_i is not summed arithmetically (not in one direction), but geometrically as in the case of an orthogonal system of vectors, i.e.:

$$\varepsilon_{\Sigma_k} = \sqrt{\sum_m \varepsilon_i^2} \sim \sqrt{\sum_m E_i} = \sqrt{E_k},$$

$i = 1, 2, ..., m$; for the longer time intervals, the total deformations are arithmetically summed in one direction, i.e.:

$$\varepsilon_\Sigma = \sum \varepsilon_{\Sigma_k} = \sum_n \sqrt{E_k} = \sum_n \sqrt{\sum_m E_i}, \tag{2.7}$$

where m is a number of events every 10 days, and n is the number of 10-day intervals in the total observation period.

Formula (2.7) was used to estimate the ordinates for the conventional strain accumulation graphs. Thus, within each 10-day period, the earthquake energies E_i proper, and the square roots of the summed energies E_k in-between were summed. This ordering in the graph plotting exercise is advantageous as it simplifies the processing of large quantities of material.

Note that with graphs plotted in this way, Benioff graphs compiled according to (2.6) do not provide quantitative values for any real strains. In fact, are only illustrative of the general variations in seismic energy release over time. In this respect, conventional strain accumulation graphs are equivalent to graphs of the seismic regime $E = E(t)$, and represent, to some extent, an integral form.

4. Graphs of variations of seismic regime parameters over time. Firstly, we discuss the functions of time t for the major parameters of the mean seismic regime, namely seismic activity A, the parameter of the recurrency curve $N(E)$, the slope γ, and the spread R of the recurrency.

When compiling graphs of this type, one should estimate the accuracy in determining the values. Otherwise, occasional fluctuations in these values due to determination errors may be taken as systematic changes in the regime. The elementary time intervals Δt for which A, γ and R can be determined, should be sufficiently long to include enough observations to achieve satisfactory accuracy; at the same time, the intervals should be small enough (compared with the total observation time) to follow variations of the regime in sufficient detail. These requirements for Δt are clearly not mutually complimentary. It is easier to meet both where the seismic activity is considerable, when the K classes of the earthquakes are more numerous (that is, for weak events), and when there are large areas S (or volumes) and observation periods. Note that the problem with time intervals also exists for the areas or volumes: areas over which the averaging is performed should not be too large in order to avoid averaging out features of the regime characteristic of individual segments.

Graphs of variations over time of the various parameters of a seismic regime reveal the occurrence or absence of characteristics prior to strong shocks. This is especially important for severe earthquake prediction.

Distribution of earthquake recurrence by energy. Recurrence curves $N(E)$ for large regions and long time spans are based on monthly earthquake recurrence numbers N for various energy classes K. Monthly average recurrencies $\bar{N} = (\Sigma N)/n$ are found for each class K, where n is the number of elementary time intervals, i.e. the number of months, during the total observation period (in this case it is about two years). K and N were taken to compile the earthquake recurrence function for energy. The sequence of points was averaged graphically by a straight line. As usual the slope of this line determined the parameters $\gamma = - \Delta \log \bar{N}/\Delta \log E$, whereas the seismic activity A was defined by the intersection with the line $K = 7$. When determining A, the estimates were made for normalized recurrence N^*.

In some cases other elementary time intervals are used for individual segments. Hence, for the Chusal site (Garm region, Tadjikistan), recurrence values for only four-hour elementary intervals were used over a total of 3 days of observation. The same techniques were used to determine γ and A in all cases.

The errors are also shown on the recurrence curves. They are presented as ellipses, the semi-axes of which are equal (on the scale of the graph) to the errors along the corresponding coordinate axes. The methods of error estimation are given below.

Fluctuations in the seismic regime. We consider the following topics; accuracy in earthquake recurrence N^*, determination for various energy classes K; temporal character of N^*, the spread in the observations; the period necessary to estimate the recurrency N^* to the desired accuracy; establishment of the spread R in earthquake recurrence and accuracy estimation for the average spread derived from mass observations; and, finally, the application of R in nomogram compilation in order to determine the necessary observational periods for earthquake recurrence of various energy classes.

1. Accuracy estimation of earthquake recurrence. To do this, a number of recurrence values $N = N_i$ ($i = 1, 2, ..., n$) are processed for each class K of energy E. The recurrences are determined in a given region for relatively short time intervals (for instance, months) into which the longer period of observations (for instance, two years) is divided. When the total period is smaller, the time intervals are reduced respectively.

The following indices were calculated for the set of values $N = N_i$; the mean

$$\bar{N} = \Sigma N/n \qquad (2.8)$$

standard (mean square) deviation of single determination:

$$\sigma_N = \sqrt{\Sigma(N - \bar{N})^2/(n - 1)}, \qquad (2.9)$$

where σ_N^2 is variance of N; the standard deviation of the mean is:

$$\sigma_{\bar{N}} = \sigma_N/\sqrt{n}, \qquad (2.10)$$

where $\sigma_{\bar{N}}^2$ is variance of $\bar{N}$; where the relative deviations of a single determination is:

$$\sigma_N = \sigma_N/N, \qquad (2.11)$$

and that of the mean $\bar{N}$ is:

$$\sigma_{\bar{N}} = \sigma_{\bar{N}}/\bar{N} = \sigma_N/\sqrt{n} \qquad (2.12)$$

Fig. 5. Distribution of earthquake recurrences for different energetic classes K, Garm region, 1955-1956.

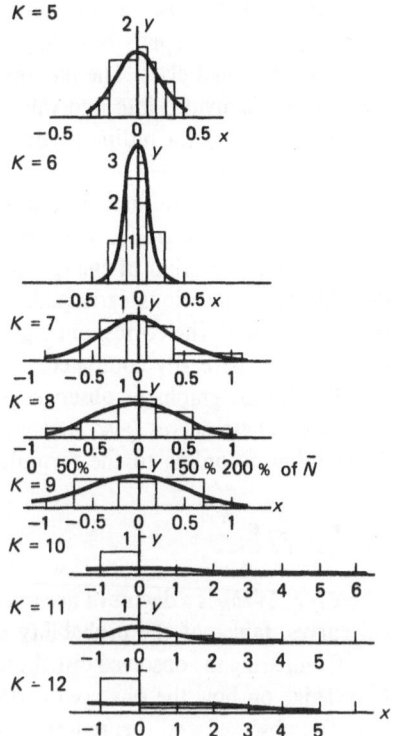

The relative deviations δ_N and $\delta_{\bar{N}}$ remain the same both for N and $\bar{N}$ and for the corresponding normalized recurrencies N^* and $\bar{N}^*$.

The value of δ_N has an important meaning. It can be regarded as the relative error in the determination of the actual long-term average recurrence N^* or $\bar{N}^*$ of earthquakes in the class and region from the data obtained for the observation period, if assuming that the average seismic regime in the region does not vary over time. In this situation, we can ignore the units used to measure the recurrence (whether it is one month or a year), the area of the region or the volume where earthquake foci are sited, viz. the seismic regime is the same over the whole region.

The error $\delta_{\bar{N}}$ depends upon total number of earthquakes ΣN from which it is estimated. Therefore, it will decrease, with increasing total observation period, area S (or volume) where the observed foci and seismic activity of the region are sited. It also depends on the spread of the recurrence from strict periodicity.

2. Comparing the monthly (or other) variances of earthquake recurrences over time to long-time averages. The goal of this research is to find out how the observed recurrence distribution deviates from the average for earthquakes of various energy classes and to compare this distribution with theoretical distributions, and also to determine the parameters of the corresponding theoretical distributions.

Graphs of the observed distribution density (histograms) for N were plotted (Fig. 5) for relatively long time spans $N = N_i(i = 1, 2, ...)$ corresponding to various energy classes K, for each individual class. The parameter $x = (N - \bar{N})/N$ is plotted along the abscissa such that the axis is divided into intervals Δx which are preferably equal. The value $y = y_n/(n\Delta x)$ is plotted up along the ordinate where n is the total number of determinations of N, whereas y_n is the number of determinations, for which N falls within the given interval of x. The lengths Δx and Δy are the base and height of the corresponding rectangle which forms the stepped distribution graph. The area under the graph is thus one unit.

This normalization of the distribution graphs provides an easy comparison between N distributions for various earthquake energy classes, the number of earthquakes varying in the same region. The corresponding data obtained in regions of different areas and seismic activities can be easily compared.

Besides the graphs of observed N distributions theoretical differential curves were also compiled. These curves (also normalized) were first plotted for the corresponding normal distribution according to the formula:

$$y = \frac{h}{\sqrt{\pi}} e^{-h^2 x^2}, \tag{2.13}$$

where $h = 1\sqrt{2}\delta_N$ is a degree of accuracy. The δ_N is determined from (2.11). To plot the theoretical curves, tables of the probability density for the normal distribution are used.

Comparing the observed distribution graphs with theoretical curves gives clear qualitative knowledge on how the observed earthquake recurrence N distribution fits the normal law.

The observed and theoretical accumulation graphs (ogives) were plotted on probability paper, on which theoretical curves for a normal distribution are transformed into straight lines (Fig. 6a). The $x = x_i (N_i - \bar{N})$ are plotted along the abscissa as usual, whereas the observed normalized accumulated frequencies of the recurrence $y = y_i = 1/n\Sigma y_{n_i}$ (that is $1/ny_{n_1}$, $1/n(y_{n_1} + y_{n_2})$, $1/n(y_{n_1} + y_{n_2} + y_{n_3})$) are plotted along the ordinate axis. The averaging straight line, representing the theoretical normal distribution, was drawn through the set of points. Note that the distance along the abscissa between the points of the straight line with ordinates $y = 15.9$ and 84.1%, which are symmetric with respect to the middle $y = 50\%$, is equal to the twice the relative standard deviation $2 \delta_N$. This may be used as a graphical determination of δ_N, irrespective of (2.8)-(2.12).

As a result of the analysis of the distribution graphs presented both in differential (Fig. 5) and integral (Fig. 6a) forms, we may make the following conclusions. For the lower energy classes K, in the example, it is $K = 6$-10, for which the recurrences N are relatively large, the observed points are plotted well along the corresponding theoretical graphs for normal distribution. For the higher classes $K > 11$, where the number of N discusses sharply. In fact during some of the time intervals into which the total observational period was divided, there were none, at all, the observed distributions become asymmetric and differ considerably from normal, as can be seen from Fig. 5. At the same time, it is difficult or sometimes impossible to show the corresponding integral data on accumulation graphs (Fig. 6a).

The observed recurrence N distribution of the earthquakes of various classes K is compared with a theoretical log-normal one. This theoretical distribution is asymmetric with respect to $\bar{N}$, the asymmetry generally being the same as observed in graphs for the higher classes in Fig. 5. The log-normal distribution for N is also more convenient as it excludes

Fig. 6. Cumulative earthquake recurrences,
Garm region, 1955-1956.

a—Representation on any paper; b—representation
on logarithmic paper

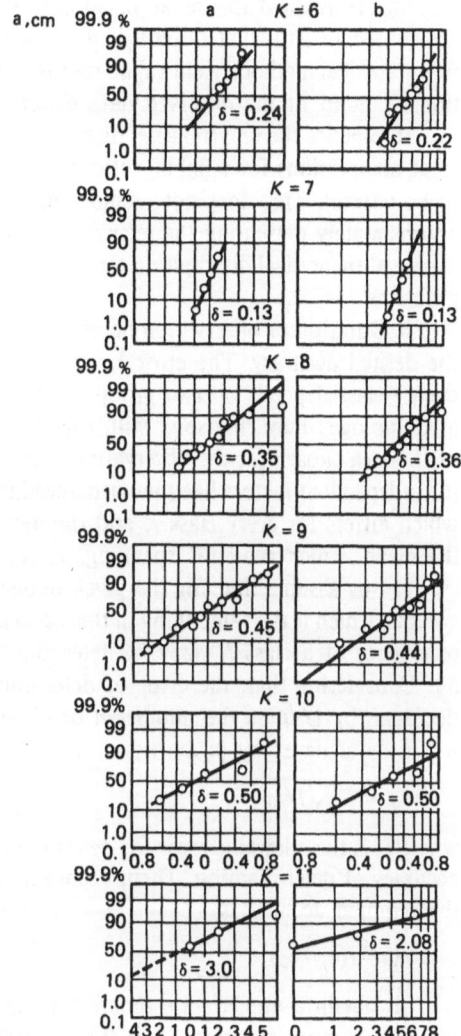

negative N, whereas with the normal distribution such values are admitted, although with
tiny probabilities. Physically, negative earthquake recurrences N are impossible.

The comparison was carried out by compiling the accumulation curves for the observed
distributions and for the corresponding theoretical ones on log-normal probability paper,
on which theoretical graphs for a log-normal distribution are straight lines. The abscissa axis
is logarithmic in these graphs (Fig. 6b). The scale $\log(x + 1)$ is marked on it, where
$x = (N - \bar{N})/\bar{N}$. As usual, the notation on this graph for x is the same as in Fig. 6a and
it characterizes relative deviations of observed N from the mean $\bar{N}$.

A comparison of the normal (Fig. 6a) and log-normal (Fig. 6b) distributions with the
observed ones indicated that for lower energy classes K, in which observations are abundant

and thus convenient for the application of statistical methods, both theoretical distributions are consistent with that observed. At higher K classes for which observed earthquakes are few, statistical methods yield vague results in both variants, although the log-normal distribution seems to fit the observed data better.

In practice the recurrence N distribution over time for the energy classes that have sufficient observations for reliable statistical results, may in the first approximation be assumed to be normal. This does not exclude the possibility and, in some cases the expediency, of approximately presenting the recurrence distribution N over time by a log-normal law. The choice of theoretical distribution mainly depends on the convenience in particular geophysical problems.

3. Duration of observation necessary to determine the earthquake mean recurrence with the desired accuracy. The error δ_N in determining the mean earthquake recurrence for an energy class depends on two groups of factors. The first group includes the factors which a seismologist may choose at will, e.g., the time t of the observations and the dimensions of the area chosen for the observations (its area S or volume V). The second group contains the independent factors like the normalized recurrence N^* of the earthquakes in a given area, which differs for every class K and the degree of natural deviation of the recurrence from the mean, which generally speaking, varies from one energy class to another.

Let us assume that for the given dimensions of the observation region, its area S (or volume, which is the same thing in this case) and seismic activity A, and the mean earthquake recurrence of a class K, may be determined in the region for a time t with relative error δ_N. Considering that: the error of determining the mean recurrence $\bar{N}$ according to (2.12) decreases by $1/\sqrt{n}$ as the number n of observations increases and that increase in number n can be achieved by proportional increase of time t of observations, we can write:

$$\delta_0 = \delta_{\bar{N}}\sqrt{t/t_0} \tag{2.14}$$

where δ_0 is the allowable error for determining the recurrence $\bar{N}$ and characterizes the desired accuracy of determination. Then, according to (2.14) the time t_0 necessary to obtain N with the required accurcy, is

$$t_0 = t(\delta_{\bar{N}}/\delta_0)^2 \tag{2.15}$$

To obtain a numerical estimate for the time t_K necessary to determine the recurrence N_K with a given accuracy $\delta_{\bar{N}} = 0.1$ ($= 10\%$) for various energy classes K in the Garm region, earthquake observations for the period from January 1955 to November 1956 were statistically processed (Table 4).

The Table shows the unnormalized monthly averages for recurrence $\bar{N}$ for the observation period (23 months) and also the relative errors δ_N for the monthly determinations calculated from (2.12) due to processing these observations. The Table also has the times (in years), calculated from (2.15) necessary to determine the recurrence $\bar{N}$ (monthly or yearly, from unnormalized or normalized recurrence $\bar{N}^*$) with a given relative error $\delta_N = 0.1$ ($= 10\%$). Similar estimates were done for the Big (area $S = 14\ 300\ \text{km}^2$) and Small ($S = 4000\ \text{km}^2$) Dushanbe regions.

Figure 7 contains these results as graphs: (a) for the Garm region and (b) for the Dushanbe region. Graphs were plotted with log-log coordinates with similar moduli on both scales. Along the abscissa $\bar{N}$ is plotted, from the right to the left, numbers $\bar{N}$ on a logarithmic scale, the values of t_0 being plotted on the same scale upward along the y axis (Table 4).

Table 4. Estimation of Time t_0 Sufficient to Obtain the Earthquake Recurrence Value $\bar{N}$ with the Error of $\delta = 10\%$ for Different Energetic Classes K, Obtained from the Observed Data of Garm Region

K	$\bar{N}$	$\delta_{\bar{N}}$, %	t_0, years	K	$\bar{N}$	$\delta_{\bar{N}}$, %	t_0, years
7	60.6	2.4	0.107	11	0.70	35.2	23.5
8	18.6	7.9	1.16	12	0.48	29.0	16.0
9	7.1	9.3	1.66	13	0.17	47.9	43.3
10	3.2	10.4	2.04				

Pairs of values $t_0 \bar{N}$ for various K classes represent the coordinates of the observed points and are indicated by double circles. We can see that the points are well distributed along a straight line. The position of the theoretical points (small circles) is determined taking into account R (see below).

We next consider the principal conclusions about the seismic regime in general.

4. Determination of the spread R of earthquake recurrence. The straight line averaging the distribution of the points with coordinates $\log \bar{N}$, $\log t_0$ in the graphs in Fig. 7 has a slope of almost $45°$ to the abscissa, which means that $t_{0K} = C/N_K$, or

$$t_{0K} N_K = C \tag{2.16}$$

for $1/N$ and t_0 values for various energy classes K; where C is a constant independent of K, the subscript K to N and t_0 shows that both belong to the same K class and vary with K.

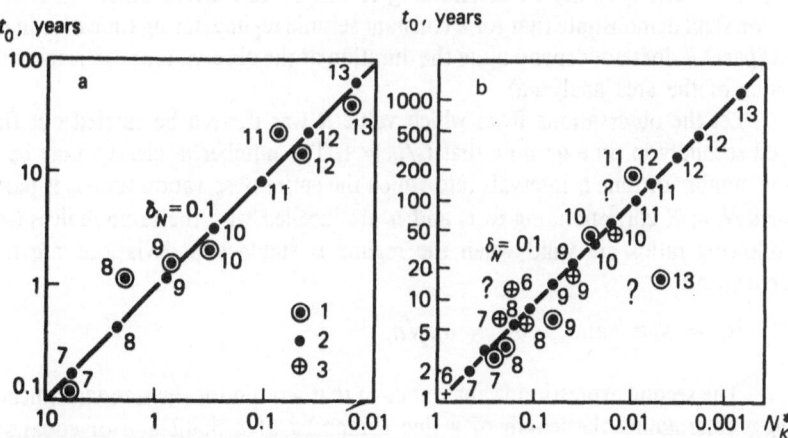

Fig. 7. Observation time interval sufficient to estimate the earthquake recurrence N with the given accuracy of 10%.

a—Garm region; b—Dushanbe region; *1*—observed values; *2*—calculated values; *3*—observed values for Close Dushanbe region

Let us consider (2.16). The term $t_{0K} N_K$ has a simple physical meaning. It equals the total number of earthquakes in the K class recorded in the time t_0 necessary to determine the recurrence $\bar{N}$ with a given error and is the same for all classes K.

Note that the approximate stability of C for all energy classes is, in this case, theoretical and does not follow from any logical construction. It seems more natural to expect that C depends on K, that is on the earthquake energy. Thus it is probably true when the energy range is wider than that analyzed and the assumption C = const is found to be inconsistent. From the available data, C is approximately constant in an energy range of 13 orders of magnitude (according to data from the Garm and Dushanbe regions, K = 7 to 13, whereas for Chusal region K = 0 to 7).

If C is the constant we can conclude that the error δ for the earthquake recurrence to be determined depends, on average, on the total number of earthquakes N_Σ from which the recurrence is determined, irrespective of the recurrences, duration of observation or importantly energy class K.

This makes us believe that the spread of earthquake recurrence in the various classes, on which δ depends, remains the same, on average, if the spread is related to the earthquake sequences in each class which have the same number of earthquakes.

Now, it is easy to find the value associated with the recurrence dispersion, which should remain approximately constant for all energy classes for which (2.16) is approximately observed. This value is the recurrence spread and has the form:

$$R = \sigma_{N_\Sigma}/\sqrt{N_\Sigma} = \delta_{N_\Sigma}\sqrt{N_\Sigma} \qquad (2.17)$$

where N_Σ is total number of observed earthquakes of a class K, for which R is determined, whereas σ_N and δ_N are the corresponding standard deviations (σ_N is absolute; $\sigma_N = \delta_N/N$ is relative).

To be exact, R is the limit towards which (2.17) tends as $N_\Sigma \to \infty$. However the application of (2.17) to observed data shows that it may also be used for relatively small N_Σ, about N_Σ = 10. The accuracy in determining R will be considered later.

We shall demonstrate that for a constant seismic regime during time t (or in other estimates, in space) R does not depend upon the duration of the observations (or, respectively, the dimensions of the area analyzed).

Let the observations from which value R was derived be carried out firstly in time t_a and secondly in time t_b; note that $t_a/t_b \gg 1$. The number n, clearly, may be assumed to be the number of time t_b intervals into which the entire observation time t_a is partitioned. Numbers N, σ, R corresponding to t_a and t_b are labelled with the same indices (a) and (b). The following ratios are valid when the regime is stable and deviations are random and independent:

$$N_a = N_b n \quad \text{and} \quad \sigma_{N_a} = \sigma_{N_b}\sqrt{n}. \qquad (2.18)$$

The second expression is analogous to that written for the standard (mean squares) error σ_{N_a}, determining the length of a line composed of a number n of equal segments, if the standard error of each segment (a random parameter) is σ_{N_b}. According to (2.17) and (2.18) we have $R_a = \sigma_{N_a}/\sqrt{N_a} = \sigma_{N_b}\sqrt{n}/\sqrt{nN_b} = \sigma_{N_b}/\sqrt{N_b} = R_b$, QED. By analogy, we can prove that certain values of R do not depend either on the dimensions of the observational area or on the seismic activity.

Table 5. R_K Values and σR_K Errors for Determining R Measures of Scattering of $\bar{N}$ Earthquake Recurrences in Various Classes of K Energy

K	$\bar{N}$	$\sigma_{\bar{N}}$	R_K	σR_K	ΔR_K
6	115.0	0.238	2.55	0.390	0.690
7	60.0	0.114	0.892	0.134	0.185
8	18.6	0.375	1.62	0.252	0.548
9	7.1	0.449	1.20	0.188	0.452
10	3.2	0.498	0.893	0.142	0.357
11	0.70	1.69	1.42	0.328	1.42
12	0.48	1.39	0.962	0.202	0.812
13	0.17	2.30	0.950	0.269	1.24

Thus, R can be determined exclusively from the spread of the earthquake recurrence typical of the process of eathquake occurrence which, as a set, comprise the seismic regime of the area.

Since, for the same N, the spread grows with increasing deviation σ_N (or δ_N), N characterizes, on average, the deviation of the earthquake recurrence from the rigorous periodicity.

Let us now consider the average spread R determined from the observed data and the accuracy of this estimate. The calculations were made using observations in the Garm region from January 1955 to November 1956 (Table 5).

In Table 5, K stands for the energy class of the earthquakes; $\bar{N}$ is the monthly recurrence average from the observational data for the total period $t = 23$ monts; $\delta_{\bar{N}}$ is the relative standard deviation for a month $\bar{N}$. Note that the minimum number of earthquakes for time $t = 23$ months, from which N was determined, was recorded for $K = 13$ and was equal to 4 ($4/23 = 0.17$). Table 5 shows the spreads R_K for various classes K as calculated by (2.17), and the corresponding standard deviations σ_{R_K} (errors). These were calculated using the formula:

$$\sigma_{R_K} = \sqrt{\left(\frac{\partial R_K}{\partial \sigma_N}\right)^2 (\sigma_{\sigma_N})^2 + \left(\frac{\partial R_K}{\partial N}\right)^2 (\sigma_N)^2}, \tag{2.19}$$

where R_K is given by (2.17). Here, are some details of the estimation. The value of σ_{σ_N} is determined by the formula for a standard deviation $\sigma_{\sigma_N} = \sigma_N / \sqrt{2n-1}$, taken from errors theory. In our case, $n = 23$ (number of months) so $\sigma_{\sigma_N} = 0.151$. Then:

$$\frac{\partial R_K}{\partial \sigma_N} \sigma_{\sigma_N} = \frac{1}{\sqrt{N}} 0.151 \sigma_N = 0.151 \sqrt{N} \delta_N$$

$$\frac{\partial R_K}{\partial N} \sigma_N = \frac{1}{2} \frac{\sigma_N}{N\sqrt{N}} = \frac{1}{2\sqrt{n}} \frac{(N\delta_N)^2}{N\sqrt{n}} = 0.104 \sqrt{N} \delta_N^2.$$

Then, finally:

$$\sigma_{R_K} = \sqrt{\frac{1}{2(n-1)} N\delta_N^2 + \frac{1}{4} N\delta_N^4} = \delta_N \sqrt{N(0.027 + 0.0109\delta_N^2)} . \tag{2.20}$$

Two things about Table 5 attract our attention. The first is the very large bias in R and the large value σ_{R_K} for $K = 6$, although this class is the most representative. The second is the very slow growth in the errors σ_{R_K} as K increases and, accordingly, N decreases. In particular, the small errors for $K = 12$ and $K = 13$ seem strange, although the numbers for these classes are so small that it was difficult to process their corresponding data statistically.

The abnormal behavior of R_K and δ_{R_K} for $K = 6$ can be explained by the poor quality of the observed data used to determine them. No doubt, systematic errors tending to underestimate them were allowed when calculating the number of earthquakes in this considered class, or possibly different observation durations were taken. The fact is that the sensitivity of modern instruments and the location of the stations in the network (the effect of station background) do not allow complete recording of weak events within the entire region. This also caused the deviation of $K = 6$ at the recurrence curve to exceed the random error.

To verify the general trend in the errors when determining R_K for $K = 6$ to 13, the errors were estimated by taking the logarithms and differentiation of them. From (2.17), we determine $\Delta R_K/R_K = \Delta\sigma_N/\sigma_N + 1/2\Delta N/N$ where ΔR_K, $\Delta\sigma_N$ and ΔN are the errors of the corresponding R_K, σ_N and N. Assuming that $\Delta\sigma_N \approx \sigma_{\sigma_N} = 0.151$ σ_N and $\Delta N = \sigma_N$. Then, $\Delta R/R_K = 0.151 + 1/2 \sigma_N/N = 0.151 + 0.5\delta_N$, and finally

$$\Delta R_K = R_K(0.151 + 0.5\delta_N). \tag{2.21}$$

The calculated errors ΔR_K yielded by this formula are given in Table 5.

Comparison of the errors σ_{R_K} and ΔR_K shows that the general trends of both are similar, but the lower reliability of high energy classes K, for statistical analysis has been mentioned. The slight increase in these errors with increasing K is typical of ΔR_K (and σ_{R_K}), which may indirectly prove the validity of the calculations of σ_{R_K} by a more rigorous method.

Let us proceed with the determination of the average spread R of the earthquake recurrence and its error estimation using the data for R_K and σ_{R_K} for a series of classes K.

The weighted average of R is written:

$$R = \Sigma R_K\omega_K/\Sigma\omega_K, \tag{2.22}$$

where the weights ω_K are $1/\sigma_{R_K}^2$. The standard error of R is:

$$\sigma_R = 1/\sqrt{\sum_K \omega_K} = 1/\sqrt{\sum_K 1/\sigma\bar{R}_K^2} . \tag{2.23}$$

If the data from Table 5 are used for all classes K, except class $K = 6$, we have $R = 1.04 \pm 0.07$ from (2.22) and (2.23); if we also exclude the data for $K = 12$ and $K = 13$ and consider only the observations within the most reliable area, we have:

$$R = 1.05 \pm 0.08. \tag{2.24}$$

This is the average spread of recurrence for the Garm region for the period 1955-1956. Note that in this case, the numerical values of R and σ_K remain almost the same when we include or exclude the observational data on earthquakes of classes $K = 12$ and $K = 13$. The numerical average of R is close to one. However, we should not assume it to be exactly equal to one as it may vary in some cases and its variations, in time and space, may be of particular concern.

5. The use of average value R to specify the duration of the period of observation. It is easy to calculate from (2.19), assuming R is a certain value (2.24) that the number N of earthquakes in any energy class necessary to make N a long-term average characterizing earthquake recurrence with a given accuracy. This has an error δ_N. Assuming various numerical values for δ_N, we have:

δ_N, %	1	5	10	20
N	11000	450	110	28

Hence, in order to determine the average recurrence of earthquakes of a given class K with an accuracy say of 10%, about 100 earthquakes must be analyzed; to increase the accuracy to 1% over 10 000 events in the same class must be considered.

We want to see how long the observation period t should be to determine the average recurrence of earthquakes of any given class K with a given error, if we know the area S of the observation region or, precisely, the region (or volume) in which the foci are sited and the seismic activity A of the region. This is a natural generalization of the problems using the graphs in Fig. 7 for particular conditions in the Garm and Dushanbe regions to provide solutions.

To solve the problem, we express the total number of observed earthquakes in class K as A and S; and then:

$$N_K = \frac{SAt_K}{100} \cdot 10^{-\gamma(K-7)}. \tag{2.25}$$

Here, area S is in km²; A, by definition, is the number of events of class $K = 7$ over 100 km² (the number 100 in the denominator) per year; and γ is a parameter indicating the decrease in earthquake recurrence with increasing energy class K, (it is approximately constant and is assumed here to be $\gamma = 0.43$). By substituting (2.25) into (2.17) we have the final formula for t_K:

$$t_K = \frac{R^2}{\delta^2} \frac{100}{SA} 10^{-\gamma(K-7)}. \tag{2.26}$$

Here, the units of measurements are the same as in (2.25).

If the seismic activity A is not calculated for the area $S = 100$ km² but for the value $V = 1000$ km³, we substitute $1000/VA$ for $100/VA$, where V will be in km³.

Formula (2.26) may be applied to any region, if the region meets the initial conditions; although for approximate estimates there is no need for the numerical values of the parameters to be exactly the same. In particular, this formula may be used to systematize the estimates made earlier for the Garm and Dushanbe regions in the graphs in Fig. 7.

To facilitate estimates using (2.26) and to make the result after the various calculations a more understandable nomograph has been compiled (Fig. 8).

The left-hand vertical scale of the nomograph is double in relation to error δ and the number N of events in a class which must be observed to determine the long-term average recurrence of a given error is δ. Going from one scale to the other gives a correlation between

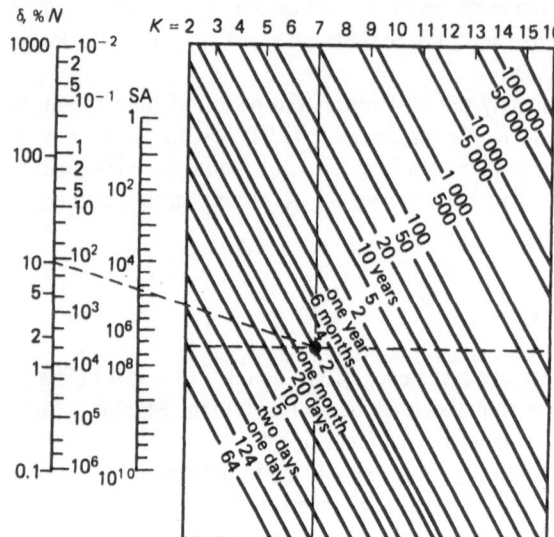

Fig. 8. Nomogram illustrating the determination of the observation time interval sufficient to estimate earthquake recurrences with the given accuracy δ (in %).

SA is the observation area (in km^2) over seismic activity $A = A_7$.

δ and N that satisfies (2.17) for $R = 1.05$. When making an estimate with the nomograph, in practice, side δ is used, side N being presented mainly for comparison.

The next vertical scale, a single one, is marked off in values of SA, where S is the area of the region in km^2 and $A = A_7$ (for $K = 7$) is the seismic activity.

Both vertical scales, together with the third one labelled $K = 7$ (this figure is circled), are to provide for the multiplication $(R^2/\delta^2)\cdot(100/SA)$ in (2.26). This part of the nomograph presents the ratio:

$$\log \frac{t_k}{10^{\gamma(K-7)}} + \log \left(\frac{\delta^2}{R} \right) = \log \frac{100}{SA}.$$

The three terms of the equation correspond to the three functional scales in the sequence: t_K (at $K = 7$); δ (and N) and SA. The multiplication is reduced to the summation of logarithms, which is done by drawing a straight line crossing all three scales.

If $K = 7$, the t_K are plotted along the vertical $K = 7$ where it intersects a series of inclined straight lines labelled by values of t_K.

To get t_K corresponding to other K, there are a number of vertical lines labelled by the corresponding numbers. Producing them from the $K = 7$ vertical is done by drawing additional horizontal straight lines. The t_K along the verticals $K = 8, 9, 10, ..., 6, 5, 4 ...$ are plotted exactly as for $K = 7$, i.e. where the vertical intersects the inclined lines labelled t_K. The slope of these lines corresponds to $\gamma = 0.43$. As an example of how to use the nomogram we use a calculation of t_K for the Garm region. Say we are interested in t_K of the observations in the region for various energy classes necessary to determine the average recurrence $\bar{N}^*$ of earthquakes in the classes with error δ not exceeding 10%. One should note that for the Garm region, the area S is 13 500 km^2, and its seismic activity A_7 is 5.0. The value of SA required for our estimates is $SA = 13\,500 \times 5.0 = 6.75 \times 10^4$.

Now, we draw a straight line through the points $\delta = 10\%$ (on the first left-hand scale) and $SA = 6.75 \times 10^4$ (on the second scale). Where this line intersects the vertical $K = 7$ we find the signs for the corresponding inclined straight line $t_K = 2$ months. This is the solution for earthquakes of class $K = 7$ under the given conditions.

To solve the problem for other values of K we can draw additional horizontal lines through the point we have not found. Where it intersects the vertical, e.g., $K = 3$, we find that it is sufficient to carry out observations for only one or two days. By analogy, the intersection of the horizontal line with the vertical $K = 16$ (the catastrophic Khait earthquake of July 1949 was $K = 16$) yields a value over 1000 years for the observation duration to determine the average earthquake recurrence of this class with an accuracy of 10%.

Note that the nomograph (Fig. 8) may help in other problems. For instance to find out what accuracy (or error δ_K) will be inherent in the average recurrence of earthquakes of some class K, if the observations in the region (with given S and A) are planned for a given time t_K.

2.2. On the Compilation Technique for Seismic Activity Maps[*]

Maps of seismic activity, epicenter maps and maps of maximum earthquake intensity are the principal graphical presentations of seismicity in the USSR. The compilation techniques for the intensity and epicentral maps have been developed for decades and, therefore they are standardized to a great extent and only the details can be improved. By contrast, the compilation of maps of seismic activity is far from being standardized. In various regions they often differ so much that they cannot be used together.

This is partly because the compilation technique for seismic activity maps was only recently proposed [Riznichenko]. The establishment of a unified technique has been hindered by the lack of a broad and detailed debate that would allow us to choose the best estimation and compilation variants.

This section is devoted to this debate. Various compilation techniques for seismic activity maps and various aspects concerning their accuracy and resolution are considered.

2.2.1. Compilation Techniques for Seismic Activity Maps

Maps of seismic activity A show the frequency of earthquake occurrence, averaged over space and time (events with a definite source value, that is, seismic energy or magnitude) per unit area. In other words, they represent the spatial density of earthquake recurrence with a definite value. Epicentral maps with indicated values for events at their sources are used as the initial material in all cases. The source depths are mainly considered as the depth range within which individual sources are sited, for instance crustal earthquakes.

Considering the general character of the initial data and the finite results, the ways in which these results may be obtained may vary. Here, we attempt to systematize some of the currently known ways. As the basic criteria for our classification we take: (1) the character of the initial values used to estimate the activity A; (2) the character of the approximating

---

[*] See [Riznichenko]. Co-author — I. V. Gorbunova.

function used to show the spatial distribution of A on the map; (3) the principle used to determine the averaging areas given the accuracy and resolution of the map A.

Initial data. The activity A is estimated at every point on the map from the calculated number of epicenters within the corresponding averaging area. Two estimation techniques exist:

(a) the number of epicenters is counted separately for each energy class of earthquakes used in the map; this may be called a *distribution technique* since it is based on the distribution density function for the earthquakes by energy class (or magnitude);

(b) the number of epicenters of earthquakes of all classes over a fixed value is totalled; it is based on a summation function (associated with the distribution function using statistical methods) therefore this can be called the *summation technique* [Riznichenko].

The distribution technique was described in [Riznichenko] before the summation technique [I. Gorbunova, Riznichenko] and is still more widely used [Masarsky et al.] even though the latter has some advantages.

The technical advantage of the summation method is that there is no need to consider separately earthquakes of different energies in order to determine the seismic activity A at each point of the map. In the distribution technique, such a consideration leads to additional estimates: either each point of the map requires its own simplified graph (or function) of earthquake distribution [Bune, Riznichenko], or a multiply stepped procedure is necessary, i.e., first, the activity maps for each class and, then, the combined maps must be compiled. The summation method reduces the compilation procedure to one step, which is as simple as compiling a map for only one class of events. In fact, compiling an activity map by the summation method becomes the same as compiling maps of epicentral density [N. Vvedenskaya], although it preserves the geophysical meaning of the seismic activity maps.

The principal advantage of the summation technique is that this technique also considers both numerous weak events and more powerful earthquakes which fall within the time and space area. In the case of the distribution technique, severe earthquakes are usually ignored when determining seismic activity, and only weak events are regarded.

Nevertheless, estimates using the summation technique, first used in 1965, only partly utilized this advantage because severe events have been recorded for much longer periods of time than weak ones. Systematic observations of the latter began lately (we speak about the "representative" earthquakes observed). Severe earthquakes which occurred before or after the period in which weak events occurred were ignored.

Testing the distribution and summation techniques for compiling seismic activity maps from observational material [Riznichenko], showed that both almost coincide within the accuracy of the activity maps, given the statistical spread. This allows us to choose the summation technique as the simpler one.

The character of the approximating function. As the result of the estimates of A in all averaging areas $i = 1, 2, 3, \ldots$ a net of discrete values of A is plotted on the map, which should then be combined into a general system covering the plane. If we assume A to be the vertical coordinate of a point (x_i, y_i, A_i), where x_i and y_i are the horizontal coordinates in the map plane, the problem is to plot the surface $A = A(x, y)$ crossing all the discrete points (x_i, y_i, A_i) at or near their general position. The relief of this surface must be drawn on the seismic activity map.

At present, there are two main ways of constructing this surface. In the first, the surface is assumed to be smooth. It is usually, but not always, constructed as a moving average, viz. the dimensions of the averaged areas are greater than the cells of the net at the nodes of which A are determined. In this case the activity maps show isoline A = const, which correspond to contour lines in topographic maps. We thus get a map in *isolines* of A. On such maps, the values of A at points between the isolines are assumed to vary smoothly, and are usually determined by linear interpolation from the isolines. In the second method, the surface is considered to be horizontally stepped. The averaging areas, in this case, are located side by side, without overlapping. The activity map comes out as a mosaic of zones with differing values of A, each value constant within the zone and changing abruptly at the boundaries. The lines on this map are not isolines, but are boundaries between the zones. Such a map may be called stepped, block or *zonal*.

The first activity maps were with isolines [Riznichenko] and they are the most widely used. Zonal activity maps have been recently developed in detail by Zakharova. However, activity maps similar to zonal appeared before. The activity A on them was indicated by isolines but the latter were more like the boundaries separating zones with different values.

Which maps—in isolines or zonal—are preferable when compiling seismic activity maps? It is a natural question. In principle, both approximations —smooth or fragemented linear (map in isolines), or fragmented constant or stepped functions (zonal maps) — are accessible. By reducing the step between the isolines or the interval between them or by reducing the dimensions of zones, the initial data may be used with almost any degree of accuracy. A technical criterion, namely the convenience and simplicity of the final result for a given accuracy may prove decisive. In the sense, the smooth or fragmented linear approximation is certainly better. Note that a sectional linear method can also be used to depict vertical steps, that is discontinuities in the function. This is associated with the possibility of showing in isolines the ruptures or vertical drops in the surface $A(x, y)$ on a map.

The boundaries of discrete activity $A(x, y)$ (if they can be discriminated and justified) are no more than one-dimension details of a two-dimensional field of activity distribution in the plane of a map (which is quasi-continuous), and a continuous distribution is certainly better represented by the isoline method. Therefore, we believe that it is usually better to present primary maps of seismic activity in isolines.

2.2.2. Correlation Between the Maps of Seismic Activity and Geological Maps

We shall consider now a less formal problem. What is the real spatial distribution of seismic activity $A(x, y)$ where seismicity has different tectonic natures, in particular in regions of block tectonics? Is it smooth or stepped? Does the function $A(x, y)$ have mainly blocked, fragmental constancy in the areas of block tectonics? We should emphasize here that even if this is true, it would not mean that the isoline techniques is betten, since a blocked, stepped $A(x, y)$ surface can also be shown easily.

It must be admitted that in areas of recent geosynclinal tectonics and areas where the material in the crust and upper mantle is inhomogeneous and macroscopically poorly differentiated, the seismicity distribution, in particular of the seismic activity $A(x, y)$, will be smooth with fuzzy transitions from the greater to lesser seismicities.

In areas of active platforms with blocked tectonics and a sharp macroscopic differentiation of materials, a different situation, at first glance, may ensue. In other words various values of seismic activity $A(x, y)$ might correspond to each block and A might be approximately constant within each individual block.

However, things are not so simple. A direct comparison between regions with different levels of seismicity and areas composed of various blocks physically does not seem wise. Seismicity, as an index of strong relative displacements of material, is not a property of individual relatively consolidated blocks (even if they are characterized by large absolute motions). It is rather a characteristic of their weakened boundary zones, that is zones where material is crushed, areas of deep living faults, i.l., seismic sutures. These zones act as a pliable plastic lubricant between solid strong blocks, moving relatively to each other. Seismicity within each block, if it really is consolidated, should be zero (that is no earthquake foci there). In fact, zones of greater seismicity need not be confined to known crushing zones, but may reach into neighboring blocks, destroying their continuity as, for instance, in the process of fracture intergrowth in technical materials.

All these suppositions testify in favor of more dispersed, macroscopically smoother spatial distribution of seismicity and seismic activity, as opposed to a probable direct correspondence between the zones with various seismicity and different inner consolidated geological blocks—known or supposed.

The opposite extreme — confinement of seismicity to block boundaries and the geometrical surfaces of deep faults only — does not seem to be correct either. All the experience of detailed observations in various parts of the USSR and abroad indicates that seismicity is related both to individual blocks and faults, and to extended systems, which may occupy considerable volume within rock masses. This does not mean that in some places earthquake foci, in particular severe events, are statistically confined to large, known faults [Bune et al.].

Despite the dispersed character of seismicity, the distribution in general, better present by smooth functions can, in some individual cases, still be drawn with clear boundaries between zones with essentially different seismicities. The simplest example is the boundary between an area with a finite seismicity and an area where seismicity is not observed at all. An example of an inert area bounded by seismic zones was given by Butovskaya et al. and Zakharova. We can draw a lot of examples (for almost every seismic zone) of the boundaries where a seismic area is adjacent to aseismic territories. However, in favorable cases, the pronounced sharp boundary can only be drawn arbitrarily. This is to some extent due to the nature of the analyzed process itself—as the latter has a discrete and dispersed character. Also, the zones of seismic silence adjacent to active zones or surrounded by the latter may be zones of temporal silence. In seismology, there are some good examples (Chile, 1960, etc.) when severe catastrophic earthquakes occurred.

The desire to draw boundaries like these on maps of seismic activity arises mainly from geological considerations. Zakharova, however, does not stick to these ideas exclusively, she also uses a quantitative seismological argument in favor of drawing boundaries on a zonal map. She proves that it is probable for considerable differences in seismic activity to occur on both sides of the boundary. That is, the observed difference is not due to casual spatial fluctuations. Since "blocked" distribution of seismicity is not an axiom, it should be shown that the stepped approximation in the observed seismicity distribution near the discriminated boundary, is higher than when a normal smooth approximation is used.

Seismicity distribution in most cases is correlated with features of the tectonic structure in regions known from geological observations and further assumptions. However, the geological picture itself is sometimes partly based on seismic data. This situation is common in seismogeological arguments. It is dangerous, then, to limit the type of seismic activity map, at its first compilation, to any geological assumption, as a vicious circle may set in.

Indeed, this sometimes occurs. On one side, we see averaging zones over-elongated, to cover the extension of a known geological structure, when calculating the activity A from epicentral maps. On the other, we see the deliberate compilation of activity maps with block structures where the blocks of the $A(x, y)$ surface are identified with structural blocks taken from geological assumptions.

Sometimes, a technique is justified in that activity maps which are made to conform in advance, with the geology, better reflect the long-term average activity, the establishment of which is the desired aim. Activity maps, in this case, become hypothetical rather than actual; it is very important to distinguish between actual and hypothetical data.

To avoid errors, it is necessary to compile primary maps of seismic activity from seismic data only. For geologists, these should be objective and independent primary material. Their geological modification and interpretation should, then, be the next stage in the seismological research.

This does not rule out the possibility of considering the general pattern of seismicity distribution on epicentral maps when compiling the maps of seismic activity. Indeed one may consider the extension of the epicentral zones and use averaging zones increased to cover these zones. By starting from seismological data alone we can delineate boundaries for step changes in activity if they can be quantitatively verified by an analysis of the seismicity variation.

2.2.3. Accuracy and Resolution of Seismic Activity Maps

In order to work from discrete epicenters to average densities in earthquake distribution in various areas of the map, we must count (by class or by sum) their number within an elementary domain of a finite size. The general requirement for these domains is that their dimensions should not be so great as to lose the detail of the spatial distribution of seismicity (in particular, the possible correlation between seismicity distribution and geological structure). However, they must be large enough for the number of epicenters, within the domains, to be statistically significant.

The size and shape of the domains determine the resolution of the final picture, the number of epicenters in the domain yielding the degree of reliability or accuracy of the map at a given place. We refer to a formal accuracy in that, an initial epicentral map with earthquake values, is true and reflects the seismicity it is supposed to characterize. However, the map is not true and this complicates the situation. Besides, the "accuracy" of the map may have another meaning. In particular, it may be combined in a new way with what we call "resolution".

To make things clearer we must introduce some quantitative definitions. The (surface) *resolution* of an activity map is the reciprocal of the area S_i of the averaging domain or a value inversely proportional to the product of its dimensions. If the dimensions of the same domain differ, the (linear) resolution of the map will be different in different directions, say along or across the strike of the seismic, geological or tectonic structure. This is anisotropy of the map's resolution. Further, for simplicity, we shall mainly consider maps with an isotropic resolution.

The *accuracy* of a map is the number N_i of epicenters from which the average density $v = N_i/S_i$ is determined for each averaging domain. To be exact, it determines the weight of the relative average epicentral density if the spread $R = \sigma_{N_i}/\sqrt{N_i} = \delta_{N_i}\sqrt{N_i}$, remains constant [Bune, Riznichenko]. Here, σ_{N_i} is the absolute and δ_{N_i} is the relative standard deviation of the average density from N_i epicenters. In fact, the accuracy in determining the average epicentral density is related to the relative deviation of δ_{N_i}. In the least squares method, the weight of the density is proportional to $1/\delta_{N_i}^2 = N_i/R^2$; for $R = $ const it is proportional and for $R = 1$ it is equal to the number of epicenters N_i. Note that for small N_i it would strictly be better for the accuracy to be defined as $(N_i - 1)$. However, for clarity, we use the approximate definition as above.

It is obvious that given these definitions of resolution and accuracy, we obtain the general expression (accuracy) × (resolution) = (density of epicenters) or:

$$n/S = v \tag{2.27}$$

Since the epicenter density v is an objective characteristic of the initial epicentral map and does not depend upon the method by which it is later processed, viz. the activity A map is compiled from it, there is a trade-off between the requirements for resolution and accuracy: the higher the accuracy, the lower the resolution, and vice versa. At the extreme, the accuracy of the definition of the overall average seismic activity over the entire region will be at a maximum, but the resolution of the spatial distribution of activity will be smallest. By contrast, when the resolution of the activity map is greatest, it will be an epicentral map. A trade-off has to be found.

To be more specific, we give some numerical estimates. Let $R = 1$, which is usually true if we consider moderate earthquakes and the period of observation does not include pronounced groups of events, for instance the aftershock sequence of a severe earthquake. If they are included, the data on them should be assigned to the time of occurrence of the severe event. If $R = 1$, the relative standard deviation δ_{N_i} characterizing the error and respectively the accuracy of the mean number N_i of epicenters is approximately $\delta_{N_i} = 1/\sqrt{N_i}$.

A quantitative relation between these values is:

N_i	3	5	8	10	15	20	100
δ_{N_i}, %	58	45	35	32	26	22	10

If we assume $\delta_{N_i} = 1/\sqrt{N_i^2 - 1}$, which would be more correct, N_i should be the number of epicenters minus one. In practice, it is not good practice to use N_i of less than say 5 or 8 events per averaging domain S in regions of scarce earthquakes. This corresponds to errors of 45 or 35%, respectively (50 or 38% using the more exact estimate).

Here, N_i is the sum of earthquakes per domain S. In the summation method it is simply N_Σ, whereas in the distribution method it is N_i, i.e. the number of epicenters of various classes K_i considered in the activity A calculations and occurring within the averaging domain. Thus, if two classes ($K_1 = 10$ and $K_2 = 11$) are used to compile a map and the numbers of epicenters over a domain S are respectively $N_1 = 5$ and $N_2 = 3$, then we have $N_i = 8$.

If for different K_i we take different dimensions for domain S_i, then N_i is the sum of the number of epicenters in these classes over an average-size domain S. For places on the map with fewer epicenters per domain S than a threshold derived from the minimum desirable accuracy, the activity A should not be calculated.

If N_i is fixed in advance, the resolution of the map should be reduced for regions of scarce earthquakes. Rational limits on the possible activity map are thus established, in accordance with the minimum admitted resolution, that is the maximum allowable dimensions of the averaging domain S.

For instance, when compiling maps with a scale of 1:1 000 000 for regions with scarce epicenters it does not seem reasonable to use domains S over 10 000 or 5 000 km^2, which means the diameters of the averaging areas should be no more than 100 or 30 km. In regions with denser distributions of epicenters, the resolution can be easily enlarged without detriment to accuracy. As the resolution is reduced, say for a scale of 1:2 500 000, the absolute resolution may be reduced leaving the relative one, which is estimated over the length of the map, approximately the same or even a little better.

Finally, in practice, we want to know for what total number of epicenters in the area analyzed is it still reasonable to compile a map of seismic activity? We can draw the following picture. Let us say that an activity map will satisfy us if it is composed of n independent elements, that is n adjacent averaging domains S. The activity is calculated from the epicenters (if a running mean is used the total number of domains S would be greater, but in some of them the activities would be calculated using epicenters which would be used in more than one element and, therefore, the activity would not be independent; the extra domains thus engendered are of no interest to us for these estimates). Let us assume, then, that we would be satisfied by an accuracy corresponding to the N_i epicenters in each independent domain. Then, the total number of epicenters in one domain for which it is still reasonable to compile the activity map is $N = nN_i$. For instance, we want a map containing $n = (3 \times 3) = 9$ independent elements, the accuracy of activity determination being no less than 32%. This corresponds to $N = 10$ epicenters in S. Thus, the minimum total number of epicenters in the area would be $N = 90$ (100 in round figures); this means the representative epicenters of various classes K_i are considered when determining the activity.

For such a total of epicenters ($N = 100$) of various classes K, the slope γ of the recurrence curve of the epicenters (which should be known whatever the method for compiling the activity map) will be accurately determined.

2.2.4. Selection of Averaging Domains

Two main ways are now used to select the averaging domains and there are two types of seismic activity maps, namely maps of constant resolution and maps of constant accuracy.

In the first case, the size of the averaging domains is constant. It should be large enough to ensure accuracy over the entire map. In this case, the number of epicenters in each domain remains moderate even in unfavorable areas of scarce epicenters. In areas of dense epicenters on the same map, the average density is determined, in this case, with too high an accuracy, even though the resolution of map (established from site dimensions) will remain the same.

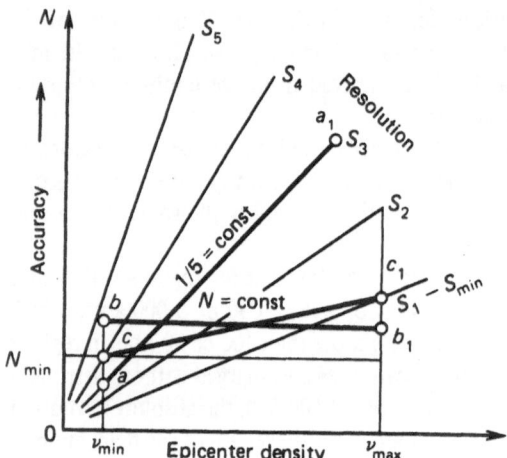

Fig. 9. Correlation between density of epicenters v, accuracy N and resolution $1/S$ of seismic activity maps:

aa_1—trajectory of constant resolution; bb_1—the same for constant accuracy; cc_1—the same for established quality

The second way is to vary the size of the averaging domain so that the same number of epicenters fall within each domain all over the map. The number of epicenters should be enough to ensure accuracy in the average density. Then, in areas of dense epicenter distribution, where dimensions of the sites are smaller, the resolution of the map will be greater than where epicenters are scarcely distributed.

The quality of a seismic activity map depends on both its accuracy and its resolution; the correlation between accuracy N and resolution $1/S$ cannot be limited by (2.27). To limit it, one, of the values, N for instance, can be bounded by a function of v; $N = f(v)$. Then, from (2.27), $1/S$ can also be derived as a function of v: $1/S = v/f(v)$.

If we assume $f(v) = N = $ const, this gives constant accuracy; if however we assume $v/f(v) = 1/S = $ const, we have the method of constant resolution. So, both methods turn out to be particular cases of a more general approach.

It is reasonable to choose the function $f(v)$ so that an increase in the epicenter density v will, within some bounds, increase both the accuracy N and resolution $1/S$ of the activity map. These boundaries, at low densities, are established by the minimum admissible accuracy N_{min} which may be less than the average accuracy N of the map. At large densities the limits are determined by the minimum averaging domain S_{min}, a reasonable size being limited by the accuracy with which the epicenter coordinates are determined, the scale of the map, and the extension and geometry of seismo-geological structures. We shall call a map compiled this way, a map of *predetermined quality*.

Figure 9 illustrates this. The trajectory as' of constant resolution $1/S$, bb' of constant accuracy N and cc' of predetermined quality are given in coordinates of accuracy N and density v of epicenters. The cc' trajectory need not be straight.

This discussion is mainly to clarify the problem in general. The compilation technique for a predetermined quality map will be more complex than either of the others and no doubt thus is a drawback. However, with certain simplifications it should not be too difficult. A continuous cc' trajectory (predetermined quality) may be approximated by segments of aa' (constant resolution) or bb'(constant accuracy) trajectories. In the first case, the activity map may then be compiled by a finite range of averaging domains S_i, e.g., three sizes: a large size for small densities v, an intermediate size for intermediate densities, and a small size for large densities. In the second case, three nomographs not one are necessary. They are the same as for the method of constant accuracy [Gorbunova], one for few epicenters N_i with small densities on the map, one for intermediate densities and one for large densities.

The different densities for the areas can be approximated by rough estimates or by first compiling a density $\varepsilon(x, y)$ map by the constant resolution or constant accuracy method.

2.2.5. Weighing Earthquakes when Determining Seismic Activity

When compiling seismic activity maps or earthquake recurrence curves for each cell of the maps or for the whole map, a traditional problem arises, namely, how to average $(K_i, \log N)$ points by a straight line. Here, K_i is the earthquake energy class, N_i is the number of events in the class; the use of magnitude M instead of energy $E = 10^K$ does not change anything. If the recurrence curve is plotted for the entire region, both parameters, namely the initial ordinate $\log A$ (A stands for activity) and slope $\gamma = \Delta \log N/\Delta K$, must be determined. On the map, the slope γ is assumed to be given and is taken from the recurrence curve for the entire region. The problem arises how to weight $(K_i, \log N_i)$ points in both cases, in particular in the case of a map. Graphs can be obviously exchanged for their formulae. A number of formulae are given in [Riznichenko].

When discussing weight, the following assumptions are made. On one hand, points representing severe earthquakes should have great weight as in many respects they are more important than weak ones. On the other hand, points corresponding to weak earthquakes are based on more events and, thus, given the statistical spread typical of seismic process, their positions on the graph are more accurate than those of severe events. Therefore, their weight should dominate. It often happens that "a golden mean" arises: the weights of all the points on the graph, one from each class K_i, are taken to be equal.

Other approaches to the problem exist. Gaisky applied Gumbel's theory of extrema to seismological problems believing that the merit of the method was its automatic preference of more severe events (weaker events simply being ignored). However, we shall only consider the case when complete information is available.

Thus, weights will always remain arbitrary to some extent. A certain formalization is necessary to make interpretation unambiguous. The formalization should correspond to the conditions and aims of the physical problem.

One concern, as usual, is to find out the most stable position of a straight line representing the earthquake recurrence graph throughout the energy range. It is natural to associate the point weight and the accuracy with which the position of each points is determined, cf. Sec. 1.2.3. Their deviations from the mean position along axis N are of principal concern. Since a logarithmic scale is used for the axis, we consider relative and not absolute deviations: if the deviations are equal they are shown by equal segments at all points along the scale. For the same earthquake spread $R = $ const, as we have already seen, the squares of the relative probable deviations from the mean are approximately proportional $1/N_i$. Consequently the number N_i of earthquakes, irrespective of their energy class, should be their weights. Note that N_i are the primary numbers of earthquakes, not the results of normalizing by time and space.

The application of the summation method [Riznichenko] leads to the same conclusion; the total number of earthquakes exceeding a specific value are calculated by this method.

Since severe and weak events are not distinguished, they are assumed to have equal weights. Considering that there is an unambiguous correlation between the assumptions of the summation method and the common method of class by class distribution, this feature of the summation method is more evidence to consider the weights of events with different scales to be equal in the distribution method as well.

However, there are conditions and aspects of the problem where this weighting method is disputable. This is primarily the case when the earthquake spread, R, is not constant but depends upon the value of the earthquakes [Bune, Katok]. Thus Katok has found from observations in the Dushanbe-Garm region, that sets of weak earthquakes tend to group more than intermediate events do (for severe earthquakes the question remains open). When events are grouped, the spread R increases [Riznichenko, Nersesov]. Since the weight is made proportional to N_i/R^2, the weights for weak events would then be less than those for severe earthquakes. A quantitative calculation may be easily done for each case by investigating R in detailed observations of the regional seismic regime.

The possible deviations of the recurrence curves (in a common K, log N coordinate system) from a straight line, particular on the right-hand side, i.e. in the area of severe earthquakes, is another such problem. We would naturally expect the curve to bend downward towards zero, for its slope γ to increase prior to the complete depletion of the seismic energy process. In practice, however, we also observe the opposite case, with γ decreasing on the right [Riznichenko]. This may be due to particular circumstances, viz. difference in the observation periods for weak and severe earthquakes in the region (severe earthquakes usually occur only within longer periods) and variations in the seismic regime with time; systematic errors in earthquake energy determination, in particular in areas of very severe events; or substantiatl fluctuations in the regime. No doubt, one should account for and try to exclude such factors. However, a study of the probable curvature $N(K)$ is very important.

The main difficulty arises because severe events are so rare. This makes the reliability of statistical conclusions (probability at particular confidence levels) smaller than those easily achieved for other segments of the $N(K)$ function with more events. Applying the same requirements (the reliability of the straight-line approximation), to the right-hand part of graph $N(K)$, as to the other parts, one usually derives a formal but unpromising conclusion that there is no evidence for rejecting the hypothesis that is, in general the straight line $N(K)$ graph over the entire range of energies, severe events included. Other ways should be sought to solve this problem.

Some of these ways were considered in the summation method [Riznichenko], viz. plotting an averaging curve through the observed points. The curve is represented by a formula, whose numerical parameters are derived from the assumption that they best conform to the observed data, as determined say by least squares or the Chebyshev method. The weights of the points in the right-hand part of the recurrence graph may temporally be assumed to be greater, so that they will be more noteworthy when calculating the averaging curve. In this case, the weights of the points are close to those in the middle of the graph (the area with the most data), where its straight-line approximation is certainly correct. When the averaging procedure is over, the points are reduced to the value determined by the representativity of the events. The result is a curve $N(K)$ with a decrease to the right in reliability of the approximation.

Another way may be to change the weights to preserve the principle of a straight-line averaging graph in its individual segments. In particular, we may debate the application of the usual method of moving window averaging with distinct or fuzzy boundaries of the delineated area. This corresponds to the procedure when the essential weights are ascribed to points (or events) captured by each interval of interest along the entire graph $N(K)$ and these weights are reduced to zero (sharply or gradually) beyond this interval. The advantage of such an approach is that it is not necessary to forecast in advance, the shape of the curve to be found, even in a general form.

Thus, it may be possible to relate earthquake weights to parameters which are themselves the subject of the research. For example, when the subject of the research is, say, seismic energy flow from earthquake foci [Riznichenko], one may wish to ascribe weights proportional to a summed earthquake energy represented by each point on the recurrence curve $N(K)$ at just this point; the conventionally accumalated strain $\Sigma\sqrt{E}$, after Benioff, is under investigation, the weights proportional to this value can be taken, etc. Such an approach (attractive at first glance) contradicts the very spirit of the statistical problem under investigation, that is the determination of the most probable numerical distribution of events characterized by any parameter directly from the value of this parameter. Such a sign, means that energy, or deformation, etc. could be used as a basis for the classification of events, but not for the establishing of their weights during the solution of the problem. Weights are finally determined, as mentioned above, by relative errors in establishing the average summation numbers N_i of events when applying the distribution technique (or of numbers N_Σ when applying the summation technique); in the simplest case when R = const they could simply be taken equal to these numbers N_i (or N_Σ).

2.3. Generalized Law of Earthquake Recurrence[*]

The value distribution $N = N(K)$ of the number N of earthquakes in terms of their value $K = \log E$ (E—seismic energy of the focus) or magnitude M was studied by B. Gutenberg and C. Richter in 1954. This study was subsequently taken up by many seismologists worldwide over. Perhaps their main concern, in discussing the "law of earthquake recurrence" was to elucidate the near-constancy or the variability of the slope $\gamma = -d \log N/dK$ of the curve $N(K)$. However, little attention was paid to the location of the right-hand end of the curve $K = K_{max}$. It was, of course, examined but not always in the correct way [Bath, C. Tsuboi].

The problem of computation of K_{max} (the maximum possible earthquake) in different regions, on the basis of indirect information—seismic and otherwise—was first formulated in concrete terms, only in 1962 [Riznichenko]. In this paper, the author suggested calculation of the mathematical correlation between the strength K of the observed earthquakes (in particular, those close to K_{max}) with other quantitative seismological as well as geophysical, geological and geodetic parameters. Initially, the solution was based on the study of the correlation between K_{max} and the seismic activity A in the area surrounding the epicenter, i.e. construction of the correlation dependence $K_{max}(A)$ (Riznichenko).

In this paper we consider the two functions $N(K)$ and $K_{max}(A)$ from a common standpoint, i.e., we construct a generalized function $N(K, A)$, that is a "generalized law of earthquake recurrence" from which the two foregoing functions would follow as particular cases. Our aim is

---

[*] See [Riznichenko]. Co-author A. I. Zakharova.

to improve the methods of computation of K_{max} further on the basis of all the data on long-term mean regional seismicity, during a limited period.

It is clear that the problem of predicting the maximum possible earthquake K_{max} in any region should be solved, considering all the available information relevant to the problem. However, the attempts to use certain other information in an objective and quantitative way, have not led to definite results [Gzovskii, Riznichenko]. Although a correlation of K_{max} with A has already proved its efficacy in a series of concrete cases [Zakharova, Riznichenko]. While not discounting the necessity of further attempts to refine the various approaches to the computation of K_{max}, clearly the present situation indicates a special need for perfecting the method mentioned above.

So far, the approach to the problem of a computation of K_{max} has been essentially this. The value K of the observed earthquakes close to the maximum was correlated with the mean seismic activity A in the region surrounding the epicenter of each considered event. If the general correlation between K_{max} and A is established from the observation of both variables in one area, it can be used to determine one of them, say, K_{max}, in another area where only the other variable A is known from observations. The assumption about the validity of the same type dependence $K_{max}(A)$ for both areas should be checked [Riznichenko].

In papers [Zakharova, Riznichenko] only data on observed earthquakes, close to the maximum, were used in correlating K and A. As for the other events which were obviously not maximal, it was noted that they should fall within the domain K, A in relation to the envelope contour $K_{max} = K_{max}(A)$. But a practical analysis of earthquake distribution in the entire A domain was not carried out. It was postponed for the future.

The first attempt to carry out such an analysis was made in [Gorbunova]. In this paper, all the earthquakes observed within the given region (maximum and non-maximum) were depicted as points on the K, A plane. However, the density distribution of these points on the K, A plane was not quantitatively analyzed. Further, it was assumed that the size of the area "responsible" for the earthquake, where the mean activity A is determined, does not depend on its value K; it remains the same both for small and large foci. Such an assumption, in our view, is not quite appropriate and it is not admissible in the context of research, using a wide range of K. Clearly, this wrong assumption accounts for the pessimistic conclusion reached in [Gorbunova] about the apparently indeterminate nature of the relationship between K and A for severe earthquakes, i.e., exactly for those earthfuakes in which we are interested. In earlier papers [Zakharova, Riznichenko], where we assumed that the dimension of this area increases with increasing K, such ambiguity did not appear. We shall therefore stick to this assumption in our further investigations.

In the present work, we move a step further along the same path. Here we carry out, for the first time, a complete quantitative analysis of recurrence N distribution, i.e., of the mean frequency of earthquake occurrences in terms of their value K and seismic activity A in the entire area of observed seismicity, not merely for the maximum earthquakes. The two-dimensional distribution $N(K, A)$ appears to be a generalized case of the familiar one-dimensional distribution $N(K)$ which is the law of earthquake recurrence.

We assume that the region "responsible" for the earthquake, where A is calculated, depends on its value K. We consider that this region is a circle with radius $R = R(K)$, as in [Riznichenko]. The principles governing the selection of parameters in the function $R(K)$, are discussed in [Riznichenko]. For earthquakes which are close to K_{max}, the diameter $2R_{max}$

of the "responsible" region usually exceeds the width of the zone, for the entire investigated seismic zone. This automatically affords a basis for determining the length and width of the zone. We tried elliptical regions, too, but we could not perceptibly improve the value of correlation density. Leaving these details aside, we confine ourselves there mainly to an investigation of the distribution $N(K, A)$.

The appearance of singularities in the earthquake distribution near the limiting contour $K_{max}(A)$ is one of the special but most interesting, aspects of this work. It is here that we find the clue to the question of maximum *possible* earthquakes K_{max} and of separating them from the near-maximum *observed* events. The present work does not offer a solution to this problem. It is only an attempt to develop one aspect of the methodology for solving the problem.

2.3.1. General Considerations

Earthquake recurrence. We know that the law of earthquake recurrence $N = N(K)$ can be represented by two main techniques: either as a summation $N_\Sigma(K)$ or as a distribution density $N(K)$, so that $N = dN_\Sigma/dK$. In practice, both functions are usually given in the form of class by class distributions, in intervals $K \pm 0.5$, where the K's are whole numbers.

When we relate N, N_Σ to unit volume or area unit, the source area (1000 km^2) and to time unit (1 year), the number $N|_{K_0 = 10} = A$ will denote seismic activity. In the usual simple case where log N linear in K, $\gamma = -d$ log N/dK = const, and both earthquake recurrence curves $N(K)$ and $N_\Sigma(K)$ (Fig. 10) are practically parallel (except for the segments close to

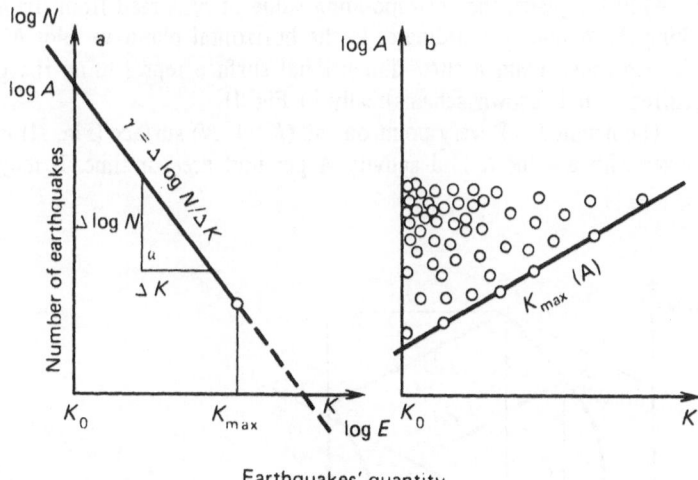

Fig. 10. The laws of earthquake recurrence $N(K)$ and maximum earthquakes K_{max}:

a—K is the earthquakes energetic value: $E = 10^K$ is seismic energy; A is seismic activity, i.e. the number of earthquakes per 1000 km^2 per year; b—$K_{max}(A)$ is the limiting seismicity contour

K_{max}); the seismic activity A and number N are then related [Anan'in, et al.] as:

$$N_\Sigma = A \frac{10^{-\gamma(K_{min} - K_0)}}{1 - 10^{-\gamma}}. \tag{2.28}$$

When $K_0 = 10$ and $\gamma = 0.5$, we have from (1.27) $N_\Sigma|_{K = 10} = 1.46\ A$.

If the epicenter map includes only representative earthquakes, N_Σ will denote the density of epicenter distribution. This holds good for A also, which when $\gamma = $ const, is distinguished from N_Σ only by a constant factor.

Maximum earthquakes. The number K_{max}, depicted by the recurrence curve of earthquake either as density distribution or as summation, is related to the value of the maximum earthquake (observed or possible), characteristic of the entire region as a whole. The earthquake recurrence curve does not provide any clue as to where exactly in the region such an earthquake can occur. This clue is given by the dependence $K_{max} = K_{max}(A)$ (Fig. 10b), which may be called the "law of maximum earthquakes". The function $K_{max}(A)$ permits us to locate the area in which we can expect the occurrence of any earthquake with K_{max}, the largest for the entire region and of other small earthquakes which are maximum in their own sub-regions; in other words, it helps us to compile a map of K_{max}, if the map of A (or the primary epicenter map) is known.

Graphs of the type shown in Fig. 10a $N(K)$ and Fig. 10b, $K_{max}(A)$ represent different functions. One cannot be directly deduced from the other (sometimes, this has caused confusion [Bath and others]. We shall now formulate a more general rule which will include both the previous values as particular cases.

Generalized earthquake recurrence. Let us take the plane K, A with coordinates K, log A as in Fig. 10a and ascribe the density N of the earthquake distribution in K to each point (K, A) in the plane; the corresponding value of N is read from the graph $N(K)$ (Fig. 10a). Taking the K and A coordinates in the horizontal plane, we plot N along a third, vertical axis. We thus obtain a three-dimensional surface representing the generalized earthquake recurrence; it is shown schematically in Fig. 11.

The height N of every point on the (K, A, N) surface (Fig. 11) is the number of earthquakes with a value K and activity A per unit area or time. It may either be represented

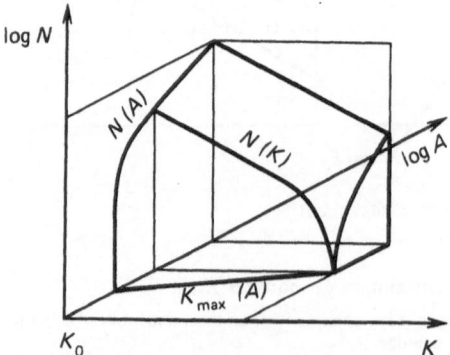

Fig. 11. Three-dimensional graph of generalized earthquake recurrence law $N(K, A)$

by a density distribution N (continuous or by class) or by a summation distribution N (continuous or by class [Riznichenko]). The volume bounded by this surface and by the planes of coordinate angle represents a cluster of epicenters of earthquakes classified according to arguments K and A. Cross sections in the K, N planes, which are perpendicular to the A-axis, are the familiar earthquake recurrence curves (Fig. 10a) and the horizontal projection of the frontal slope of the cluster is the graph of the maximum earthquakes K_{max} (A) (Fig. 10b).

Thus, the function $N(K, A)$ (Fig. 11) includes the function $N(K)$ (Fig. 10a) and K_{max} (A) (Fig. 10b) as particular cases. Thus, it contains more information than the two curves $N(K)$ and K_{max} (A) and is over the entire region: $N(K, A)$ contains all the graphs $N(K)$ in the region at various A. It also includes all the right-hand breaks or bends, i.e., areas of maximum possible earthquakes K_{max}.

In the simplest case, when the recurrence curves $N(K)$ can be considered linear (in K, log N, coordinates) with constant slope γ, and with a bendless break at K_{max}, the $N(K, A)$ surface (Fig. 11) becomes part of an inclined plane. The slope of this plane to the horizontal in a cross section A = const, equals $-\gamma$ (decreasing towards greater K); in the cross section with the coordinate plane $K = K_0$, the slope equals $+1$ (increasing towards greater A), since A is linearly related to N (2.28). In the simplest case, the equation of the plane can be written as:

$$\log N = \log N_0 + (\log A - \log A_0) - \gamma(K - K_0) \qquad (2.29)$$

where N_0 is the value of N when $A = A_0$ and $K = K_0$. In the general case, the $N(K, A)$ surface is curved and its shape should be determined from observations. It can be approximately described either by (2.29) or by other more suitable interpolation formula.

The surface described by (2.29) cannot itself be used to obtain K_{max} unless we have information about the limiting contour K_{max} (A). The $N(K, A)$ surface, however, can yield K_{max} at least in the probability sense, if it has a sufficiently steep frontal slope. In addition, the statistical character of the solution is, in fact, inevitable because all the relations between long-term mean indices are obtained from inaccurate observations; these indices, including, in principle, their mean scatter, are not completely determinate and contain a considerable random component.

2.3.2. Analysis of Observations in Central Asia

The catalogue of crustal earthquakes of Central Asia, between $\varphi = 36° - 45°$N and $\lambda = 65° - 81°$E for the period 1962 to 1966 [Riznichenko] was used to make an initial analysis of a general earthquake recurrence law $N(K, A)$; the catalogue contains more than 1500 earthquakes with $K \geqslant 10$, and contains some corrections from [Zakharova].

The usual graphs of earthquake recurrence for this region over the period 1962-1966 in the form of class by class distributions $N(K)$ and summations $N_\Sigma(K)$ are shown in Fig. 12. We see that beginning from $K = 10$ both graphs are practically straight and parallel over the range $K = 10$-15 which are the limits observed during this period. Points for $K = 9$ diverge noticeably downwards. This and considerations of the recording distance, leads us to suppose that earthquakes weaker than $K = 10$ are not all contained in the observations; in view of this we have not taken them into consideration. The total number of earthquakes with $K \geqslant 10$ used for our computations was 1538.

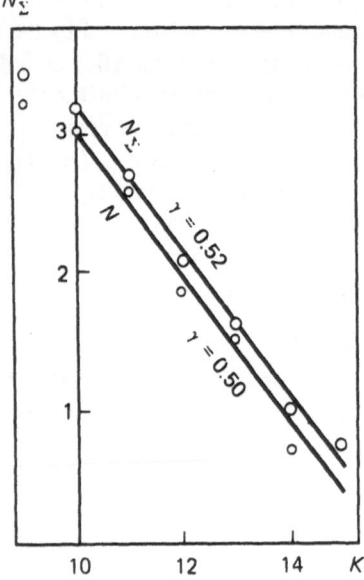

Fig. 12. Earthquake recurrence in Central Asia for the whole region, 1962-1966.

N-distribution; N_Σ—summation

For a practical construction of the $N(K, A)$ function from the observational data, we had to fill in the cells in the table with the numbers $N(K, A)$ (see ahead). Every cell of the table must be filled by numbers N_{ij}, adjusted to unit area and time, with parameters within limits $K_i \pm 0.5$; $\log A_j \pm 0.1$; $j = 1, 2, 3, ...$, were found for the different local area elements. To do this we first determined the individual "earthquake recurrence law" for each area element on the epicenter map; we obtained a series of numbers $N(K_i)$, $i = 1, 2, 3, ...$ for a fixed value of A_j, characterizing this area element. Consequently, these values fall into a row of the table A_{ij}. We did this for a number of local area elements and thus filled in the series of rows A_j, $j = 1, 2, 3, ...$ of the table and in each cell there appeared, in general, certain values of N_{ij} from different areas. Finally, we averaged, in a certain way, all the N_{ij} falling in the same cell and thus obtained the value of the function $\bar{N}_{ij}(A, K)$. Details of this procedure are explained below.

Determination of local earthquake recurrencies. A circle was drawn with radius $R = R(K)$ on the epicenter map around each epicenter of an earthquake with $K \geqslant 12$, R increasing with increasing K. The dependence $R(K)$ was taken to be the same as in [Riznichenko, et al.], their exact values being

$K = 12 \quad 13 \quad 14 \quad 15$
$R = 21 \quad 27 \quad 34 \quad 46$ km

The number of epicenters of earthquakes of different classes $K_j \geqslant 10$ was counted within the boundaries of every circle to establish the local relation between the recurrences $N(K)$ and $N_\Sigma(K)$ and to determinte the local activity A from (2.28). Graphs of $N(K)$ and $N_\Sigma(K)$ were not actually constructed. The computation was done numerically. The numbers N and N_Σ were normalized for one year and for an area of 1000 km^2.

The local recurrences were generalized for cells of the table in columns $K_i = 10, 11 ...,$ 15 (± 0.5) and rows $\log A_j = 1.0, 1.2, 1.4 ... 0.8, 1.0$ (± 0.1). Numbers $N_{ij}(K, A)$ in the same cell were averaged after weighting. The weights ω of the observed local numbers $N_\Sigma(K)$— and not their logarithms—were taken to be proportional to the total number of epicenters of all energy values $K \geqslant 10$ in a given local area element, $\omega = N_{\Sigma|K=10}$. Since $A \cong \text{const}$ for each row of the table and since A is proportional to $N_{\Sigma|K=10}$, the assumption of proportionality is tantamount to the assumption that in the limits of each row the weights ω are proportional to the areas $S = \pi R^2$ of the local area elements where the numbers $N_{\Sigma_t} K$ were determined. The mean value in each cell was computed from the formula:

$$\bar{N}_\Sigma(K, A) = \frac{\Sigma \omega N_\Sigma(K, A)}{\Sigma \omega} = \frac{\Sigma N_{\Sigma|K=10} N_\Sigma(K, A)}{\Sigma N_{\Sigma|K=10}} = \frac{\Sigma S N_\Sigma(K, A)}{\Sigma S} \tag{2.30}$$

So far the area S is concerned, this formula is only valid at the limits of each row $A = \text{const}$, but is valid everywhere for the numbers $N_{\Sigma|K=10}$, $N_\Sigma(K, A) = N_{\Sigma|K_i}$.

In essence, this method of weighting enables us to obtain, in each row $A = A_j$ of the table and to an accuracy of $\log A_j \pm 0.1$, an integrated earthquake recurrence correlation $N(K_i)|_{A=A_j}$ over the total area ΣS of all those local area elements S of our region, where the activity is equal to A_j.

We note that the contribution of some area elements S to the recurrence $N_{\Sigma|K_i}$ of earthquakes with certain values of K_i, in particular when $K_i \rightarrow K_{max}$, may be zero. There is nothing wrong in this; it only means that during the observed period, such earthquakes have not taken place. As against this, they might occur in other area elements S having the same A_j. This is all reflected in the overall mean (2.30), which is statistically significant for all the area elements.

Displacement. Since the interval $\log A \pm 0.1$ is finite, the mean numbers $N_{\Sigma|K=10}$ obtained in a given interval after weighting may not be exactly equal to those which conform, according to (2.28), to the rounded value $A = A_j$. In order to eliminate this small discrepancy, the observed mean weighting numbers $\bar{N}_\Sigma$ for different K's in each row A_j were changed proportionately to ensure equality. In practice, the same additive correction $\Delta \log N_\Sigma = \log \bar{N}_{\Sigma|K=10^-} - \log N_{\Sigma|K=10}$, is introduced for each row A_j, in each logarithm of the observed mean $\log \bar{N}_\Sigma$, for different K_i. This is equivalent to a parallel transition of each cumulative graph of earthquake recurrence with parameter A_j, along the $\log N$ axis through a distance $\Delta \log N_\Sigma$ so that this curve begins from the point $(K = 10, \log N_{\Sigma|K=10})$, which strictly conforms to equation (2.28) for a given A_j.

The accuracy of the determination of the mean weighted $\bar{N}_\Sigma$ for different cells K_i of every row A_j of the table $N_\Sigma(K, A)$, was estimated for undisplaced values of $\bar{N}_\Sigma$ by the usual formula for the mean square deviation:

$$\sigma = \sqrt{\frac{\sum_{i=1}^{n} \omega_i \delta_i^2}{(n-1) \Sigma \omega_i}}, \tag{2.31}$$

where ω_i is the weight of an individual determination of N_{Σ_i}, proportional from (2.31) to the earthquake summation $N_{\Sigma_i|K=10}$ in the local area element or to its area S_i; δ_i is the deviation of the observed number N_Σ from undisplaced mean weighted $\bar{N}_\Sigma$; and n is the number of determinations (number of local area elements S), on the basis of which, the weighted mean is found.

The generalized mean earthquake recurrences $N_\Sigma(K)$ in Central Asia for the period 1962-1966 are reproduced in Table 6. Each row of the table contains, in effect, the distribution $N_\Sigma(K)$ over the total area of all the area elements where the activity is equal to the given $A = A_j$. In the last column of the table (the weight) is the total number of earthquakes, on the basis of which the distribution is obtained; this is the number N_Σ before normalizing it to unit time T or unit area S for $K = 10$. The number N_Σ, per unit T and S for $K = 10$ is also characterized by the same weight. The weights of the other numbers in the row, for K_i other than 10, are related to the weight of this number as are the numbers themselves.

Table 6. The General Values of Earthquake Recurrencies $\bar{N}_\Sigma(K, A)$ Related to a Year and to 1000 km^2 During Observations in Central Asia in 1962-1966

log A	$K = 10$	11	12	13	14	15	Weight
0.8	9.24	2.70	0.696				53
0.6	5.82	1.98	0.495	0.069	0.069		162
0.4	3.68	1.18	0.387	0.136	0.0525	0.00973	342
0.2	2.32	0.699	0.187	0.0578	0.0215	0.00960	685
0	1.46	0.530	0.179	0.0716	0.00640		199
$\bar{1}.8$	0.922	0.252	0.0660	0.0203	0.00495	0.00244	
	±0.019	±0.028	±0.0150	±0.0058	±0.00989	±0.00370	719
$\bar{1}.6$	0.582	0.350	0.192	0.0356	0.0250	0.0125	54
$\bar{1}.4$	0.367	0.112	0.0312	0.00934	0.00260	0.00106	1347
$\bar{1}.2$	0.232	0.0896	0.0320	0.00689	0.00152	0.000600	265

The weight is the total number of earthquakes $\Sigma N_{\Sigma|K=10}$ of all classes $K \geqslant 10$ and over all area elements S of the region where the seismic activity A is the range log $A \pm 0.1$.

The standard deviations and possible errors, computed from (2.31), are given, as examples, for the row log $A = \bar{1}.8$. We see that with increasing K_i, the error rapidly increases and when $K = 14$, it exceeds the measured value itself. This behavior of the error is characteristic of the distribution $N(K)$ and it is associated with the rarity of events with large K.

The earthquake recurrence graphs to which the table relates are given in Fig. 13. The parameter characterizing the family of curves is the activity A, more precisely log A. We see that, in the first approximation, these graphs are straight and have the same slope as the primary graphs $N(K)$, and $N_\Sigma(K)$ for the entire region (see Fig. 12). On closer look, however, we see that the graphs in Fig. 13 are not quite straight. This is especially noticeable in graphs with the parameter log $A = 1.6$, which has a total weight of 54, a value which is very small when compared to neighboring values (see Table 6). We know from experience that constructing graphs $N(K)$ from less than 100 observations is uncertain since the accuracy will be small [Riznichenko].

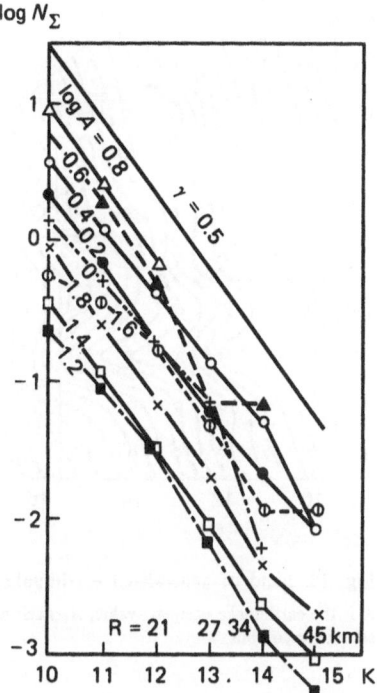

Fig. 13. Earthquake recurrence in Central
Asia for areas of given seismic activity A_j,
1962-1966, within limits $\log A_j \pm 0.1$;
$A_j = 1.2-0.8$

The curvature and nonparallel nature of the graphs in Fig. 13 do not, however, reveal any perceptible systematic tendencies. The slope γ is almost independent of the seismic activity A. On the right-hand side, the graphs curve downward as often as upward. However, since all these inferences relate to only one case it is not possible to generalize over much.

The generalized earthquake recurrence correlation based on the given observational material is depicted in Fig. 14 in the form of an N_Σ-field in the K, A plane. The field is mapped as isolines of $\log N_\Sigma$. We notice that the surface $\log N_\Sigma(K, A)$ over the range of arguments, approximates to an inclined plane. Its microrelief (especially well-defined for $\log A = 1.6$) is not evident, which can perhaps be explained by the adequacy of the statistical material as well as by random deviations from long-term mean relationships.

Isolines of $\log N_\Sigma(K, A)$ are only shown by continuous lines within the area enclosed by the contour, where they are computed from actual data for the observed period 1962-1966, in Central Asia. Outside this contour, the isolines are extrapolated on the assumption that the $\log N_\Sigma(K, A)$ surface is approximately a plane; such isolines are shown by dotted lines.

At the lower right-hand corner of Fig. 14, we have line $K_{max}(A)$, drawn on the basis of earlier summarized observational data [Riznichenko and others] on earthquakes of Central Asia, over a long period. Some of them, particularly those for which we can get a mean seismic activity A in the surrounding region from the data for 1962-1966, are depicted in Fig. 14 as circles. These are the earthquakes of Sarez 1911, Argankul 1934, Andizhan 1902, Kurshab 1924, Ulugchat 1955, Khait 1949, Chatkal 1946, Kashgar 1902, and Karatag 1907. We see that all these earthquakes fall outside the area of the graph of Fig. 14, defined by

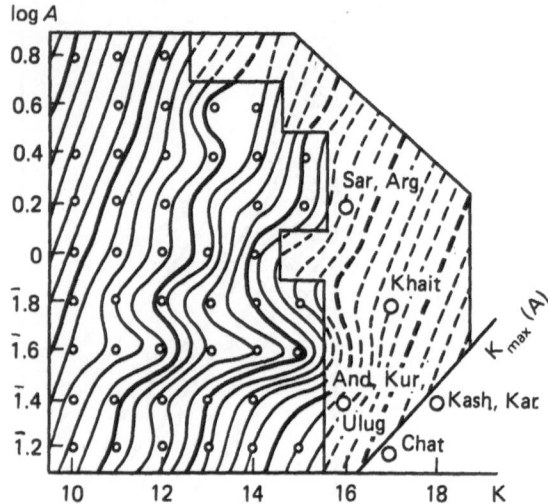

Fig. 14. Field of generalized earthquake recurrence N_Σ (K, A) for Central Asia, 1962-1966.

K is the earthquake energetic value; A is seismic activity; *isolines* represent the value log N_Σ; $K_{max}(A)$ is the limiting seismicity contour

the observational data for 1962-1966. The actual recurrence of such severe earthquakes is not known. It can only be estimated from the extrapolated field.

For example, we see from Fig. 14 that log $N_\Sigma = \overline{4}.7$, i.e., $N_\Sigma = 0.5 \times 10^{-3}$, for the Andizhan earthquake $(K = 16)$, which is the mean number of such earthquakes in an area of 1000 km^2 per year. The mean recurrence period T of such events in this area is given by $T = 1/N_\Sigma = 2000$ years. We shall also evaluate the mean frequency of such events in the "responsible" area (according to the computation in [Riznichenko]). When $K = 16$, the radius of this region is taken to be $R = 72$ km. For this area, we obtain a frequency of:

$$N|_{R=72} = \frac{\pi R^2}{1000} 0.5 \times 10^{-3} = 8.17 \times 10^{-3} \text{ year}^{-1}.$$

From this it follows that the recurrence period of such events in the "responsible" area is $T|_{R=72} = 1/N|_{R=72} = 120$ years. It is not surprising, therefore, that during the observed period of only 5 years (1962-1966) we did not get any direct observed evidence on the frequency of similar events from this region (Fig. 14).

We note that the Andizhan earthquake, judging by its interior location relative to the contour $K_{max}(A)$, is not the maximum possible one in the region. The Chatkal earthquake is located on the contour, i.e., it should be considered the maximum possible. We shall now evaluate its recurrence. From Fig. 14 we see that log $N_\Sigma = \overline{5}.8$ for this earthquake and therefore $N_\Sigma = 0.63 \times 10^{-4}$, so the mean recurrence period for such events in an area of 1000 km^2 is $T = 16\,000$ years. In the area of radius $R = 144$ km "responsible" for this earthquake, with $K = 17$, the recurrence is $N_{\Sigma|R=144} = 4.1 \times 10^{-3}$ year^{-1}, which corresponds to a period of $T|_{R\,=\,144} = 240$ years. Unfortunately, we cannot directly check this figure from our observed data; it can only be found from extrapolation.

The events during this period (up to $K = 15$) were certainly smaller in magnitude than the maximum possible earthquake in their corresponding regions (up to $K_{max} = 17$), making it impossible for us to obtain here the actual distribution $N(K, A)$ close to the limiting contour $K_{max}(A)$. However, we can evaluate the recurrence for a series of severe earthquakes in the past, even though we can only do it for the present by extrapolation.

The method of constructing a "generalized earthquake recurrence law" $N(K, A)$, which we have developed here, can be used to process observed material from a number of uniform seismic zones.

2.4. Interpretation of Earthquake Recurrence in Terms of Energy[*]

Present-day methods for the quantitative mapping of seismicity employing seismic activity A are based on an empirical statistical correlation of earthquake recurrence. A further clarification of its physical essence is of interest. Earthquake statistics [Riznichenko, Nersesov et al, Gutenberg], seismo-acoustic observations of rock bursts and the shattering of rocks in mines due to rock pressure [Borisov et al, Riznichenko etc.] and laboratory experiments on the destruction of rock specimens and other materials in presses [Vinogradov, Riznichenko] demonstrate that, in all these instances, the distribution of fractures with respect to the energy ($E = 10^K$) of the elastic waves generated by the fractures is monotypic within wide limits:

$$N = A \cdot 10^{-\gamma(K - K_0)} \qquad (2.32)$$

In this *earthquake recurrence correlation* $N = N(E)$ is the space (volumetric or surface) distribution density of the frequency of the fractures with respect to $K = \log E$; $A = N|_{K = K_0}$ is seismic activity [Riznichenko]; $K_0 = $ const is an assigned number; γ is an empirically determined and provisionally constant parameter that usually fluctuates in fairly narrow range of 0.4-0.6, where E is related to the surface of a sphere of fixed radius (10 km) surrounding the source, a range of 0.6-1.0, where E is related to the surface of the source and there is some energy absorption in its vicinity.

The exponential form of (2.32) signifies no more than a similarity between the phenomena over a wide range of variable values. Several attempts have been made to explain the numerical value of γ on the basis of various concepts concerning the connection between energy E and the geometry of the fractures or the volumes of the separate solid bodies thus formed.

An attempt is made here to approach the matter from a more general standpoint. The value of the parameter γ that ensures that the elastic energy of the process is at a minimum is sought on the basis of (2.32) in general form. Certain assumptions about the nature of the process and the properties of the material are made. It seems likely that this general approach may also be applied to more complicated models than those considered below.

Rock deformation during tectonic motion accompanying earthquakes is treated as macroscopic continuous flows of large volumes of rock masses, i.e., as a "tectonic flow" whose elastic-discontinuous component forms "seismic flow" [Riznichenko]. In essence, it is only this part of the process of tectonic flow, that will be dealt with in calculations given below.

---

[*] See [Riznichenko].

2.4.1. General Properties of the Integrals Under Consideration

We start with some auxiliary calculations. Calculations of the seismic energy flux carried by elastic waves from earthquakes $w = (1/VT)\Sigma E_i$ in a given space (V)-time (T) area, of the strain release accumulation values $w_B = (1/VT)c_B \Sigma \sqrt{E_i}$, in the spirit of Benioff, and of some other quantities relating to similar seismological questions in a continuous macroscopic interpretation [Riznichenko], lead us to integrals of the form:

$$w_m = \int_{K_1}^{K_2} NE^m dK, \tag{2.33}$$

where $m > 0$. Here, $m = 1$ for earthquake seismic energy flux w, and $m = 0.5$ for strain release. The limits K_1 and K_2 in (2.33) are assumed to be assigned values. When calculating the energy flux w for $\gamma < 1$, K_1 may be assumed to be very small, with a limit of $K_1 = -\infty$, and $K_2 = K_{max}$; this latter proviso corresponds to the seismic energy $E_{max} = 10^{K_{max}}$ of the maximum possible earthquake.

Let us demonstrate that integral (2.33) has a minimum of $m = \gamma$, where γ is the exponential factor in the earthquake recurrence correlation (1.35). For this purpose we substitute N from (2.32) into (2.33) and, assuming $E = 10^K$, we obtain:

$$w_m = \frac{A \cdot 10^{\gamma K_0}}{(m - \gamma) \ln 10} [10^{(m-\gamma)K}]_{K_1}^{K_2}. \tag{2.34}$$

If we assume $K_1 = K_0 - \Delta K$ and $K_2 = K_0 + \Delta K$ ($K_0 + \Delta K < K_{max}$) (K_0 does not necessarily coincide here with the value at which activity A is determined), this equation may be reduced to:

$$w_m = \frac{A \cdot 10^{K_0} \cdot \Delta K}{\ln 10} \frac{10^x - 10^{-x}}{x}, \tag{2.35}$$

where $x = (m - \gamma)\Delta K$. The indeterminate nature of (2.35) when $x = 0$ is readily established by L'Hopital's rule:

$$\lim_{x=0} \frac{10^x - 10^{-x}}{x} = 2 \ln 10.$$

Equating the derivative to zero:

$$\frac{d}{dx} \frac{10^x - 10^{-x}}{x} = 0,$$

we find the unique root of this equation at $x = 0$. The trivial solution of the equality that flows from this is:

$$\frac{1 + x \ln 10}{1 - x \ln 10} = 10^{2x}.$$

When $x = 0$ we obtain $\gamma = m$. It may be established by ordinary methods that this corresponds to the minimum of (2.35), and consequently also to the minimum of the integral (2.33).

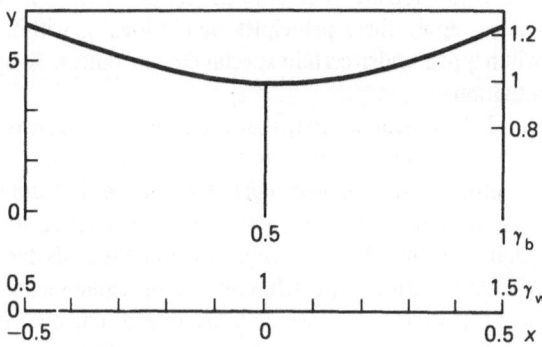

Fig. 15. Graph of the function $y = (10^x - 10^{-x})/x = \omega_m$. The minimum of this function (when $x = 0$) corresponds to the minimum seismic energy flow ω when $v_\omega = 1$ or with released strain, as defined by Benioff, when $\gamma_B = 0.5$

Figure 15 is a graph of the relative values of w_m as a function of $x = (m - \gamma)\Delta K$. From it we may infer, in particular, the depth of the minimum of the integral (2.33) when $x = 0$, i.e. when $\gamma = m$. The depth of the energy minimum governs the stability of the processes regulated by it.

Let us make a further general comment on integrals of the form of (2.33). One can see from (2.34) that the value $\gamma = m$ is a certain critical value: when $\gamma < m$ the integral (2.33) diverges towards the upper limit and converges towards the lower limit, so that we may freely assume $K_1 = -\infty$, $K_2 = K_{max}$ without the risk of obtaining an infinite value. When $\gamma > m$ this integral diverges in the opposite direction, and it should no longer be assumed that $K_1 = -\infty$. At the critical value $\gamma = m$ (which, as we have seen, also ensures the minimum) the integral in (2.33) diverges to both sides. In this case it is less risky to take it in finite limits of K_1 and K_2.

Now let us pass directly to a consideration of the seismological questions in which we are interested.

2.4.2. The Possibility of Using the Energy Minimum Principle When Considering Seismic Flow

The principle of minimum strain energy, the principle of stationary action, and other analogous variational energetic principles known from mechanics relate to an isolated system. The area, where tectonic flow accompanied by seismicity takes place, is not isolated, but is related to surrounding regions of the Earth that move relative to each other. Our internal system is set in motion by the forces of its interaction with the outer part, the energy being derived from the processes of radioactive decay and heating, gravitational differentiation and isostatic compensation of the Earth. The mechanical principles which have been indicated should be applied, in general, to the whole system, including the region of flow and the region of external factors.

Nevertheless, since we have sufficiently reliable quantitative data only for the local system of the flow itself, or to be more precise for its seismic features, and for the parameters of the seismic regime, including the seismic energy flux of earthquake sources, we consider it best to apply these principles purely locally, within the flow area. This, if possible at all, will happen under certain special circumstances. To specify these, we idealize the geophysical situation.

Let us assume, first, that the tectonic process is macroscopically steady throughout the system. The use of the term macroscopic here implies that we shall consider parameters of the process averaged over fairly large space-time domains; individual earthquakes will be incorporated in the microstructure. The justification for this assumption is that significant variations will only become apparent over intervals far greater than the periods of recurrence of even the strongest earthquakes in the same region. However, it would be possible for our purposes to restrict considerations to areas narrower in space, in time and in earthquake energies, where the stability of the process could be assumed on the basis of seismic observations to be statistically established. This type of situation is frequently encountered in seismological practice.

We assumed that a steady, tectonic energy is uniformly injected into the flow mechanism and uniformly expended by it. Part of the incoming energy is expended in the area of flow on inelastic processes and is dissipated in the form of heat. Another part passes through the phase of elastic potential energy of the flow process before dissipation. The elastic energy reserve in the medium remains unaltered. It is distributed, in particular, in many earthquake foci of various magnitudes in various stages of preparation for release. This energy is released and dissipated during the earthquake as the seismic energy of the foci. When the process is in steady state, the seismic energy flux of the foci is constant and equals the potential elastic energy flux contained in the foci before rupturing.

Let us add a corollary to the second important condition, namely the absence of mutual effects between the flow parameters in the domain under consideration and the forces acting on this domain from outside the system. For ease of argument, let us assume that the whole domain of flow is homogeneous. The process will then occur as if we had a large isolated volume of macroscopically homogeneous, structurally heterogeneous material brought to a state of homogeneous flow by a system of constant forces applied to its boundaries and so distributed as to ensure homogeneity. The general energy principles of mechanics are clearly applicable to such a system as a whole, and so they are also applicable to any macroscopic part of the volume, i.e., they are, in a sense, locally applicable. It is similarly assumed that these principles are locally applicable in technology, when considering the energy features of the process of brittle fracture of metal in a press, if the press operates with a large energy reserve, for example pneumatically from a reserve of large volume [Drozdovsky et all].

Let us now outline the properties of the medium in the flow area.

We assume that elastic energy is here the only form of potential energy. This frees us from a number of circumstances whose role is apparently secondary for sources of the most widely encountered shear type. They include conversions of energy into the potential energy of the position of masses in the gravity field. They may develop in three-dimensional elastic and other deformations, related to temperature and pressure variations. They also include possible energy conversions to chemical energy and to crystal lattice energy in phase transformations and metamorphic processes in rocks. Let us also disregard the surface energy of the fissures that appear and disappear during flow.

Finally, let us make specific assumptions concerning the textural composition of the medium. Let the medium consist in the simplest case of domains of a viscous fluid, where no potential energy accumulates, and of domains of an ideally elastic material subject to distortion and brittle fracture that are earthquake foci i.e., which contain elastic energy. During fracture, this energy is liberated either wholly or in part and is converted into the dissipating energy of seismic waves.

For all these conditions let us apply to the region of tectonic flow, the principle of the steady state:

$$\delta \int L \, dt = 0.$$

Here L is the difference between the kinetic and potential energies of the system, t is the time, δ is the symbol of variation for all variables defining the process. In our case (stability) neither of these energies is dependent on time. Moreover, the kinetic energy of the tectonic flow may be disregarded in comparison with its potential elastic energy. Let us assume that the seismic energy flux per unit of time is a fixed fraction of the reserve of potential elastic energy present in the flow. The problem then becomes one of finding the steady-state conditions for the seismic energy flux from earthquake sources.

The integral of the seismic energy flux will only be varied with respect to parameter γ of the earthquake recurrence correlation (2.33) in which we are interested. We assume, provisionally, that the parameter is independent of the other parameters A and K_{max} of the correlation (2.33), and that they can be considered as fixed. The whole problem then becomes one of finding a simple extremum, in particular the minimum of the seismic energy flux integral with respect to γ.

All this reasoning is highly schematic, but is an approach that may be justified by the correctness of the results.

2.4.3. The Condition of the Potential Energy Minimum for Total Liberation From a Source

Whatever properties we may agree to ascribe to a medium when considering it structurally, either microscopically and continuously, or macroscopically, we shall always be concerned with the seismic energy flux in earthquake foci, which may be determined to the accuracy of the observations:

$$w = \frac{1}{VT} \Sigma E_i = \int_{K_i}^{K_{max}} NE dK. \tag{2.36}$$

Here w is the seismic energy flux radiated by the sources per unit volume V of the medium in unit time T.

Let us assume that this flux represents the entire potential elastic energy contained in the earthquake focus, before fracture in the sense indicated above. Then, assuming that the integral (2.36) corresponds to the minimum elastic energy of the flow for a value of γ in the earthquake recurrency correlation (2.32), we immediately establish by comparing (2.36) and (2.33) that $\gamma = 1$.

In fact, at the finite limits K_1 and $K_2 = K_{max}$ (2.36) should coincide with the integral in (2.33) where $m = 1$. We saw in Section 2.4.1 that this corresponds exactly to the minimum of such an integral for $\gamma = 1$. It was for this reason that we were careful not to write

$K_1 = -\infty$ in (2.36), as could have been done if we had been certain beforehand that $\gamma < 1$, as is usually found from observation.

This contradiction may be explained, at least in part, by the fact that the value $\gamma \approx 0.5$ usually obtained from seismological observations relate to calculations of seismic energy E_i at the surface of a sphere of constant radius R (the usual value is $R = 10$ km) surrounding the hypocenter, rather than to the surface of the focus as a zone where inelastic fracture and plastic movements are significant. If E_i is related to the surface of the focus, γ closer to 1 can be obtained, the actual value being dependent on what is actually taken as the focus boundary, and also on the value adopted for the absorption of elastic wave energy between this boundary and the surface of the sphere R: such values are often approximately 0.7-0.8 (cf. e.g. [G. Gurevich, Riznichenko]). Values of approximately $\gamma = 0.8$ were also obtained in tests on specimens (Vinogradov and other papers). It is possible that different values may be obtained for γ, especially values closer to unity, and possibly even exceeding it, by certain other approaches to the calculation of the elastic energy dissipated near and within the focus. Particular attention will then have to be paid to limitation of the energy integral (2.35) in the direction of K_1.

The large difference between the value $\gamma = 1$ obtained by minimizing the integral (2.35) and the values observed in seismology is less surprising than the proximity of this "theoretical" value to the observed one given the extremely schematic approach. It may be that seismic flow, in fact, plays a far more significant role in tectonic flow than might appear at first glance. The crushing of material and the liberation of potential elastic energy in fractures may, in large measure, be a property of the deep microstructure of plastic flow itself.

2.4.4. The Condition of the Elastic Energy Minimum in the Accumulation of Strain Releases in the Benioff Sense

Following an idea put forward by Reid, Benioff considered the movement of rock masses along a major deep-seated fault such as the San Andreas (California, USA), as a sequence of elementary processes of slowly increasing continuous elastic strains in individual sectors of the fault zone, alternating with acts of rapid discontinuity of the material and strain release (Fig. 16). According to Benioff, the maximum elastic energy that accumulates under conditions of simple shear in a unit parallelepiped of earth masses (Fig. 17)[*] is:

$$E_i = \frac{\tau_{max} \, \varepsilon_{max}}{2} \, V_i = \frac{\mu \varepsilon_{max}^2}{2} \, V_i, \qquad\qquad (2.37)$$

where $\tau_{max} = \mu \varepsilon_{max}$ is the shear strength (breaking point), ε_{max} is the corresponding shear stress, μ is the elastic shear modulus, V_i is the volume of the parallelepiped. During the process of fracture (earthquake) both parts of the parallelepiped once again enter an unstressed state and are re-deformed, and the energy (2.37) is fully liberated, transforming into seismic wave energy. A residual shear stress forms in every fracture, and on the average equals the maximum elastic strain of the entire parallelepiped

$$\varepsilon_i = c\sqrt{E_i},$$

---

[*] Although, as noted in [Gurevich], this is inaccurate, the order of magnitude of the values remains, and this suffices for our estimates.

Fig. 16. Diagram of the accumulation (A, B) and release (C, D) of elastic deformation for an earthquake in the zone of a deep-seated fault, according to Reid and Benioff [Maruyama]

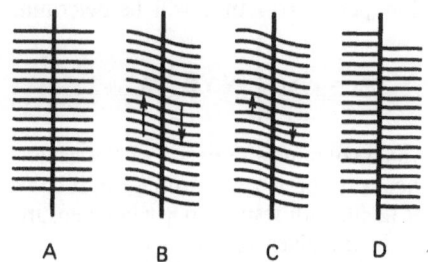

A   B   C   D

where c = const is defined by the properties of the material. This strain accumulates as there is a repetition of similar elementary processes:

$$\varepsilon_\Sigma = \Sigma \varepsilon_i = c\Sigma \sqrt{E_i}, \tag{2.38}$$

forming, in macroscopic terms, a tectonic flow, or to be more precise, its seismic part.

This outline of the elementary process has been improved and more detail added by other researchers, including Soviet ones, but the orders of magnitude of the values remain if the analysis is still made in terms of a body subjected to linear elastic deformation and brittle fracture [Gurevich].

When normalizing deformation to unit of volume and time one may write in accordance with (2.38) that the mean elastic shear strain in seismic flow equals:

$$\varepsilon_0 = c_0 \int_{K_1}^{K_2} N \sqrt{E} \, dK, \tag{2.39}$$

where the proportionality factor c_0 = const is defined by the properties of the material. It should be noted that in (2.39), as in (2.38), we follow Benioff in assuming that the deformations produced by various earthquakes are simply arithmetically summed, although this assumption can not be said to be clear. Another view on addition of the deformations produced by a set of many earthquakes of different sizes has been developed in [Riznichenko]. Here, nevertheless, we shall provisionally retain Benioff's approach.

It may be assumed for the entire seismic flow that it occurs at some constant mean stress τ_0. In this case the accumulation and release of the elastic energy $w_0 = \tau_0 \varepsilon_0$ of the

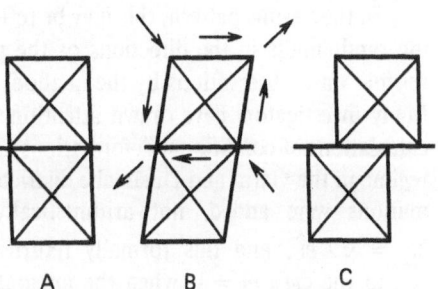

Fig. 17. Diagram of accumulation (B) and release (C) of elastic deformation in a unit parallelepiped (A) according to Benioff [Maruyama].

The *arrows* denote the stresses, tangential and normal (not exactly in accordance to Benioff)

A   B   C

flow per unit of time will be determined by the equation:

$$w_0 = c_0 \tau_0 \int\limits_{K_1}^{K_2} N \sqrt{E} \, dk, \qquad\qquad\qquad (2.40)$$

which contains an integral of the general form as in (2.33) for $m = 0.5$. On the assumption that τ_0 is not dependent on γ (or, if convenient, that γ is not dependent on τ_0), the minimum of (2.40) with respect to γ will be ensured, as we already know from Section 2.4.1 at $\gamma = m$, i.e., in the given case at $\gamma = 0.5$.

Therefore, even with this, even more schematic approach an estimate of a reasonable order of magnitude is obtained for γ.

Let us now turn our attention to the fact that in the expressions for specific deformation (2.39) and for specific energy (2.40) the integrals diverge in the neighborhood of K_1 from the actually observed values $\gamma \approx 0.5$ and above (when energy w is related to the surface of the focus).

On the assumption that they remain finite within the finite intervals K_1, K_2, we shall obtain infinitely large deformation ε_0 and energies w when integration is continued towards $K_1 = -\infty$, i.e. towards weak earthquakes, but this is physically impossible.

Continuing to reason in the same terms, a way out of the difficulty may be sought by suggesting firstly that the earthquake recurrence law ceases to be defined by a straight line in log E, log N coordinate for sufficiently weak earthquakes with small K, i.e., that the recurrence graph bends downwards not only to the right (near $K_2 = K_{max}$), but also to the left (near $K_1 = K_{max}$). Secondly that simple arithmetic summation of the deformations ε_i by Benioff's outline (2.38) becomes invalid in the region of small values of K.

Although both possibilities are feasible in principle, the second must be seen as the more significant. Real deviations of observed frequency graphs from a straight line have not been observed for small values of K in the range of variation of this quantity reviewed either under natural conditions (earthquakes, rock bursts), or under laboratory conditions (experiments with specimens). In relation to the arithmetic summing of deformations according to Benioff, it probably remains permissible only in an extremely limited range of the severest earthquakes of approximately equal magnitude and for the same deep-seated fault. Displacements in severe earthquakes, which involve the motion of large volumes of rock masses, should in fact reflect the general mean tendency of tectonic flow in the fault zone. Weak earthquakes may be characterized by a more random distribution of the directions of displacement.

Morphologically this is manifested by the fact that faults of the first or principal order of magnitude are usually accompanied by second-order faults that approach them at oblique angles and form "feathering". Second-order faults may similarly have their own feathering and so on.

In the seismic pattern, this may be reflected for increasingly weak earthquakes by increasing randomness in the directions of the principal stresses and movements calculated from seismic wave observations by the methods used in investigation of the fault plane solutions. Many investigators have drawn attention to this. It was allowed for in the author's earlier calculations of conditional deformations according to Benioff, in investigations of the seismic regime in the Garm and Dushanbe districts of Tadzhikistan: for weak earthquakes the deformations were added, not arithmetically (2.38), but as an orthogonal vector system $\varepsilon_{\Sigma'} = \sqrt{\Sigma \varepsilon_i^2}$, and this formally returns us again to an integral of the form of (1.4), i.e., to the case $m = 1$, when the integral converges downward for $\gamma < 1$.

Equation (2.40), in which it is assumed that stress is constant and may be assigned irrespective of deformations in the foci, and that it does not influence γ, may also be interpreted in the following manner. Let us assume that all stresses and strains accumulated in a focus do not revert to zero in each fracture, i.e. that only a small part, and not all the accumulated potential elastic energy in the focus is released. The stress will then change little, and may be assumed not to be dependent on the changes of deformation that develop. At the same time the total deformation of the seismic flow is made up precisely of these elementary changes of strains. This case may correspond to an expression for seismic energy flux in the form of (2.40), whose minimum is reached when $\gamma = 0.5$.

This view permits a new approach to the observed values of γ. In fact, if it is held that the case of complete release of the potential elastic energy of foci during fractures corresponds to $\gamma = 1$, and the case of very small release corresponds to $\gamma = 0.5$, it would appear that the relative amount of elastic energy released by foci could be assessed from observed values of γ. It is known from experiments that the value of γ may vary perceptibly in different places and at different stages in the seismic process. This offers wide scope for tempting geophysical comparisons and arguments.

It should, however, be noted that the parameter γ may also be dependent on other circumstances for other, more intricate models of a medium with tectonic and seismic energy flow.

An energy interpretation has been attempted with the object of increasing physical comprehension of the empirical "earthquake recurrence law". The slope γ of the recurrence graph is arrived at by minimizing the potential elastic energy accumulated in unit volumes of a medium involved in seismic flow and released in ruptures as seismic wave energy.

Direct calculation of the seismic energy flux of earthquake foci yields the theoretical value $\gamma = 1$; a theoretical value of $\gamma = 0.5$ is obtained from consideration of Benioff's slightly artificial outline of strain release accumulation. Both these values, despite the extremely schematic nature of the actual formulation of the problem, agree, within an order of magnitude, with the actually observed values of γ for earthquakes and rock bursts and with laboratory experiments on specimens.

This makes it possible to consider the present energy approach a promising one meriting further examination and consideration in greater detail. This could be especially in the sense of considering not only the elasticity and brittle fracture of the material, but also its other rheological properties. This should make it possible to relate more clearly the parameters of seismic regime, especially A, γ and K_{max}, to the properties of the material of rock masses and to the characteristics of its deformation in the process of tectonic movements.

3. The Maximum Possible Earthquake

3.1. General

Large earthquakes... Seismic danger... Prediction of earthquakes... Seismic zoning...
Lately these problems have become popular topics with writers, broadcasters, and movie makers. It may seem that the prediction of earthquakes is a magic key to the avoidance of seismic danger, even that the topic is the main scientific and economic problem in modern seismology.

To see if this is so, we have first to specify what earthquake prediction is and how seismic danger is defined.

Then we can address the description of the maximum possible earthquake. Having done so, we shall look at its relation to seismic danger and various other important seismological problems, in particular seismic zoning.

In seismology, earthquake prediction is the prediction of the time, place and size of an individual large earthquake, and is thus the natural analog of a weather forecast.

If we agree with this way of understanding of earthquake prediction, it is unreasonable to talk about a prediction when calculating the maximum possible event in a given place on the Earth's surface without giving a time for its probable occurrence. Such an estimate would be say, a "climatic" attribute of the seismic regime in a given region, and it is usually done by considering its constant average parameters over time. This assumption may be justified by the fact that geological tectonic processes, to which earthquakes are confined, do not vary much in the time periods which are of practical interest to humanity.

Sometimes, geophysicists "predict" certain seismic climate elements instead of "determining" where, on the Earth, these elements are still poorly investigated. By analogy, we may "predict" mineral deposits in poorly studied territories. We do not criticize geological terminology here. In seismology, such a free use of the term does not seem possible. It would be true, if we were concerned with the prediction of possible changes in the seismic climate over time due to changes in tectonic processes, just as we may simulate the meteorological climate due to increasing air pollution, deforestation, or irrigation. Sometimes, similar problems may arise in seismology: the possible increase in seismicity due to nuclear tests, the intensification of mining, or the loading of a surface as a large reservoir is filled. Such facts are known, however, they are only second-order phenomena. On the whole at present, the natural seismic climate on the Earth remains approximately the same. So, our "prediction" of the seismic climate for the near future does not go beyond this simple statement. To determine the climatic elements of seismicity with respect to a point on the Earth's surface, assuming their stability over time, is the aim of zoning for a long-term average seismic danger, or in short, "seismic zoning" (Fig. 18).

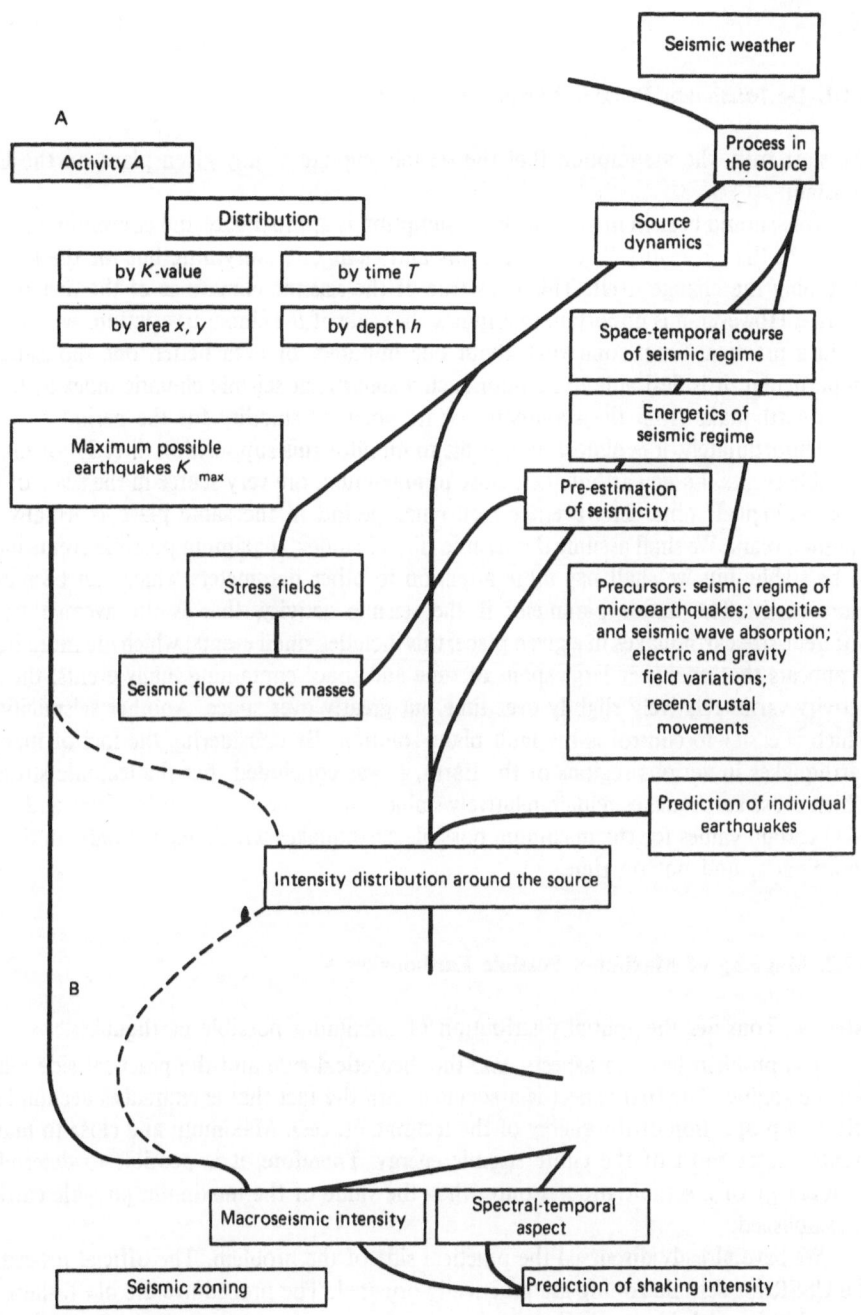

Fig. 18. Schematic representation of the main problems in seismology.

The scheme consists of two parts. The upper part (A) includes seismicity, source activity, the process of rupture formation in rock masses and seismic wave generation. This is part of the dynamics of the Earth's interior, a specific kind
of tectonic of orogene process. The lower part (B) includes shakeability; seismic danger denotes the demonstration
of seismic processes on the Earth's surface, its influence on the people and the consequences i.e. on constructions,
dams, bridges, houses. The scheme can also be divided into two parts horizontally, the *left* part relating to seismic
climate, the *right* one relating to seismic weather. The climate typical for a given place on the Earth is then defined
by the totality of local weather at a given moment, either meteorological or seismic

3.1.1. Do Maximum Possible Earthquakes Exist?

We start from the assumption that the seismic climate at any given place on the Earth is constant.

One should bear in mind that this assumption is approximate and conventional. We have no doubt that in reality "everything flows and changes", everything but for the stable laws governing the change itself. This is as true of the seismic climate as of the meteorological climate. However, it is important to estimate the scale of the climatic variations we are interested in a practical time: somewhat about one hundred, or even better, one thousand years. In particular, it is desirable to monitor such a significant seismic climatic index as the maximum earthquake from the standpoint of its practical stability for the period.

Unfortunately, it is almost impossible to monitor this supposition directly for maximum possible events. Large earthquakes, close to maximum, are very scarce in the scale of human (not geological) time. The average recurrence period in the same place is usually several hundred years. We shall assume the seismic climatic index, maximum possible events included, to be stable, but we shall pay more attention to other parameters which can be monitored more easily. The major parameter is the seismic activity, that is the average frequency different-size earthquakes in a given place; this includes small events, which are more frequent. It appears that over very large spans of time and space containing many events, the seismic activity varies relatively slightly over time but greatly over space. Another seismicity index, which is easier to control is the fault plane solution. By considering the foci of many large earthquakes in various regions of the Earth, it was concluded that the tectonic stress fields which cause earthquakes remain relatively stable. These, and various other facts and concepts yield certain values for the maximum possible earthquakes which depend only on the spatial coordinates, and not on time.

3.1.2. Mapping of Maximum Possible Earthquakes K_{max}

Now, we consider the spatial distribution of maximum possible earthquakes.

The problem has two aspects, viz. the theoretical side and the practical side related to seismic zoning. The first aspect is associated with the fact that earthquakes accumulate and release a proportion of the energy of the tectonic process. Maximum and close to maximum events release most of the entire seismic energy. Therefore, it is possible to determine the total energy of a seismic process only when the value of the maximum possible earthquake is established.

We have already discussed the practical side of the problem. The official procedures in the USSR for seismic zoning are practically oriented. The procedures are old fashioned and are to be changed. However, K_{max} is also necessary for the system of seismic zoning derived from seismic shakeability, which is to be substituted for the former.

Methods for calculating seismic shakeability, as well as establishing the maximum possible earthquake, have been the topic of research for a number of years by the author and a large team of seismologists in the USSR. They work in central and peripheral institutes of the Academy of Sciences and other institutions directly involved in studying quantitatively seismicity, and seismic danger. These studies have confirmed that the definition and mapping of K_{max} is extremely complicated.

Several approaches are possible. All are based on observed events close to the maximum extrapolated to the expected possible events. We shall list the most significant approaches.

The direct method involves a search through the historical data for large earthquakes which have occurred in a region over the maximum possible time period. In the USSR, the best region for this is the Caucasus, which has been inhabited by civilized peoples with ancient recorded histories. Abundant historical data are available here for a period of about 1000 years. However, this is still too poor to characterize the seismicity of the territory to the required resolution. The data only also concern certain individual sites. Vast territories which were previously uninhabited have no such data. Besides, even where there is a long recorded history, the data on large earthquakes might have been lost as a result of war or other tragic events. Paleoseismological data provide useful contributions to our knowledge of large earthquakes. Large earthquakes sometimes leave landforms on the face of the Earth, e.g. thrusts, shears, rock-falls and land-slides, which have been preserved for thousands of years. These scars may be dated by geological or geomorphological methods, while their size indicates the scale of the events. However, this method works mainly in rock regions in which the basic rocks contain primary seismic dislocations and the covering materials consist of unconsolidated sediments. These are less stable than in the lowland areas, and so are exposed more often. In addition, the consequences of a sequence of lesser events may be mistaken for that of one large event.

Historical and paleoseismic data can only be applied to a few individual places or "points" on the map. These data are generalized and extrapolated to neighboring territories. Usually the procedure is to dissect the region into presumably, homogeneous seismotectonic segments. The maximum observed seismic event is located within each of them. An event of a similar size is assumed to be possible at any other point in the territory.

It is clear that this technique is qualitative and intuitive in character. It is difficult, if not impossible, to tell which parts of a territory are homogeneous: in nature, no two parts are the same. It is extremely risky to state that the maximum observed earthquake in each area is in fact the maximum possible event there. However, today's standard seismic zoning maps are based on this logical assumption.

It seems expedient to use other techniques, to find out more definite, constructive, and algorithmic methods for assertaining the maximum possible earthquakes. It seems natural to base these methods on general regularities to be found in the seismic process itself and their correlation to other interior processes studied using geological, geomorphological, geodetic methods, and to various physical fields in the Earth (gravity etc.).

The second approach is based on empirical correlations between the seismic and other phenomena and their values. The first approach is more fundamental, but it requires more time. The second is more superficial, but takes less time. In addition, the first approach cannot be used efficiently without a sufficient preliminary, fundamental analysis of the second type but physical theory cannot be built without accumulation and systematization of actual data.

The first possible thing to do is to establish and use correlations between the parameters of the seismic process itself, earthquake value K and seismic activity A (Figs. 10, 11)

It appeared that K_{max} as a function of A is rather general: it is approximately the same for areas with remarkably different tectonics, e.g. regions with Alpine folding like the Caucasus, Crimea, Apennines (Italy), regions with predominantly block tectonics (the Baikal Rift), and regions with intensive recent tectonics such as Japan (see Sec. 2.5), the Kuril Islands and Kamchatka (Fig. 19).

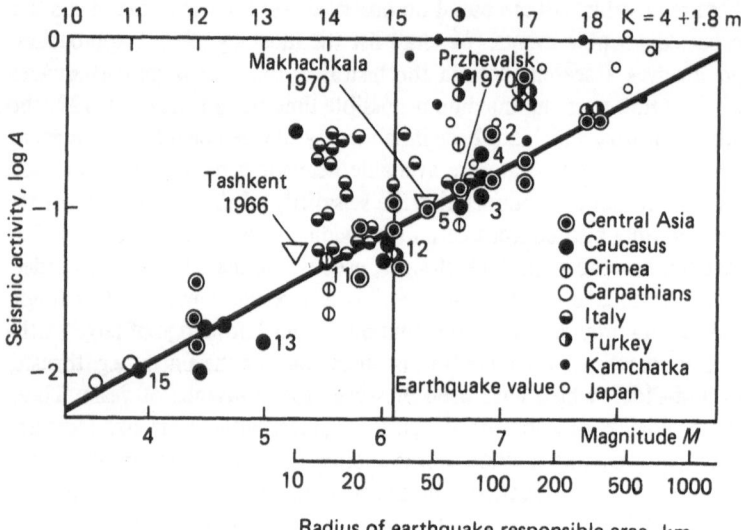

Fig. 19. Data summary on the dependence of seismic activity on the earthquake energetic value K in various regions of Eurasia

The correlation between seismic activity A and the maximum possible earthquake K_{max} was used to estimate the seismic shakeability in certain regions of the USSR. Despite the very schematic correlation, the estimates of the shakeability, including the data on K_{max}, were on the whole in good agreement with the observed shakeability where historical data are abundant; only there is such a comparison possible.

3.1.3. Complex Solution of the K_{max} Problem

Estimation techniques for assessing the maximum possible earthquakes should in the near future be comprehensive and based on seismological and other geophysical, geological and geodetic information.

Seismicity has for a long time been compared with geological and other data, but this has been done mainly in a descriptive and qualitative way. Now, the problem is to make this comparison in a quantitative way using statistical methods developed for such comparisons.

Two major features of such comparison should be first the necessity to organize it in such a way as to obtain over the widest area the densest grouping of observed points as round empirical averaging curves or surfaces (multi-dimensional in general) that can be derived. The second is the necessity to find a neat procedure not only of obtaining a simple statistical correlation (for instance, a regression analysis) but also to establish the probable position of the boundary delineating the maximum possible events (earthquakes) from the observed set of events. We have presented the principles of this technique when comparing earthquake values with seismic activity.

In the near future, the maximum possible earthquakes will be assessed from comprehensive seismological-geological-geodetic and geophysical data in each data type. The solution of this problem in space should be extended to the space-time domain. Then, the problem of finding the parameters of the seismic climate will be united with the problem of attaining the seismic weather, the problem of earthquake prediction. Empirical correlation methods will then be supplemented by physical consideration, theoretical simulations, and a physical model of the process. Further research is obvious and the field does not seem to be exhausted.

3.1.4. What About Earthquake Prediction?

We have considered maximum possible earthquakes as a part of the seismic climate and its practical application to estimating the long-term average seismic danger, seismic zoning. We have tried to show that the problem is a "hot-spot" in the chain of seismological problems and is vital in both its scientific and practical aspects.

What then should our attitude to earthquake prediction be? We do not mean a general term to which the entire problem of seismic hazard may be ascribed casually, including the maximum possible earthquake, seismic zoning etc. but rather a scientific prediction of the time, place and size of each large earthquake. Some interesting parallel aspects concern the role of such a prediction in the development of seismology as science and the practical economic application of seismology to mitigate the losses due to a seismic hazard.

Earthquake prediction is a facinating idea. It has stimulated seismological science and the related geosciences, and is now a more and more promising branch of research. The system of short-term earthquake prediction that is beginning to appear, will make it possible to avoid most of the human and property losses; for instance, by preventing the fires which usually accompany large earthquakes. The construction of a system for the long-term prediction of seismicity has also been outlined (on the basis of the energy model of seismic regime). It may be used to delineate time-dependent seismic zoning.

A complete understanding of earthquake prediction would mean the compilation of an accurate schedule of expected large seismic events for many years ahead. It would, in principle, mean the establishment of the seismic climate as the set of possible seismic events, and hence would yield, the long-term average seismic hazard, and a system of seismic zoning. However, at present, this is only a dream.

It is and will be, for the foreseable future, seismic zoning that can contribute to the economy. Buildings and other constructions should be able to withstand all aspects of the seismic weather (just as they do meteorological) for years and even centuries. We are therefore concerned with the regularities in the seismic climate as a set of events. These bulk, statistical regularities are simpler, than individual ones; they can be easily solved and used. This is in fact, being done, if rather poorly because only the maximum events are considered. The situation can be improved by considering the probable recurrence of events of any size, that is shakeability.

The scientific aims are clear and should be realized in the near future. We believe this is a major task in modern seismology as it addresses economic needs.

Earthquake prediction is the most important problem in seismology. The search for efficient methods of prediction should be parallel with major research, despite the fact that its practical application, in the near future, is limited and not so clear.

3.2. Seismic Activity and the Energy of Maximum Earthquakes[*]

The basic notion in this research is the quantitative definition of seismic activity A [Riznichenko] derived from studies of Tadjik earthquakes, a joint project by Tadjik and Moscow seismologists (the Tadjik Integrated Seismological Expedition—TISE—of the Physics of Earth Institute, Acad. Sci. USSR and the Tadjik Institute of Seismic Engineering and Seismology. Tadjik territory has always attracted the attention of our seismologists because its high seismic activity is the greatest of all the continental regions in the USSR.

The main aspect of this research is the discussion of the quantitative relations between the seismic activity A and the other indices of tectonic activity in the regions, and the maximum possible earthquake in a particular region [Riznichenko].

The estimation of maximum possible earthquake is a vital problem in seismology, whose practical significance for seismic zoning cannot be overestimated..

3.2.1. Seismic Activity

Earthquakes can differ in size, described as the seismic energy E radiated from the focus as seismic waves. This energy is usually normalized for a sphere with a definite radius, say 10 km, from the focus and can be determined from instrumental observations of earthquakes at seismic stations.

It is known from observation that earthquakes are distributed by seismic energy rather as the stars are by their visible brightness: the greater the value, the less numerous they are. The frequency of occurrence (or recurrence) of earthquakes distributed by energy E, $N = N(E)$ is shown schematically in Fig. 10a. As first approximation, it may be considered a straight line in logarithmic coordinates.

The seismic activity A of an area characterizes its seismicity in the earthquake region; events are close to class K_0 ($K_0 = 10$). To derive a complete frequency distribution $N = N(E)$ given this approximation, we need two more values, viz. γ and K_{max} (or E_{max}).

To make thing clearer we shall make some comparisons. A seismic activity of $A = 0.01$ is considered weak, whereas $A = 1$ is severe. An earthquake of class $K = 10$, that is of seismic energy $E = 10^{10}$ J, corresponds approximately to the seismic effect of the underground blast produced by a nuclear bomb the size of that dropped on Hiroshima. According to modern standards, the Hiroshima bomb was comparatively small. At a focus depth of 10 to 15 km, typical of Tadjikistan, the blast of such a bomb under the surface would have corresponded to shaking of 5 or 6 degrees at the epicenter; such event can be felt but it is not destructive.

The catastrophic Khait earthquake of 1949 (Tadjikistan) was of 9 degrees or class $K = 16$ or 17. Its seismic energy E was 10^6 times greater than that released by the bomb and equals approximately the annual energy generated by the Bratsk hydro-electric power plant[**], one of the greatest in the world. The almost instantaneous release of such an enormous energy under the ground would cause a catastrophic earthquake.

---

[*] See [Riznichenko].

[**] In 1964, the Bratsk hydro-electric plant produced 12.5 billions kW · hr of electric energy, that is 4.5×10^{16} W · s—*Author's note*.

If the seismic energy E of the event and the depth of its focus are known, we can determine its seismic effect on the surface—to find out the corresponding degree of intensity. A knowledge of the frequency N of earthquakes of various energies will allow us to derive the earthquake frequency of various degrees on the Earth's surface. Thus, a study of the long-term average seismic activity A and the other parameters γ and K_{max}, that is the law of earthquake recurrence and a determination of focus coordinates, gives an objective quantitative basis for seismic zoning.

3.2.2. Is the Estimation of the Maximum Earthquakes Possible in Advance?

Reliable and precise seismic zoning is hindered because we only have short-term instrumental or even non-instrumental seismic observations at our disposal. By recording small earthquakes say $K = 10$ or less, which occur often, we can reliably plot some of the recurrence curve in the area close to $K = 10$ for a relatively short time (in Tadjikistan, under the normal seismic regime, it is several years). However, for severe events, which are scarce, this is almost impossible. So, in order to establish the average recurrence of events like the Khait earthquake of 1949 directly from observations, one might have to wait several thousand years or more depending on the extension and seismicity of the region under analysis.

A linear extrapolation of the observed events from $K = 10$ events on the recurrence graph to greater energies would yield a hypothetical frequency for earthquakes with any value of K, if we assume that all earthquakes are possible in the given region. But the question as to whether they are in fact possible remains open.

Note that despite the obviously one-sided and incomplete character of this method of estimating seismic danger, it is of some practical importance because it can be used. To establish an upper limit for the maximum possible hazard, *a priori* we overestimate the hazard, assuming it to be infinite as far as maximum power is concerned, but very rare in frequency. It is at least an estimate. When, for instance an event with a shaking intensity 9° MSK is estimated to occur (if at all) in a region less often than once every hundred thousand years, it does not seem reasonable to take it into practical consideration.

Nevertheless, it is clear that such a solution should be improved upon. Therefore, we should learn how to establish maximum values $K = K_{max}$ or energy E_{max} for the maximum possible event in a given region. This should be done both by direct observations of such earthquakes, and from indirect data, which are more abundant and can be obtained in a shorter time.

In the past this was done by a pure qualitative consideration of geological data and their comparison with seismic and other data, arising mainly from intuitive suppositions. If an outstanding seismo-geologist with good intuition attempts to persuade the scientific community that his estimates of seismic danger and his scheme of seismic zoning are correct, it is really hard or impossible to verify the reliability of these ideas, which have arisen from particular objective data. One has to wait for the estimates to be realized (if they ever do). In such a situation it would be desirable to obtain more definite, justifiable, and accessible method that can be objectively verified. Such a method should be exclusively quantitative.

In principle, two major approaches may be considered. The first is to search for internal physical relations between the observed facts and to establish some functional relations between them. The second is to study external statistical correlations between the phenomena and approach the problem from the spirit of probability that is search for the most probable average ratios between a quantity E_{max} in our case, (find its expected reliability), and its probable deviation from the average.

In both cases, the solution should be based on geological, geomorphological and geodetic data, some geophysical data (tiltmeters, strainmeters, gravimeters), and certainly seismic data about weaker but more frequent events.

These data should be presented in quantitative form so they can be processed jointly. Some data have been already "treated", while others have yet to be "treated" in an appropriate way. This, in particular, concerns the geological data, was interpreted only qualitatively.

However, the quantitative approach is not completely alien to classical geology, e.g., the vector roses of cleavage fracturing, which have been used for a long time. Recently, movement towards the inclusion of more quantitative consideration has occurred, the determination of the average velocity and velocity gradients of the vertical displacements of rock masses over geological time. No doubt, it will be possible, in future, to transfer pure qualitative geological ideas vital for seismology into quantitative indices by methods currently used to estimate the resolution and contrast of photographic images, or the characteristics of the maps as isolines, for instance, as in topographic maps to indicate the gradient, detail, extent, and spatial distribution of structures.

The regularities we are interested in, functional or correlational should be established from data on long studied regions. Then these relations may be used and refined by applying them to larger, less well studied areas. Finally they can be used to make predictions.

Seismicity is a complex process. It depends upon many factors which are difficult to monitor. This hinders a direct study of seismic quantities as functions of other parameters, though some topics are amenable to such a study. To achieve fast results for several significant general seismic parameters, the statistical approach seems more promising. Certainly, it is desirable and possible to combine it with the functional approach.

As a first step we shall only consider statistics of seismic data. To evaluate maximum earthquakes we recall the fact that severe earthquakes usually occur in regions where moderate earthquakes are frequent and that as the earthquake potential decreases, so too does the average frequency (that is seismic activity) of the moderate earthquakes.

To be exact, we shall consider the correlation between the long-term average seismic activity A and energy $E_{max} = 10^{K_{max}}$ of the maximum earthquakes in some places.

We should be aware of the difficulties and limitations of this approach. A direct A and E_{max} correlation is known to hold good only on average and over rather large spatial-time areas. Seismologists know many cases when the direct correlation is violated over limited time intervals and within some bounded areas. However, such cases are scarcer than where the direct correlation holds. This is what makes this approach so promising.

This approach might provide some useful results that can be practically applied if the correlation is close. The closeness of fit defines the degree to which observed, particular cases in the correlation deviate from the average, or rather from the bulk correlations to be exact, with respect to the value and frequency of deviations from the average. When using a statistical correlation for prediction, these deviations act as possible errors in the calculation technique. It is desirable to reduce them in number and magnitude.

Meantime, the fitness of the analyzed correlation can sometimes be controlled and improved by selecting more suitable conditions for comparing the data. In the case of A and

E_{max}, these conditions are dimensions of the space time domain within which the comparison is to be performed.

When choosing the dimensions of these domains we face the following dilemma. On one hand, in order to get a better fit we should use larger domains, in both space and time. On the other hand, in order to get a good resolution of E in a prediction, we should stick to small areas of comparison.

First, we shall consider the temporal aspect of the problem. Starting from the simplest assumption that the seismic regime is an approximate steady state over time, we get an upper boundary for the time interval beyond which seismic regime will significantly change in its long-term average features. Then we can begin to show the systematic, secular variations, due to general variations in the tectonic process as manifested by seismicity. One may suppose that such variations occur over geological time intervals, viz., several thousand years or more. The observed seismicity variations, during historical time in the regions, may be attributed to fluctuations in the seismic process considering the tendency of a seismic event to clustering in time. A vivid example of this is the temporal increase in local seismicity after a powerful earthquake. It is significant that this tendency to cluster remains approximately the same over time.

The reliable seismic data we have at our disposal have been collected usually for very short periods. Instrumental observations of moderate events are only available for about ten years at best and, at worst, only one year perhaps. The information on severe earthquakes is better. Instrumental data about these events cover about 50 years, whereas non-instrumental data, from which it is also sometimes possible to determine earthquake "magnitude" the energy, cover greater periods. However, they also very seldom exceed one thousand years. Note that the reliability of these data decreases with age.

Geological methods give us an opportunity, in principle, to enlarge the range of time intervals to obtain reliable seismic data. This will probably help when considering, in future, systematic variations in the seismic regimes. However, at present, no reliable way of monitoring exists. We have to be satisfied by the assumption that the process is steady state.

In this situation, the procedure is reduced to using reliable seismic data for the longest possible time. For this time, the "long-term average" seismic activity A and energy E_{max} of the maximum earthquakes are determined. Note that for these reasons, the time intervals for which each of these values is determined may not coincide.

Now, let us consider the spatial aspect. Here, the choice of conditions for the A to E_{max} comparison is wider.

Each earthquake is obviously attributed to a process which occurs in some extended space, which is in direct proportion to the level of the event, rather than at the point symbolizing its hypocenter or focus. We do not mean the area of the earthquake focus proper, where the medium fractures and deforms plastically medium over short time of the event, but the larger area with a longer life time responsible for preparation of the event. There, most of the tectonic stresses which cause the earthquake, and the related continuous and discontinuous seismic deformations, both preliminary and subsequent, are concentrated. It is also the source of most of the elastic energy which is released and for the very short time during the main shock.

It is natural to assume that the preparation area of each powerful earthquake has some long-term geological structure or change in the tectonic processes which resulted in earthquakes in the past and may cause re-occurrence in the future. These characteristics may affect the long-time indices of the seismic regime with respect to weaker events, in particular the long-term average seismic activity. It is therefore reasonable to compare average value of seismic activity A with energy value E_{max} of each maximum earthquake just in the corresponding responsible area. We shall consider a particular example when the dimensions of this "area of conformity" were determined as a function of the size of the earthquake, viz., its seismic energy.

3.2.3. Seismic Activity and Energy Correlation for Maximum Earthquake

The data were obtained from observations in the Eastern Tien Shan, Dzhungaria, and Altai-Sayan mountains. Good maps of seismic activity are available for these regions. They were compiled by Gorbunova from instrumental data for 16 years, 11 years or fewer years [Gorbunova]. The data on the severest events which occurred there between 1887 to 1957 were taken from "Earthquakes in the USSR" [see Bibl.], and for more recent time from [Gamburtsev].

Table 7 presents the list of most powerful events. To get a correlation with the activity A, only the severe events located inside the area for which A was determined were used.

Table 7. Major Earthquakes in Eastern Tien Shan and Altai-Sayan Region

No.	Name	Date	K
1	Kemin	3. I 1911	17
2	Chilik	12. VII 1889	16.5
3	Verny	9. VI 1887	16
4	Kemin-Chu	20. VI 1938	15
5	Son Kul'	13. X 1958	13.5
6	Dzhungarian	21. XII 1958	15
7	Dzhungarian	19. VII 1962	15
8	Mondy	4. IV 1950	13.5
9	Altai	25. VII 1922	15
10	Barguzin	26. V 1939	14.5

For the less active Altai-Sayan area five more earthquakes of weaker intensity ($K = 12$ including) were used. The latter appeared to be the maximum for the corresponding regions of low seismic activity. The magnitudes M given in [Masarsky et al] of the severe events were converted to energy values K by the Rautian formula: $K = 4 + 1.8 \, M$.

The spatial comparison conditions for E_{max} and A were chosen as follows. It was assumed for all the severe events that the conformity area was in the shape of a hemisphere with center O at the earthquake's epicenter (Fig. 20), while the volume of the hemisphere $V = (2/3) \, \pi r^3$ is related to the energy E_{max} by the equation

$$E_{max} = kV = cr^3 \tag{3.1}$$

Fig. 20. The area of conformity of correlating seismic activity with maximum earthquake energy.

O—Earthquake epicenter; S—the area of conformity with the Earth's surface

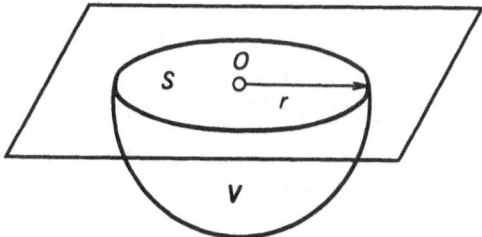

where k and $c =$ are constants irrespective of E_{max}. This permits us to assume that the bulk
density $k = E_{max}/V$ of the energy E_{max} is distributed over the conformity area and remains
the same for all earthquakes. The average seismic activity A was determined as the average
over the area $S = \pi r^2$ of the circle along which the volume area of the conformity is attributed
to the Earth's surface.

We have just considered shallow continental seismicity as observed in the regions in question. The seismic activity plotted on the maps is a surface one, that is per unit surface
area. In the case of deep seismicity, i.e., continental or oceanic, distributed in a volume,
which extends both horizontally and vertically, the activity should be considered as a bulk
parameter.

Note that the assumptions we made above are not the only ones possible. Instead of
the earlier assumption we could presume that the spatial areas responsible for all severe crustal
earthquakes are shallow and go down to similar depths. Thus, instead of (3.1), we should
assume the correlation $E = c'r^2$. Then, assuming that the areas in which severe earthquakes
and their foci develop, are predominantly linear and we can introduce the function $E = c''r$
instead of (3.1).

Then, we can question the assumption that $K = E_{max}/V$ and, accordingly, the coefficient
c in (3.1) is constant. The physical strength of macro-volumes of the material of the crust
and upper mantle, where earthquakes are generated, depends mainly upon the occurrence
of defects, or weak zones there. The destruction of these seems to start the destruction of
a larger volume (earthquake), we suspect that an earthquake's total effective strength should
decrease as the volume V increases. This would cause the reduction of k and c with increasing
E_{max}. On the other hand, for each individual focus, the relative amount of released elastic
energy may increase with focus diameter. This is because an increase, in diameter causes the
energy loss due to rigid or viscous friction near the rupture surface to increase in proportion
to the square of the diameter, while the total energy accumulated within the volume rises
in proportion to the cube of the diameter. Hence there might be an increase in k and c with
increasing E_{max}.

However, we shall not analyze all the possibilities and remain with (3.1).

Now, the problem is how to set the values of the constants k and c in (2.21). As the
relation between them is simple $k = 3c/2\pi$, it is enough to find, say, c.

It would be nice to derive c from physics. However, we should need extensive quantitative
information on the properties of the Earth's mass material and conditions under which they
deform; alas, we do not have this information at the present time. Therefore, in this case
we shall use a simpler, pure statistical method of research.

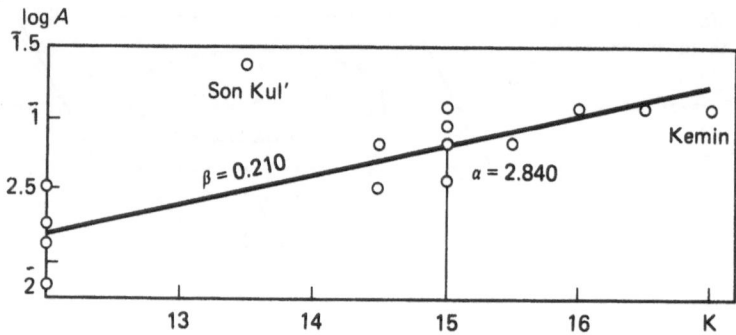

Fig. 21. Correlation dependence between K, the logarithm of seismic energy of maximum earthquake, and log A, the logarithm of average seismic activity within the surrounding area

We shall test a sequence of values for the constant $c = c_i$, $i = 1, 2, 3 \ldots$. At various values of c_i we shall average activity A for various areas S_i around the epicenter of each maximum earthquake. Correspondingly, we shall have various averages of A_i. At each fixed value of c_i, we correlate A_i and E_{max} for the whole set of maximum earthquakes. For each particular correlation (with parameter c_i), we shall estimate the discrepancy ξ_i between the set of observed pairs and the average function. Finally, we can select correlations for which the discrepancy is least. The value of c_i used in this correlation will be assumed to be the final one. This solution is optimal in the sense that there is the greatest possibility of its practical application.

We conducted the correlation between A and $E_{max} = 10^{K_{max}}$ for Central Asia and Siberia by linear regression of log A over K_{max}, so that K_{max} were assumed to be exactly known, whereas only a "random" dispersion of values A was considered (Fig. 21). During a more detailed study, one should also consider the inaccuracy in K_{max}.

Figure 21 shows the case for the closest correlation. It corresponds to $1/c = 0.315 \cdot 10^{-10}$ joules$^{-1} \cdot$ km^3, for which the energy density $K = 3c/2\pi = 1.5 \times 10^{10}$ joules$^{-1} \cdot$ km^3. For the Chilik earthquake of 1889 ($K = 16.5$), the radius of the area S of the optimal conformity equals to 100 km. This radius is 15 km for events with $K_{max} = 14$, and for $r = 3.2$ km it is $K_{max} = 12$. These dimensions only slightly exceed the estimates given earlier for the area of the earthquake focus proper. The latter was then considered to be the area from which most of the seismic energy of the event radiated and where rupturing and other essentially non-elastic deformations occur [S. Medvedev]. However, our area of optimal conformity and the area of a physical source need not coincide. We mentioned earlier that the physical mean may be remarkably greater than the latter.

To characterize the spread of the observed points (K_{max}, log A) about the straight line average in the graphical analog to Fig. 21, the sum of absolute deviations of the points from the straight line was taken $\xi = \Sigma \mid \Delta \log A \mid$. Figure 22 illustrates the way this value depends upon the parameter $1/c$ in the interval $0 < 1/c < 1.3 \cdot 10^{-10}$ Joules$^{-1} \cdot$ km^3 for Eastern Tien Shan and Dzhungaria. Because $1/c = r^3/E_{max}$, a zero $1/c$ in Fig. 22 corresponds to zero dimensions of areas $S = \pi r^2$ of the averaged activity A round the epicenters of maximum

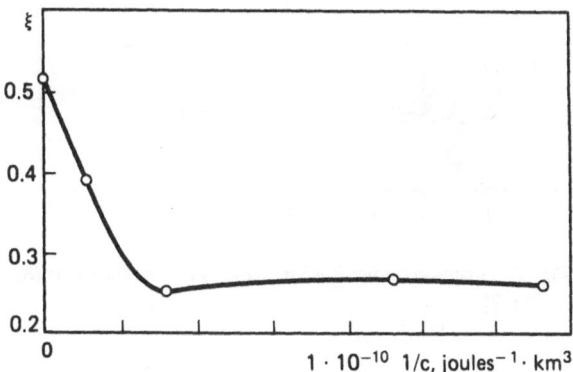

Fig. 22. The dependence between scattering of *dots* ξ on graph representing the E_{max} correlation with A and the average amount of activity in areas around epicenters, $1/c = r^3/E_{max}$

earthquakes E_{max}, whereas large $1/c$ correspond to large dimensions. Figure 22 shows that in the case of small areas of averaging, starting from the situation when A and E_{max} are correlated at the points of the epicenters of E_{max}, the quality of this correlation is poor and the spread is great. Later as the spread first decreases and reaches a minimum (in our case, at $1/c = 0.315 \cdot 10^{-10}$ joules $\cdot$ km^3) as the area of averaging increases, i.e. the correlation becomes optimal; then it increases slightly and becomes stable.

In other test cases [Riznichenko], this relation sometimes has a distinct minimum at approximately the same $1/c$. Sometimes, the minimum disappeared and, starting from the same value of $1/c$, the rate of decrease lessens and there is a certain stabilization within the same variations of $1/c$. For the Altai-Sayans mountains, the graph proved to be a horizontal straight line. This may be due to the averaging area, namely 30 000 km^2 [Massarsky, Gorbuno-va] from which the map of seismic acvitity was compiled in advance. The radius of the circle for this area is about 100 km, which is almost the optional radius for severe earthquakes like Chilik. Under such conditions, the additional variation of the averaging areas did not result in any considerable change to the average values of A in the epicentral areas with E_{max}.

For an optimal correlation between K_{max} and $\log A$ (Fig. 21), the averaging straight line in the graph is described by the following regression equation:

$$\log A = \log \alpha + \beta (K_{max} - K_\alpha) \tag{3.2}$$

where α, β, and K_α are constants, and at $K_\alpha = 15$ the ordinate $\log \alpha = 2.84$ with a slope $\beta = 0.21$.

The spread of the observed points K_{max}, $\log A$ (which means their deviation from (3.2)) is expressed by the distribution function in Fig. 23. We have a symmetric distribution which resembles the Gaussian one, except for one point—viz., the Sonkul' earthquake—which strays from the general pattern in Fig. 21.

The deviations of the points do not exceed $\Delta \log A = \pm 0.2$ in over 70% of the consid--ered earthquakes of E_{max} (10 out of 14 cases). The horizontal deviation $\Delta K_{max} = \pm 1.0$ corresponds to the vertical deviation from the averaging straight line slope β. This gives us a

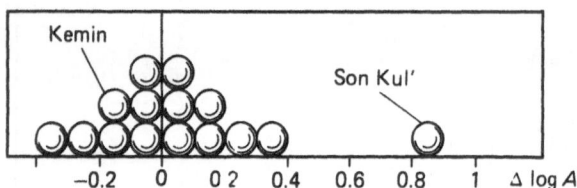

Fig. 23. Diviation of distribution $\Delta \log A$ of observed *dots* (K_{max}, $\log A$) from the averaged line presented in Fig. 21

formal estimation of the accuracy with which K_{max} and the correspondingly energy $E_{max} = 10^{K_{max}}$ can be determined from (3.2) for a given value of A:

$$K_{max} = K_\alpha + 1/\beta \log (A/\alpha) \tag{3.3}$$

at the 70%-confidence level.

At this stage in the research, this result is satisfactory, indeed it is unexpectedly good. However, we should warn against its wide application, in particular against any practical conclusions, until the same analysis has been performed for other well-studied regions, Tadjikistan in particular.

We note one feature of this solution to the problem. When correlating values A and K_{max}, we assumed that A characterizes the long-term seismic activity for a steady state evolution of the process, whereas K_{max} determines the maximum possible earthquake in the same area. Meantime, we were to consider both values observed only for a limited period of time. The fluctuations in both values and also the secular variations in the regimes which were not considered, might change the spread of the points compared to the spread which might be obtained over substantially longer periods of observations for a steady state process.

Since we are always bounded by the limited periods of observations, we could try to improve the technique by considering temporal variations of seismic regimes. However, this goes beyond our present discussion.

Let us suppose that the data we possess adequately describes the long-term average seismic activity, but that the observation time was too short to reveal all the possible maximum events and, instead, only the maximum observed earthquakes were used in individual cases. The points for these earthquakes (weaker than the maximum possible) would be located at our correlation (Fig. 21) to the left of the place where the maximum possible events should have been. The stray Sonkul' earthquake may be one such example.

Following this we may, at least in principle, get rid of the assumption that all the strong earthquakes are the maximum possible. In fact, for a correlation graph like Fig. 21, all the observed earthquakes, both small and strong, may be indicated by (K, $\log A$) points. They will form a cluster of points in the left-hand upper corner of the graph. They will include the points belonging to really maximum possible earthquakes. These being near the right-hand lower boundary of the cluster. This boundary, which has the same role in the solution of the general problem, seems to coincide with the position of our straight line.

On the distribution graph (Fig. 23) such an approach would not change considerably its left-hand corner and would fill with points to the right and upward from the average vertical ($\Delta \log A = 0$). The position of the line, say a straight one which symbolizes the conventional boundary of the cluster of discrete points on the correlation, might be subordinated to the condition of maximum steepness of the left cluster slope in a graph similar to Fig. 23, the steepness being a measure at a fixed level of the slope.

This latter assumption is only one example of how to improve the approach. The main development of statistical methods to predict the maximum possible earthquakes for seismic zoning should consider the joint numericla account of a variety of information: seismic and other (geophysical, geodetic, geological).

3.2.4. Determination of the Energy Flow of Earthquake Sources[*]

To determine the average specific power or "seismic energy flow" ω of all earthquake sources n released over a time T in a volume V or area (Earth's surface) S, find the sum $\omega =$ $= (1/VT) \sum\limits_{i=1}^{n} E_i$, or $\omega = 1/ST \sum\limits_{i=1}^{n} E_i$, where E_i is the seismic energy of each individual earthquake in the area. This way of determining ω and its mapping seems acceptable, but only for small-scale general maps. This approach, however, meets with great difficulties if we are to compare ω with the geological-geophysical setting of the regions and to study the temporal variations of the seismic regime, because ω is mainly determined from scarce earthquakes and fluctuates greatly.

To reduce such fluctuations the following function was proposed

$$\omega = \int\limits_{-\infty}^{K_{max}} NE \, dK \tag{3.4}$$

where N is distribution density of the earthquakes in time, space, and K is determined from a recurrence law, e.g. (1.3). In this case

$$\omega = \frac{10^{\gamma K_0}}{(1 - \gamma) \ln 10} A \cdot 10^{(1 - \gamma) K_{max}} \tag{3.5}$$

The difficulties are still not overcome as K_{max} is consiered independently in each particular area and so, still fluctuates remarkably. However, the problem may be solved if we relate K_{max} to a general function which fluctuates less, primarily associated with the seismic activity. We shall use the correlation between K_{max} and A determined in section 3.2 and shall write (3.2) as

$$A = A_M 10^{\beta (k - K_M)} \tag{3.6}$$

Here $A_M = \alpha$ and K_M is the value of K_{max} corresponding to the detemined parameter A_M. Eliminating K_{max} from (3.5) and (3.6), we have

$$\omega = \frac{10^{K_M + \gamma (K_0 - K_M)}}{A^{1 - \gamma/\beta} (1 - \gamma) \ln 10} A^{1 + (1 - \gamma)/\beta} \tag{3.7}$$

---

[*]/ See [Riznichenko].

which provides the solution required. Its better stability compared with the definition of ω by summing the observed values E_i, is due to the averages (2.32) and (3.6) which are obtained from more voluminous statistics and are used in addition to the local data.

We present some numerical comparisons. If (3.7) according to (3.6), we assume that $\gamma = 0.5$, $K_0 = 10$, $K_M = 15$, $\log A_M = 2.8$, and $\beta = 0.2$, we get $\omega = 2.74 \times 10^{15} \times A^{3.5}$ joules $(1000 \ \text{km}^2)^{-1}$ year^{-1}. Then, for regions with moderate ($A = A_1 = 0.01$), high ($A_2 = 0.1$) and extremely high ($A_3 = 1$) seismicity, the value will correspondingly be $\omega_1 = 8.7 \times 10^{-9}$, $\omega_2 = 2.8 \times 10^{-5}$ and $\omega_3 = 8.7 \times 10^{-2}$ joules $\cdot$ m^{-2}s^{-1}. If we assume $\gamma = 0.6$, we shall have from (3.7) for the same values of the other parameters $\omega = 2.72 \times 10^{14}$ joules $(1000 \ \text{km}^2)^{-1}$ year^{-1}. Then, $\omega_1 = 8.6 \times 10^{-9}$, $\omega_2 = 8.6 \times 10^{-6}$ and $\omega_3 = 8.6 \times 10^{-3}$ joules $\cdot$ m^{-2}s^{-1}, which differs from the previous result by no more than one order.

We shall compare these values with the heat flow of the Earth. The mean heat flow through the Earth's surface is $Q = 1.2 \times 10^{-6}$ cal $\cdot$ cm$^{-2} \cdot$ s$^{-1} = 5.0 \times 10^{-2}$ joules $\cdot$ m$^{-2} \cdot$ s^{-1}. So only the energy flow transmitted by elastic waves of earthquakes in the regions of extremely high seismicity approximate to the order of the heat flow. Naturally, it would be more interesting to compare the seismic energy flow and heat flow in some local areas, but we lack detailed thermometric data.

We suppose that a ratio like (3.7), which relates seismic activity A with seismic energy flow ω and which is based on two observed ratios, namely the recurrence law (2.32) and the correlation between maximum earthquakes and activity (3.6), will make it possible to map the seismic energy flow in detail. Then it may be compared with other mapped values, mainly the geodetic and geophysical data on slow deformation process of the Earth. This would lead to the development of a general physical theory of tectonic motion "seismic flow" [Riznichenko], the energy of which is debated.

3.3. Calculation and Mapping of K_{max} for Japan[*]

The correlation technique (Sections 2.3 and 2.4) for determining the maximum possible earthquake K_{max} values from data on average seismic activity A and then based on other geological-geophysical factors Φ_i (see [Riznichenko] and Chapter 3) was developed and applied for continental Eurasia, characterized by moderate seismicity. The principal aspect of the technique for any earthquake size is the choice of the dimensions of the area "responsible for the event". These come from the factors Φ_i for a particular value A, averaged for a correlation with the observed value K, and finally with maximum possible earthquake K_{max}. These factors may be (1) seismological, viz., the activity A, thickness of seismogenic layer, seismic energy, indices of the fault plane solution, seismic moment and the seismic flow of the rock mass; (2) other geophysical, viz., functions, gradients and other transformators of the gravity, magnetic, electromagnetic, and heat fields; (3) data on the topography and on properties (density, velocity of elastic waves) of various topographic and deep geological and geophysical (e.g. density and velocity) surfaces and boundaries of the corresponding layers which reflect neotectonics, young and recent tectonics—the slopes of deformed levelling surfaces, the gradients of motion velocities or strain rates, faults, etc.

---

[*] See [Riznichenko]. A. M. Bagdasarova is a co-author.

This section is devoted to the characteristics of the correlation technique for determining K_{max} from A and for severe earthquakes. Japan is a classical example of such seismicity, and so was chosen for our consideration.

3.3.1. Selection of the Averaging Areas Responsible for Earthquakes

The correlation technique for determining the maximum possible earthquake K_{max} is based on the assumption that the long-term average seismic regime (K_{max} is one of its parameters) is steady state. We can explain it by slow changes in the tectonic process, which generates seismicity as compared to the period of human activity, during which the average seismicity and its spatial distribution have been calculated.

With such an approach, temporary deviations in the realization of seismicity from the long-term average should be avoided. This can be achieved, firstly, if one accounts for and excludes the effects of excess earthquake grouping as compared to a "purely random" Poisson process by ignoring aftershocks and swarms. Secondly, by averaging and smoothing the Poisson data in spatial-temporal areas. The direct determination of K_{max} in each elementary cell with dimensions comparable to the focus size of the earthquake of K_{max} is not possible due to scarcity of events close to K_{max}. Experience, however, indicates that the approximation of the long-term average activity A in this area is possible. Then the problem can be reduced to the determination of the most accurate average of A in each elementary cell and to the establishment of the correlation function $K_{max}(\overline{A})$ in a considerably larger spatial area. Using this function and knowing the local values of $\overline{A}$, we can determine K_{max} in each elementary cell. In addition to A, other geological-geophysical factors Φ_i instead, may be used for the same purpose. A knowledge of the factors may be transformed into K_{max} either directly or by tentative activity $A(\Phi_i)$ from which it is easy to transfer to K_{max} in known seismological way.

The accuracy of the average $\overline{A}$ is directly proportional to the number of observed earthquakes and, at the given natural density of events, to the dimensions of the space and time averaging volume. In fact, this assumption is based on the admissibility of the ergodic principle, as known in statistical physics [Gaiskii]. The duration of seismic observations is limited so the volume of the area may be only enlarged by increasing its spatial dimensions. However, this increase contradicts the requirements for greater resolution in the definition of the spatial distribution of seismicity—the principle of indeterminacy in seismology [Riznichenko]. We have to be satisfied by a practical acceptable compromise. When using other factors Φ_i which in essence have a long-term average nature (for instance, the velocity gradients of neotectonic motions), areas of their spatial averaging may be not so great. This makes it possible, in principle, to increase the resolution in determination of the long-term average parameters of seismicity in addition to that achieved when utilizing seismological data. Another difficulty may arise as the periods of time within which the seismo-tectonic process can be assumed to be a steady state may be violated. Therefore, additional monitoring is desirable for temporal variations of neotectonic deformations, namely their evolution.

When comparing Φ_i with seismological factors, in particular A, we should use spatial averaging areas with dimensions of least as big as those for A. In all these correlations, we should also consider a number of geological-geophysical states we shall now consider.

The accuracy of determining the focus coordinates. In present-day detailed regional studies of seismicity, which we are considering here, the errors in determining the coordinates of the epi- or hypocenter of the event are usually about ± 2 to ± 5 km; for very detailed studies they may be reduced to ± 0.5 km, but they may reach ± 10 km and more for less detailed work. The latter estimate is true for the material from Japan, which we discuss. When determining the activity A the dimensions of the averaging areas should obviously not be less than these errors.

Linear dimensions of a focus [Riznichenko]. For small earthquakes $K = 10\text{-}15$. These dimensions are usually less than the average errors when determining the coordinates of epi- or hypocenters, consequently dimensions of the focus may be ignored. For large events the focus dimensions considerably exceed these errors; therefore, when analyzing such earthquakes, an area equal to the focus region should be taken as the minimum spatial cell, in which the correlated values are to be averaged.

Accuracy and resolution of seismic mapping. For seismic activity mapping this aspect was discussed in [Riznichenko]. Maximum resolution can be achieved for minimum allowable accuracy. In the case of a Poisson process, a 50% average error is obtained when there are approximately three epicenters within the averaging site. It does not seem reasonable to use a lower accuracy and hence greater resolution. Where epicenters are scarce, it is convenient to fix the accuracy. However in areas with denser epicenter spacing it is often more convenient to fix the resolution by associating the form and geometry of the averaging sites with the geographical network, and to determine the dimension of sites from the errors in determining the focus coordinates. In practice regional studies with medium resolution usually take the minimum dimensions of the averaging sites for determining the activity A, to be about 10 to 20 km. When A is calculated from catalogues, a moving window technique with double overlapping is used for each direction, that is fourfold over the square.

Sometimes the summed density of the seismic energy flow from earthquake sources ΣE or values $\Sigma \sqrt{E}$ of the Benioff type are mapped. These energy values are predominantly determined by large earthquakes, which are scarce. Therefore, for an equal resolution, the activity A maps turn out to be more accurate than either ΣE maps (usually 10 times) or $\Sigma \sqrt{E}$ maps (3 times). For equal accuracies, ΣE and $\Sigma \sqrt{E}$ maps should be less detailed than maps of A. If one ignores this condition, ΣE and $\Sigma \sqrt{E}$ maps would not reflect long-term average conditions, and instead reflect the transient situation, viz., the random fluctuation manifested during the observation period.

Thus, in order to reflect the long-term average seismicity we should use maps of seismic activity A. The energy ΣE or the parameter $\Sigma \sqrt{E}$ may be calculated and, if necessary, mapped from A, γ, $K_{\max}$.

Shape of averaging zones. When mapping A, we should use isometric averaging areas to avoid a subjective approach in the configuration and orientation of zones with different intensity. Thus, for instance, not to produce a false impression of their systematic extension along the strike of major geological structures, which may be not active at the time or at a given site. The sites should be in the isometric form of a circle, square, rectangle or a trapezium. It is also true when mapping $K_{\max}$ at least to a first approximation. At the following stages it may be possible to use a distinct orientation of the seismic zones and to stress it when averaging it by extending the averaging sites along the strike for A when correlating with K. However, when the lack objective optimum criteria, it is risky to introduce these distortions onto the observed seismic pattern.

It is reasonable to choose a circle geometry for the averaging site for activity A when correlating it with K to determine $K_{max}(\overline{A})$ [Riznichenko, etc.]. It is significant to correlate K_{max} both with the activity A at the point, at which K_{max} is being sought, and with its neighborhood, viz., the width and extension along the strike of a seismic zone where K_{max} is being determined. The total length of an extended seismic zone does not seem meaningful when compared to the dimensions of the averaging site, for instance, the moderate values of K_{max} in narrow seismic zones of the mid-oceanic ridges and in some continental structures of various extension (sometimes it may be very long).

Let two seismic zones have a similar activity A, but different widths. It would be natural to suppose that the greater K_{max} (all things being equal) should be ascribed to the zone which is wider. If we assume that K_{max} is directly related to A, its correlation with the width of the zone should be made, naturally, using a circular averaging area, the diameter of which would lightly exceed its width. However, such a correlation cannot be achieved if the equivalent averaging area of an elongated form is entirely located within the zone and does not depend upon its width. A circular area or an elliptical one, makes it possible to consider properly the continuity or discontinuity in the extension of a seismic zone in the region of K_{max} determination.

Possible temporal seismicity variations. The dimensions of the area where is a gap in the temporal seismic activity in the region of small earthquakes prior to a large event are compatible with and are approximately twice as great as the linear dimensions of the focus. The phenomenon of a seismic gap was first recognized when analyzing rock bursts [Riznichenko] and then true tectonic earthquakes [N. Borovik]. To avoid underestimating the long-term average seismic activity in such areas they should be covered by observations in time and space. For large events, it is hard to do this in the spatial area without a loss in the desired resolution of the study. Hence, the duration of observations is of a particular importance. From estimates by Borovik, the seismic gap in the Baikal region for the events with $K = 16$ ($M = 6.5$) is about 7 years, for smaller events it is shorter, for larger events it seems greater. The duration of reliable seismic observations often exceeds these time periods. Thus, the overlap of the quiet regions, on the time scale and the space coordinates before earthquakes can usually be obtained.

Areas of temporarily reduced seismic activity may appear after a substantial liberation of potential seismic energy by severe earthquakes with their seismic train of aftershocks and the subsequent revival of seismicity due to a weakening of the medium before it "heals" [Riznichenko]. Such areas constitute a greater danger. It seems that the local region of reduced seismic activity within a broad strip of high seismic activity, cited by Gutenberg, was of this kind. This local region was overlapped by the disastrous 1960 Chile earthquake with its pronounced seismic train. Possibly the region of the famous 1885-1889 Northern Tien-Shan earthquakes was of this type; after these earthquakes, the seismic activity A observed at the present time is still lower than the average long-term seismic activity, a situation typical of such powerful earthquakes. It is possible that the zone of the 1907 Karatag earthquake (intensity 10°) in Tadjikistan is still characterized by a reduced seismic activity A; according to macroseismic data, this earthquake had a magnitude $K = 18$ ($M = 8$). The $K_{max}(\overline{A})$ estimates made by Yakovleva, with our method, indicate that the present average activity A in the region of that earthquake corresponds to the severest possible earthquake with $K_{max} = 16.3$, i.e., the present average activity is lower than the published value of $K = 18$ by $\Delta K = 1.7$. The difference ΔK is twice as great as the normal maximum differences which are tolerated when the method is used. It is possible, however, that the estimate $K = 18$ for this earthquake was too great, because a great focal depth of $H = 40$ km was assumed[*]. The earthquakes which now occur in this area are characterized by focal depths $h = 2\text{-}20$ km. The intensity $I_0 = 10°$ observed in 1907 on a small area is possible even for $K = 16$ when the above depth figures are corrected.

Many confusing points exist; but the situation is, in essence, as follows. One must attempt to eliminate as much as possible the influence of time-dependent seismicity fluctuations. This is in order to reach sufficient accuracy in the determination of the long-term average value so that the area of averaging can be increased. But at the same time the area must be kept small so that adequate resolution is preserved. This dilemma is especially aggravated for severe earthquakes. It seems that one can only escape this by defining and satisfying certain optimum conditions which are justified.

3.3.2. Determination of the Parameters of the Law of Maximum Earthquakes

The earlier estimates of reasonable dimensions of the region in which the parameters Φ_i and particularly the parameter A are averaged for the determination of K_{max} with the correlation method, were only based on general conclusions or at best on inequalities. But our goal is to specify the problem completely and to derive expressions, in the form of equations, for its solution.

The functions for Central Asia. In the first papers [Riznichenko et al.] for the crustal earthquakes of Central Asia, the equation

$$r^3 = 0.315 \cdot 10^{K-10}, \tag{3.8}$$

was used. The radius r was for a circular averaging area of A for the determination of $K_{max}\,(\overline{A})$, where r is expressed in kilometers and the seismic energy $E = 10^K$ is expressed in joules. For small K, Eq. (3.8) results in r-values which are smaller than the minimum (r_{min}) obtained from considerations of accuracy when determining the focal coordinates or from considerations of both accuracy and resolution of the seismic activity mapping. In this case the values $r = r_{min}$ are used. According to Eq. (3.8) regions with $r_{min} = $ const and $r = $ var are conventionally connected via a smooth transition. The following equation for the limiting contour K_{max}

$$\log \overline{A} = \overline{2.84}\,(K_{max} - 15), \tag{3.9}$$

was obtained for Central Asia.

Equations (3.8) and (3.9) for Central Asia were checked against the numerous data which have been obtained in the detailed seismological investigations which have been made in many areas of Eurasia, including Japan and Kamchatka [Riznichenko]. The equations were usually agreed with the observations in all the areas considered. Equations (3.8) and (3.9), based on the data for Central Asia, hardly had to be modified when they were applied to other regions. Attempts to modify the equations were made but were not dictated by a genuine need for

---

[*] In the new catalog of severe earthquakes for the USSR, the energy value of Karatag earthquake was estimated to be from $M = 7.3$ to 7.4 ($K = 17$)—*Ed. note.*

modification. The relations remained valid not only for the crustal earthquakes of the various regions but also for the subcrustal earthquakes in the Balkans (region of the Crete arch) and in the Carpathians (Vrancha region) [Riznichenko], i.e., they were valid for all regions we have considered to date. A few deviations from the general laws resemble the above described deviations and can be explained, in most cases, by specific details which are often of a purely technical nature. These include shortcomings of both the observation system and the data evaluation methods.

Thus, equations like Eqs. (3.8) and (3.9) describe a stable statistic of the strongest possible earthquakes, which characterizes the average long-term seismic regime in general. Thus, in a first approximation, this statistic is independent of the local geological conditions such as structure and tectonics. Under the condition $r(K)$ we can term our $K_{max}(\overline{A})$ relation the law of the maximum earthquake by analogy with the $N(K)$ law of earthquake recurrence.

Naturally, no absolute validity must be ascribed to these laws. In the ensuing approximation one must attempt to find statistically significant deviations from the established average laws, correlate the deviations with specific conditions, treat the deviations qualitatively, and determine correlations and relations in terms of the physics of the parameters of the deviations as functions of specific details.

Equations (3.8) and (3.9) were used to construct maps of the maximum possible earthquake (K_{max}) in many regions. In a region with moderately severe earthquakes ($K_{max} \leqslant 15$), the maps are not contentious. But for major earthquakes $K_{max} > 15$ the maps often seem to be too general and smoothed and the K_{max} isolines covered zones that were too large. The obvious reason for this was that the radii r obtained with Eq. (3.8) for the averaging areas were too large for particularly large earthquakes.

Most of the data from continental Europe and the Mediterranean area, which were used to derive and check (3.8) and (3.9), are for earthquakes with energies $K = 10\text{-}15$. It was hard to check the equations beyond this interval. Earthquakes with $K \geqslant 15$ are frequent in Japan and the maximum magnitudes of the Japanese earthquakes reach the greatest values known for the globe. We again considered the statistics of the Japanese earthquakes in order to improve the law of the maximum earthquake for large K or M.

3.3.3. Defining the Law of the Strongest Earthquake from Data for Japan

In [Riznichenko et al.] we mapped the seismic activity A for Japan and correlated K with A; we aslo checked the possibility of whether the law of the severe earthquake could be applied to Japan using (3.8) and (3.9). So far, observations in regions of moderate seismicity were used mainly to derive the appropriate limit curve $K_{max}(A)$. Now our main task is to select parameters for (3.8) and (3.9) which are consistent with the data for severe earthquakes both for the limiting contour and the mapping of K_{max} with maximum possible resolution.

Here, as in [Riznichenko, Bogdasarova], the same initial data catalogues (for a list see [Riznichenko et al.]) were used to study shallow earthquakes with focal depths of up to 60 km in Japan; the epicenters of the earthquakes are located on the islands, the continental shelf and in the deep-sea trough. These earthquakes make up the majority of the earthquakes in that region. It is not reasonable only to consider crustal foci there because the seismogenic layer extends directly into the upper mantle.

Table 8. Major Earthquakes in Japan with $M > 6.5$ ($K > 16$), 1926-1965 by Means of Which K_{max} Map Is Drawn

No.	Year	Coordinates		M	K	h, km
		N	E			
1	1933	39.1	144.7	8.3	19.2	0-20
2	1946	33.0	135.6	8.1	18.9	30
3	1944	33.7	136.2	8.0	18.7	0-30
4	1964	38.3	139.2	7.5	17.8	40
5	1953	34.3	141.8	7.5	17.8	40-60
6	1947	43.8	141.6	7.0	16.9	0-30
7	1940	44.1	139.5	7.0	16.9	0-20
8	1940	36.2	132.2	6.8	16.6	0-30

The basic data to be analyzed are the representative earthquakes with a value $K \geqslant 15$ ($M \geqslant 6$) of the 1926-1965 period. The largest of these earthquakes with $K > 16$ ($M > 6.5$) are listed in Table 8.

The seismic activity to be compared with the energetic value K of these earthquakes is $A = A_{15}$ in the case under consideration, i.e. the number of earthquakes in the interval $K_0 = 15 \pm 0.5$ per year, and in an area of $10^5 \sqrt{10}$ km^2. The activity A_{15} (which is convenient when the seismicity is very high as in this case) and the more convenient activity A_{10} (useful in the case of moderate seismicity, such as the continental seismicity of Central Asia) are normalized. This means that for $\gamma = 0.5$ (a value close to usual average observations) the numerical values of the activities coincide, i.e. $A_{15} = A_{10}$. In determinations of the activity A in [Riznichenko et al.], the effects of the earthquake grouping (aftershocks and swarms) were excluded. In this manner, the results approached the long-term average. The map of the activity was compiled by the fixed quality method [Riznichenko] with a total number N_Σ of epicenters of representative earthquakes in the area of averaging depending upon the epicenter density and, accordingly, upon the activity A: $N_\Sigma = 3$ for $A < 0.1$; $N_\Sigma = 4$ for $A = 0.1$-1.0, and $N_\Sigma = 5$ for $A > 1.0$. This corresponds to a 50% standard error in the determination of A in the first case and to a 20% standard error or a smaller error in the latter case, Poisson scattering being assumed.

When we varied the r (K) and K_{max} (A) resembling (3.8) and (3.9) for Central Asia, we had to reject our previous assumption of similarity in terms of energy at a constant energy density $E = 10^K$ with the volume r^3 of the area of averaging. The condition that the correlation must be optimal in the construction of the contour K_{max} (A) was not dropped. We introduced the additional condition that zones with large K_{max} should be enveloped, when mapping, by zones with lesser K_{max}. In the case of Central Asia, for example, the latter condition need not be emphasized because there the condition is usually automatically satisfied for moderate maximum K_{max} values and moderate r values.

Figure 24 shows several versions of the function r (K) for Japan. The previous function of (3.8) for Central Asia was extrapolated and is labelled 1; the average radius R of the earth-

Fig. 24. Investigated versions of the dependence of radius R of A average activity on the
earthquake energetic value K or magnitude
M.

1—Extrapolated function for Central Asia;
2—mean source radii (aftershock zone), after Utsu-
Seki; *3, 4*—intermediate versions; *5*—the final version for Central Asia and Japan, Eq. (3.12)

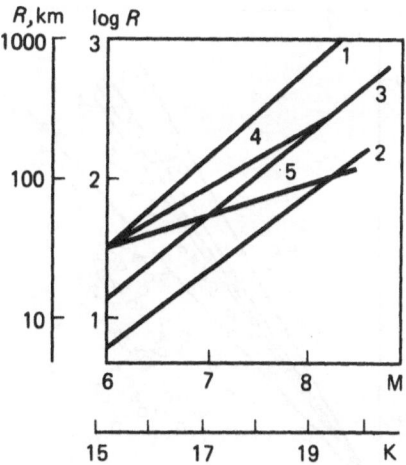

quake focus is labelled *2* [Riznichenko]; the average radius R was estimated from areas of
$S = \pi R^2$, of equal size in the aftershock zone; the procedure is done according to Utsu-Seki;
and *3-5* denote intermediate versions between *1* and *2*.

The Utsu-Seki formula, which was obtained from 38 earthquakes in Japan, is

$$\log S = 1.02\, M - 4.01 \tag{3.10}$$

We checked the applicability of this formula on 30 earthquakes in Japan and their aftershocks observed between 1926 and 1965. The earthquakes had magnitudes of $M \geqslant 6.8$. We
obtained

$$\log S = 1.01\, M - 3.65$$

This is not very different from (3.10) so we may use the more popular Utsu-Seki formula
(3.10), which we then really used.

Figure 25 shows the various forms of the $\log \overline{A}$ versus K correlations, which correspond
to versions *1-5* of the function $r\,(K)$. The limiting contours $K_{max}\,(A)$ are shown as straight
lines through the point $\log A = \overline{2}.84$, $K = 15$; this point also belongs to the previous curve
of (3.9) for Central Asia. The contour remains valid for earthquakes with $K \leqslant 15$. The data
obtained from observations of severe earthquakes with $K \geqslant 15$ in Japan are consistent with
it (data shown in Fig. 25). These data were used to compile the K_{max}-maps for each version.
We only show the map resulting from version *5*. The other versions are discussed in the text.
Let us consider each of versions *1-5* separately.

Version 1. We used the functions of (3.8) and (3.9) for Central Asia. The observed points
for this version on the correlation field of Fig. 25 are indicated by open circles. These circles
appear over a wide front.

All these points should be and are in fact located under the limiting contour *1* for Central
Asia though the limiting contour was extrapolated far from the limits with $K = 16$, i.e. far
from the maximum magnitude of the Central Asian earthquakes, from which the limit curve

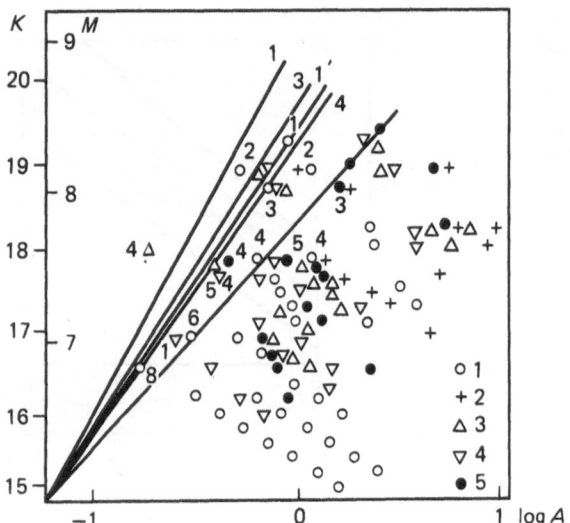

Fig. 25. Correlation of average seismic activity $\overline{A}$ with the largest earthquake value K observed within the corresponding area.

Points *1-5* and the corresponding limiting contours $K_{max}(A)$

was derived. When the Japanese data are used (Fig. 25), the limiting contour should probably have the form of straight line *1'* running through points *1, 3,* and *8* (the number at a point corresponds to the sequential number of the large earthquakes in Table 8). Only point *2* was beyond the limit and this may be by chance. Furthermore, this detail is not very important, because we reject this version for other reasons.

The regular sequence of the points in the lower part of the field is caused by technical details which are unrelated to our subject. In this case the activity $\overline{A} \ (= \overline{A}_{15})$ to be compared with K was determined by

$$\overline{A} = (1 - 10^{-\gamma}) \cdot 10^5 \sqrt{10} \ \frac{N_\Sigma}{\pi R^2 T} \tag{3.11}$$

where T denotes the observation period (in years); r is expressed in km. When a minimum number N_Σ of epicenters on the averaging area is used (e.g. $N_\Sigma = 1$). Equation (3.11) shows that the activity $\overline{A}$ is inversely proportional to r^2. Only this formal detail indicates the regular sequence of points. However, this is a minor detail. The correlation field as a whole does not provoke critical remarks with this version.

However, the corresponding K_{max} map of this version is unacceptable. The areas with the greatest values ($K_{max} = 19$) are so large on this map that they overlap aseismic neighboring regions. It is intuitively wrong to treat these regions as seismic regions responsible for an earthquake. There is an even better criterion of inconsistency. According to physics or geophysics, zones of large K_{max} must be inside zones with smaller K_{max}. In this case the inverse situation was observed: the $K_{max} = 18$ isoline is situated inside the contour of the $K_{max} = 19$ isoline which envelops too large on area. Mainly because this "inside-out" situation the K_{max} map that was obtained for severe earthquakes, version *1* must be rejected.

Version 2. The radii r of the averaging areas were calculated on the basis of (3.10) by Utsu-Seki $\pi r^2 S$. The observed points (denoted by crosses) form a narrow strip which does not reliably define the limiting contour. The limiting contour was tentatively drawn through points *1* and *3* (line *2*); point *2* remained beyond the limiting contour. The $K_{max} = 19$ isoline on the K_{max} map appropriately encompasses sections in which the major earthquakes and large earthquake swarms of Japan occurred. The shortcoming of this version is that the r values are small for moderate earthquakes with $M = 6\text{-}6.5$. For example, $r = 6.4$ km is obtained for $M = 6$. In such an area even a single earthquake may yield a high average activity $\overline{A}$ and the corresponding point on the correlation field is then so far out of the limiting contour that the limiting contour cannot be fixed for earthquakes of this magnitude. The construction of the K_{max} map for such earthquakes then becomes uncertain.

Version 3. Intermediate radii between the two preceding versions were used for the radii. On the correlation field of Fig. 25 the front of the corresponding points broadens near the limiting contour (see points *2* and *5*) when a comparison with the preceding version is made. This means that this curve is defined with greater certainty. Like version *1*, though to a lesser degree, the $K_{max} = 19$ isoline on the K_{max} map still encompasses an excessively large area and at some sites overlaps the $K_{max} = 18$ isoline. The $K_{max} = 18$ isoline falls inside the $K_{max} = 19$ isoline, which should not occur.

Version 4. As far as the magnitude of r is concerned, this version is intermediate between versions *1* and *3*. Version *4* proved to be intermediate between versions *1* and *3* with regard to the shortcomings of the correlation field and the K_{max} map.

Version 5 is the final version. The radii of the averaging regions were selected so that at the lower limit $K = 15$ of earthquake magnitude value for Japan, the radius r coincides with the radii given by (3.8) for Central Asia; at the upper limit $K = 19$ ($M = 8.5$), the radius r coincides with the average focal radius or the aftershock zone according to Utsu-Seki. Both the correlation field and the K_{max} map are satisfactory. The equations for the radius $r(K)$ of the averaging area and for the limit curve $K_{max}(\overline{A})$ are in this case

$$\log r = 1.505 + 0.111 \ (K - 15) \tag{3.12}$$

$$\log \overline{A} = \overline{2}.84 + 0.39 \ (K_{max} - 15), \tag{3.13}$$

for $K = 15\text{-}19$. However, the r value (expressed in km) can be calculated with this formula also for smaller K. Differences between (3.12) and (3.8) for small K are immaterial because the dimensions of the averaging area A for the determination of K_{max} are, in this case, given by the size of the averaging areas used for compiling the map of the seismic activity A, as indicated above. The differences between (3.12) and (3.8) for severe earthquakes ($K = 16\text{-}19$) become more important as K increases and strongly affect the slope (and, hence, the position) of the $K_{max}(A)$ curve for severe earthquakes (see (3.13)); when compared with the slope of (3.9) for weak earthquakes with $K \leqslant 15$.

Length, width and average diameter of the focus. An inspection of the aftershock zones of 30 Japanese earthquakes with $M \geqslant 6.6$ in the 1926-1965 period has shown that the form of the zones usually considered as the focus area, is almost elliptical with a length ratio of about 2 for the axes; the major axis usually lies along the extention, i.e. parallel to the "ridges"

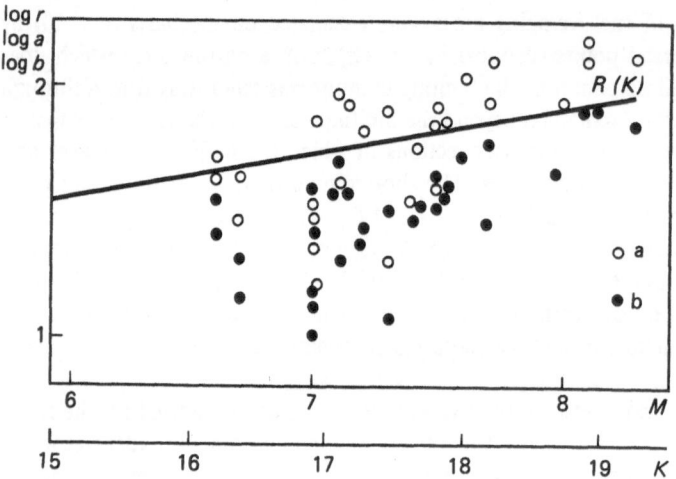

Fig. 26. Comparison of radii R of averaged zones r with major a and minor b semi-axes of source areas (aftershock zones)

of the seismic activity A. We attempted to use elliptical averaging areas with the above axis ratio instead of circular averaging zones. However, no substantially new limiting contours $K_{max}(A)$ or K_{max} maps were obtained and the $K_{max} = 19$ isoline was unaffected (see below). Since it is much more difficult to work with an elliptical net and to orient it properly, a circular net was always preferred.

In order to obtain a clear idea of the extent to which the circular averaging regions with radius r (according to Eq. 3.12) can broaden the K_{max} zones relative to the focal dimensions of earthquakes with the corresponding magnitudes K or M, the R values in Fig. 26 were compared with the lengths of the semi-major a and semi-minor b axes of the foci (see also [Riznichenko et al.]). In the case of the largest earthquakes for both Japan and the world $M \geqslant 8$ or ($K \geqslant 18.8$), our r values are intermediate between the semi-major and semi-minor focus axes and generally tend to the semi-minor axis. This should not noticeable broaden the K_{max} zones for these earthquakes. But in the case of earthquakes with $K = 15\text{-}16$ ($M = 6\text{-}6.5$), the r values tend towards the focus semi-major axis, which is also acceptable. We note that for $K = 15$ the new radii according to (3.12) coincide with the radii according to (3.8) for Central Asia and for other regions of the Eurasian continent.

3.3.4. Map of the Maximum Possible Japanese Earthquakes

The map (see Fig. 27) was constructed using the general method described [Riznichenko, et al.] but in accordance with the new relationships obtained in the form of (3.12) and (3.13). The isolines on the map are loci of the points at which the relationships hold for earthquakes of a particular magnitude. The condition that the number of epicenters on the averaging area with the radius r [Eq. (3.12)] should be minimal was also considered: the number of epicenters was not smaller than the number with which the activity map A was constructed.

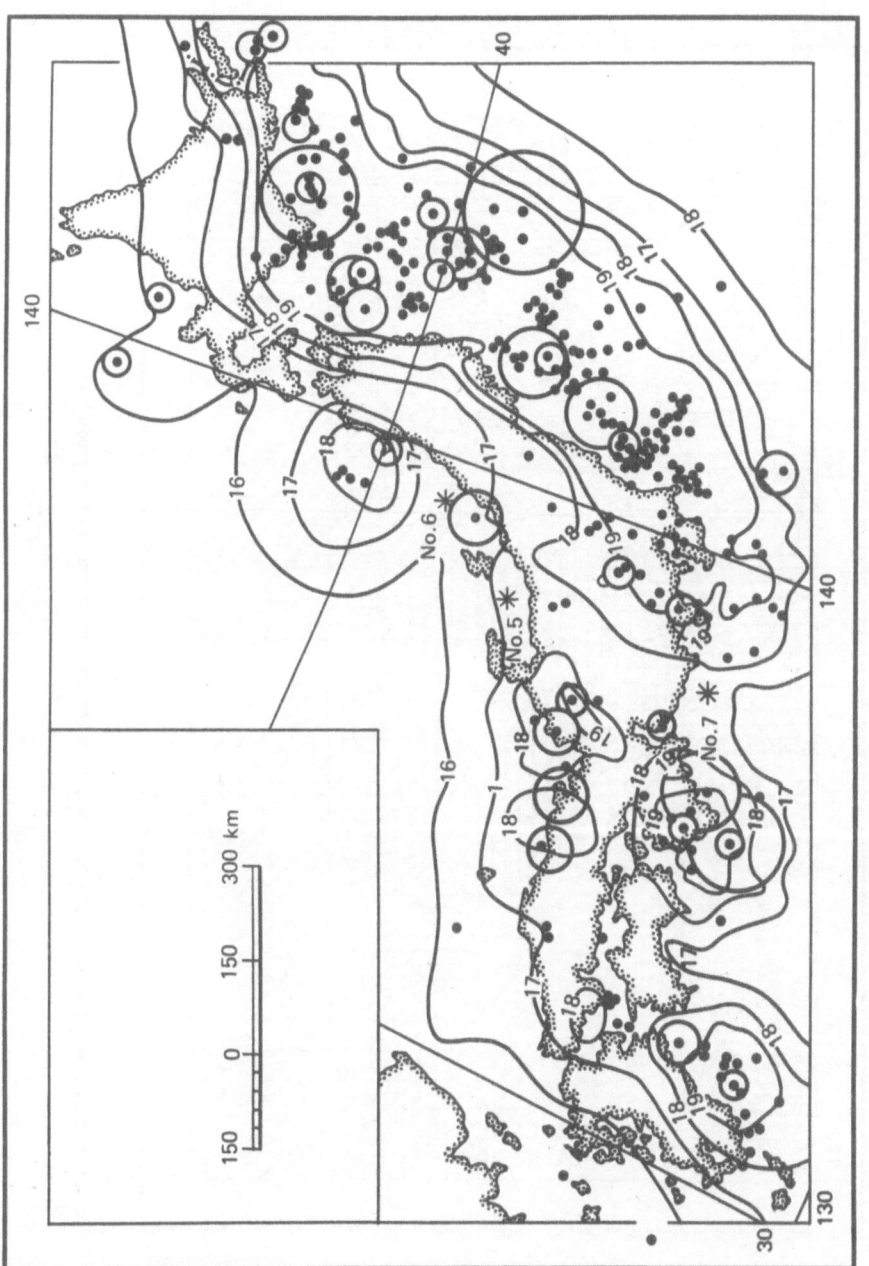

Fig. 27. Map of maximum possible earthquakes K_{max} of Japan.

The map was compiled with a $K_{max}(\overline{A})$ technique from observations of earthquakes with the energetic value $K \geqslant 16$ ($M \geqslant 6.5$) for 1926-1955

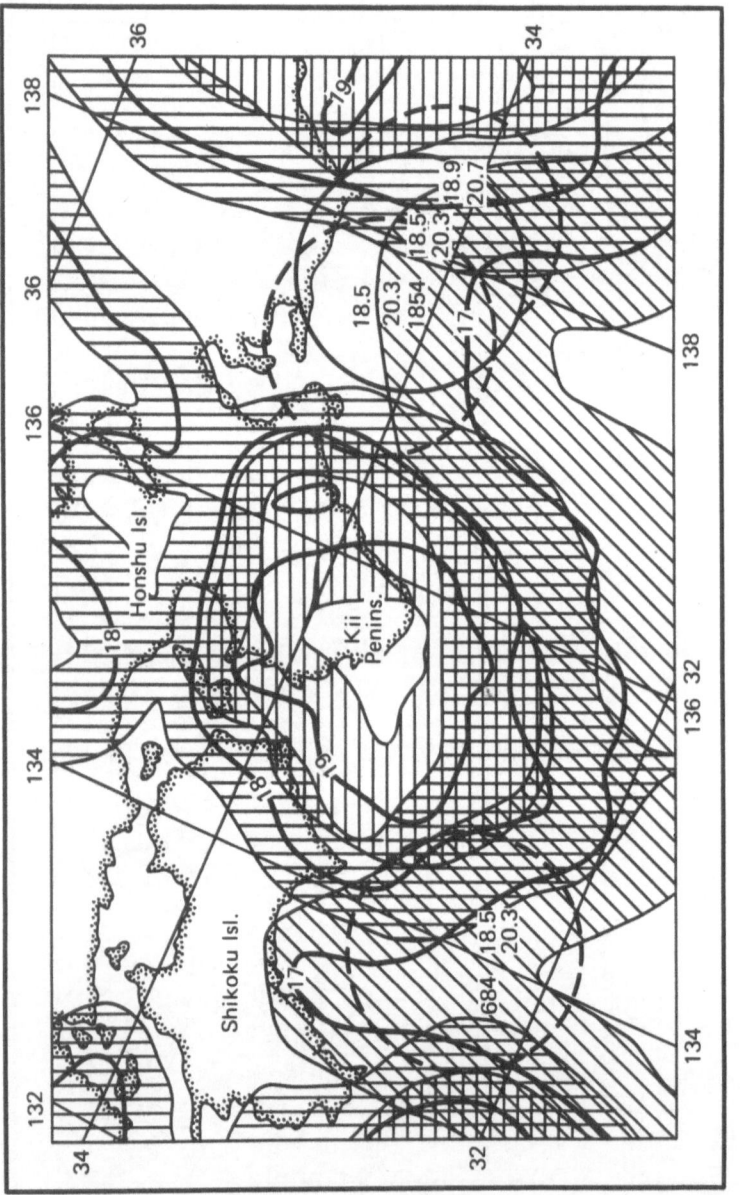

Fig. 28. A fragment of a K_{max} map of Japan with verification of the inverse prediction

Table 9. Major Earthquakes of Japan with $M > 7$ $(K > 17)$ in the 684-1925 Period; the Earthquakes Occurred at the Corresponding Positions on the K_{max} Map

| No. | Year | Coordinates | | M | K | K_{max} |
		N	E			
1	745	35.5	136.6	7.9	18.5	18
2	818	35.2	139.3	7.9	18.5	19
3	830	39.8	140.1	7.4	17.6	17
4	869	38.5	143.8	8.6	19.8	19
5	887	33.0	135.3	8.6	19.8	19
6	1361	33.0	135.0	8.4	19.4	19
7	1586	36.0	136.8	7.9	18.5	18
8	1605	34.3	140.4	7.9	18.5	18
9	1611	38.2	143.8	8.1	18.9	19
10	1677	38.7	144.0	8.1	18.9	19
11	1703	34.7	139.8	8.2	19.1	19
12	1707	33.2	135.9	8.4	19.4	19
13	1843	41.8	144.8	8.4	19.4	19
14	1854	33.2	135.6	8.4	19.4	19
15	1891	35.6	136.6	8.4	19.4	18
16	1894	42.4	146.3	7.9	18.5	19
17	1900	39.0	141.0	7.3	17.4	17
18	1905	34.2	132.3	7.6	18.0	18
19	1908	33.7	138.5	7.7	18.2	18
20	1909	32.1	133.1	7.9	18.5	18
21	1911	43.0	144.0	7.8	18.3	18

Under this condition (3.12) for r had to be used for plotting the $K_{max} = 19$ isoline of the most powerful earthquakes. The other isolines were constructed directly in accordance with the A-map and the correlations stated in (3.13).

The epicenters of the 1926-1965 earthquakes are indicated on the map in Fig. 28. Dots indicate earthquakes with $M = 6.0-6.9$; circles which have the radii of the sources according to (3.10) by Utsu and Seki refer to earthquakes with magnitude $M \geq 7.0$. Both the positions and magnitudes of the severe earthquakes which occurred during that period are in good agreement with the isolines except for earthquake No. 4 (see Table 8) which is beyond the limiting contour of Fig. 25. The observed energy value of this earthquake was $K = 17.8$, yet the earthquake appears on the $K_{max} = 17$ isoline on the map so that the error is $\Delta K = 0.8$ $(\Delta M = 0.4)$. This error, which is observed in only one case, is not excessively large.

The good agreement between the K_{max} map and the observed K values in the observation period (1926-1965) is not surprising; it only confirms our deductions. It is more interesting to check how the severe earthquakes which occurred in this region outside the 1926-1965 time interval, conform to the map and to assess in this manner the value of the map as a "forecaster".

We checked the earthquake catalogue for 1966-1971 and did not find a single event which was inconsistent with the map in Fig. 27. However, this is not very significant, since no partic-

Table 10. Major Earthquakes of Japan with $M > 7$ ($K > 17$) in the 684-1925 Period; the Earthquakes Did Not Occur at the Corresponding Positions on the K_{max} Map

No.	Year	Coordinates		M	K	K_{max}	r	L_1	L_2	K_{max}	ΔL
		N	E								
1	684	32.5	134.0	8.4	19.4	16	107	90	85	19	—
2	880	35.4	132.8	7.4	17.6	16	33	60	10	17	27
3	1096	34.2	137.3	8.4	19.4	17	107	100	90	19	—
4	1498	34.1	138.2	8.6	19.8	18	135	100	90	19	—
5	1614	37.5	138.0	7.7	18.2	17	47	90	80	17	43
6	1833	38.7	139.2	7.4	17.6	16	33	80	70	17	47
7	1854	34.1	137.8	8.4	19.4	17	107	110	100	19	3
8	1894	39.2	139.5	7.3	17.4	16	30	40	25	17	10
9	1896	39.5	140.7	7.5	17.8	16	37	60	30	17	23
10	1896	33.0	131.0	7.1	17.1	16	23	30	25	17	7

ularly outstanding events occurred during this short time interval. It was more important to consider the rich data of the historic past.

Backward prediction. We marked on our K_{max}-map the epicenters of the earthquakes with $M \geqslant 6$ which occurred between 684 and 1925. The total number is about 120. Tables 9 and 10 list data on the strongest of these earthquakes with $M > 7$ ($K > 17$) and their locations. The agreement with the K_{max} map is obviously satisfactory when the earthquakes are situated in zones surrounded by the corresponding K_{max} isolines or when the earthquakes are beyond these zones within the acceptable magnitude (ΔK or ΔM) or distance (ΔL) errors. We note that it is not only necessary to pack all events into the limits of the zones of corresponding K_{max}, but that this would be actually undesirable because it would show that, within the error limits, a better localization of the K_{max} zones (which is always desirable) is actually possible.

It follows from Table 9 that all the 21 earthquakes listed in the table have their proper positions on the K_{max} map. Leaving aside those earthquakes, let us consider, in detail, the 10 earthquakes which are listed in Table 10 epicenters beyond the limits of the corresponding contours. In order to determine the significance of the deviations, we compare these earthquakes with possible errors made in the determination of the earthquake parameters. The estimates are as follows. We can assume for 1926-1965 that $\Delta M = \pm 0.25$-0.35 and $\Delta L = \pm 25$-50 km; and for 684-1925, that $\Delta M = \pm 0.5$ and $\Delta L = \pm 50$-70 km; in view of the lateral position of the points of observation on the islands relative to the main seismic zone extending in the ocean parallel to the deep-sea trough, the ΔL errors are reduced perpendicularly to, and increased in parallel with the strike of the main structures.

The disagreement between the earthquakes listed in Table 10 and the K_{max} map of Fig. 27 originates from the fact that the indicated K values of earthquakes were greater than the K_{max} of the zone in which the epicenters of these earthquakes were located (columns 6 and 7 of Table 10). Data that allow evaluation of the importance of this discrepancy are listed in columns 8-12. Here r denotes the radius of the averaging area from (3.12); for the major earthquakes, the radius is equal to the average radius of the focal area; L_1 denotes the smallest distance between the epicenter and the point on the map at which $K_{max} = K$ for that particular earthquake; L_2 denotes the distance between the epicenter and the nearest K_{max} isoline surrounding the zone in which this earthquake should occur; K_{max} denotes the K_{max} value at the point of the map which the r contour of the area of averaging or of focus $\Delta L = L_1 - r$ denotes the distance by which the r contour must be shifted to reach the point at which $K_{max} = K$ for the earthquake on the map.

It follows from Table 10 that though the epicenters of earthquakes Nos. 1, 3, 4, 7, and 10 do not coincide with the corresponding points at which $K = K_{max}$ on the map, the r zones overlap the points (in the case of earthquakes Nos. 1, 3, 4) or almost these points (in the case of earthquakes Nos. 7 and 10). For the remaining earthquakes, the section ΔL of the "missing shift" is greater though it nowhere exceeds the possible error $\Delta L = \pm 50$ km in the determination of the epicenter.

Let us consider the situation in the focal region of earthquakes Nos. 5 (1614) and 6 (1833) for which the ΔL discrepancy in Table 10 shows the greatest values (43 and 47 km). The epicenters of these earthquakes are denoted by asterisks on the map of Fig. 27. Earthquake No. 5 with $K = 18.2$ ($M = 7.7$) coincides with the point at which $K_{max} = 17.2$ on the map so that $\Delta K = 1.0$ or $\Delta M = 0.6$. This is almost consistent with the acceptable limits $\Delta M = \pm 0$. Shifting the epicenter over an acceptable distance $\Delta L = 50$ km to the East would lead to the point at which $K_{max} = 17.8$, misses the predicted value $K_{max} = K = 18.2$ by only $\Delta K = 0.4$ or $\Delta M = 0.2$. Earthquake No. 6 with $K = 17.6$ ($M = 7.4$) should coincide with the point for $K_{max} = 16.7$ on the map. The error $\Delta K = 0.9$ ($\Delta M = 0.5$) is equal to the acceptable error even when the epicenter coordinates are not modified. Thus even the two worst cases in regard to ΔL were not beyond the possible error limits.

To check the forecasting capability of our K_{max} map (Fig. 27), further we drew ± 25 km confidence strips on the continents and ± 50 km confidence strips in the oceans parallel to the isolines. These values are approximately equal to the accuracy with which the epicenter coordinates are determined. The focal areas [R of Equation (3.12)] of the major earthquakes known for the 684-1925 period were marked on the map a fragment of which is shown in Fig. 28. It was found that except for one, all focal areas appear in the proper position on the map when the possible error in the determination of the magnitude, $K_{max} = 0.5$, is taken into consideration. Let us dwell on the only exception as it is very instructive.

We refer to earthquake No. 7 in Fig. 27 and in Table 10 (1854; $M = 8.4$ and $K = 19.4$); the closest position of its focus is shown on map in Fig. 28. The epicenter of this earthquake coincides with a point characterized by $K_{max} = 17.5$ so that $\Delta K = 1.9$ (or $\Delta M = 1.1$). The circular area r (Equation 3.12) of its source only touches the confidence strip of the $K_{max} = 19$ isoline. The foci of two other very severe earthquakes of the past (No. 3 of 1096 and No. 4 of 1498) are located in approximately the same region. All the 1096-1854 earthquakes of global scale are located on the axis of the "main ridge" of seismicity, viz., in the region of the "saddle" in the ridge as given by the 1926-1965 observation data. It is possible that the saddle results from a long-lasting quiescence which began after the liberation of the potential energy of this group of earthquakes. A somewhat similar is observed in the focal area of earthquake No. 1 (Table 3; 684, $M = 8.4$; see Fig. 28, left side). However, in this case the focal area of the earthquake comprises to a greater extent regions of high seismicity to the right and to the left of the narrow "valley" of seismicity in its "main ridge". As long as there is no doubt about the extreme magnitude K of known severe earthquakes of the past, there is obviously no reason for disregarding their focal areas as potentially dangerous in the future or for ascribing these earthquakes to the $K_{max} = K$ zone, although the $K_{max}(\overline{A})$ correlation does not provide a formal basis for this.

The observation could be interpreted in another way, namely by assuming that after a severe earthquake a long quiescence must be expected and that therefore K_{max} must be reduced in this region and should be assumed in the form of the $K_{max}(A)$ correlation as indicated in Fig. 27. But this approach is dangerous for reasons we partially explain below.

Recurrence of the maximum earthquakes. As far as large earthquakes close to the local maximum possible earthquakes are concerned, it is often assumed that these earthquakes "disassociate" due to the temporary liberation of seismogenic stress. In the course of time this effect should manifest itself in greater periodicity of these earthquakes relative to the randomness of a Poisson process. In space coordinates, the effect should imply non-overlapping focal areas in time intervals smaller than the period T of the "seismic cycle", which was estimated at $T = 140 \pm 60$ years by Fedotov for both the Kuril-Kamchatka and Northern Japan regions. Let us check whether this tendency can be confirmed for the largest Japanese earthquakes with $M \geqslant 8$ ($K \geqslant 18.8$) when the observations from the 684-1971 period are used.

It follows from Tables 8-10 and the 1966-1971 catalogs that the epicenters of these earthquakes never occurred at the same point within this period. This is not surprising for these rare events even if they are completely random Poisson events. But there are four sites at which the focal areas of the earthquakes overlap considerably in this period (see Table 11).

The time intervals T observed between any two subsequent events are listed in column 7 of Table 11. Obviously, the time intervals vary widely, ranging from 2 to 742 years. The average recurrence period $\overline{T}_i$ of the events (column 8) naturally varies to a smaller extent in each zone but the variation is still considerable probably because the small number of events in each zone. In order to increase the volume of our statistical data, we assume that the recurrence laws of the largest earthquakes of Japan are the same in all zones considered. Let us calculate the average value $\overline{T}$ from all T listed in column 7 and estimate the deviation. We obtain $\overline{T} = 272 \pm 220$ years. We have indicated the mean square deviation which proved to be close to the average value $\overline{T}$ proper.

This fact suggests that the above set of events might be appropriately described by a flow of independent events, i.e. by a Poisson process rather than by a trend towards "disassociation" and, hence, a trend towards periodicity in the sense of [Fedotov]. In order to check this conclusion, we compared the observed distribution of the T values in the form of a histogram, Fig. 29 (column 7 of Table 11), with the corresponding theoretical exponential Poisson distribution of the time intervals between the events

$$P(T) = (1/\overline{T}) \exp(-T/\overline{T})$$

represented by curve *1*. We also compared the observed distribution of T with a normal Gaussian distribution reflecting the trend of clustering of T around an average $\overline{T}$ (curve *2*). It was assumed for this curve that $\overline{T} = 272 \pm 116$ years, where the relative mean square deviation σ/T was assumed to be equal to that of the Fedotov case with $T = 140 \pm 60$ years. It follows from Fig. 29 that the Poisson distribution of independent events is in much better agreement with the observations than the normal distribution with a clear trend to periodicity.

Table 11. The Greatest Japanese Earthquakes with $M \geqslant 8$ ($K \geqslant 18.8$) Whose Source Areas Overlapped During the Period of 684-1971

No.	Year	Coordinates		M	K	T_i	$\bar{T}_i$
		N	E				
1	887	33.0	135.3	8.6	19.8		
2	1361	33.0	135.0	8.4	19.4	474	
3	1707	33.2	135.9	8.4	19.4	346	
4	1854	33.2	135.6	8.4	19.4		
5	1944	33.7	136.2	8.0	18.7	147	212
6	1946	33.0	135.6	8.1	18.9	90	
1	1096	34.2	137.3	8.4	19.4	2	
2	1498	34.1	138.2	8.6	19.8	402	
3	1854	34.1	137.8	8.4	19.4	356	379
1	869	38.5	143.8	8.6	19.8		
2	1611	38.2	143.8	8.1	18.9	742	
3	1677	38.7	144.0	8.1	18.9	66	354
4	1933	39.1	144.7	8.3	19.2	256	
1	1843	41.8	144.8	8.4	19.4		
						109	109
2	1952	42.2	143.9	8.1	18.9		

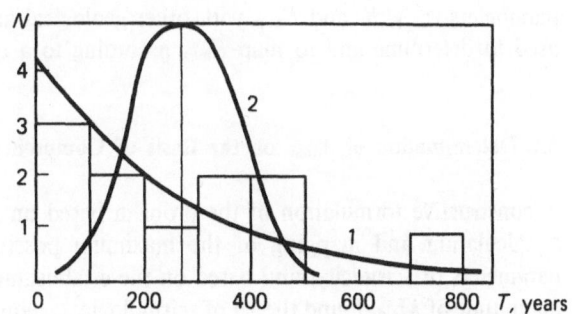

Fig. 29. Distribution of the greatest earthquakes in Japan with $M > 8$ ($K > 19'$) occurring in their own places during 684-1971 over T periods of their recurrence.

The histogram shows: *1*—a Poisson pure random process; *2*—a Gaussian process with a trend to periodicity

This confirms our reservations about the conclusion that after each near-maximum value of earthquake the probability of another event of about the same magnitude at the particular point is systematically reduced for some time smaller than T. This almost was practically constant, at least in the example under consideration.

In order to avoid misunderstandings and wrong interpretations, we emphasize that we do not consider the largest earthquakes to be events which are truly independent physically. We only assume that no purely time-dependent model of the seismic process, in the form of a simple seismic cycle, can adequately describe the real behavior of a seismic region with dimensions similar to the focal dimensions of the maximum possible earthquakes, if this region is surrounded by other similar regions. The concept of a simple cycle is only applicable for an isolated source area. In a general case, when secondary effects cannot be ruled out, the time dependence of the energy relations and other details of the seismic process, including the recurrence of the maximum possible earthquakes at a particular point, must be determined as space- and time-dependent functions. This approach was used for the first time in [Riznichenko].

Investigations of the space and time dependence of seismicity will, in future, help in approaching the physical interpretation of the external empirical correlations we have now learned to recognize. These can be utilized for quantitative investigations and mapping of the basic parameters of the average long-term seismicity (including K_{max}). Thereafter, better estimatime of seismic hazard is possible in terms of the probabile parameters of seismic shakeability.

Later the combined relationships $r(K)$ and $K_{max}(\overline{A})$ for Central Asia and Japan were checked and proved to be valid for the continental seismicity of Central Asia and several other regions such as the Caucasus, the Crimea, and the Carpathian-Balkans. We may assume that this indicates a stable, general law of seismicity which, in a first approximation, does not depend upon the local details of either tectonics or regional features. Further investigations must be concerned with the regional and local variations of the law and their interpretation.

It is supposed that the optimal dimensions which were determined in the correlation of K with $\overline{A}$ for the regions $r(K)$ "responsible for an earthquake" (i.e. the regions within which factors to be compared were averaged) are also valid for the comparison of the seismic parameters A, γ, K, and K_{max} with other geological and geophysical or geodetic factors Φ_i used to determine and to map K_{max} according to a quantitative correlation technique.

3.4. Determination of K_{max} on the Basis of Comprehensive Data[*]

A constructive formulation of the problem based on a quantitative determination, i.e., on a calculation and mapping of the maximum possible earthquake as one of the main parameters of seismicity, and based on the establishment of their energy value K_{max} or the magnitude of M_{max} found the set of seismological, geophysical, geological, geomorphological and geodetic data was started only in 1962.

---

[*] See [Riznichenko et al.]. E. A. Dzhibladze is a co-author.

Reference [Riznichenko] treats two possible operational solutions to this problem. The first is based on the establishment and use of physical cause and effect relationships. The second is based on the establishment and use of external patterns and correlations only. Unfortunately, the first, at present is less promising than the second. While accepting the superiority of the first method, and leaving it for the future, we can follow the second, which gives a clear-cut perspective for a quick solution of the problem to a first approximation. The first step in carrying out the integrated program formulated in [Riznichenko] was to evaluate the possibilities for qualitative use of pure seismological information (see Sec. 3.2).

3.4.1. Quantitative Methods for Determining K_{max} from Seismological Data

First, earthquakes observed as a maximum at a given place, K, were compared with the mean seismic activity A in the area "responsible for the earthquake" and surrounding the epicenter. The greater the activity, the higher the value of K. The value of K_{max} was determined by the limiting contour on the plane of the K versus A correlation field. This limits the area where (within the limits of allowable error and possible dispersion associated with parameter fluctuations of the seismic regime) almost all such events fall within the region under consideration. The results of these experiments have been considered repeatedly [Gorbunova, Riznichenko and others]. This method was tested and has been used to compile maps of K_{max} in many regions of Eurasia from the Apennian region to the Japanese islands. On the basis of these data on K_{max} and on other seismicity parameters, the probability factors of seismic danger were calculated for many territories in terms of seismic shakeability $B = B(I)$, i.e. the mean recurrence of earth shaking of various intensities [A. Drumea et al, Zakharova, Riznichenko etc.].

Another, also purely seismological correlation method for determining and mapping K_{max} (or M_{max}) was proposed by N. V. Shebalin. This is based on a comparison of the observed maximum earthquake K with the thickness h of the active seismic layer in the Earth's crust and the length l of the seismic zone. Also here, the determination of K_{max} is based on a certain limiting contour $K_{max}(h, l)$ which is analogous to the contour $K_{max}(A)$. With respect to its physical meaning, this approximation is close to the previous one. In both cases, the extent of the zones responsible for the earthquakes is considered with the first method, on the basis of its area, i.e., its length and width, while with the second method, using its length only. On the other hand, the value of h in method two has the same role as A in the first method. In [Riznichenko, et al.], the conditions are indicated where both methods coincide: this is the stability of volumetric seismic activity. The experience accumulated by trying to use both methods together on data observed in specific regions has shown that under conditions of reliable data and suitable selection of the parameters, the determinations of K_{max} by means of both methods are either close to each other or effectively the same.

A third seismological approach to the same problem is more formal. It is based on Gumbel's statistical theory of extreme values, which is frequently used to assess the calculated characteristics of floods and hurricanes. It is assumed in the seismological works in this field [V. Gaiskii] that no definite finite value for K_{max} exists and that $K_{max} \to \infty$ asymptotically. Furthermore, a well known assumption states that high values for K occur only rarely, more rarely than the lower values. Thus, by a low probability P, we may always choose a high value for $K = K(P)$ which may be identified with a suitably chosen limit of the unknown value of K_{max}. As a rule, such constructions are made separately for each previously chosen "quasi-homogeneous" zone.

It may be proved that this approximation, in essence an agnostic one with respect to the existence of an ultimate K_{max}, is untenable physically. In the main K interval of reliable observations, the parameter $\gamma = -d \lg N/dK$ characterizing the decrease in the number N of earthquakes with K is always less than one, i.e., $\gamma < 1$. It was shown in [Riznichenko] that when $K \to \infty$, the integral density of the seismic energy of earthquakes per time and space (volume area) unit will tend to infinity, which is absurd. In local zones, where large earthquakes are observed, the parameter γ tends to decrease rather than increase, which only confirms the inequality $\gamma < 1$, i.e., it continues to widen the energy integral, and thus causes a deepening of the absurdity. Later it was proved [Kogan, Shakirova], that the first limiting Gumbel distribution, which was used to calculate K_{max}, is invalid in the proximity of K_{max} with a sufficiently general statistic: it is here that its constant parameters start to change substantially. Such a possibility seemed earlier to be valid with regard to the statistics of the maximum earthquake for the entire world. Now, we may apply it to a number of individual regions. Thus, we must admit that a final limit on K_{max} actually exists, and that, in principle, it may be established from observation. Thus, it makes sense not to try to establish it according to Gumbel as a conditional $K(P)$ value, but to search for it and to find the actual value of K_{max} under specific conditions on the basis of other physical considerations.

Within the limits of a seismological method, this may be done by plotting the observed K values and establishing the limiting contour K_{max} by comparing them either with the factors already mentioned. These include: $\overline{A}$ (for given l and ω), or h (for given l and/or ω), or other characteristics of the A-field such as grad log $\overline{A}$, the transformants of the field of the integral density of the seismic source energy, the seismic moment M_0 of all the earthquakes, the parameters of the focal mechanism and seismic flow of rock masses; the spectral characteristics of earthquakes, and the attenuation of seismic energy with distance [Ananien, Riznichenko and others]. It also makes sense to use several factors together. In this way, the possibilities of seismological approximations are not exhausted.

3.4.2. Qualitative Approximations for Determining K_{max} and I_{max} on the Basis of Integrated Data

In reviewing the history of this problem, we should acknowledge that the first attempt at a constructive formulation was Gamburtsev's program of investigation concerning the prediction of earthquakes, including seismic zoning. It has to be said that these papers were published at a time (1955) when neither the modern correlation approach to the determination of K_{max} had been formulated nor did a qualitative concept of seismic activity A, only formulated in 1958) exist [Riznichenko].

Gamburtsev's basic ideas about seismic zoning, essentially the determination of K_{max} and correspondingly I_{max} (I is the macroseismic intensity), were formulated as three main hypotheses: (1) "on average, the seismic regime is stable"; (2) "the foci of intensive earthquakes are attributed to zones of deep ruptures in the earth's crust, i.e. active seismic faults"; and (3) "a severe earthquake whose focus is attributed to a place of an active seismic fault is a sign for possible seismicity of the fault as a whole, as well as of the other faults connected with it". While acknowledging the importance of geological data in seismic zoning ("geological and geomorphological criteria of seismicity", "contrast of recent vertical movements", "transection of structures", etc.), Gamburtsev also pointed out that "recently, the possibility of using geological methods for the solution of tasks of seismic zoning has clearly been overestimated" [Gamburtsev]; that all geological and geophysical indexes of seismicity have a "relative" and "qualitative" character [Gamburtsev]; that "not every seismic fault may be regarded as being seismogenic at the present time"; and that "except for information about severe earthquakes in the past, there exists no other reliable parameter on the degree of seismic activity of a seismic fault for the present time". It is interesting that regarding the K_{max} and I_{max} topic Gamburtsev wrote that the hypothesis on the stability of seismic regimes "leaves open the question whether each seismic region has its own earthquake of maximum force or whether the difference among regions only consists of the frequency of occurrence of earthquakes of the same intensity of the most powerful earthquakes. On the basis of the data at hand, the first hypothesis seems more probable. However, if the second hypothesis were true, this would also lead to a maximum magnitude for each region in practice. In fact, to design actual antiseismic measures, one can hardly take earthquakes into consideration whose probability during 100-200 years is negligibly low (for example, an earthquake which may occur once in thousand years)". We have already mentioned our present positive solution of the problem of the existence of K_{max}. Data concerning the recurrence A of earthquakes and shakeability B are complements rather than substitutes for data on K_{max} and I_{max}. A discussion of the hypotheses proposed in [Gamburtsev] is given in [Riznichenko].

Gamburtsev's ideas were supported by a large circle of geologists and geophysicists dealing with problems on seismicity and seismic hazard in the country. Their influence is still felt. The following approach to the establishment of seismically dangerous zones and determining I_{max} (or K_{max}) for each zone is a specific corrollary. "Homogeneous" zones are selected as a rule, due to deep faults. Within the limits of each, the maximum observed seismic shaking I_{max} is found. It is this shaking (earthquake) that is regarded as being the maximum possible at any place of this zone. In essence, this was the approach for the seismic zoning of the USSR[*], including its later refinement in 1969 (Building Code SNiP 12-69). Officially, this is still in force, after 20 years[**]. This is a very long time considering the rapid development of the geological and geophysical sciences and of the instrumentation which has taken place during this time, which also testifies to the progressive character of the approach in [Gamburtsev].

We mention only two logical imperfections of the previous approximation to K_{max}. The first is the identification of an event (earthquake K_{max}, or shaking I_{max}) observed as a maximum in every supposedly homogeneous zone with that which may be observed as a maximum within the zone. The second is the lack of a definition for the term "homogeneity" of a zone. Gamburtsev knew about both imperfections, but could see no way out at the time; at present, a solution is clear and we cover it below. The general development of the data follows.

To develop Gamburtsev's ideas seismologists required refinement and numerical expressions for "seismicity" and "seismic hazard". This led to the determination and characterization of these terms by means of the long term seismic activity A, the maximum possible

---

[*] Seismic Zoning of the USSR by S. Medvedev (ed.), Moscow, 1968 (in Russian).

[**] The 2-69 Building Code is in force at present.

earthquake K_{max}, and the seismic shakeability B. Similarly, the development led to existing geological criteria for seismicity being refined and new ones developed. The largest contribution was made by M. V. Gzovskii and his co-workers. A great number of criteria and their modifications were suggested but most attention was directed: (1) to the lines and strips along the strike of deep fractures and, in particular, to their intersection nodes; (2) to the horizontal gradients of vertical tectonic movements, or more exactly, the modulus, of the horizontal gradients of the velocity of vertical displacements or, (the same) the velocity of deformation of a simple dislocation along the vertical.

These gradients are quantitative experessions of the criterion of "contrast of motion" we discussed earlier. Nevertheless, they seemed new to seismologists. Moreover, they at first seemed to be the key of the final solution to the K_{max} problem. However, accumulated data and comparisons rapidly dispelled the optimistic expectations. The gradients were found to be indices of seismicity as "relative" and "qualitative" as the faults themselves.

In investigations carried out most recently, features of traditional qualitative approximation are found in [Bune] (for the Caucasus Region). Although the gradients of neotectonic motions are considered, the general treatment of the geological and other criteria is of the same descriptive character as it was years ago.

A. A. Borisov and G. A. Shenkareva were possibly the first seismologists to acknowledge the ineffectiveness of using traditional descriptive geological techniques for solving the problem of "seismicity and seismic hazard" with respect to K_{max} and I_{max}, and to look (1972) towards statistical correlation methods for a quantitative solution. Now we can repeat their main conclusions from [Borisov].

The first conclusion was: "No one of the tectonic", neotectonic, and geophysical features, taken separately, or any combination for various zones can be regarded as an absolute indicator of seismicity. Under various geotectonic conditions, the geophysical characteristics of seismogenic structures in seismically active regions may be quite different, both with regard to the individual features and to their combination." The second conclusion was: "Only a combination of a set of different independent features, a multivariate vector may characterize a seismogenic structure and a seismically dangerous zone with the required reliability. It is this multivariate vector that should be determined for various regions, and within the boundaries of a given region, for its various zones."

These statements are certainly strongly critical in which respect they are close to [Gamburtsev]. But creatively, i.e. in the development of a new method for studying seismicity, only the first steps were taken in [Borisov et al]; namely the multidimensional character was shown and a technique was proposed for the formulation of multivariate vectors (in dimentionless units) which were earlier used to solve structural problems. The Table given in [Borisov] for 52 quasi-homogeneous zones (selected on the basis of about the same seismicity as well as "mainly morphological and tectonical features") and for 25 features (various characteristics of the gravitational and magnetic fields and relief of deep boundaries) were debated. Generally speaking, the Table was difficult to survey, a fact also reflected in the conclusions.

The approximation in [Borisov] is, in general, promising. It is a start, although it leaves something to be desired. For example, the integrated specific power of the seismic energy was used to express the seismicity quantitatively although it is known that this quantity fluctuates widely and, therefore, reflects the implied long-term mean conditions less well than, let us say, the seismic activity A. Neither is it clear whether a great number of quasi-heterogeneous zones can be selected *a priori* in view of a general lack of definition for this term and an incomplete formalization of the principles for their selection. The classification system of the zones is too coarse for some practical aims, since they are divided into three categories according to seismicity: severe, medium, and weak. The proposed algorithm itself for finding the required seismicity parameters K_{max} and I_{max} for each of the zones on the basis of a multivariate vector was not clarified, although it was the basic idea in the paper.

3.4.3. Correlation Method for Determining K_{max} from Integrated Data

The idea of using indirect geological and geophysical information for establishing the long-term average seismicity coefficients A and K_{max} is based on the assumption that the geological and geophysical fields were formed over an extended period of time. They should, therefore, represent a long-term average tectonic activity together with its manifestation, seismicity. It may, therefore, be expected that the indirect determination of the theoretical parameters of the long-term mean seismicity $\tilde{A}$ and $\tilde{K}_{max}$ on the basis of factors Φ_i of the geological and geophysical fields may be useful for checking and correcting the seismological values A and K_{max}. These are established from relatively short-duration observations, and may therefore be subject to considerable fluctuations.

Our approach to the common use of some Φ_i factors in the determination of $\tilde{K}_{max}$ [Riznichenko, etc.] may be regarded as a natural generalization for the multidimensional case of the known determination method of K_{max} on the basis of either the factor A [Riznichenko] or the factor h (Shebalin) only.

The value of $\tilde{K}_{max}$ is determined by the limiting contours of the correlation fields $K(\Phi_i)$ i.e. either by a series of bidimentional contours for each factor Φ_i with a subsequent determination of the mean with respect to all the factors or immediately by the multidimensional contour, thereby considering some factors simultaneously. Here, K is the observed energy value of the earthquakes; above all, the values which are close to the maximum at given places. In this process, the regions within whose limits the mean values of the factors Φ_i are determined are correlated with the respective values of K; the higher K, the larger the averaging area. For the time being, the limiting contours are simply constructed graphically; suitable statistical techniques should be carried out on this procedure.

The value of K_{max} may also be determined, indirectly predicting the value of A established by the correlational method on the basis of geological, geophysical and other factors $\tilde{A} = \tilde{A}(\Phi_i)$. The transition from $\tilde{A}$ to $\tilde{K}_{max}$ follows relationships established for A and K_{max} seismologically [Riznichenko]. The correlation Φ_i and A is carried out, in the simplest cases, in pairs, i.e., graphically with a later determination of the average $\tilde{A}$ on the basis of a number of factors, or analytically by simple or multivariate regression analysis.

It should be noted that our new method is essentially quantitative in character and constructional, formalizing our approximation of the determination of K_{max} on the basis of complex geological and geophysical data, and it differs from the methods applied earlier in several important features.

We do not attempt to establish the function $K_{max}(\Phi_i)$ individually for each "quasi-homogeneous" zone. This would lead, in essence, to the substitution of the maximum possible earthquake in every zone by the maximum earthquake observed in the given zone. On the contrary, we have tried to find a general function $K_{max}(\Phi_i)$ for a possibly large combination of zones. The more universal this function, the better. "Success" depends on the choice of factors which should have a direct physical relationship to tectonics and seismicity [Riz-nichenko]. If any of the factors correlates well with A and K_{max} in a narrow localized zone only, its prediction value will be doubtful and it should not be considered even within this zone. For every zone, checking is needed to see whether the general solution is in contradiction with the actual data or not. Only in the presence of statistically significant contradictions, should the corresponding zone or combination of zones be subject to a special investigation and, possibly, discrimination. All these will eliminate the most essential, fundamental draw-backs in the earlier method, i.e., the identification of the maximum observed event in a zone with the maximum possible one in it.

The second drawback is the lack of a definition for "heterogeneity" or "quasi-heterogeneity" of a zone. This drawback is now eliminated because we do not need any *a priori* selection of such zones. The zone is not the elementary cell where the values of A, K, Φ_i are compared and for which the solution A, K_{max} is obtained, but every point on the map of the region or, more exactly, the area ascribed to it. In practice, the points of examina-tion Φ_i and the points of the obtained solutions $\tilde{A}$ and $\tilde{K}_{max}$ are usually located on the map at the nodes of a fairly dense two-dimensional net. In an almost continuous situation all the numerical fields could be generalized and represented as a "smoothed" map in isolines and not as a "fragmentary homogeneous" zonal map.

Nevertheless, the occurrence of quite extended zones is not excluded. In essence, any con-tinuous map may, if we wish, be coarsened and approximately represented in a zonal, frag-mented homogeneous, and horizontally gradual form. This relates to maps both of the initial and resulting values. Moreover, separate treatment of the data on the zones (not as a substitute for the common treatment of the data in the region as a whole, but as a supplement to it) may be useful for various geological, geophysical, and statistical conclusions. In this case, the zones may be selected on the basis of seismological, tectonic, or other considerations, depending on the aim of the investigation. The number of selected zones should be small so that the data present on each zone are sufficient for reliably establishing the patterns. It is required, in particular, that the values being compared in each zone vary within a large range. Thus, it is not required that these zones be quasi-homogeneous with respect to every feature, which was the aim earlier.

Sometimes, the region is divided into parts on the basis of formal considerations, one part being used to obtain correlations, while the other is used to check them. However, fre-quently the same region is simultaneously the territory for establishing and for using the established relationships. The situation in an individual region is obtained by transferring the results in a better investigated region to another less well investigated region. Examples of this kind for the determination of $K_{max}(A)$ are given in [Riznichenko, etc].

These standpoints have crystallized from analyzing some, although not many, results ob-tained using these methods in various regions. The first qualitative correlation of geological, geophysical and seismological data in this sense was carried out in the Baikal region [Riz-nichenko]. Here, the seismic activity A was correlated with the height h and with height gra-dient (graph h) of the average relief representing the tectonics of the region for the Neogene-Quaternary period (25 million years), and the isostatic gravity anomaly parameter Δg. In this investigation the net of examination points was 15×15 km^2 and the dimension of the averaging area around each point varied from 30×30 km^2 to 90×90 km^2 (the resolu-tion should be greater with respect to regions studied). Earlier, non-Soviet, qualitative compar-isons of seismicity with gravitation [I. Ananien, Tsuboi], did not complete comparisons with such high resolution. In [Tsuboi] the character of comparison is close to ours [Riznichenko], but the averaging areas were very large (5 × 5°) which, of course, does not make it possible to solve regional scale tasks needed for seismic zoning. No special investigations for zoning seem to have been carried out outside the Soviet Union.

In our first report [Riznichenko], low values were obtained for the correlation coefficient r between the seismicity index and other factors; in comparing A with (grad h), the best value was $r = 0.36$-0.39. Later, comparisons of A, h, and Δh were continued by Yu. A. Zorin and M. R. Novoselova in the same region. Instead of A, they took log A, which made it possible to increase the correlation coefficient of log A (with grad h, Δg) and (grad Δg) up to $r = 0.8$; this is already a fair result. The convenience of working with log A in order to bring the distribution closer to normal and the regression equation close to linear was already felt in [Riznichenko]. It is also very important to choose suitable measures for the averaging areas and, naturally, for the parameters of the logical, geophysical, and other fields. The physics of this problem is discussed in [Riznichenko].

Later seismicity parameters were considered with neotectonic parameters (h, grad h and etc.) on the basis of a similar technique by S. Sherman for the Baikal region, by A. Drumea for the Eastern-Precarpathian region, by M. Artem'yev et al. with the gravitational anomalies for the Crimean-Caucasus region, and by V. N. Gaiskii et al. for the Altai-Sayan territory.

The velocity boundary in the earth's crust, like the surface of the seismological "granitic" layer, which moves with a longitudinal wave velocity of 6 km/s, may serve as an indication of recent tectonic movements, and therefore, potential seismicity. From data obtained in a detailed study of such a boundary, using E. M. Butovskaya's method, a number of correlations of h (grad h) and other seismicity parameters (A, K_{max}) can be found in [E. Butovskaya and Riznichenko].

Multidiscipline investigations of the correlations between seismicity and tectonic, deep structure, and geophysical fields such as (gravitational, magnetic and thermal) parameters were carried out for [Azizbekov]. Analogous quantitative comparisons are being carried out increasingly frequently in various regions of the USSR and in neighboring countries. There is no doubt about the rapid accumulation of experience in this field.

4. A Model of Space-Time Evolution of Seismicity

4.1. A Model of the Energetics of a Seismic Regime

While the construction of theoretical and physical models of an individual earthquake focus has already become traditional, the simulation of a seismic regime as a space-time totality of earthquakes of various energies [Riznichenko] is still in an embryonic state. The first attempts in this direction were the attempts by Benioff and later Kuznetsova, Knopoff, and Kasahara to compare such regimes to the processes occurring in mechanical systems with discrete components symbolizing, as it were, the microstructure of the medium. In [Knopoff] a physical device for realizing such a model is also described. All these spatially zero-dimensional or one-dimensional discrete models, although they help in visualizing the internal mechanics of the process and in reproducing certain external features of the behavior of nature, still remain distant from the possibility of real use in seismology.

In the present work, an attempt is made to outline the means of solving another problem: leaving aside the microstructure of the medium and the process, to develop bases for constructing a phenomenological model which is completely formalized, has no internal contradictions, and is externally physically plausible, and which, in principle, would allow us to simulate the seismicity observed under specific conditions. For this purpose, we shall consider, probably for the first time, a fundamentally new, continual, spatially two- or three-dimensional theoretical model, close in spirit to the well-known energy concepts of the seismologists Tsuboi, Fedotov, and others. The relationship between this model and the discreteness of the process is also discussed, but in the present work it plays a secondary role.

When matching the action of this model to the observed seismic regimes in specific regions it passes directly to the solution of the problem of predicting earthquakes in these regions. Thus it is possible that the proposed model will become one of the links in the integrated solution of this current problem.

4.1.1. Continual Model

The structure and character of the action of the seismic-regime model proposed below is postulated simply. The discussion of its features, in comparison with what is known from nature, should be regarded not as proof that the process in nature closely follows the rules indicated for the model, but merely as an explanation of why the author chose a particular method of formalization. Nevertheless, this does not mean that the model has no relation to nature at all. It was constructed considering a number of phenomena known in seismology and related sciences. Some were clearly understood only recently, while the description of others has barely begun. The functional dependences in the calculation method are only schematic and general in form. Making them more specific is a future task. The purpose of this paper is to indicate precisely what values are needed for calculations and what their interaction is in the overall system. But the main task is actually outlining this system as a whole and we are trying to do it here.

With the aid of this model we propose to obtain a very simple first approximation to nature. To what extent this approximation satisfies the rational requirements of the present day will be shown only by the results of actual simulations. We can hope that the construction of a model, and an experimentally consistent theory of the space-time variation of a seismic regime, will be accomplished in the near future.

Proceeding to the actual description of the model, let us now consider, for simplicity and specificity, a spatially two-dimensional case (crustal seismicity). Let there be given on the x, y plane (in the seismic region being considered) an energy source function of the seismic regime $\omega_0 = \omega_0(x, y, t)$, which represents a flow of potential units to time and area, basically the elastic energy reproducing the seismicity, i.e. the density of the power delivered to this region from without (from the mantle) to sustain the seismic process in it. This is a source of replenishment of the reserve of potential elastic energy in the medium [Riznichenko] and of the seismic energy flow, liberated from this reserve, of the set of earthquakes comprising the regime.

As our initial approximation we assume that ω_0 does not depend on time, i.e., $\omega_0 = \omega_0(x, y)$. The approximate assumption that ω_0 is constant in time is justified by the fact that the tectonic processes, of which seismicity is one, may vary appreciably in character, in particular, energy intensity, only over much longer time intervals (millions of years) than those of interest to us when studying contemporary seismicity (thousands or tens of thousands of years).

Let us introduce into the discussion two energy functions of space and time, which will be used to monitor the regime in our model, namely, the potential energy ε stored in the medium and the energy "strength" ε_{max}. More exactly, $\varepsilon = \varepsilon(x, y, t)$ is the density of that part of the potential energy (mainly the stressed state in the medium) which is capable of reproducing the seismicity; ε_{max} is the maximum density of the energy ε, at which the rupture of the medium assumes an avalanche-type character, i.e. a large earthquake occurs.

We are speaking here of the "seismogenic" part of the potential energy ε. The remainder, for example, the elastic energy of triaxial compression under conditions of hydrostatic equilibrium in the absence of appreciable displacements in the gravity field, the thermal energy of uniform heating of rocks located at deep levels, the potential chemical energy in the absence of reactions, etc. does not play an active role in the seismic process. The rest of the potential energy is expended on various continuous-flow processes, but any energy, not only elastic energy, which is capable of reproducing discontinuous (or quasidiscontinuous) motion, i.e., seismicity, is included in ε.

Energy "strength." During a short-term (minutes) liberation of elastic energy from the focal zone of an individual earthquake it may still be possible to speak cautiously of an effective maximum strength of material, that is supposedly not dependent on time. However during a seismic regime lasting a long time (millennia) this is clearly inadmissible. It is necessary to introduce into the analysis a time-dependent energy strength.

The concept of energy strength in a time-independent "instantaneous" aspect dates back to Maxwell (subsequently, the "theory of maximum energy of form change" of Beltrami, Huber, and others). In addition to considering the time-dependent aspect of the strength, bearing in mind the interests of seismologists, we use the term "critical state" to mean the transition of rock material to a state of avalanche-type destruction, with the formation of discontinuities or high-plasticity zones concentrated along certain surfaces, which enable individual blocks of the initially continuous solid mass to move rapidly relative to each other. Engineers, by contrast, usually understood critical state as the beginning of plastic flow of

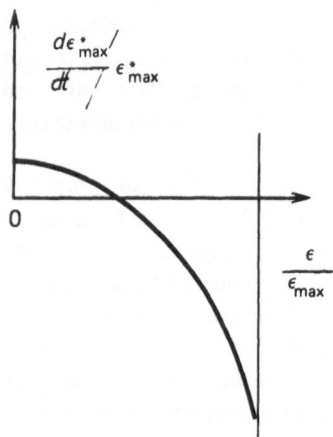

the material, perhaps even without the formation of discontinuities, and all the more so without total destruction: it is sufficient for the product to be noticeably irreversibly deformed for it to be distorted.

We assume that in the absence of a large earthquake the energy strength ε_{max} decreases with time t if the relative stress energy $\varepsilon/\varepsilon_{max}$ is large, and increases if $\varepsilon/\varepsilon_{max}$ is small. With time this increase tends towards saturation: $\varepsilon_{max} \rightarrow \varepsilon_{max\ max}$.

The effect of a strength decrease in time, an effect which also depends on the stress state, is well-known in engineering physics [Zhurkov]. It is related to the various forms of flow of solid material, which in turn are related to the displacement, coalescence, and concentration of dislocations in the crystals, and to the formation, growth, and coalescence of microfractures preceding total destruction. In mining and geophysics [Vinogradov], in particular when considering a seismic process, it is customary to class as weakening phenomena the formation, growth, and coalescence of macrofractures or local zones of higher plasticit. Also included are those defects in the continuity of a solid mass which show themselves seismically, producing small earthquakes constituting a "background seismicity" but not yet leading directly to avalanche-type, catastrophic, total destruction of large volumes of the medium, i.e., to large earthquakes.

The effect of an increase in the strength of rock masses under conditions of a low tectonic stress, on the other hand, symbolizes those processes of deep metamorphism which lead to the consolidation of the rock material in zones of previously inactive faults, large and small, and to a decrease in the amount of crushing and cracking occurring in them, i.e. these faults being healed. We may perhaps include as a process: intrusions of melts into cracks and their subsequent hardening; pneumatolithic and hydrothermal processes which cement crushed rocks and fill the fractures with vein material—in total, all those multifaceted physical and chemical processes which gradually convert back into a monolith those parts of a rock mass which have been destroyed by faults and earthquakes can be included.

We assume, therefore, that the variation in strength with time t occurs in accordance with the functional values shown schematically in Figs. 30 and 31. Let the relative velocity $a = d\varepsilon_{max}^{*}/\varepsilon_{max}^{*}dt$ of the time variation of the "calculated" strength ε_{max}^{*} (this term is explained below) depend on the relative value of the stress energy $\varepsilon/\varepsilon_{max}$ (Fig. 30). Moreover, let the

Fig. 31. Saturation of the energy strength ε_{max}, ε_{max}^* is the calculated strength

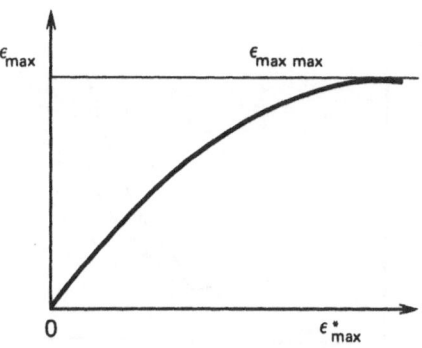

quantity a be positive (strengthening the material) at small stresses and negative (weakening the material) at large stresses. In the particular case where $\varepsilon/\varepsilon_{max} = \text{const}$ we obtain

$$\frac{d\varepsilon_{max}^*}{dt} / \varepsilon_{max}^* = \frac{d \ln\varepsilon_{max}^*}{dt} = a = \text{const}$$

thus, integrating, we obtain

$$\varepsilon_{max}^* = \varepsilon_0 e^{a(t-t_0)},$$

where ε and t_0 are constants characterizing the initial conditions. Thus in this particular case our calculated strength varies exponentially in time. In the case of the variable $\varepsilon/\varepsilon_{max}$ the dependence of ε_{max}^* on t is, naturally, more complicated.

A decrease in strength with time, perhaps even to zero, in the presence of large stresses ($\varepsilon/\varepsilon_{max} \to 1$), will obviously not cause any physical objections, so that under these conditions the calculated strength may be regarded as the actual strength. However, for small stresses ($\varepsilon/\varepsilon_{max} \to 0$), as the time t increases, our calculated strength ε_{max}^* increases without limit, which is unacceptable for a real strength: the strength should move towards some finite limit ε_{max}. We are considering this by introducing a relation between the calculated strength ε_{max}^* and the real strength ε_{max} as shown schematically in Fig. 31.

The calculation of the time variation of the strength ε_{max} is formalized finally as follows. Let the stress energy ε and the strength $\varepsilon_{max}|t$ be given at a point (x, y) at a moment t. To determine the new value of the strength at the subsequent moment of time $t + \Delta t$ from the ε_{max}, let us find, from Fig. 31, the value of the calculated strength ε_{max}^*. From Fig. 30, we find the rate of growth of ε_{max}^* with time that corresponds to the given ε and ε_{max}, and, multiplying it by Δt, we obtain the increment $\Delta\varepsilon_{max}^*$. From this we obtain a new value for the calculated strength $\varepsilon_{max}^*|_{t + \Delta t}$. Finally, with the aid of Fig. 31, we return to the unknown value of the real saturating strength $\varepsilon_{max}|_{t + \Delta t}$.

We shall discuss the remaining behavior of the strength, in particular, during and following a large earthquake, when we consider the processes related to such an earthquake.

Two forms of destruction and seismicity. Let us assume that the destruction can be of two types: total and partial. In relation to seismicity, total destruction corresponds to a large earthquake. The immediate aftershocks of a large earthquake are also ascribed to it. Partial

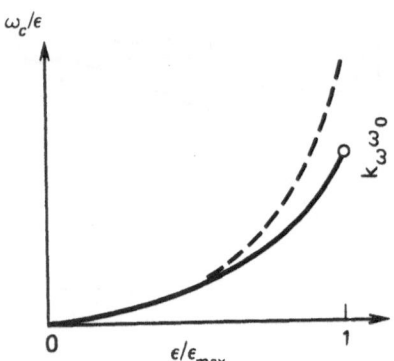

ω_c/ϵ

$k_\omega \omega_0$

0

ϵ/ϵ_{max}

1

Fig. 32. Dependence of the background energy flux ω_c seismicity on the relative stress
energy ϵ/ϵ_{max};

ϵ is the potential stress energy; ϵ_{max} is the strength

destruction corresponds to the "background" seismicity, which occurs in the periods between large earthquakes. We shall class the increase in seismicity before a large earthquake as partial destruction and, consequently, as background seismicity.

Let total destruction occur when the stress energy reaches the energy strength $\epsilon = \epsilon_{max}$. It constitutes a process of avalanche rapidity; in the given model it is an instantaneous process. Partial destruction is a permanent process. Let the seismic-energy flow ω_0 liberated in this case (the background seismicity) depend on the relative stress energy ϵ/ϵ_{max} and increase with an increase in the latter, as shown (in two variants) in Fig. 32. In this case $0 < k_\omega < 1$. An increase in the background seismicity as ϵ approaches ϵ_{max} delays the moment of unleashing of a large earthquake, but leads to an increase in its dimensions, thus giving the surrounding regions time to move closer to the critical state. If the potential elastic energy is not sufficient or rapidly concentrated, it may be completely expended on an almost critical amplification of the background seismicity (an earthquake swarm), so that no single large earthquake results.

A large earthquake. Such an earthquake is initiated at the point where the moment (t_0) when the condition $\epsilon = \epsilon_{max}$ is first attained (Fig. 33). In this case the subscript -0 denotes the stress energy ϵ and the strength ϵ_{max} at the moment $t_0 = 0$ immediately before the earthquake. The values of the functions at the moment $t_0 \neq 0$ immediately after the earthquake are denoted by the subscript $+0$.

We shall distinguish two zones within the focus: an internal zone—the focus proper— a zone predominantly of destruction D, and an external, predominantly elastic zone E. Let the boundary B between these zones be determined by the place where $(\epsilon_{max-0} - \epsilon_{-0})/\epsilon_{max-0} = k$ is a fixed number, $0 < k < 1$. Inside this boundary, let the new values of the stress energy ϵ_{+0} and strength ϵ_{max+0} be determined by the stress and strength conditions before the earthquake, i.e. they depend on $(\epsilon_{max-0} - \epsilon_{-0})/\epsilon_{max-0}$; on the boundary, let $\epsilon_{+0}/\epsilon_{-0} = k_2$ and $\epsilon_{max+0}/\epsilon_{max-0} = k_{\epsilon_{max}}$; $0 < k_2, k_{\epsilon_{max}} < 1$, where k_ϵ and $k_{\epsilon_{max}}$ are fixed numbers. Outside the boundary, let the values of ϵ_{+0} and ϵ_{max+0} be determined by the relative distance l/L from this boundary, asymptotically approaching ϵ_{-0} and ϵ_{max-0} with increasing distance from it. In this case L is the diameter of the focus. When the argument is changed, the functions ϵ_{+0} and ϵ_{max+0} do not undergo any discontinuity on the boundary of the focus.

Fig. 33. Source of a major earthquake:

D—destruction zone; *E*—elastic zone. *Above,* model of the phenomenon; *below,* behavior of the strength function ε_{max} and the potential energy function ε

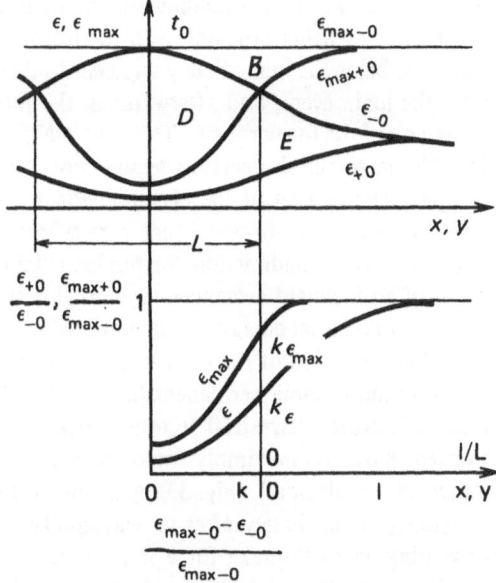

This formalization of the functions ε and ε_{max} in the focal zone is justified by the following physical considerations. In the inner focal zone, the destruction process causing the change in strength and the liberation of elastic energy is entirely determined by internal factors: the properties and the state of the medium in the focus itself. In the model adopted, these properties and state are characterized by two functions ε_{max} and ε. It is these functions which are taken as the arguments characterizing what is occurring.

In the outer zone, on the other hand, the variations in the properties and state of the medium occur mainly as a result of what has happened at the focus. Moreover, the outer zone remains mainly elastic: the destruction and weakening processes in this case, at least at first, are not significant. The spatial distribution, in the outer zone, of the elastic energy liberated from this zone may be calculated even on the basis of the theory of seismic dislocations [Vvedenskaya], if the focal mechanism is known. In the simplest case, it is simulated by a displacement over part of a plane—a fracture surface. The geometrical dimensions of this surface and the character of the displacement along it determine the function of liberation of elastic energy in the surrounding space. In turn, it is customary in seismology to correlate the amplitude of a displacement along a fault with its length. The total seismic energy of a source is also correlated with this length. Thus the spatial length of the focus L is a convenient main argument for characterizing external effects. Finally, in the elastic region the phenomena are similar to each other. This enables us to make constructive use of the relative distances l/L. This is what we have done.

Let us turn to some further details. In nature, the destruction zone and the elastic zone do not have a sharp boundary; there are only gradual transitions between them, as can be seen from the dispersed distribution of the aftershock sources. This is included in the assumption that the functions ε_{+0} and ε_{max+0} are continuous. Furthermore, in the destruction zone the elastic energy may not be completely liberated. Pieces of partly crushed rock masses in the focus and in the surrounding area may still retain some elastic energy and expend it only gradually on further crushing with the liberation of free energy [K. Mogi]. The weakening wave (see below) is immediately adjacent to these processes. All this is shown by aftershocks after the main event, and afterwards in the form of a long-lasting increase in seismicity in the region of the former focus. This is formalized by the fact that in the focal zone L (Fig. 33), in addition to the destruction region proper D, part of the elastic region E remains. The complete elimination of the elastic region from this zone, i.e., the assumption that in this case $\varepsilon_{+0} = \varepsilon_{\max+0}$, or even that somewhere in the focus, if only at one of its points, $\varepsilon_{+0} = \varepsilon_{\max+0}$, is inadmissible for fundamental reasons. The latter equality is, by convention, proof of an incipient avalanche-type, total destruction. Owing to the constant action of the sources of potential energy ω_0, the point where $\varepsilon_{+0} = \varepsilon_{\max+0}$ would become a site of boundlessly long retention of the critical state: the strength and potential-energy "gap" in the focal region would become permanent (like the Red Spot of Jupiter or the volcano Stromboli on Earth). However, terrestrial tectonic earthquakes clearly occur under different conditions. Consequently, we are simply forced to assume that the destruction zone D of the focus is protected "on all sides" (Fig. 33) by a zone of liberation of elastic energy. In this latter zone, the stresses immediately after an earthquake fail to reach the strength $\varepsilon_{+0} < \varepsilon_{\max+0}$, thus preventing an earthquake for a long time.

In a particular case we can admit that the boundary B of the focus is the locus of points where $\varepsilon_{\max+0} = \varepsilon_{-0}$. Then at points on the boundary itself we obtain $k_{\max} = 1 - k$. At first sight, this assumption may appear attractive from a physical standpoint, but does not prove to be constructive. Therefore we shall not introduce it.

It remains for us to justify the introduction of a small (and attenuating with increasing distance from the focus) decrease in the strength of the medium in the outer zone — a transition from the curve for $\varepsilon_{\max-0}$ to the curve for $\varepsilon_{\max+0}$ in Fig. 33 in this zone. This is related to the consideration that a strong elastic wave propagating from the source can cause partial destruction and weaking in the surrounding medium, either directly by its shock action, or as a result of a trigger mechanism activating defects previously prepared or in a state of development. Similar phenomena are known in engineering and seismic prospecting. The weakening phenomenon under consideration arises at the "moment" of action of the focus itself. It differs from the weakening wave discussed below, which develops afterwards and much more slowly.

Let us assume that the seismic energy E_c of a large earthquake includes the energy liberated from both source zones: the inner zone D and the outer zone E.

Weakening wave. From observations it is known that after a large earthquake and its immediate aftershocks, the seismicity is usually intensified in an expanding area around the source; "divergence of the earthquakes" is observed [Katok]. The linear rate of propagation of the front of this area may range from several tens of kilometers per day in the case of large catastrophes (the Alaskan earthquake of 1964) to several kilometers per year (centimeters per day) in the case of smaller magnitude phenomena (the Khait, Tadjikistan, earthquake of 1949). It is natural to assume that this occurs mainly as a result of expansion of a low-strength zone which arose initially at the focus. Similar phenomena were observed in other regions. Thus in mining a slowly propagating (from centimeters to tens of meters per day) and rapidly damping (tens of meters) wave transporting a bearing pressure from the mine face into the rock mass and related to a partial loss of strength of the rocks (cleavage and

Fig. 34. Distrengthening wave:
$\bar{\varepsilon}_{max}$ is the strength; r is the distance along the vector
line grad

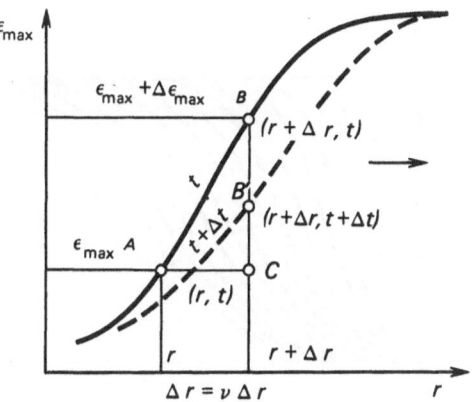

other phenomena) near the zone of total relaxation to the mine working has been observed [Riznichenko]. On the shores of new reservoirs or during the pumping of water into wells, owing to a change in the stresses, a propagating system of faults, which are sometimes even accompanied by a slight seismicity, arises. A similar system of fractures and, generally speaking, weakening is observed in the vicinity of channels and downcuttings in young river valleys. During impacts and explosions, in addition to elastic waves, plastic waves — both longitudinal and transverse are observed. Their propagation velocity decreases to a point considerably below the elastic-wave velocity (according to [Vasiliev], by ten times and more) with increasing distance from the source.

Let us introduce such a weakening wave into our model. This wave will cause a seismicity-amplifying wave in the model, owing to the mechanism formalized above of background seismicity. The weakening wave itself is calculated as follows.

Let it propagate in the direction of the vector line grad ε_{max} means the direction r in Fig. 34. Let us denote its velocity:

$$v = \Delta r / \Delta t,$$

where $\Delta r = \sqrt{(\Delta x)^2 + (\Delta y)^2}$.

If this wave were not damped, then after a time Δt point A in the wavefront would move to point C with the spatial coordinate $r + \Delta r$. Here the strength at the point $r + \Delta r$ would have decreased during this time to $\Delta \varepsilon_{max} = BC$. If, on the other hand, damping occurs, the decrease in strength will amount to a smaller value, BB', so that $B'C = \varkappa BC$, where $0 \leq \varkappa \leq 1$. The parameter $\varkappa$ may be called the damping coefficient of the weakening wave.

Let us find a differential equation for this wave. For this purpose, let us write the expression $B'C = \varkappa BC$ in the following form (see Fig. 34):

$$\varepsilon_{max} (r + \Delta r, t + \Delta t) - \varepsilon_{max} (r, t)$$
$$= |\varkappa[\varepsilon_{max} (r + \Delta r, t + \Delta t) - \varepsilon_{max} (r, t)].$$

Expanding the terms of this equation in series in powers of the small quantities Δr, Δt and limiting ourselves to the first order powers, we obtain:

$$(1 - \varkappa) \frac{\partial \varepsilon_{max}}{\partial r} \Delta r + \frac{\partial \varepsilon_{max}}{\partial t} \Delta t = 0.$$

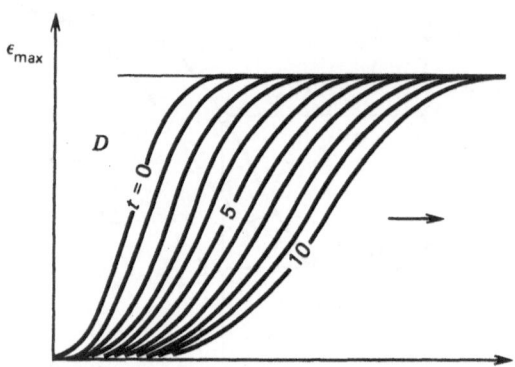

Fig. 35. Displacement of the front wave of weakening with the passage of time t. It is assumed that the wave damping coefficient $x = 0.5$.

D—the destruction zone

Introducing the notation $v = \Delta r / \Delta t$, we finally obtain

$$(1 - x) \frac{\partial \varepsilon_{max}}{\partial r} + \frac{1}{v} \frac{\partial \varepsilon_{max}}{\partial t} = 0, \qquad (4.1)$$

or in another form:

$$(1 - x)|\text{grad } \varepsilon_{max}| + \frac{1}{v} \frac{\partial \varepsilon_{max}}{\partial t} = 0. \qquad (4.1')$$

The general solution of this equation has the form:

$$\varepsilon_{max} = \varphi \left(\frac{1}{1 - x} r - vt \right), \qquad (4.2)$$

where φ is an arbitrary function. It resembles one of the two solutions to the wave equation for a plane wave, with the sole exception that in the present case there is a factor related to the damping x associated with the spatial argument. For $x = $ const the damping only leads to a decrease in the propagation velocity of this wave, without any change in its shape; for $x = 1$ the wave stops.

In deriving Eq. (4.1), it was tacitly assumed that the parameters v (the velocity of the weakening wave) and x (its damping coefficient) do not depend on ε_{max}. However, let us assume that they can be functions of the strength ε_{max}, and also of the stressed-state potential energy ε, or, more exactly, the ratio between them $\varepsilon / \varepsilon_{max}$ (Figs. 36, 37). Then, instead of Eq. (4.1), we would have to write a more complex equation. We shall not do this here. However, the procedure for plotting the weakening wave from initial data (a Cauchy problem) given above with the aid of finite differences, shown in Fig. 34, remains valid in this case too. Thus, a numerical solution of the problem, in terms of finite differences, can be carried out for this more general case too.

When v and x are variable, the weakening wave generally changes shape during propagating. Figure 35 shows where the propagation velocity of large amplitudes with a variation of ε_{max} (the lower part of the strength range) is less than for small amplitudes (the upper part of the well). The plastic wave behaves in this manner. Actually, Fig. 35 was obtained by a graphical construction, according to the scheme in Fig. 34, with a spacing Δr of finite

Fig. 36. Dependence of the velocity V of the weakening wave on the relative energy of the stressed state $\varepsilon/\varepsilon_{max}$

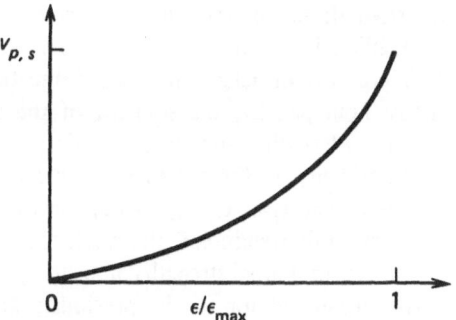

length and $x = 0.5$. In this case the wave configuration varied, owing to the finiteness of Δr. This reminds us once again of the danger of cumulative errors resulting from multiple iterations with finite differences.

It is assumed that the weakening wave is superimposed on the time variation of the strength under conditions of background seismicity. Under appropriate conditions (when the stress energy ε is close to the critical energy ε_{max}) the movement of the weakening wave may be greatly accelerated. In certain cases (where $\varepsilon \to \varepsilon_{max}$) this wave can serve as a trigger mechanism for a new large earthquake (large twin earthquakes are an example). When $\varepsilon \to \varepsilon_{max}$, the amplitude of the weakening wave may increase, and the damping coefficient x may become negative.

The question may arise: is the weakening wave to be regarded as an accompaniment only of large earthquakes or is it an inevitable feature of the background seismic activity in general? Formally, we can adopt either of these suggestions. Physically, however, only the second one is likely.

Indeed, the previously formalized concept of time-dependent energy strength (Figs. 34 and 35) does not explicitly contain the spatial relation between the strengths of neighboring parts of the medium. Nevertheless, such a relation unquestionably exists. This relation involves passing fractures and other defects of the solid monolithic mass in a neighboring area. These defects may be regarded as the phenomenological microstructures of the medium, while

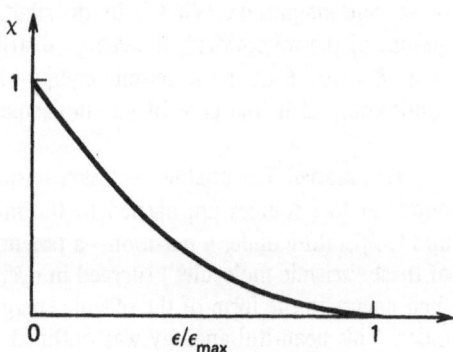

Fig. 37. Dependence of the damping coefficient x of the weakening wave on the relative energy of the stressed state $\varepsilon/\varepsilon_{max}$

the strength may be thought of as their macroscopic form. When the macroscopic state of the medium is homogeneous, there are no appreciable strength gradients in it, and the process of intergrowth of defects in opposite directions does not lead to a disturbance of the macroscopic homogeneity; the strength of the medium decreases equally everywhere, and the presence of spatial strength connections remains hidden. Only in the case of macroscopic inhomogeneity, in the presence of an appreciable strength gradient, do these connections manifest themselves clearly in the form of a weakening wave.

From this standpoint, the weakening wave introduced here constitutes an addition to the process of mutual strength connectivity between neighboring parts of the medium, already partly accounted for in the previously adopted behavior of the time-dependent energy strength, which is related to the presence of a strength gradient.

Approaching from more general positions, we would have to introduce the concept of space-time strength, the fundamental premise of which is the weakening wave, whose parameters depend on the stressed state. Then the time-dependent strength will be regarded as a particular case of space-time strength, namely, the case where the strength gradients in the medium are small. But this is a subject for another discussion.

For the special question of a seismic regime which we are considering, it is of practical importance that the weakening wave significantly affects the process only under conditions of large gradients of ε_{max} and in the presence of small strength stability of the medium (large values of $\varepsilon/\varepsilon_{max}$), i.e., exactly in the focal regions of large earthquakes that have just started. This enables us to regard it in a first approximation as a formal attribute of these earthquakes.

With this the formalization of the continual energy processes of a seismic regime—its space-time variation—is seen to be complete.

4.1.2. Relation to the Discreteness of the Process

The model described above enables us to calculate, under certain conditions, the variation in the liberation of seismic energy in space and in time in the form of two functions: (1) the continuous flux ω_s of seismic energy of the background seismicity, which also includes the activation and divergence of the seismic process after large earthquakes and the rise in seismicity before large earthquakes, and (2) the simultaneous liberation of considerable portions of energy E_c from the foci of large earthquakes, which also includes the energy of their immediate aftershocks. We have regarded both of these forms of energy liberation as continual. In nature, on the other hand, seismic energy is liberated, as is known, in discrete portions of varying magnitude. What is its distribution over these macroquanta? In contrast to the quanta of the microworld, the energy distribution of earthquakes with respect to energetic value $K = \log E$ (E is the seismic energy of an individual earthquake) may be regarded as continuous, as in the case of gas molecules in classical physics.

Gas model. The analogy between seismicity and a gas can be extended. In Fig. 38 the potential foci sources are likened to the molecules of a gas situated at a certain pressure and temperature under a partition—a potential barrier ($\varepsilon_{max} - \varepsilon$). Only the more "energetic" of these "seismic molecules" succeed in overcoming this barrier, bursting out, and liberating their energy in the form of the seismic energy E of individual earthquakes of various magnitudes. This beautiful analogy was outlined to me by E.M. Butovskaya, who tried to apply

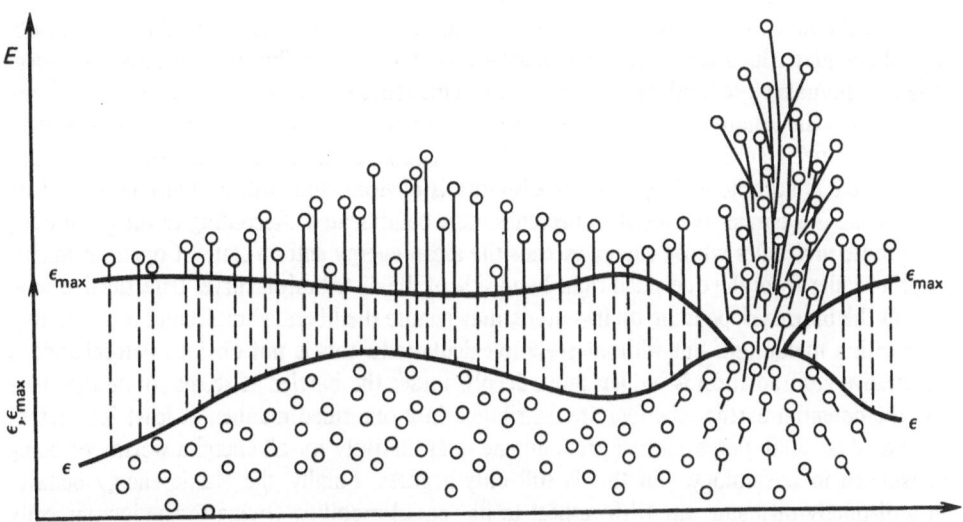

Fig. 38. Gas model of the seismic regime.
Left—background seismicity; *right*—a large earthquake

the theory of fluctuations to a quantitative determination of the long-term mean energy distribution of both potential and "triggered" earthquake foci. The distribution of gas molecules is a particular conclusion of this theory. We shall extend this analogy, however, only for purposes of illustration and to bring out the time-dependent aspect of the phenomena.

The situation shown in Fig. 38, on the left, illustrates the discrete aspect of the seismic process in our continual model in the background-seismicity phase. In this phase, according to the assumption made (Fig. 36), the seismicity ω_s is lower where the potential barrier $(\varepsilon_{max} - \varepsilon)/\varepsilon_{max}$ is thick and higher where it is thin. With the passage of time, according to the law adopted (Fig. 34), the strength ε_{max} at the thin places will decrease more rapidly than at the thick places. On the other hand, the reserve of potential energy ε is constantly replenished as a result of the inflow ω_s from outside. Consequently, the potential barrier at a thin place may wear away completely, and the seismic gas will gush out through the gap, as is shown on the right-hand side of Fig. 38. This will be a large earthquake. Its flow, naturally, also consists of individual molecules possessing various energies E. The first molecule to jump out through the opening may possibly be the most energetic molecule—the main tremor (although it does not necessarily have to be so, it usually does turn out this way with earthquakes). This molecule is followed by other—the aftershocks. The analogy turns out to be actually entertaining.

However, this simple gas model of the seismic process is still not valid quantitatively. It does not contain a mechanism for stopping an avalanche-type liberation of potential energy at a focus: for large dimensions of the gaseous seismic region, although still under conditions of constant influx of seismic gas from outside ω_s, the spurting of the gas, once started, should continue forever. Even more important is the suspicion that the gas molecules are distributed in energy in the same way that the earthquakes are distributed. The characteristic distribution of a gas is one (Maxwellian) with a maximum number of molecules at a certain mean energy. With earthquakes, on the other hand, at least for surface earthquakes, in the range of observed energies which can have seismological meaning the distribution is clearly one-sided: small earthquakes predominate quantitatively. This remains true even for shocks which are weaker than the usually observed seismic events: in the case of rock bursts or during laboratory tests of specimen destruction [Vinogradov]. For large earthquakes with their aftershocks, where there is no potential barrier, the disparity is obvious, but with background seismicity when there is a barrier the problem becomes more complicated. According to the gas model, only those molecules whose energy exceeds the mean energy enter a state of open seismicity, i.e., molecules from the one-sided right-hand side of the distribution curve. True, the deceleration at the barrier, depending on its nature, may restore the liberated molecules to a distribution with a maximum, but without specific calculations, this is not obvious beforehand. A qualitative contradiction is hidden more deeply, under the barrier. It seems to me that here too the potential earthquake foci are distributed in a one-sided manner, at least for surface earthquakes: small potential foci predominate quantitatively for all energies worthy of being considered in seismology. But this is still only a guess. Finally, the elastic energy behaves in a distinctly different way with respect to its spatial mobility. Owing to the low viscosity of the gas, the elastic energy easily flows together with the gas into a region of lower pressure. The elastic energy in a solid medium, on the other hand, is attached, as it were, to certain places, remaining almost stationary. It is here that the roots of the discrepancy for the particular problem of stopping of a gas outflow and an earthquake lie.

Thus, without any special reworking, the gas model of seismicity seems to me to be quantitatively unsuitable, although it is a good illustration of the discrete aspect of the process. It is this molecular aspect of it which remains attractive, and, perhaps, with further development, this idea will yet be used in simulating seismicity.

Here, however, we shall proceed along another, purely phenomenological line, since it seems, iniatilly to be simpler and more constructive. To break down the energy in the above-described continual model we make direct use of those statistical laws of energy distribution of earthquakes which are known from seismological observations.

Considering the energy distribution of the earthquakes. The main factor is the "recurrence law of earthquakes" $N = N(K)$ (see Chapter 2). In the usual elementary, linear approximation this law is written as follows:

$$\begin{aligned}\log N &= \log A - \gamma(K - K_0) \quad &\text{for } K \leqslant K_{max} \\ N &= 0 &\text{for } K > K_{max}.\end{aligned} \qquad (4.3)$$

In assumption (4.3) and for the usually justified condition $\gamma < 1$, the total-energy flow of earthquakes is determined by the expression (Riznichenko):

$$\omega = \int\limits_{-\infty}^{K_{max}} ENdK = \frac{10^{(\gamma K_0)}}{(1 - \gamma) \ln 10} A^{(1 - \gamma)K_{max}}. \qquad (4.4)$$

This establishes a relationship between the continual total energy or energy-flux power of earthquakes ω and the parameters A, γ, and K_{max} of the probability distribution of the earthquakes as discrete events with respect to their energetic value $N = N(K)$, $K = \log E$.

Naturally, instead of an elementary, linear approximation (4.3), we could use a more complex, curvilinear one, if it is justified by observations and sensible calculations.

In the case of background seismicity, we must take $\omega = \omega_0$ in (4.4). Here by N we are to understand the time-dependent density, i.e., the recurrence rate of the earthquakes. In the case of a large earthquake with its aftershocks, on the other hand, which we formally considered a simultaneous event, we are to understand by N in formula (4.4) not the recurrence rate, but simply the number of earthquakes—the main shock and aftershocks—during the entire activity time of the focus, and then substitute E_c, instead of ω, in (4.4).

We shall confine ourselves at this time to these general remarks.

4.1.3. Adjustment to Observations and Prediction of Earthquakes

Adjustment to observations. The above model of a seismic regime includes a number of values not yet been specified (Figs. 30 to 33, 36, 37). These must be made from observation. The results of laboratory and mine investigations and, in particular, field observations of earthquakes proper can be used. In the latter case, it is necessary to use mainly those situations where a particular value occurs in its purest possible form.

The next step should be an attempt to reproduce, in a model, the complex space-time situations resulting from the combined action of all factors, under conditions of both "fictitious" and observed seismicity. For a comparison with nature, we should choose seismic regions that are as large and self-contained as possible, such as Central Asia (possibly dividing it into subregions), the Baikal region, the Caucasus, and the Crimea and its environs.

We propose to use as the initial approximation for the initial field (defined in the model) of potential-energy sources $\omega_0 = \omega_0(x, y)$, vitalizing the entire seismic process, the field of liberation of long-term mean seismic energy ω_s, calculated by a well-known method [Riznichenko] and from the results of seismological instrumental and noninstrumental (historical) observations. However, the usual lack of information means that in order to get reliable judgments of the long-term averages, especially with respect to the total seismic energy, further approximations in the choice of the appropriate focal function are required.

To begin an actual calculation of the space-time variation of a seismic regime with this model, it is necessary to give, in addition to the initial focal field $\omega_0(x, y)$, the initial conditions with for following two space-time functions: the potential energy $\varepsilon = \varepsilon(x, t)$ and the energy strength $\varepsilon_{max} = \varepsilon_{max}(x, y, t)$ controlling the regime. These functions are still difficult to find from observations, even now, although in recent years certain approaches to this problem have been outlined. Let us assume the worst: that these functions are not known beforehand. Then we can propose the following calculation procedure.

Let us begin the calculation of the regime variation for a specific region for a remote "prehistoric" initial moment of time t_i, for which the functions $\varepsilon(x, y, t_i)$ and $\varepsilon_{max}(x, y, t_i)$ are given arbitrarily. Owing to the presence of strong dissipative processes (the energy drift into background and focus seismicity) in the model, we can assume that with time the process in the model will depend less and less on its prehistory, i.e., in particular, on the initial energy situation. After a fairly long time, t_i, it will be determined with ever increasing accuracy, only by a constantly acting factor—the focal function $\omega_0(x, y)$. The beginning of this "guaranteed" period of time in the calculation—the establishment of a "seismic climate" in the model—may be chosen experimentally by a number of trial calculations for various initial conditions. Theoretical estimates of it may also be made.

Next, when calculating the variation of the process in the model for a fairly long period under conditions of the steady-state climate, let us choose the part of the variation which best reflects the seismic regime in nature, i.e., the "seismic weather" for the historical period at our disposal.

Fitting the process in the model to the process in nature in a given region must proceed by a further directed variation in the source function $\omega_0(x, y)$. This may be accomplished using self-adaptive systems. The process of matching the action of the model to nature may be perfected simultaneously for tectonically similar regions by varying the functions controlling the calculation (Figs. 30 to 33, 36, 37): these functions may differ somewhat in different regions. In this way we can solve the physical problem haunting present-day seismologists: the determination of the internal mechanism of the seismic process, at least in its energy aspect.

The main difficulty lies in the inadequacy of differentiated information—we are usually capable of only observing the combined action of many factors. In this connection, the combined effect most accessible to a quantitative study is the phenomenon we are now studying—a seismic regime.

Forecasting. The transition from fitting the action of the model with observed seismicity variations in the past to predicting this variation for the future does not require any explanations: it occurs automatically in the given model, as soon as the time in the model passes through the "present" into the "future".

A clear survey of the results of a calculation of the space-time seismicity variation in the particular region for the past and the future may be made with the aid of motion pictures.

This investigational trend, if successful, may prove to be useful both in forecasting earthquakes. Also perhaps in refining our theories concerning long-term mean seismicity indices (maps of seismic activity, maximum possible earthquakes, etc), and also in improving the calculations (based on these indices) of the indices of the long-term mean shakeability [Riznichenko], the latter should become the geophysical basis of new statistical quantitative methods of seismic zoning for earthquake-resistant construction standards.

We have considered a logical scheme for the construction and action of a continual theoretical energy model of a seismic regime.

We have considered in general outline the relation between the seismic energy calculated in a continual model and the discreteness of the seismic process observed in nature.

We have discussed ways to fit the functions controlling the calculation in the model to seismic observations, and ways of using the model in forecasting earthquakes.

The model may also prove to be useful in refining the calculation of the long-term mean indices of seismicity and seismic hazard, i.e., the seismic shakeability.

The author is clearly aware that the proposed model is schematic. Its main arbitrariness lies in the assumption that the energies of the seismic process can be considered stationary, constant in seismicity and that there is no specific dependence on other processes actually interacting with the seismicity.

4.2. The Development of a Model of the Energetics of the Space-time Course of Seismicity[*]

A model of the energetics of the seismic regime [Riznichenko] is in principle a deterministic system of functions and the connections between them which is intended for an approximate, general description of the course of energy of the seismic process in space and time. Its purpose is to make the extrapolation of this course into the future possible, i.e. to the long-term seismicity forecast.

It was assumed earlier [Riznichenko] for the model of seismic energy that the medium may sometimes locally reach a state of complete disintegration with $\varepsilon = \varepsilon_m$, and the occurrence of a "major earthquake". In the case of a subcritical condition $\varepsilon < \varepsilon_m$ however, there is a continuous "background seismicity". In order to account for the expansion of the region of decreased strength ε_m, especially noticeable after large earthquakes (the expansion of the aftershock area, and afterwards a wider area of increased seismicity, which can lead to the general migration of seismicity), a special mechanism "weakening waves" was introduced.

However the shortcomings of breaking down the value of the strength ε_m into separate time (with the variable parameter $\varepsilon/\varepsilon_m$) and spatial coordinate weakening wave functions was acknowledged above, and the possibility of introducing a space-time energy strength was expressed. This is accomplished here. The energy strength satisfies a diffusion-type equation which includes both a dependence of ε_m on $\varepsilon/\varepsilon_m$ and a space-time dependence governed by a diffusion coefficient D.

In this way we eliminate the need for a separate consideration of two forms of seismicity, viz. background and that of larger earthquakes, and the need for a special mechanism for the behavior of the latter. In the new model, the single unified form of seismicity is considered to be continuous, much as background seismicity was. The need for a special consideration of the weakening waves is also eliminated. This type of wave will become, as was earlier, needed, a natural consequence of the general laws, to which the space-time-dependent energy strength is now subject.

We now proceed to a concrete exposition. We make the assumptions that (a) the behavior of the cracks in the medium can be compared to the behavior of dislocations in solid bodies, and (b) the energy strength decreases with increasing concentration of dislocations, or, in accordance with the first assumption, cracks, and increases with a decrease in the concentration of dislocations. These assumptions imply a limitation on the size of the representative region.

The size of the representative region must be much larger than the length of the maximum crack in the medium, i.e. $L \gg l_{max}$, where L is the size of the representative region and l_{max} is the length of the maximum crack. This stipulation permits us to use statistics.

---

[*] See [Riznichenko]. A. M. Artamonov is a co-author.

4.2.1. The Space-time Dependent Energy Strength

We will return to the concept of energy strength. In technology the concept of strength, σ_m, expresses the ability of the material to sustain mechanical stress $\sigma < \sigma_m$ without signi-ficant destruction, in which case this ability is decreased. Our concept of energy strength too, ε_m expresses the ability of the material to retain a certain reserve of potential energy $\varepsilon < \varepsilon_m$ without its rapid loss into seismic energy ω_0, viz. fracture of the material.

There are many different definitions of strength in technology and physics, taking into account the different aspects of fracture. The simplest is the "limit strength" as a constant characteristic of a given material. However even in technology this concept is only meaningful in the very special case of "instantaneous" loads. The change in the properties of the loaded material with time is significant over long durations, and it is necessary to take into consideration some type of "time-dependent strength". This is also necessary in geology and geophysics, particularly for seismic processes, see [Azizbekov and Ananien].

The time-dependent change in strength of materials under load is associated with the presence and development of defects of different scales: from atomic and molecular (dislocations in crystals) to macroscopic (cracks). The dislocation approach is widely used in seismology (see [Vvedenskaya]); however it has usually been discussed only in a general form. It has been used constructively to a limited extent in the examination of the earthquake focus itself. The strength characteristics of the surrounding medium have not been touched upon; it was assumed they remain perfectly elastic. Here we proceed differently: we assume that the characteristics of the entire medium in which a seismic process or, more accurately, its energy, develops are determined by the presence in the medium of a large number of defects (dislocations, cracks, faults) whose behavior may be compared with that of dislocations in a solid.

The energy strength of a volume L^3 is that density of potential energy which corresponds to the energy which this volume must obtain in order for complete instantaneous disintegration to take place (the manner of disintegration, whether one large crack appears which cuts through this volume, or whether an echelon of fine united cracks appears, is unimportant). Complete disintegration will be understood to be the formation of a discontinuity surface covering an area equal to or greater than that of a cross section of the volume in the original formation. The formation of a surface greater in area than a cross section of the volume must be understood in the sense that the surface need not be flat. We examine the behavior of the energy strength. We emphasize that the energy density which corresponds to the energy stored in a volume L^3, is small in comparison with the energy strength of this volume, i.e., $\varepsilon/\varepsilon_m < 1$. Otherwise, as $\varepsilon/\varepsilon_m \to 1$, cracks comparable in length with L would occur in L^3; this would take us outside the framework of the assumptions made concerning the similarity of cracks and dislocations. Thus, we are dealing with the integral macroscopic effect of the complete destruction of small volumes (the change in the effective value of the energy strength, as we call it, relative to the large volume L^3).

4.2.2. Derivation of Equations Connecting the Change in the Energy Strength with the Potential Energy Density

The behavior of dislocations in a solid body is described by the diffusion equation

$$\frac{\partial \nu}{\partial t} = D_i \frac{\partial^2 \nu}{\partial \xi_i^2},$$

where ν is the concentration of dislocations, D is the diffusion coefficient, ξ is the coordinate, and i is its number, $i = 1, 2$. Consequently, on the strength of the assumptions made concerning the similarity of cracks occurring at the time of an earthquake to dislocations, we may also write the same equation for the concentration of cracks:

$$\frac{\partial \varrho}{\partial t} = D_i \frac{\partial^2 \varrho}{\partial \xi_i^2},$$

where ϱ is the concentration of cracks. It is possible to show that with some simplifications for the energy strength ε_m the equation:

$$\frac{\partial \varepsilon_m}{\partial t} \cong D_i \frac{\partial^2 \varepsilon_m}{\partial \xi_i^2} \tag{4.5}$$

will be satisfied.

It is necessary to formalize the dependence presented graphically in [Artamonov]. For the energy strength we take:

$$\frac{d\varepsilon_m}{dt} = \left[a - b \left(\frac{\varepsilon}{\varepsilon_m} \right)^2 - b_1 \left(\frac{\varepsilon}{\varepsilon_m} \right)^4 - \dots \right] \frac{\varepsilon_m}{c} (c - \varepsilon_m).$$

Further, since $\varepsilon/\varepsilon_m < 1$, we discard terms over a second order. Then we obtain (for Figs. 30 and 31):

$$\frac{d\varepsilon_m}{dt} = \left[a - b \left(\frac{\varepsilon}{\varepsilon_m} \right)^2 \right] \frac{\varepsilon_m}{c} (c - \varepsilon_m). \tag{4.6}$$

Here c is the limit value of the energy strength ε_m. We explain the meaning of the coefficients a and b. If the energy density per unit of volume were equal to zero ($\varepsilon = 0$), then, under conditions of constant ε the energy strength ε_m would increase by a factor e over the time interval $T_1 \approx 1/a$ (e is the base of natural logarithms), while ε_m remains much smaller than its limit value c. If, however, $\varepsilon/\varepsilon_m$ is close to unity, then assuming the ratio $\varepsilon/\varepsilon_m$ to be constant, the energy strength would decrease by a factor e over the time interval $T_2 \approx 1/(b - a)$. In reality, such a decrease would take place over a time less than T_2, since b is the coefficient of the second-degree term in the expansion into a series of the function $\left. \frac{d\varepsilon_m^*}{dt} \right|_{\varepsilon_m^*} \left(\frac{\varepsilon}{\varepsilon_m} \right)$ while we neglect the other terms.

We represent the seismic energy ω_s by the formula (in accordance with Fig. 32)

$$\omega_s = k \left(\frac{\varepsilon}{\varepsilon_m} \right)^\beta, \tag{4.7}$$

where k and β are constants.

In accordance with the above, the rate of change of the energy strength is defined by the equation

$$\frac{\partial \varepsilon_m}{dt} = \left[a - b \left(\frac{\varepsilon}{\varepsilon_m} \right)^2 \right] \frac{\varepsilon_m}{c} (c - \varepsilon_m) + D_i \frac{\partial^2 \varepsilon_m}{\partial \xi_i^2}. \tag{4.8}$$

The first term of the right-hand part of equation (4.8) is an analytical expression of the dependence in Figs. 30 and 31, while the second term of the right-hand part is obtained from (4.5).

To construct an equation giving the change in ε, the density of the energy stored in the medium, we may use following considerations. Let us suppose that in the volume dV, where $dV \sim L^3$, the energy density changed by $d\varepsilon$. This change could have come about due to an influx of ω_0 (energy of the foci) and due to the liberation of energy in the form of ω_s (the seismic energy flux) i.e.,

$$d\varepsilon \cdot rdV = \omega_0 \, dVdt - \omega_s \, dVdt.$$

Dividing both parts of the equation by $dV \cdot rdt$ and taking (4.7) into consideration, we obtain:

$$\frac{\partial \varepsilon}{dt} = \omega_0 - k\varepsilon \left(\frac{\varepsilon}{\varepsilon_m} \right)^\beta. \tag{4.9}$$

Thus, we have a system of differential equations between the change in ε (energy density) and ε_m (energy strength):

$$\frac{\partial \varepsilon}{\partial t} = \omega_0 - k\varepsilon \left(\frac{\varepsilon}{\varepsilon_m} \right)^\beta$$

$$\frac{\partial \varepsilon_m}{\partial t} = \left[a - b \left(\frac{\varepsilon}{\varepsilon_m} \right)^2 \right] \frac{\varepsilon_m}{c} (c - \varepsilon_m) + D_i \frac{\partial^2 \varepsilon_m}{\partial \xi_i^2}. \tag{4.10}$$

Here ω_0, D, β, a, b, c and k are parameters of the medium.

We make several assumptions to simplify the solution of the system. We will examine a one-dimensional problem, i.e., we suppose that $i = 1$. We will let ξ denote the space coordinate. Then ε and ε_m will be functions of two independent variables, the coordinate ξ and the time t:

$$\varepsilon = \varepsilon(\xi, \, t), \quad \varepsilon_m = \varepsilon_m(\xi, \, t).$$

Earthquakes at a seismic suture are an analog of the one-dimensional problem. Further, let the power of the foci be constant in time and along the coordinate, i.e., $\omega_0 = $ const. We assume $\beta = 2$ in Eq. (4.7) for ω_s. We make the following substitution of variables:

$$x = \frac{\varepsilon}{c}, \quad y = \frac{\varepsilon_m}{c}, \quad y = y, \quad p = \frac{x}{y}.$$

Then system (4.10) will take the form:

$$y \frac{\partial p}{\partial t} + p \frac{\partial y}{\partial t} = \frac{\omega_0}{c} - kyp^3$$

$$\frac{\partial y}{\partial t} = [a - bp^2]y(1 - y) + D \frac{\partial^2 y}{\partial \xi_i^2}. \tag{4.11}$$

4.2.3. Solution of the System of Equations and an Analysis of the Solution

It is possible to show that the solution of (4.11) may be sought in the form:

$$p = \sqrt{\frac{a}{b}} + v(t) \sin \frac{2\pi}{\lambda} \xi$$

$$y = \frac{\varepsilon_m}{c} + u(t) \sin \frac{2\pi}{\lambda} \xi, \qquad (4.12)$$

where $v(t)$ and $u(t)$ are periodic functions.

We will use isoclinic lines to find the form of the functions $v(t)$ and $u(t)$. It is possible to show that the integral curve will have the form shown in Fig. 39 in the u, v coordinate. If, at the beginning the system is located at the point with the coordinates $u = 0$, $v = 0$, then any fluctuation will take the system out of unstable equilibrium and in the course of time the system will move away from this point, so that as $t \to \infty$, the integral curve will approach a cycle limit from within. If, however, the system is removed from the equilibrium position by some sort of shock, so that it falls outside the limit of the cycle but sufficiently close to it (we consider small deviations from the equilibrium position), then the integral curve will approach the limit cycle from the outside. Thus, in any case, as $t \to \infty$, the integral curve hits the limit cycle and no longer deviates from it. After a certain time T the integral curve arrives again at its starting point. In other words, as $t \to \infty$, the functions $v(t)$ and $u(t)$ become periodic with period T.

If we construct the functions $u(t)$ and $v(t)$ in the coordinate systems (u, t) and (v, t), then curves are obtained which are described satisfactorily by

$$v = A \cos (\omega t + \varphi)$$

$$u = B \cos \omega t, \qquad (4.13)$$

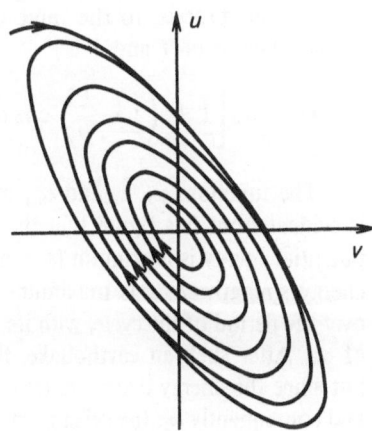

Fig. 39. The integral curves of the system of equations for $u(t)$ and $v(t)$

where $\varphi \approx \pi - \pi/50$, i.e. it is possible to consider $\varphi = \pi$, and thus:

$$v \approx -A \cos \omega t$$

$$u \approx B \cos \omega t,$$

where

$$A = \left(1 - \frac{k}{b}\right)\sqrt{\frac{a}{b}}; \quad B = \bar{\varepsilon}_m \frac{k}{b}.$$

From this

$$\varepsilon_m \cong \bar{\varepsilon}_m \left(1 + \frac{k}{b} \cos \omega t \sin \frac{2\pi}{\lambda} \xi\right)$$

(4.14)

$$\varepsilon \approx \bar{\varepsilon}_m \sqrt{\frac{a}{b}} \left[1 - \left(1 - \frac{k}{b}\right) \cos \omega t \sin \frac{2\pi}{\beta} \xi\right],$$

Here $\bar{\varepsilon}_m$ is the energy strength of the region under consideration averaged over time and along the coordinate. The angular frequency ω and the wavelength λ of the seismic process are determined from a linear approximation of (4.11)

$$\omega^2 \approx \frac{4\omega_0 \sqrt{ab}}{\varepsilon_m}$$

(4.15)

$$\lambda^2 \approx \frac{4\pi^2 D}{a}.$$

(4.16)

It is evident from (4.15) that the frequency ω of the change in the energy strength ω_m and the potential energy ε (4.14) will be inversely proportional to the average strength $\bar{\varepsilon}_m$ of the medium and directly proportional to ω_0, the power of the foci. An obvious conclusion is that from (4.14) the greater $\bar{\varepsilon}_m$ (the average stength of the medium), the higher the level of ε (the density of the potential energy stored in the medium capable of reproducing seismicity):

$$\varepsilon_0 = \varepsilon_m \sqrt{ab},$$

(4.17)

where ε_0 is the potential energy density ε averaged over time and along the coordinate analogous to $\bar{\varepsilon}_m$.

We now proceed to the value of most interest to us, viz. ω_s, the seismic energy flux. Its dependence on t and ξ is:

$$\omega_s \approx \omega_0 \left[1 + \left(1 - \frac{k}{b}\right) \cos \omega t \sin \frac{2\pi}{\lambda} \xi\right]^3 \left(1 - \frac{k}{b} \cos \bar{\omega} t \sin \frac{2\pi}{\lambda} \xi\right).$$

(4.18)

The functions ω_s, ε, and ε_m, of time t for a fixed coordinate ξ is shown in Fig. 40. It is evident from this figure that the seismic energy flux ω_s is at its maximum not when the potential energy is maximum ($\varepsilon = $ max): but rather when the potential energy relative to the energy strength $\varepsilon/\varepsilon_m$, is maximum and the maximum earthquake (K_{max}) at given location over the period of the cycle, with its foreshocks and aftershocks, correspond to the maximum of ω_s. After such an earthquake, the energy strength continues to decline for some time, but since the energy decreases even more rapidly during this time, the ratio $\varepsilon/\varepsilon_m$ decreases, and consequently ω_s, the seismic energy flux, also decreases. The ratio between the maximum

Fig. 40. The dependence of the functions ω_s
and ε_m on time t with the fixed coordinate ξ

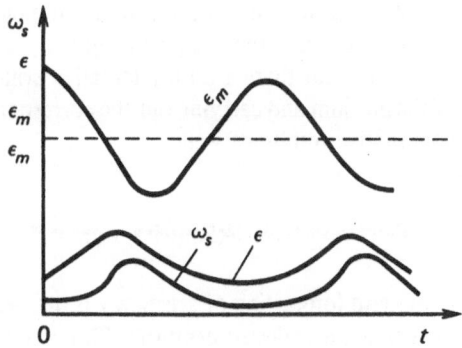

seismic energy flux and the minimum over the period T is approximately:

$$\frac{\omega_{s\,\max}}{\omega_{s\,\min}} = \frac{(1 - k/b)[1 + (1 - k/b)]^3}{(1 + k/b)[1 - (1 + k/b)]^3}. \tag{4.19}$$

Thus, this ratio increases with an increase in $\bar{\varepsilon}_m$, the average strength of the medium, and in the parameter a, and decreases with an increase in the parameter b. We remember that a and b are parameters of the time-dependent variation of the energy stength (4.6).

It is possible to determine ω_0, the power of the foci in the solution, if we make use of the following consideration. We arbitrarily consider only the seismogenic energy of the sources, i.e. that part of the tectonic energy which is finally completely transformed into seismic energy. It is possible to regard (4.10) as transforming ω_0 (the constant power of the sources) into ω_s (a signal of alternating power) without loss, with an efficiency equal to unity. Then we may write:

$$\omega_0 = \frac{\int\limits_0^Z \int\limits_0^T \omega_s \, d\xi \, dt}{TZ} \approx \frac{\sum\limits_j E_j}{TZ}.$$

Here $\sum\limits_j E_j$ is the sum of the energies of all earthquakes, whose foci are located in the sector under consideration over the time T (the period of the process); Z is the length of the sector.

Now we can draw certain conclusions. Our model explains the existence of a "seismic cycle", i.e., a periodicity in the occurrence of large-scale earthquakes in the space-time region. The time T, over which the cycle takes place (the period of the cycle) is proportional to the mean square root of the energy strength $\bar{\varepsilon}_m$. The larger $\bar{\varepsilon}_m$ is, the longer the accumulation time of ε, the potential energy in the medium, preceding the arrival of the maximum seismic energy flux, including the maximum possible earthquake. Here it follows from (4.19) that the maximum of ω_s will be larger for a large strength $\bar{\varepsilon}_m$ than for a small strength $\bar{\varepsilon}_m$, i.e., the earthquakes will be larger in the first case than in the second. Furthermore, the energy strength ε_m in the fixed moment of time is not identical along the entire length of the seismic suture ξ, but varies from point to point according to a sine rule at the first approximation. The rule according to which the energy strength ε_m varies over the time t for a fixed coordinate ξ also has a sinusoidal character within the limits of the assumptions made. We may also say the same for the variation in ε, the potential energy density; moreover, the phase shift over time between the energy strength ε_m, and the potential energy is somewhat less than π.

At present these are only tentative evaluations, but, having established the parameters of the medium and carrying out the corresponding calculations, it would be possible to obtain quantitative results as well.

4.3. On Long-term Seismicity Forecasting[*]

Long-term forecasting of seismicity is defined as an estimate of its expected variation during the coming decades or centuries. To do this, we use data concerning its variation during a similar or greater period in the past. Also, other geological-geophysical and geodetic data concerning the structure and properties of the medium and on the movements of the medium including neotectonics and contemporary movements, of which seismicity is a particular manifestation can be used.

Below we present results from a continuation of a study [Riznichenko], where for the first time we formulated the main ideas of an energy model of the space-time seismicity — a theoretical deterministic macroscopic continuous model, which was subsequently developed in [Artamonova, Riznichenko]. It differs from a number of other models which were proposed later [Kuznetsova, R. Burridge, Knopoff] and which have a basically stochastic and discontinuous character.

4.3.1. Model of Energetics

The main quantity studied and predicted is an energy parameter the "seismic weather" [Riznichenko]. It is the space-time density ω_s of the total seismic energy, which depends on the space and time coordinates x_i and t ($i = 1, 2, 3$), i.e., the spatial density of the power of the set of earthquakes, the foci of which lie in each elementary space-time volume with its center at any point, at any time (x_i, t). The dimensions of this elementary volume in space are no less than the volume of the focus of the maximum possible earthquake with energetic value K_{max} at a particular spot. The dimensions in time constitute the same, of the order 10%, of the average recurrence period at a given point x_i, to large earthquakes close to the maximum possible, let us say, with an energy value of $K_{max} = 1$ or greater. In practice, the dimensions of the "characteristic" volume are of the order of ten or a hundred kilometers in space and of the order of ten years in time. The expected degree of detail of the results is roughly the same. At present this approach is directed not so much toward the long-term prediction of individual large earthquakes (although this is not excluded) as toward an analysis and forecast of the overall variation of a seismic process which includes a large number of earthquakes of different magnitudes.

Three energy density functions, which control the behavior of the seismic weather ω_s, are introduced: a function of the energy sources of the long-term average seismicity of "seismic climate" ω_0; the potential seismicity energy ε, which characterizes the degree of stress in the medium; and the energy strength of ε_m. Let us consider them in turn.

---

[*] See [Riznichenko].

Seismic climate sources ω_0. It is the long-term average spatial density of the seismic energy power, which, in the final analysis, is converted into the seismic energy of a set of earthquake foci. In principle, to determine the seismic climate, we could start from a tectonic energy power ω_T slowly varying in time and assume that $\omega_s = \omega_T k_{T\text{-s}}$, where $K_{T\text{-s}}$ is a coefficient of the "seismic efficiency" of the tectonics and is similar to the efficiency of a machine. Judging from observations on samples [Vinogradov], in mines [Vinogradov] and observing the seismic flow of rock masses under field conditions [Riznichenko], this coefficient is usually only a fraction of a percent. But, since ω_T and $k_{T\text{-s}}$, have not yet been studied much and are still inaccessible to mapping, it is more constructive at present to proceed directly from ω_0. It may be determined [Riznichenko] from known parameters of long-term average seismicity: the slope of the earthquake recurrence graph $\gamma = -d \log N/dK$, the seismic activity A and the energetic value of the maximum possible earthquake K_{max}, which are calculated and mapped for the entire USSR on the basis of seismological and complex seismological-geological-geophysical and other data.

Next it is assumed that in a study of the seismic weather $\omega_s(x_i, t)$, the seismic climate remains stable, i.e., the quantity ω_0 depends only on the spatial coordinates x_i and not on time t. Under this condition, the seismic weather function $\omega_s(x_i, t)$ and the seismic climate function $\omega_0(x_i)$ over a prolonged period of time are related as:

$$\omega_0 = \lim \frac{1}{\Delta t} \int_{t}^{t+\Delta t} \omega_s \, dt. \tag{4.20}$$

The climate function ω_0 can also serve as an integral function to control the calculation of the weather function ω_s.

The degree of medium stress—the spatial density ε of the potential seismicity energy. This is a measure of weather and depends greatly on time $\varepsilon = \varepsilon(x_i, t)$. It represents that part of the potential energy which is stored in the medium as a result of an inflow of energy from the sources (ω_0) and will be expended by earthquakes and be removed from the process in the form of seismic wave energy (ω_s). In this sense this is the seismogenic part of the potential energy of the tectonic process.

Basically, ε is the potential elastic energy of the medium. Other forms of energy—gravitational, thermal, chemical—as a rule only play an indirect and secondary role in the generation of earthquakes. The elastic energy of triaxial compression is not included in ε (if we assume that the principal focal mechanism of the earthquakes is a shear type fault). Not included in ε is the large part of the tectonic energy expended on slow movements of the medium: continuous, plastic-flow type, leading to folding, and discontinuous creep type, which often alternate with seismic movements, but do not produce them.

The energetic functions of the seismic weather (ω_s), the seismic climate (ω_0) and the degree of stressing of the medium (ε) figure in the main energy balance of the model:

$$\frac{\partial \varepsilon}{\partial t} = \omega_0 - \omega_s,$$

or in integral form:

$$\Delta \varepsilon = \int_{t}^{t_2} (\omega_0 - \omega_s) \, dt.$$

Here $\Delta\varepsilon = \Delta\varepsilon(x_i, t_1, t_2)$ is the increase in the potential seismicity energy density ε at any point x_i of the medium during a time interval t_1, t_2.

The energetic strength ε_m. This is the maximum value of the potential energy ε in a characteristic element of the medium, which if reached, would cause complete failure of the element. Partial failure, however, can occur even when $\varepsilon < \varepsilon_m$. By failure we understand here the rapid avalanche-type destruction, which seismicity is related to.

The strength ε_m, like the stress state ε, is a measure of weather, $\varepsilon_m = \varepsilon_m (x_i, t)$. It is assumed that in the absence of a noticeable strength gradient, ε_m varies in time at a rate which depends on the degree of stress in the medium, i.e. on ε. At large values of $\varepsilon/\varepsilon_m \to 1$, the value of ε_m decreases, and of the medium loses strength, while at small values of $\varepsilon/\varepsilon_m \to 0$ it increases, approaching a certain limit as $\varepsilon_m \to \varepsilon_{mm}$, and the medium strengthens. The dependence of strength on time under stress is known from experiments on samples [Zhurkov] and is generally accepted in physics and modern engineering. The strengthening postulated in our model at small loads, at the limit of zero ($\varepsilon/\varepsilon_m \to 0$), considers the phenomenon of rock consolidation in the Earth's crust under the action of temperature and pressure, which is well-known from geology. Also known is the phenomenon of material transport, for example, "healing" of cracks during the precipitation of cementing minerals from hydrothermal solutions. This "relaxation strengthening" differs from the strengthening in engineering and physics at high stresses, close to destructive stresses, which precedes the sharp decrease in strength during the actual destruction. In this model the phenomenon of strengthening at high stresses is not considered. However, in principle, this can be done and then we could simulate the phenomenon of a "lull" before large earthquakes, well-known from field seismological observations and noted earlier during physical modeling of earthquakes on samples and during rock-bursts in mines [Vinogradov, Riznichenko].

In the presence of a large strength gradient $(\partial\varepsilon_m/\partial x_i)$, especially near strength "wells" caused by triggered sources of large earthquakes, a decrease in strength diffuses, as it were, into the surrounding medium—a weakening wave is formed which can cause the migration of seismicity. In the first variant of our model two forms of seismicity were considered arbitrarily: a "background" form, which comprises small earthquakes, and a "large earthquake". In this case the weakening wave was considered only in relation to the second form, where it shows itself most strongly. In the second variant both forms of seismicity are merged into one, and the phenomenon of a weakening wave is formalized by the introduction of a generalized concept of a space-time dependent energetic strength ε_m, governed by a diffusion type or heat-conduction equation (4.10).

The first term on the right-hand side of this equation realizes one of the possible formalizations of the temporal behavior of the strength as a function of the stressed state of the medium. The second and remaining terms, ensure space-time relations of the strength, including the weakening wave. Here a, b, p and ε_{mm} are constant parameters of the temporal behavior of the strength, while D is a parameter of its diffusion in space.

Let us return again to the main seismicity function—the density ω_s of the seismic energy power liberated by earthquakes. It is assumed that this quantity is all the greater, the greater the stress in the medium (ε) and the closer the stress is to the failure threshold (ε_m). Let us write these conditions as

$$\omega_s = k\varepsilon\left(\frac{\varepsilon}{\varepsilon_m}\right)^n, \qquad\qquad\qquad\qquad (4.21)$$

where k and n are constants. Their values, like the parameters in (4.10), must be determined by comparing theoretical results with observations. This can be done in various ways. Preferably using some procedure for optimizing the solution, for example, in the sense of the minimum sum of the squares of the deviations between the ω_s calculated from the model and the observed seismicity power obtained for a fairly large number of elements of a fairly large space and time region.

4.3.2. Results of Analytic Study

The model was analyzed for a spatially one-dimensional case simulating the seismicity behavior of an individual "seismic suture", i.e., an isolated, linearly extended seismic zone. The calculation was carried out under the simplest assumption of a homogeneous distribution of the seismic climate sources $\omega_0 = \text{const}$ on a segment of finite length x.

A solution to the system of (4.20) and (4.21) is obtained by the isocline method. It is shown that after a sufficiently large time has passed from the initial moment $t_0 = 0$, the solution becomes independent of the initial conditions $\varepsilon(x, t_0)$ and $\varepsilon_m(x, t_0)$.

This model yields an explanation of the Fedotov "seismic cycle" [Kogan etal.], the tendency toward periodicity of large earthquakes (which in this model are related to maxima of the seismicity function ω_s). This periodicity occurs not only in time but also in the space-and-time domain. Thus it follows from (4.15) that the cycle period T is all the greater, the greater the average strength of the medium $(\bar{\varepsilon}_m)$, and the smaller the power of the seismic climate sources (ω_0). Also it follows from (4.16) that the seismicity wavelength λ is all the greater, the greater the strength diffusion coefficient D.

Although (4.14) to (4.16) gives standing waves characterizing the steady variation of the self-oscillations in seismicity (ω_s) on the seismic suture of finite length x, we can also estimate the possible rate of seismicity migration, if we take it to be the quantity $v = \lambda/T$. According to (4.15) and (4.16) this velocity should be all the greater, the greater ω_0 and D are, the smaller $\bar{\varepsilon}_m$ is.

Obviously, these results do not challenge the intuitions of seismologists.

4.3.3. Comparison with Observation

To compare the results with observation, Artamonov used catalogs of earthquakes for four linearly extended seismic zones: on the coast of the Adriatic Sea in the Balkans, the Kurile-Kamchatka zone, the southern Tien-Shan zone, and the Arctic zone from the North Pole region to the Verkhoyansk ridge (see Table 12).

Each zone was divided into a number of sections of equal length Δx and on each of these we calculated the observed seismic energy power density for a number of identical time intervals Δt.

$$\omega_{s\ \text{obs}} = \frac{1}{\Delta x \Delta t} \sum_j E_j, \tag{4.22}$$

where the energy E_j of each earthquake was calculated from its magnitude M_j according to the Gutenberg-Rautian formula, $\log E = 4 + 1.8\,M$, for E in joules. The numerical values

Table 12. Parameters of Seismic Zones

Zone	λ, km	T, years	v, km/year	$\bar{\tau}$, bar	$\bar{\tau}_m$, bar	$\bar{\tau}/\bar{\tau}_m$, %	$\hat{\tau}/\bar{\tau}_m$, %
Balkan	600	30	20	3	35	9	30
Kuril-Kamchatka	1100	90	12	20	150	13	52
Southern Tien-Shan	1100	110	10	10	93	11	36
Arctic	900	60	15	2.5	34	7	22

Note. λ—wavelength of seismicity; T—period of oscillations; $v = \lambda/T$—migration velocity; τ—mean effective stress in a zone; $\hat{\tau}$—maximum stress in a zone; τ_m—mean effective strength

of some of the quantities obtained from a comparison of the model with observations in these regions are presented in the table.

In this table λ is the seismicity wavelength, T is the seismicity self-oscillation period, $v = \lambda/T$ is the seismicity migration velocity, $\bar{\tau}$ is the average effective stress in the zone, and $\bar{\tau}_m$ is the average effective strength. The stress and strength characteristics are given in the table in stress units (bars) in accordance with the dependences:

$$\varepsilon = \frac{\tau^2}{2\mu}, \qquad \varepsilon_m = \frac{\tau_m^2}{2\mu}, \tag{4.23}$$

where ε is the elastic potential energy density during pure shear strain, ω is the shear elasticity modulus, and ε_m are τ_m and the corresponding strengths: energetic and in shear stress.

The last two columns of the table contain the relative values (in %) of the average $\bar{\tau}$ and maximum $\hat{\tau}$ stresses in the zone in comparison with the average strength $\bar{\tau}_m$ in it. The Kurile-Kamchatka zone was found to be in a stage closest to the "strength limit".

The average shear strengths $\bar{\tau}_m$ obtained with the aid of the given model were found in all zones to be of the same order of magnitude as those obtained in the usual manner from completely different considerations, on the basis of seismic wave spectra and the stress drop $\Delta\delta$ in earthquake sources.

The present energetic model of the space-time variation in seismicity is, unquestionably, extremely schematic. But the reasonableness and physical sense of the results obtained on the basis of it enable us to assume that it may be regarded as a promising empirical scheme for long-term prediction of seismicity. Also it may have a use as a formalized constructive system of concepts of a physical and geological-geophysical nature, which fairly correctly depicts the essential features of reality.

5. The Mass Determination of the Coordinates of Local Earthquakes and of the Velocitites of Seismic Waves in the Focal Areas[*]

A classic problem in instrumental seismology is to establish the focal coordinates of local earthquakes and to determine the velocity V of seismic waves in a given medium. Every seismic study in an area begins with these purely geometrical and very simple but very important problems.

The literature gives numerous methods for solving problems with known V = const with unknown V = const with V-var [Savarensky and others] and new ones still emerge. While we do not wish here to review or discuss the various methods, it should be pointed out that the most effective techniques are those which (a) allow us to process simultaneously all observed data of the same type, including excess data, without arrangement first into arbitrary groups; (b) allow us to exploit data to their limit of accuracy; (c) permit an easy evaluation of the compatibility of all data so as to recognize and discard those containing obvious error; (d) give a clear evaluation of the consistency and accuracy of the results; (e) are simple and not labour intensive. This latter point is of great importance in processing large quantities of information.

Present methods do not fulfill all these requirements. Graphical methods usually do not comply with (a) to (d). Analytical methods based on difficult calculations of the least-squares method are impractical and unworkable; therefore, this does not comply with (e).

This paper describes some methods which fulfill all the requirements and which, at the same time, do not need computers. The methods are based on the comparison of observations with theoretical graphs which can be worked out conveniently as master curves.

Every relevant geophysical characteristic is given in these master curves: the known parameters pertinent to the location of the focus (hypocenter) and the composition of the medium (velocity); the geophysical effect, or, in this case, the arrival times, surface or linear travel time curves. The preliminary steps, i.e., preparing the graphs, are rather tedious. The reverse processes, however, like determining the parameters of the location of a focus and the velocities of the medium, are easy and fast with the master curves. This does not require any computations whatever. The difference in the effort required for preliminary calculations and ultimate application of this system satisfies the requirements for processing large quantities of data.

We describe the following methods based on this system: (1) isochrone master curves of the surface travel-time curves for determining focal coordinates; (2) master curves of the theoretical linear travel time curves for determining both the focal depths and the seismic velocities in a medium; (3) the vertical travel time curve method, which allows us to combine a series of individual focal depth determinations with velocity data in the surrounding medium, in order to obtain an average velocity at depth, or, in other words, to find the relationship between average velocity $\overline{V}$ or velocity in specific formation (V_b) and depth in particular area.

---

[*] See [Riznichenko].

The first and second techniques resemble the methods described by N. A. Vvedenskaya; they differ, however, in their use of a computation system and in providing a more complete solution to the problem.

All three methods are based on the assumptions and procedures developed by us for seismic prospecting [Riznichenko]. Their application in seismology was necessitated by the mass of observational data on earthquakes. This was done in connection with the start of detailed studies (under the supervision of G. A. Gamburtsev) of the seismic regime.

5.1. Determination of Focal Coordinates by Isochrone

Outline of the method. We assume that the propagation velocities V_P and V_S of the seismic P (longitudinal) and S (transverse) waves are known, and that they are functions of the depth. Basically, P and S in this case are direct or penetrating refracted waves travelling from the source toward the seismic station. The station is located near the epicenter, whereas the focus may be located in any given layer of the continental crust or even beneath the crust. We are here discussing this type of P and S waves because they are most distinct, as a rule, in local earthquakes and consistently stand out on seismograms. Other types of waves, e.g., refracted or reflected head waves may be used with this method if their systematic observation is feasible. The technique of isochrone master curves may be applied to local shocks and also to distant earthquakes (by means of a globe).

Let us assume that the coordinates of the focus (hypocenter) are x_0, y_0, and $z_0 = H$, and that the stations are located on the Earth's surface $z = 0$ at different azimuths α_i and different distances $r = r_i$, from the epicenter; $i = 1, 2, 3, \ldots, n$; n = number of stations.

In this case the usual travel time curves for P and S waves can be computed and plotted for any value of H (focal depth); $t_P = t_P(r)$, $t_S = t_S(r)$, as well as travel time curves for imaginary (S-P) waves: $t_F = t_S - t_P = t_F(r)$, the propagation velocity of which can be determined from the relation

$$V_F = \frac{V_P, V_S}{V_P - V_S}. \tag{5.1}$$

In the simplest case (constant velocities) all travel time curves are hyperbolas, as shown in Fig. 41.

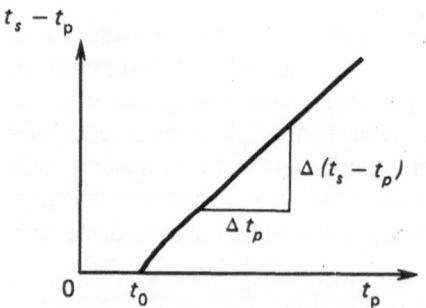

Fig. 41. Linear travel time curve $t = t(r)$ of the seismic waves.

For a local earthquake the focal depth is H (*upper half*) and the surface view (*lower half*). O— Epicenter; *circles* are isochrones of the surface travel time curves shown on the master curves; A, B, C, $\ldots$—seismic stations; aa'—their average location line

Fig. 42. Wadati curve for determining the time t_0 in the focus.

Slope of curve $\tan \varphi = \Delta(t_S - T_P)/\Delta t_P = V_P/V_F$; velocity relation $V_P/V_S = \tan \varphi + 1$

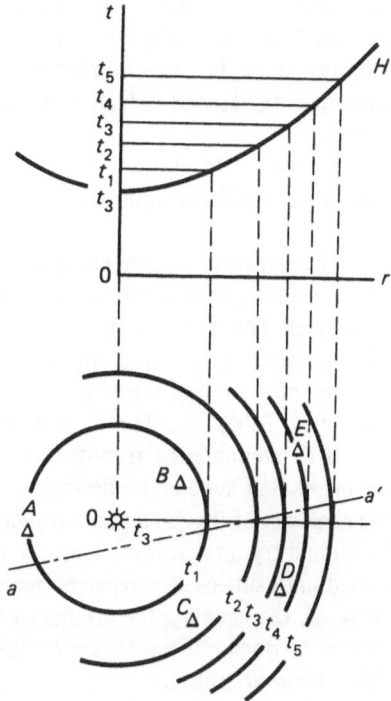

It is not difficult to plot travel time curves related to the surface $z = 0$ (on which the stations are located) from the corresponding linear travel time curves (the special character of the location of a station—surface relief; low velocity zone—should be considered and corresponding residuals could be added to the observed arrival time). The isochrones of the surface travel time curves are then a group of concentric circles the center of which coincides with the epicenter of the earthquake. H will be the parameter of the isochrone group for each wave type (P, S, or S-P). The "theoretical" surface travel time curves, based on the isochrones of the various wave types, are our "isochrone master curves" from which the position of the focus of a recorded earthquake can be determined.

We discuss below primarily isochrone master curves of imaginary S-P waves, thus eliminating the time element t_0 at the hypocenter, a factor which cannot always be determined with sufficient accuracy. It is more expedient, as a general rule, not to use unnecessarily any values requiring additional definitions. We may use the graphs separately for P and S waves if we remember that t_0 can be determined in advance by any known method ([Riznichenko] and others), e.g., by the Wadati method which shows the relationship of $(t_S - t_P)$ to t_S, or of $(t_S - T_P)$ to t_P (see Fig. 42) graphically.

It is basically possible to determine the coordinates of a hypocenter without the use of S waves and to base all the calculation on the first arrivals of the P wave. This is, however, unnecessary because good records show in most cases a very distinct S phase, the arrival time of which can be determined. Besides, the exclusive use of P waves (or for that matter, S waves) would entail a decrease in consistency of the results, and, therefore, less accuracy in practical application. And finally, whereas using both P and S waves requires data from at least 3 stations, using P waves or S waves alone would require data from four or more stations. With the same number of stations ($n \geqslant 4$), the first version provides, thus, more data than the second, which is important for obtaining additional control and greater accuracy.

We prefer, therefore, the simultaneous use of P and S wave master curves in processing data about local earthquakes.

The isochrone master curve technique boils down to this. The locations A, B, C, ..., of the stations, and the travel times t_i of the imaginary S-P waves for each station are put on a map. The map has then a transparent isochrone master curve for a certain value of the parameter H. The map and graph have the same scale. By moving the isochrone graph over the map we find the position where the theoretical and recorded (t_i) travel times agree best at A, B, C, ... The time is then determined from the isochrones by interpolating.

If we compare the recorded and the theoretical times for a first arbitrary value of H, we might find systematic discrepancies due to an incorrectly chosen H. After determining the character of the discrepancy (shallow foci give "deeper" travel time curves, and vice versa) we should try an isochrone master curve with a different value for H. This should be continued until sufficient agreement between recorded and theoretical travel times (t_i) has been achieved, i.e., equaling the accuracy of the recordings at every station A, B, C, ... (for this reason the isochrone master curves should be selected carefully so as to correspond with the observation accuracy).

This done, we transfer the center point of the isochrone master curve onto the map. This marks the location of the epicenter of the earthquake. The focal depth can be found by reading off the value for the parameter H of this master curve. The above method enables us to determine the coordinates of a hypocenter in a constant-velocity medium.

The consistency and accuracy of the results should be evaluated during the data processing. This means that distinct limit values for the determined parameters exist within which the differences between recorded and theoretical times do not exceed the admissible limits. A change of results may ensue also through variations in the observed values.

The isochrone master curve technique is based on the velocity depth relationship in 1955, and was used later by the Tadzhik Complex Seismological Expedition in processing earthquake records. During this practical application T. G. Rautian developed additional ways of using the technique by extending it to cases with a laterally heterogeneous medium. These improvements increase noticeably the applicability and accuracy of the isochrone master curve technique under certain circumstances and deserve a detailed discussing which is, however, beyond the scope of this section.

5.2. Theoretical Travel Time Master Curves for the Simultaneous Determination of the Focal Depth of an Earthquake and of the Average Velocity of Seismic Wave Propagation

Outline of the method. We assume that the velocities $V = V_p$, V_s, or V_F of the P, S, or imaginary S-P waves in the medium are unknown, but are constant for a given focal depth H and a given series A, B, C, ..., of stations ($V = \text{const}$). All seismic rays will then be

Fig. 43. Travel time curves (*above*): *dashed lines*—theoretical travel time curves; *A, B, C*,...—points of the recorded travel time curve; cross section (*below*): *N*—hypocenter; *O*—epicenter; *M*—arbitrary observation point

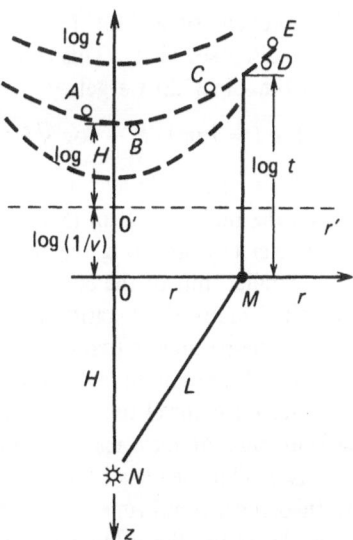

straight lines, and the common linear travel time curves of all waves in the *r, t* system will be hyperbolas (see Fig. 43):

$$t = L/V = 1/V\sqrt{H^2 + r^2},\tag{5.2}$$

L being the distance from the focus *N* to any given point *M* on the surveyed surface ($z = 0$); *r*—distance of *M* from the epicenter; *t*—travel time of the various waves along *L*.

We assume that the location of the epicenter *O* has been predetermined by means of some other method, e.g., with isochrone master curves based on approximate velocities in the medium.

Basically it is possible to determine all three focal coordinates simultaneously from the velocity *V* when processing the data of each earthquake individually. This is sometimes done with other known methods but leads to rather inconsistent results. It is more convenient and more precise to determine the coordinates separately. Finding the epicenter coordinates of an earthquake usually does not present any difficulties—even with incorrect (but fixed) values for velocity and depth, and gives fairly accurate results. If the velocity is a function of depth only, and if the density of control points (stations) is sufficiently high, the determination of the epicenter can be based on the circular symmetry of the surface travel time curves. Besides, a predetermined epicenter location simplifies the problem and results in more consistent and accurate values for *H* (depth) and *V* (velocity)[*].

In the *r, t* coordinates system where for both *r* and *t* linear scales are used, the shape of the travel time curve (5.2) depends on two parameters (*V* and *H*). We cannot compile,

---

* It is not necessary, when using the above method, to predetermine the location of the epicenter if all used stations are located more of less on a straight line. In that case the values for *V* and *H* will be generally related to the inclined plane which passes through this straight line and the focus. If the epicenter also is located on the same straight line, this plane will be vertical.

therefore, one single "theoretical" graph which would contain all the varieties of travel time curves discussed above. This is, however, possible in a semilogarithmic plot $(r, \log t)$. Taking logarithms in (5.2) we get

$$\log t = \log 1/V + \log \sqrt{H^2 + r^2}. \tag{5.3}$$

From (5.3) it follows that in the r, $\log t$ plot the shape of the theoretical travel time curve, the radical term in (5.3) depends on the parameter (H) only. A change of the other parameter (V), affecting the first expression on the right side of the Eq. (5.3) only results in a parallel shift of the entire set of theoretical travel time curves (expressed by the radical function) along the logarithmic axis of the ordinate of the graph (see Fig. 43).

A one-parameter (parameter H) set of "theoretical travel time curves", as expressed by the radical function in (5.2), form the set of master curves (Fig. 44). These can be compared with recorded travel time curves during the simultaneous determination of V and H from seismograms of local earthquakes.

A similar idea has been used for a long time in seismic prospecting, where it is applied to theoretical travel time curves of reflected waves [Riznichenko]. The main objective there is to determine the velocities and the depths of the reflecting horizons, and, at the same time, the dips of these horizons. In seismic prospecting, point N, Fig. 43, is the mirror image of the shot point (located near the Earth's surface) on a flat reflecting surface. The distance $ON = H$ is twice the depth of the reflecting horizon, if the latter is horizontal. Seismic prospecting knows over a dozen different ways to solve this problem. A paper [Riznichenko] compares the advantages of the theoretical travel time curve technique with other methods. These advantages become obvious under conditions similar to those mentioned in the introduction of this section. The seismological master curves (Fig. 44) differ from master curves used in seismic prospecting [Riznichenko] and others by its wider range of r/H, by a different meaning and different values (twice the depth) for H, and by other, less important, features.

The master curves, Fig. 44, were prepared for the processing of recording of earthquakes located within the continental crust ($H = 1 \div 50$ kilometers), at an epicentral distance up to $r = 100$ kilometers. By changing the scale (see for example [Riznichenko]), however, it may be applied to greater depths and epicentral distances (the grid in Fig. 44, is diagrammatic).

The practical application in seismology of the theoretical travel time curve technique is, in short, as follows. A straight line (like aa', Fig. 41, lower half) showing the average distribution of stations A, B, C, ..., is drawn (or only imagined) on the map with the locations of the stations. Points A, B, C, ..., and the epicenter O are then projected onto this line. This is done to find out whether it would be more feasible to show the measured values of the epicentral distances (r) and the times (t) of the points A, B, C, ..., to the right or left of O on the recorded travel time curve, Fig. 43. In the r, $\log t$ plot, using the same scale as the master curves for the theoretical travel times, the observed travel time curve is then plotted from points A, B, C, ... For this, the semilogarithmic plot in Fig. 44 is used.

The master curves for theoretical travel time curves are then used as an overlay on the graph of the observed travel time curve (one of the plots should be transparent). Moving the overlay back and forth, while keeping the corresponding axes parallel, we find the same theoretical curve which is best fitting the average distribution of A, B, C, ... on the master curve for the recorded travel time curve.

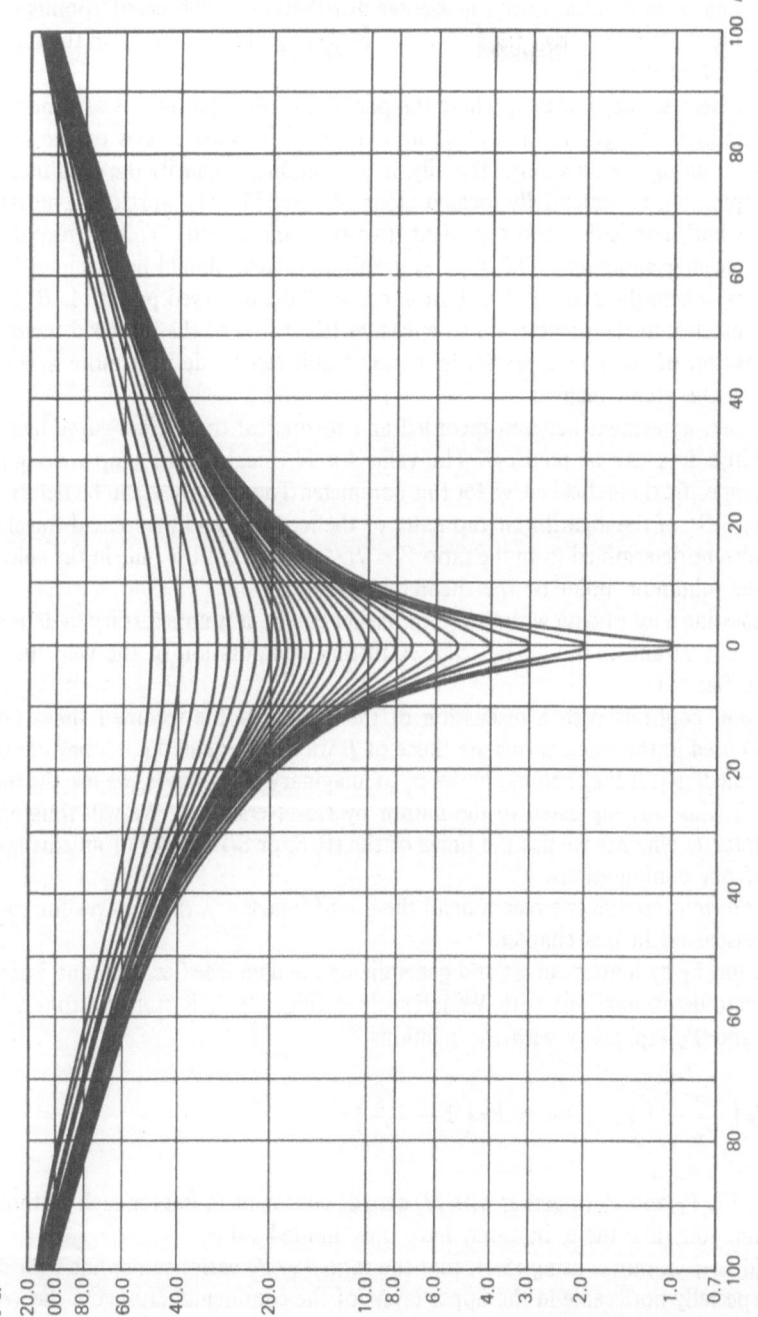

Fig. 44. Master curves for theoretical travel time curves of the direct waves of local earthquakes used to determine the local depth H and the propagation velocity V of waves. H is read on either side at the minimum point of the travel time curve

The consistency and accuracy of the results is checked as in the previous cases (see 5.1). It should be noted that "inside" errors only can be discovered in this way. "Outside" errors can be found from discrepancies between independent solutions and locally recorded data. In practice this can be recognized from the average distribution of "observed" points on the vertical travel time curve (see below, 5.3), the construction of which is based on the recorded data of a series of earthquakes.

With the present set-up, however, where the position of the epicenter is assumed to be accurately determined, the graphs should be moved along the vertical axis of the overlay, i.e. free movement along one axis only. Actually, if the stations, primarily the ones most distant from the epicenter, are essentially located along aa' (see Fig. 41), horizontal movement of the graphs is both admissible and expedient (two-stage movement). This corresponds to a shift of the epicenter along aa'. This type of graph movement should be executed (under above conditions) where the average distribution curve of the observed points A, B, C, ... (Fig. 42) is asymmetric to the predetermined epicenter (the r axis of the recorded travel time curve). The position of the epicenter can be checked and can be defined more accurately if we proceed in the above manner.

When the best agreement between recorded and theoretical travel time curve has been found, the desired data can be read off. The value for H (focal depth) simply designates, on the plot, the specific theoretical curve for this parameter. The velocity V can be determined by the intercept OO' of the logarithmic ordinates of the recorded and theoretical travel time curves. It can also be determined from the ratio $V = H/t_e$, where t_e is the time in the epicenter indicated by the minimum point of the theoretical curve.

While processing a lot of data with the set of master curves, it is unnecessary to determine V every time. Only H and t_e should be read. A further computation of the velocities will be discussed in Sec. 5.3.

We shall now continue with a discussion of the results. If the recorded times (A, B, C, ..., Fig. 43) used in the calculations are those of P waves, V equals V_P; if they are those of S waves, V equals V_S; if they belong, however, to imaginary (S-P) waves, we use the imaginary velocity V_F (this was suggested to the author by I. L. Nersessov). We will thus obtain the same value for H whether we use the times of the (P, S, or S-P) waves of an earthquake as the basis of our computations.

As in the previous section, we recommend the use of imaginary, (S-P) waves for solving the problems discussed in this chapter.

Having found V_F by master curves and generalizing the data (see Sec. 5.3), and knowing V_P/V_S from mass determinations with Wadati's curve (Fig. 42), it is simple (from 5.2) to determine V_P and V_S separately with the relations

$$V_P = V_F\left(\frac{V_P}{V_S} - 1\right), \quad V_S = V_F\left(1 - \frac{V_S}{V_P}\right). \tag{5.4}$$

If desired, V_F, V_P and V_S (together with H) can, of course, be found for each earthquake. Single determinations like these, however, have only limited value.

We should keep in mind, using (5.4), that the ratio V_P/V_S varies somewhat with depth (H). This is especially noticeable in the upper layers of the continental crust. On the Wadati graph, Fig. 42, where $\log \varphi = (V_P/V_S) - 1$, this appears as a bending of the early portion of the mostly straight line which expresses the relationship of $(t_S - t_P)$ to t_P.

5.3. The Velocity of Seismic Waves as a Function of Depth Determined from Vertical Travel Time Curves Based on Recordings of a Series of Earthquakes

Outline of the method. Let us assume that we have processed a series of earthquakes with various focal depths (H_i). Also, we have determined these depths as well as the travel times t_i (equal to t_e) for the P (resp. S or S-P) waves in the epicenter, by using master curves for the theoretical travel times (Sec. 5.2). We will then have a series of pairs of "observed" values (H_i, t_i), $i = 1, 2, 3, \ldots, k$, where k = total number of processed earthquakes.

We plot a graph $t = t(z)$ using the points $t = t_i$, $z = H_i$. This is a "vertical travel time curve" (see Fig. 45, lower right).

According to the kinematic principle of reciprocity, the travel time curve remains the same if the focus (hypocenter of an earthquake) is located at O on the Earth's surface, and the seismic stations were beneath the surface along a straight vertical axis (z) at various depths (H). The slope of the straight line OM from the zero point of the coordinates to an arbitrary point on the curve (dashed lines, Fig. 45) gives the average velocity $\overline{V} = (z/t)$ of the respective waves along their path from the hypocenter to the epicenter, i.e. from the depth z on the vertical axis to the Earth's surface $z = 0$. The slope dz/dt of the tangent to $t(z)$ at M gives the true velocity of the waves at depth z on the vertical axis. We shall call the average true velocity, limited to a certain depth interval in the vicinity of M, the layer velocity for that depth z.

Figure 45, left half, depicts the curves $V(z)$ and $\overline{V}(z)$ which show the layer V and average $\overline{V}$ velocities as functions of depth z. They correspond to the vertical travel time curve $t(z)$. The upper right quadrant of Fig. 45 shows the related $\overline{V}(t)$ curve. It is probably unnecessary to plot here a $\overline{V}(t)$ curve too. Vertical travel time curves are widely used in seismic prospecting. For a description of this method see [Riznichenko] and (more detailed) [I. Berzon]. Practical experience with this method in seismic prospecting suggests that the recorded data should be used to draw first a vertical travel time curve $t(z)$, but not any of the other curves shown in Fig. 45, even though the curves are closely related. Therefore any one of them could serve

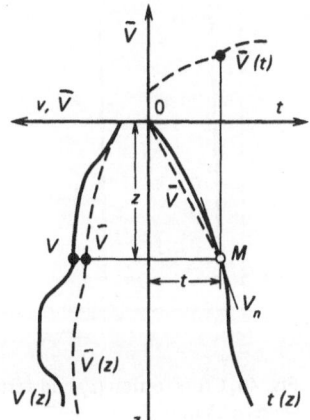

Fig. 45. Vertival travel time curve $t(z)$ (*lower right*) and the corresponding curves for average $\overline{V}$ and layer V velocities as related to depth z (*lower left*), and the travel time $t = t_e$ in the epicenter (*upper right*)

as an initial curve. We shall not explain here the reasons for this important practical application, except that it is justified and adapted to seismological conditions.

The practical application of the vertical travel time curve technique in processing many earthquake records proceeds as follows. For reasons explained above, it is more feasible to use the travel times $t_F = t_S - t_P$ of imaginary S-P waves for the construction of the curve $t(z)$. The axis of the vertical travel time curve is scaled in t_F values. To convert to P and S wave data more easily it is sufficient to scale the t axis in t_P and t_S values, which means that this axis will have a triple scale. The t_P, t_S and t_F ratios are determined from the Wadati graph (Fig. 42).

The plotting of all points (H_i, t_i) recorded in the area of interest on the same vertical travel time curve graph will give clusters of points arranged in a certain manner along a line of average distribution $t(z)$. Figure 46 gives an example of a vertical travel time curve compiled according to this principle by I. L. Nersessov from recordings taken in 1955-56 by the Tadzhik Complex Seismological Expedition in a highly-seismic region of the Peter-The First Range in northwestern Pamirs.

Due to the limited number and scattering of the point clusters, some liberty may be used in drawing the average distribution line $t(z)$. The line may be shown curving gradually, indicating a gradual velocity (V) change with depth (z); or, as in Fig. 46, breaking and thus agreeing with a velocity step change with depth. The average distribution line is, above all,

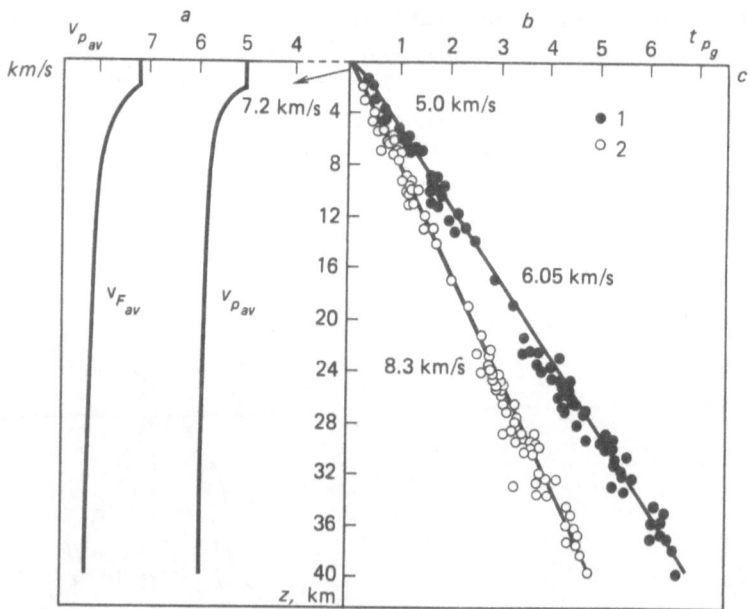

Fig. 46. Cross section (*a*) and vertical travel time curve (*b*) recorded in the Peter-the-First Range in northwestern Pamir:

1—for P waves; *2*—for S-P waves

determined by its best position relative to the recorded data. This alone would be insufficient, however, in individual cases with a single solution. Here then other factors should be considered such as the similarity of data in adjacent and geophysically identical regions; deep seismic probing, also basic geophysical concepts about the variability of the continental crust and subcrustal zones. The interpretation of recorded vertical travel time curves should add detail to our present general ideas on this subject, and this is the main reason for the seismological application of the method described here.

Below follows a more detailed discussion of the specific methical aspects of interpretating vertical travel time curve without using additional data and outside concepts.

If the average distribution curve $t(z)$ is drawn as a broken line, i.e., if we assume that the velocity change at the boundary of two layers is sudden, it would be pointless to subdivide the line into numerous segments, and, accordingly, the medium into many layers. We should restrict this to the minimum number of layers deducible from the general position and average distribution of the point (H_i, t_i) clusters. Based on the point clusters in Fig. 46, only two different segments of the straight lines of the vertical travel time curve are discernible as such, and consequently only two layers with substantially different velocities can be present within the depth interval covered by these recordings.

If the point distribution over all or part of the graph $t(z)$ is averaged by a gradual curve, i.e. if we assume that the velocity V changes gradually with depth z, the relation $V = V(z)$ can be replaced by some empirical equation like:

$$V_0 = V_0 f(\beta, z),$$ (5.5)

where the parameters V_0 and β are constant; f is a specific function given in general form. If, for instance, f is a linear function, (5.5) takes the form $V = V_0(1 + \beta z)$.

The following method may be used in order to obtain more adequate values for V_0 and β. From the basic relation $V = V(z)$, a general equation for the vertical travel time curve $t = t(z)$, may be found from

$$t = \int_0^z \frac{dz}{V(z)}.$$ (5.6)

A set of theoretical travel time curves is then compiled for various values of V_0 and β. If (5.5) is used for $V = V(z)$, the set of vertical travel time curves can be constructed with a coordinate system $(z, \log t)$ that will eliminate the effect of V_0 on the shape of the curves of the vertical travel time. The values for both V_0 and β can be determined from the recorded data (H_i, t_i) by a method identical to the one given in Sec. 5.2. A paper [Vinogradov] gives an example of a similar process applied to problems in seismic prospecting.

It would be possible to select the parameters of the empirical relation $V(z)$ by a different method viz. a gradually curving average distribution line $t(z)$ is drawn freehand through the cluster of travel time points, then this line is graphically differentiated in order to define the curve $V(z)$ and to determine the empirical equation for it. Solving the problem in this manner is difficult due to the arbitrarines in drawing a curve for scattered points, and because of the low accuracy of graphical differentiation.

We discuss finally the following theoretically important consideration. Each H_i, t_i point of the vertical travel time curve was determined, in Sec. 5.2, by the assumption that the layer $0 \leqslant z \leqslant H_i$ was homogeneous, and that the velocity $V_i = H_i/t_i$ in it was constant. It might be necessary, however, during the interpretation of similar point aggregates on the vertical travel time curve, to subdivide the layers into zones of different velocities. The question arises here, as to how much the "effective" velocity V_e, determined with theoretical travel time master curve for constant velocity conditions, may differ from the true average velocity V in a vertically heterogeneous and, possibly, horizontally layered medium $V = V(z)$ with continuous or stepped layer velocity changes as a function of depth z.

This question was carefully analyzed in seismic prospecting. A paper [Riznichenko] discusses the case of a layer with vertical velocity step changes, evaluates some possible errors $(V_e - \bar{V})/\bar{V}$, and comes up with suggestions for correcting them. Errors of the same type, occurring in cases with gradually changing velocities $V(z)$, are discussed in another paper [Berzon]. The equations given in [Riznichenko] can be applied directly to similar computations in seismology. We shall only mention that in most field cases with horizontally bedded media, errors of this type are small, usually not exceeding the error in velocity determinations from scattered points on vertical travel time curves constructed as given above.

It is anticipated that the techniques will be applied to processing local earthquake records. Such recordings are being made systematically by the regional network of seismic stations operating continuously in many parts of the Soviet Union.

This section also presents an average velocity determination method based on the simultaneous processing of recordings of a series of earthquakes with foci at different depths. The vertical travel time curve technique provides more accuracy and detail than previous methods for the determination of layer velocities at the depth ranges of earthquake. The application of this method, in a number of different areas, will provide new and more accurate data concerning the composition and properties of the continental crust and, in some areas, of subcrustal zones. These, in turn, will give us a broader foundation for discussions of these significant geophysical problems.

It appears reasonable to use the resultant layer velocity data from the crust. The data will help us to define more accurately the results of the deep seismic sounding (DSS) of the continental crust by refraction shooting according to the G. A. Gamburtsev method (it is very difficult to arrive at accurate determinations of the layer velocities with the present layer method).

The author is deeply grateful to I. L. Nersessov for friendly criticism of all the basic problems discussed in this section.

6. Seismic Magnitudes of the Underground Nuclear Explosions*

The main diffuculty in detecting underground nuclear explosions by seismic techniques is the necessity to distinguish between the seismic effect of the explosion and that of the natural earthquake. The seismic effect of the explosion and of the earthquake can be conventionally expressed by magnitude. Obviously, the lower the magnitude of the nuclear explosion, the greater number of false earthquakes could be recorded and, therefore it is more difficult to recognize such a nuclear explosion.

To estimate the efficiency of the International System of Control we need to know the total annual number of world earthquakes corresponding to or, to be exact, of greater magnitude than the underground nuclear explosion of the given power.

To solve the problem of detecting underground nuclear explosions we used the materials of seismic observations of the explosion series carried out at the test area in Nevada (USA) in 1957-1958.

6.1. Magnitudes of Explosions

At present, we have several scales of magnitudes for earthquakes. They are compiled from the values of amplitudes A or A/T ratios, where T is the period. These scales differ in the types of oscillations used, areas of application and, generally speaking, in the numerical results obtained. A certain correlation has been established between different scales of magnitudes**.

In practice, three scales of magnitudes are widely used.

1. **Local magnitudes M_L.** The estimation of such magnitudes is performed in a "local" zone that starts in the vicinity of the epicenter and continues for the epicentral distances of $x = 1000$-1100 km, up to the known "shadow zone". To calculate M_L we use maximum amplitudes of oscillations registered by standard Wood-Anderson torsion seismometers. These amplitudes correspond to "shear" — transverse or surface — waves which are not distinguished in the local zone near the epicenter.

2. **Magnitudes M or M_S,** as they are written sometimes, are determined by *surface* waves. These waves can be traced independently in the local zone not very close to the epicenter and can be then observed in the "shadow zone" (zone of diffraction) and in the subsequent "distant" zone. They allowed Gutenberg to use the scale M to correlate the two other scales:

---

* See [Riznichenko].

** The process of the introducing new scales of magnitudes in seismological practice and establishing correlations between them is still going on — (*Editor's note*).

magnitude determined in the local zone up to the shadow zone and teleseismic magnitude m, determined in the distant zone, beyond the shadow zone.

3. Teleseismic magnitudes m. They are determined in the distant zone which, according to Gutenberg, starts from $x = 1700$ km. For explosions the "distant" zone is taken (as we can see below) from about $x = 2500$ km from the epicenter. The start of the area for which the teleseismic scale exists can be related to the continuation of the diffraction zone which starts at the shadow zone ($x = 1000$-1100 km), including the latter. Further on it covers the zone of the focused seismic signal that continues approximately for $x = 2500$ km. To calculate teleseismic magnitudes m one uses amplitudes and periods of body—transverse and longitudinal—waves. Gutenberg, using this scale, has elaborated a *unified* scale of magnitudes. So, that teleseismic magnitudes mentioned below, coincide with the unified ones.

The unified scale of magnitudes m seems to be most theoretically sound and connected with the major physical values characterizing the seismic effect of the focus, namely with its seismic energy E. In addition, it is extremely important for the present problem that of the correlation between explosions and natural earthquakes—that the statistics used—Gutenberg's statistics of the total annual number of earthquakes worldwide (determining their magnitudes) were based on the unified magnitudes m.

Certainly, it is expedient to determine not only the conventional value—magnitude, but to determine directly a clearer physical value, i.e., seismic energy of the focus related to the sphere of a given radius surrounding the focus. Also, it is important to consider the frequency spectrum of the focus. It is of special concern when we compare seismic effects of earthquakes and explosions, the latter having higher frequency sources of oscillations as a rule. The author considers that the energy-spectral trend of investigation is more progressive. Since this work is aimed at international use we would preserve here a more widely used expression of earthquake value by magnitude. Moreover, one can easily transfer it into the seismic energy of the focus and then correlate it with the average frequency spectrum of the earthquake.

When the usual technique in determining earthquake magnitudes was first applied to underground explosions a number of peculiar phenomena arose which seemed to be paradoxical. They were mainly thought to result from a lack of proper consideration, when calculating magnitudes, to different effects in frequency spectra of the seismic wave sources. In addition the effects associated with different attenuation of seismic waves with distance from the épicenter to the zone of diffraction. One cannot ignore the role of the different distribution of the complete seismic energy of the explosion or earthquake in the waves of various types.

Initial data on magnitudes of underground nuclear explosions. The first underground nuclear explosion—Reinier—was carried out in September 1957 at the Nevada test area, USA; its TNT equivalent was 1.7 kT. It was a part of the Plumb bob operation. The depth of the explosion was around 300 m, in country rock of layered volcanic tuff. The magnitude M_L of the Reinier explosion was estimated by American specialists as equal to 4.25 ± 0.4 [Bailey, Romney]. This value was taken as a basis for calculation performed by experts in Geneva in summer of 1958. The first conference of experts resulted in mutual agreement in all main problems; the recommendations of the conference were approved by governments of the USSR, USA, Great Britain and other participating countries.

A series of five nuclear explosions, a part of Hardtack II operation was executed shortly after this, in October 1958, at the same test area in Nevada [C. Romney]. This series contained two larger underground explosions, namely Blanca with TNT equivalent of 19 kT (the first estimation was 23 kT) and Logan with TNT equivalent of 5 kT. The depth of these explosions was the same as in the Reinier test, around 300 m. The remaining three explosions were markedly smaller, less than 0.1 kT each.

Within all the mentioned underground nuclear explosions only three fall into the main power interval from 1 to 20 kT. This interval was discussed by experts in Geneva to establish the possibility of control and control measures. One of the small explosions of the Hardtack II series named Tamalpais (0.072 kT, depth of about 100 m) could still be used for comparison with explosions in the main power range. Two other explosions were useless for this. The Evans explosion (0.055 kT, depth of about 300 m) had no essential difference in power from Tamalpais, the causes of this remain unexplained by the Americans themselves [Romney], produced a very low seismic effect. Its oscillation amplitudes were less than 0.1 compared with Tamalpais. This event was called by them the "Evans Mistery". The Neptun explosion (0.090 kT) executed at the depth of 30 m only produced a large outburst into the atmosphere, so that this explosion cannot be, considered as an underground one.

Explosions of the main power range should be able to be treated, in a seismological sense, as well camouflaged, i.e., exactly *underground* ones. One of them, however, the strongest, the Blanca event was accompanied by some atmospheric pollution: for more than a dozen seconds, after the moment of explosion, an escape of gas and dust was observed over the epicenter. It was supposed however that this was produced by a rock collapse into the cavity formed in the partial vacuum after the explosion. The main severe seismic shock was connected with the very moment of the explosion, and radiated seismic waves that were registered distantly. The value of the shock did not seem much reduced as a result of the incomplete camouflage of the explosion.

The summary of data on measured amplitudes for these explosions was presented by American specialists to the Second Expert Conference held December 14, 1959. We reproduce in Tables 13 and 14 the data from this summary concerning the explosions in Reinier, Logan and Blanca. The data on M_L values (Table 13) and m values (Table 14) form the bas is of all our following calculations related to these explosions.[*]

Local magnitudes M_L (Table 13) were calculated [Romney] with the use of maximum "shear" wave amplitudes in the first local zone, within the interval of epicentral distances from 180.7 to 583 km. The records of torsion Wood-Anderson seismometers were used; precisely for those seismometers the initial scale and data processing for magnitude determination was elaborated by Richter. The same processing was used now in determination of M_L [Romney].

Using these values of M_L we calculated the corresponding values of m according to the unified Gutenberg scale. The transformation from one scale to the other was done according to known formula [B. Gutenberg, C. Richter]

$$m = 1.7 + 0.8M_L - 0.01M_L^2. \tag{6.1}$$

---

[*] When compiling Tables 13 and 14 we omitted only one figure (Dec. 14, summary), namely the particular value of local magnitude of Blanca explosion obtained by St. Nicolas station, 540.0 km from the epicenter. This station was the only which has not reported data on magnitudes for two other explosions. This impedes the effective use of this value in the calculation system. Considering the omitted value we would change the average value of local magnitude for the Blanca explosion for −0.03 only, but this has no practical meaning — (*Author's note*).

Table 13. Local M_L and Unified m Magnitudes of Blanca, Logan and Reinier Explosions as Taken from the Observations in the 1st Zone

Serial, No.	Station	Distance, km	Magnitude					
			Blanca, 19 kT		Logan, 5 kT		Reinier, 1.7 kT	
			M_L	m	M_L	m	M_L	m
1	Tinemaha	180.7	5.1	5.5	4.8	5.3	4.2	4.9
2	Woody	289.1	4.2	4.9	3.9	4.7	3.6*	4.4
3	Riverside	370.8	4.7	5.2	4.4	5.0	3.8*	4.6
4	Pasadena	382.2	4.8	5.3	4.4	5.0	4.0	4.7
5	Mt. Hamilton	482.8	5.4	5.7	5.0	5.4	4.7	5.2
6	Barret	502.8	3.9	4.7	3.7	4.5	3.3*	4.2
7	Palo Alto	530.4	5.2	5.6	4.8	5.3	4.4	5.0
8	Berkeley	540.5	4.9	5.4	4.5	5.1	4.1	4.8
9	San Francisco	556.6	4.9	5.4	4.5	5.1	4.2	4.9
10	Mineral	583.0	4.8	5.3	4.4	5.0	4.3	5.0
11	Average magnitude		4.79	5.30	4.44	5.04	4.06	4.77
	σ_{M_L}, σ_m		±0.45	±0.31	±0.40	±0.28	±0.41	±0.30
	$\sigma_{\bar{M}_L}, \sigma_m$		±0.14	±0.10	±0.13	±0.09	±0.13	±0.10

* This value was not used in the initial determination of magnitude for Reinier explosion, 1958.

The transformations of M_L into m was used to express all magnitudes obtained in the local zone (Table 13) and in the distant zone (Table 14) by the unified system of units. Also it gave us the possibility of comparing the explosion data with Gutenberg's earthquake statistics [Gutenberg].

Teleseismic magnitudes m presented in Table 14 were calculated in USA from P-wave — amplitudes recorded by Benioff seismometers at the stations (Zone 2, Nos 1 to 8, Zone 3, Nos 1 to 4); calculations were completed according to rules and tables from [Gutenberg]. Three final magnitudes were obtained at very distant Soviet stations (Zone 3, Nos 5 to 7). Here the amplitudes and periods of P-waves recorded by CVK-M seismometers were used.

All magnitudes m in Table 14 correspond completely to the unified Gutenberg scale and do not need further recalculation. The same units of m were used in his statistic.

Discussion of magnitude data. The initial magnitude of Reinier, the first underground nuclear explosion was calculated by D. Carder and W. Cloud from the seismic energy determined using near-field observations. The result was $M_L = 4.6$-4.7. D. Carder and W. Cloud used the square dependence of $\log E$ on M in the form

$$\log E = 9.4 + 2.14M_L - 0.054M_L^2$$

which is valid for the local magnitude [Gutenberg et al]. In the book [Richter] the similar formula is reproduced with slightly different coefficients

$$\log E = 9.9 + 1.9M_L - 0.024M_L^2$$

Table 14. Teleseismic (Unified) Magnitudes m of Blanca and Logan Explosions in the 2nd and 3rd Zones

Serial, No.	Station	Distance, km	Magnitude	
			Blanca, 19 kT	Logan 5 kT
		Zone 2		
1	Temporal	1706.7	4.5	
2	Ditto	1803.7		3.8
3	Ditto	1842.0	4.7	
4	Ditto	1902.1		4.1
5	Ditto	2011.2	4.1	
6	Ditto	2111.3	4.5	4.0
7	Ditto	2208.8	4.7	
8	Ditto	2305.0		4.9
	Average		4.50	4.20
	magnitude		$\sigma_m \pm 0.24$	± 0.48
			$\sigma_{\bar{m}} \pm 0.11$	± 0.24
		Zone 3		
1	Temporal	2506.0		4.5
2	Ditto	2665.3	5.1	
3	Ditto	3017.4	5.0	
4	Ditto	4020.6	5.1	4.9
5	Tixie	6890	5.2	4.9
6	Temporal	8320	5.1	
7	Ditto	10080	5.2	
	Average		5.12	4.76
	magnitude		$\sigma_m \pm 0.08$	± 0.23
			$\sigma_{\bar{m}} \pm 0.03$	± 0.13

which are in better agreement with the formula (6.1). This, however, does not change the essence of the matter. It is important that D. Carder and W. Cloud were true in the discrimination between local magnitude M_L and unified magnitude m, which, as it is well known [Gutenberg], has linear relation to $\log E$:

$$\log E = 5.8 + 2.4m$$

The authors then decided that, considering the high absorption of seismic waves at short distances, from 2 km, the magnitude $M \approx 4$ would be more suitable for comparison with earthquakes. We shall not consider the reasons for the assumption concerning high absorption.

Let us turn now to the data on the Reinier magnitude announced at the First Expert Conference in Geneva, 1958. The data on local magnitude in the near-field zone only including readings from seven of ten stations mentioned in Table 13 were presented for this event, at that time a unique underground atomic explosion. The readings from stations Riverside, Woody and Barret (marked with an asterisk) were not available. The reason explained by the American delegation later on, during the Second Conference in December 1959, was because of "extremely low accuracy caused by small amplitudes" of Reinier records from these stations. It is interesting that all three additional stations give lower magnitudes (less than

4) in comparison with other seven stations (4 and more). Magnitudes from Barret and Woody are expecially small.

The average magnitude calculated omitting data from three stations for Reinier explosion was $M_L = 4.27$. Approximately this value, namely $M_L = 4.25 \pm 0.4$, was officially presented by the American delegation in 1958. It was taken as a basis for calculation at the First Expert Conference, while the difference between M_L and m seems not to have been considered.

The first official document[*] on "New Seismic Data" did not contain the figures marked by an asterisk in Table 13. It was, however, said that the magnitude determinations for Blanca and Logan explosions with the use of all ten stations have demonstrated that "observations of Reinier explosion were carried out by only those stations which recorded anomalously high magnitudes" and that, in accordance with such conclusions it is necessary to reconsider the Reinier magnitude and to reduce it to the value of $M_L = 4.1 \pm 0.4$.

Similar magnitudes were published by C. Romney (namely either 4.0 or 4.05 or 4.1) and presented in his principal report at the second expert conference. This quantity is largely in agreement with the following slightly lesser average value obtained later on from all American data (Table 13), including the three stations with lower magnitudes:

$$M_L = 4.06 + 0.13 \tag{6.2}$$

Here the value ± 0.13 is the standard (mean square) deviation of average value of M_L; in two previous estimates ± 0.4 is the standard deviation of an individual determination.

We show later that instead of "anomalously high" magnitudes at seven stations it would be better to mention the "anomalously low" magnitudes from three additional stations, especially from Barret and Woody. In the meantime we shall assume that the value (6.2) for the Reinier magnitude is correct and will use it in further estimations.

Now we turn to discussion of the magnitudes of the Blanca and Logan explosions. Observations of these explosions were carried on not only in the 1st local zone (as for Reinier explosion), but also in the 2nd zone. This is the zone of diffraction, covering distances from 1100 km to 2500 km, and even more distant, in the 3rd remote zone, up to 10 000 kilometers from the epicenter. This provides as with rich material for analysis and comparison of seismic effect from explosions and earthquakes.

Let us follow first the logic the American delegation members applied to the question under consideration.

The average local magnitudes from the 1st zone were calculated first separately for the Blanca and Logan explosion when determining the average explosion magnitudes used in the "Working Document" and later on in the paper [Romney]; the following figures were obtained (according to latest summary of Dec. 14, 1959, see Tables 13 and 14):

$$\text{average local magnitudes } M_L \begin{cases} \text{Blanca} & 4.76 \pm 0.13 \\ \text{Logan} & 4.44 \pm 0.12. \end{cases}$$

The next calculation was made for both 2nd and 3rd zones:

$$\text{average teleseismic magnitudes } m \begin{cases} \text{Blanca} & 4.84 \pm 0.11 \\ \text{Logan} & 4.44 \pm 0.21 \ . \end{cases}$$

---

[*] Atomic Energy Commission Release on Hardtack Bomb Tests. 1959, No. 2, March, 39 p.

Here the standard deviations of average values are shown. The consideration then was as follows (see [Romney]).

Two average values for the 1st and the $(2 + 3)$d zones and for each explosion separately demonstrate that average magnitudes of these explosions determined by two different methods according to two different scales M_L and m almost coincide. From Gutenberg's viewpoint such a coincidence should be considered as unexpected one, for the earthquake at least; his investigations show, in fact, that the magnitude estimations in scales M_L and m should differ by magnitude 4.5 in approximately one unit, as was demonstrated by C. Romney. "This difference, however, could be of less importance because of uncertainty in correlation between the two magnitude scales" [Romney].

The decisive solution is then made: one assumes that local (Table 13) and teleseismic (Table 14) magnitude scales form together a homogeneous sequence of values. One assumes further on the absence in this sequence of any systematic deviations (all deviations are considered as random ones), equal weights are ascribed to all individual observations. As a result, the magnitudes are determined as general averages for all $(1 + 2 + 3)$ zones, being expressed in "any" scale.

The figures are taken in [Romney] for all-zone averages or, say, "out-of-scale" magnitudes, which are in agreement with Table 13 (for M_L) and Table 14 (for m):

$$\text{"Out of scale" magnitude} \begin{cases} \text{Blanca} & 4.8 \pm 0.1 \\ \text{Logan} & 4.4 \pm 0.1, \end{cases}$$

where ± 0.1 is the standard deviation of the average.

It is interesting to note that the small but necessary (in view of Gutenberg's investigations), correction in transition from M_L to m, as shown in [Romney], was in good co-existence in the official position of the American specialists. Those specialists paid great attention precisely to the reduction of Reinier magnitude from 4.25 to 4.1 (see [Romney]), in 0.15 only. The difference in *signs* of these corrections together with their consequences played perhaps a role: the negative correction of 0.15 corresponds to *increasing* the number of earthquakes exceeding (by magnitude) the explosion of given power by a factor of ca 1.5. It seems to testify the greater difficulty of discriminating between earthquakes and explosions. At the same time the positive correction in 1.0 would lead to the *decreasing* of this number by a factor ca 10, i.e. to the conclusion explosions would be markedly easier to distinguish.

The refusal of American experts to analyze the causes of coincidence of local and teleseismic magnitudes which was "unexpected" for themselves could be explained at best by the hypnosis of the fact proper in numerical coincidence of values of different nature. Such a coincidence is in fact surprizing, as it will be shown below: in the case of different distribution of temporary stations, recorded Blanca and Logan explosions between 2nd and 3rd zones (or even only within the 2nd zone where the strong systematic trend in amplitudes was observed), the coincidence could not occur. In such a case the analysis presented below would be, perhaps, performed by the American specialists themselves.

6.2. Analysis of Explosion Magnitudes

Systematic trend of explosion magnitudes. Let us consider the dependence of initially measured magnitudes M_L and m for Blanca and Logan explosions upon the epicentral distance x. We shall use in this case the average values of M_L and m calculated in Tables 13 and

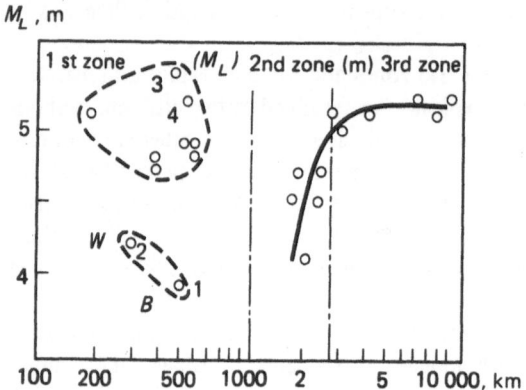

Fig. 47. Dependence of measured magnitudes M_L and m on epicentral distance x for Blanca explosion:

1—Barret; *2*—Woody; *3*—Mt. Hamilton; *4*—Palo Alto

14 for three separate zones. We shall consider also Figs. 47 and 48 where the dependences of magnitude M_L, in the 1st zone, and magnitude m, in the 2nd and 3rd zones, are presented in graphical form. These data are conventionally presented in spirit of American viewpoint as if all initial magnitude data would be quite comparable.

It is a good opportunity to remind here that the earthquake magnitude is a value characterizing the seismic effect in the earthquake source proper and not depending, in its own essence, upon the epicentral distance x. Seismologists proceed, when compiling the magnitude scales, just from this idea. The magnitudes being calculated at different *epicentral* distances and being expressed in the *unified unit system* should be equal provided they are calculated with the use of these individual scales or by its mutually correlated combination. This relates with certainty to the systematic trend of values only, as random deviations of measured values from the average ones are possible. Such random spread of magnitudes are discussed below. This means that the average magnitudes, in particular those measured in 1st, 2nd and 3rd zones, should be equal to the limits of random spread. When subtracted from the known differences between earthquakes and explosions as sources of seismic waves, one can expect preferable the explosions to show the same independence of magnitudes with distance and the same average values in different zones; it is surely preferable if these magnitudes are presented everywhere in the same unit system. This is expressed in Figs. 47 and 48 by random deviation only of observed point from the horizontal lines which mark the average magnitudes at the graph ordinates.

One can observe, however, from Tables 13 and 14 that for both explosions, Blanca and Logan, the average "out-of-scale" (after [Romney]) magnitudes show remarkable differences in the three mentioned zones. In the shadow zone (2) the average magnitude is 0.5 units lower than in the remote zone (3). In the near-field zone (1) these magnitudes also appear to be smaller by ca 0.3 of a unit than in the remote zone.

The mentioned differences are pronounced more clearly at curves (Figs. 47 and 48). Some important details in distribution of corresponding dots could be also seen there. A systematic trend of deviations in the 2nd zone, the zone of diffraction, is most pronounced: the magnitudes decrease with penetration into this zone from the remote 3rd zone, i.e., from right to left.

Although the difference in observed magnitudes m in the 2nd and 3rd zones, as demonstrated in Figs. 47 and 48, is convincing ecnough by sight, we shall perform the simplest quantitative analysis of this difference. We conventionaly assume that, in both zones, the deviations of individual values from the average ones are random, and average values are either equal or different. Necessary calculations were performed by V. F. Pisarenko.

Let us assume, according to statistical routine, that actually measured magnitudes in the 2nd and 3rd zones are a limited sample of some general population corresponding to an infinitely large number of observations. The sample average of magnitudes from the 2nd zone is indicated as $\bar{x}_2$, from the 3rd as $\bar{x}_3$. Let the corresponding average values of general population be m_2 and m_3. We know that magnitudes in the 2nd and 3rd zones are normally distributed with parameters (m_2, σ_2) and (m_3, σ_3). The values σ_2^2 and σ_3^2 are the known dispersions of general populations for the 2nd and 3rd zones. We can assume for our purposes that

$$\sigma_2 = S_2, \qquad \sigma_3 = S_3 \tag{6.3}$$

and S_2^2, S_3^2 are sample dispersions. The other supposition is considered below instead of (6.3). When assuming (6.3) the magnitude difference will be normally distributed with parameters $(m_2 - m_3, \sqrt{S_2^2/n_2 + S_3^2/n_3})$, where n_2 and n_3 are the numbers of observations in the 2nd and 3rd zones.

Let us check now whether it is possible to suppose that the numerical values of magnitude observed in the 2nd and 3rd zones during Logan and Blanca explosions are in agreement with the hypothesis $m_2 = m_3$. To say in other words that magnitudes measured for these explosions are equal in both zones within the limits of random spread.

We obtain for the Logan explosion:

$$\bar{x}_2 = 4.16; \ \bar{x}_3 = 4.78; \quad \sqrt{S_2^2/n_2 + S_3^2/n_3} = 0.24$$

$$\frac{\bar{x}_2 - \bar{x}_3}{\sqrt{S_2^2/n_2 + S_3^2/n_3}} = -2.58$$

From the other side the 98% interval of confidence is $(-2.33; +2.33)$. We can see that observed value -2.58 lies outside of this interval. Thus the hypothesis $m_2 = m_3$ has a significance level not exceeding 2%, it has a low probability and should be rejected.

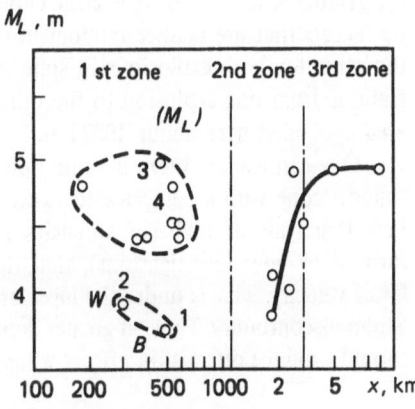

Fig. 48. Dependence of measured magnitudes M_L and m on epicentral distance x for Logan explosion:

1—Barret; 2—Woody; 3—Mt. Hamilton; 4—Palo Alto

For the Blanca explosion following values are found:

$$\bar{x}_2 = 4.5; \quad \bar{x}_3 = 5.1; \quad \sqrt{S_2^2/n_2 + S_3^2/n_3} = 1.05$$

$$\frac{\bar{x}_2 - \bar{x}_3}{\sqrt{S_2^2/n_2 + S_3^2/n_3}} = -5.71$$

The 99.9% interval of confidence is $(-3.3; +3.3)$. Observed value -5.71 lays out of this interval. So, in this case the hypothesis $m_2 = m_3$ has the significance level 0.1%, even less than the previous case, and should undoubtedly be rejected. A more certain result for the Blanca explosion could be partly explained by a greater number of observations (11 stations) in comparison with the Logan explosion (7 stations).

Instead of assumption (6.3) as another version one could suppose that

$$\sigma_2 = \sigma_3 \tag{6.3'}$$

In such a case the appropriately normalized value $(\bar{x}_2 - \bar{x}_3)/\sqrt{n_2 S_2^2 + n_3 S_3^2}$ has the Student's distribution. The calculation shows that, in this case, the significance level for the hypothesis $m_2 = m_3$ is rejected, is 2% for the Logan case and 0.1% for the Blanca case, i.e. similar to the results of first calculation version with the supposition (6.3).

Thus, the quantitative statistical estimation leads to the conclusion that the difference between 2nd and 3rd zones in measured magnitudes for both Logan and Blanca explosions could not be explained by a random spread of data.

A systematic decrease of magnitudes for Blanca and Logan explosions in the 2nd zone in relation to the 3rd zone is not connected with the choice of magnitude scale: in both zones the same scale m was used.

But the assumption is left that the observed effect is connected:

(a) either with the "random" local peculiarities of seismic stations that occurred in this zone;

(b) or with regular, systematic differences between waves from earthquakes and from explosions, caused by different geophysical conditions in the formation of this zone, (the specific diffraction zone) and by differences in explosions and earthquakes as different sources of seismic waves as well.

Little material can be found to discuss (a). The question could be clarified better if not only explosions but also earthquakes were observed at the same stations. This, however, was not so, and that was an important defect of seismic observation methodology applied to the Hardtack test series. It is clear enough that the observed trend of magnitudes is caused by factors that are neither random nor very local. This trend appears to be approximately the same for both explosions in spite of changes in mutual disposition of the majority of stations from one explosion to the other. The trend is pronounced over a sufficiently large area expanded over about 1000 km.

One can express more definite opinion about (b). On one side, the existence of present shadow zone with a complicated fuzzy seismic signal is well known in seismology. Distinct first P-arrivals of higher frequencies are almost absent in this zone. Gutenberg believed (proved by later investigations), that the cause is in the existing of a layer with relatively lower velocity. This is under the layer of higher seismic wave velocity directly underlying the Moho discontinuity. Thus, at greater depths velocities begin to increase again. Such a structure should produce diffraction effects which can be observed at a distance from the source. Let us, however, not to go into detailed geophysical explanations. What is important for us, is the existence of the zone of diffractional signal and the impoverishment of the signal in this zone by high frequency oscillation components.

From the other side, it is a well-known fact that the explosion produces relatively more high-frequent oscillations than the earthquake of the same seismic focal energy. This is connected with more localized and shorter duration of impact for an explosion source compared with a corresponding earthquake force. We apply the term "source" to that zone of intensive breakage and nonlinear motion from which stresses and strains in seismic waves are mostly of continuous and linear character.

It follows directly from the comparison of these two statements that in the diffraction zone (especially unfavourable for passage of high frequencies) the seismic effect of relatively more high-frequency explosions should be suppressed in comparison with the effect of corresponding earthquake. This has a direct relation to the magnitudes as well.

The specific behavior of explosion magnitudes in the 2nd zone has thus a quite natural geophysical explanation. What is more, it would be extremely strange if these differences resulting from variations in frequency spectra of explosions and earthquakes were not observed in this zone.

Just such a strange thing is found when considering magnitudes in the 1st, near-field zone and agreeing with the American experts with the opinion that local M_L and teleseismic m magnitudes, could form a "homogeneous" population $M_L \equiv m$.

In fact, the explosion waves should attenuate with the distance more quickly than the earthquake waves, as the explosion has a higher frequency than the earthquakes. The explosion signal should be more suppressed in the remote 3rd zone than the earthquake signal because, in this zone, all observed oscillations are relatively low-frequency ones. This is caused by the earlier absorption of high frequencies, So, if both signals are equalized in the 3rd zone in their intensity, within the 1st zone, the level of explosion signal should be relatively higher. The same should be related to magnitudes. We could see, however (see Tables 13 and 14, Figs. 47 and 48), that for "out of scale" magnitudes of the Blanca and Logan explosions an inverse relation is observed.

This apparent contradiction disappears if one ignores the difference between local and teleseismic magnitudes known from Gutenberg's papers and reduces all magnitudes to the unified scale m (see corresponding columns in Table 13). We can see that explosion magnitudes m in the 1st zone became slightly greater than the magnitudes in the 3rd zone, as was expected. The difference is small and equal to 0.2 of a magnitude unit only; this is two times higher than the standard deviations of average magnitudes estimated separately in the 1st and 3rd zones. The given analysis testifies to the approximate correctness of Gutenberg's relation between scales M_L and m [Gutenberg et al.]. It confirms the incorrectness of Romney's statement concerning "uncertainty" [*] in order to justify his procedure of total averaging.

The generalized dependence of unified magnitude m on the epicentral distance is presented in Fig. 49. The data on the Logan and Blanca explosions are conventionally combined here and reduced to the Logan level. The procedure was as follows: the Logan magnitudes were kept without variations, and Blanca magnitudes are diminished by 0.26, the difference between average values for both explosions in the local zone, where the same ten stations were

---

* Romney, C. F. Amplitude of Seismic Body Waves from Underground Nuclear Explosions.—J. Geophys., 1959, vol. 64, No. 10, P. 1489-1498.

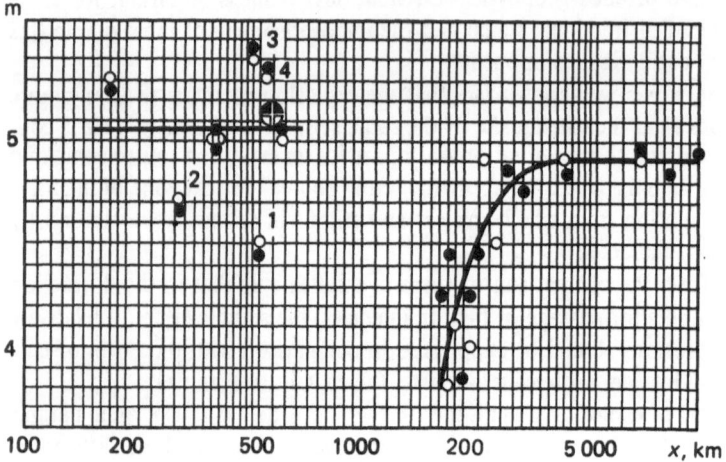

Fig. 49. Generalized systematic dependence of unified magnitudes m on epicentral distance for Logan (*open circles*) and Blanca (filled circles) explosions:

○—Logan; •—Blanca

used for observations. One can see from Fig. 61 that, in this case, the dots (x, m) for both explosions in the diffraction zone and in the remote zone $(x = 1700\text{-}10\,000$ km) forman automatically united sequence. It is interesting that such attempt at combination, performed earlier for magnitudes M_L and m, without reducing them to the single scale, was unsuccessful. It was found that compared with combination of dots (x, M_L) in the local zone, the dots (x, m) in two other zones show a systematic shift, especially pronounced in the remote zone: the Logan dots are higher than the Blanca dots. This fact can be considered as an additional proof in favor of Gutenberg's relation between scales M_L and m and its approximate application to the explosions.

By the way, the averaging solid lines in Fig. 49 could be used to determine the corrections for measured magnitudes m of explosions in order to obtain the unified evaluation of magnitude, reduced to the level of any given zone, local for example.

We notice that the essence of the problem in comparison with different sources cannot be solved completely by the consideration of spectral differences between explosions and earthquakes, in particular, by their local magnitudes. The differences between explosions and earthquakes do exist also in the distribution of energy between different types of waves at the source, viz. between longitudinal and transverse waves, and between body waves and surface waves; the difference exists also in the distribution of energy between different components of oscillation.

It is natural to expect that for explosions the relation of longitudinal wave energy to the transverse wave energy is larger than for earthquakes. Further, the energy of high-frequency surface waves should be larger for explosions than for earthquakes, when the summed energy of their sources is equal (because of high-frequency character and small depth of the explosion); the energy of low-frequency surface waves for explosions should be smaller (because of low level of low-frequency part of the explosion spectrum). Calculations of relative magnitudes of explosions and earthquakes in the local zone with the use of "shear" oscillations, transverse and surface waves including, are then very formal and hardly comparable with the calculation of magnitudes from longitudinal waves for an approach not purely empirical.

Wave forms of explosions should be generally simpler than earthquake waves. The cause is short duration of explosion proper and its shallow setting hampering the generation of waves reflected from the Earth's surface in the epicentral area which would not be superimposed on direct waves.

A difference in relation between transverse and radial oscillation components for explosions and earthquakes should exist, especially for transverse waves, both body waves and surface waves: for explosions this relation should be smaller because of near-axial symmetry of the source. This would be more noticeable for sufficiently low frequencies when the influence of local horizontal unhomogeneities of the medium is smoothed. In particular, the relation of low-frequency Love and Rayleigh waves should be much smaller for explosions than for earthquakes.

There are a lot of observational data concerning all these differences. However, they need special and careful investigations in order to elaborate a complex criterion for separating explosions from the earthquakes. This criterion would be more effective than the direction of the first arrival.

The comparison of level of seismic effect, magnitudes in particular, for explosions and earthquakes should be performed as the basis of those waves and components which are to be used for their discrimination. When separating explosions and earthquakes by the direction of the first arrival of longitudinal waves only, it would be reasonable to determine the magnitudes from these waves, not only in the local but also in other zones. The local magnitude scale for first longitudinal waves is then needed.

So, we see that explosion magnitude being calculated with the use of earthquake-orientated technique and utilized for comparison of seismic effect from explosions and earthquakes appear to have a value depending on those indexes which are used for such comparison. If we are limited himself by: the comparison of "shear" wave magnitudes in the local zone; of amplitudes and periods of longitudinal waves in the diffraction zone and in the remote zone; with corresponding calculation of M_L and m, as was done when processing seismic observations of Reinier, Logan and Blanca explosions; or if we expressed all measured magnitudes in the single system of units m according to Gutenberg's unified scale; the result is that magnitude of explosion depends, initially, on the epicentral distance. The highest numerical value is in the 1st local zone; in the remote 3rd zone the magnitude seems to be slightly smaller; in the zone of diffraction (2nd zone) it is markably diminished and varies with the position within the zone.

The scattering of explosion magnitudes. We have considered already the "systematic" aspect of the explosion magnitude problem. Let us turn to consideration of the "random" aspect, the scattering of individual calculated magnitudes in relation to average values. From Fig. 49 we see that such an approach is possible without reserve in the 1st zone ($x \leqslant 1000$ km) and in the 3rd zone ($r_x \geqslant 2500$ km) separately, as in these zones the lines averaging the magnitudes are horizontal. But in the 2nd zone, where a strong systematic trend is observed, the base should be measured not from a horizontal line but from the averaging rather steep curve.

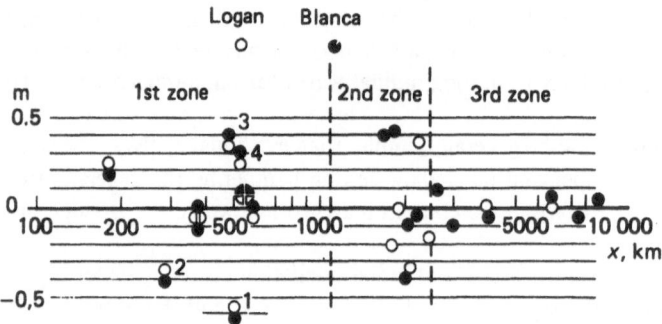

Fig. 50. A random spread of calculated magnitudes *m* from the averaged lines for Logan and Blanca explosions (see Fig. 49).

Numbers indicate station numbers, see Figs. 47 and 48

The spread of points *x*, *m* from the averaging lines shown in Fig. 49 is presented in Fig. 50 for the Logan and Blanca explosions. A large difference in spread between the 1st and 2nd zones taken together and the 3rd zone, with essentially lower spread, is evident. The same can be found from the values of standard deviations σ_m and σ_{M_L} for individual magnitudes *m* and M_L in the 1st and 3rd zones (see Tables 13 and 14). The deviations for the 2nd zone given in Table 14 have no random spread as they include a systematic component as well.

Spread of magnitudes M_L and, correspondingly, *m* in the 1st zone is approximately equal to ±0.4:

	M_L	m
Blanca	±0.45	±0.31
Logan	±0.40	±0.28
Reinier	±0.41	0.30

The value ±0.40 was indicated in the "Working Document" also for the whole set of magnitudes over all three zones without differentiation between M_L and *m*. There is no other way except to consider this as an occasional coincidence.

The standard deviation of ±0.4 obtained during the American observations of underground explosion magnitudes appears to be too large, noticeable larger than the deviations usually obtained from good stations when observing earthquakes. One could expect the inverse relation between the value of explosion spread and earthquake spread: the explosion has more symmetrical source than the earthquake, and thus more homogeneous spatial distribution of seismic effect could be expected for the explosion. This strange event was noticed as far back the beginning of 1959 during the discussion on the "New Seismic Data" held at the session of the Disarmament Subcommission, the Senate's Commission of Foreign Affairs, under the chairmanship of Senator Hubert Humphrey. Analogous deviations for earthquakes at good seismic stations in the USA and other countries, according to the evidence of L. Murphy, the head of Seismological Division, US Geodetic Survey, usually are equal to ±0.1-±0.2. The same level of deviations is mentioned in the example from the Richter's book [Richter]. It is interesting that in this example for an earthquake, as in our case for underground nuclear explosions (see Table 13, Figs. 47 to 49), the lowest magnitudes M_L were observed at stations Barret and Woody (see points *1* and *2* in Figs. 47-50).

A problem appears as to why the spread of calculated explosion magnitudes is larger in zones which are closer to the source than in a remote zone. This seems to be a paradox: it would be useful to be closer to the event in order to estimate it better. However, this does not work in our case. Observational data, under the conditions of equally reliable observations in local and remote zones, should be more subjected to fluctuations in the 1st zone; the fluctuations here are connected with details of structural; heterogeneities in the Earth's upper layers, the waves arriving in the zone passing through them. It is known that structural heterogeneity of the Earth decreases with depth. Besides, the fluctuations of waveforms increase close to the epicenter because of more high-frequency seismic oscillations in this zone in comparison with the 3rd zone. Thus, correspondingly because of less smoothing by structural features of the medium where seismic waves propagate: it is known that the smaller the frequency (it means that the period is bigger), the higher the smoothing.

The condition of equal reliability was understood as the observations used in calculations were not spoiled by microseismic noise, which is more important in the remote zone where the useful signal is weaker. In our case microseisms had not hindered much in magnitude determination.

The decrease of spread in earthquake magnitudes in comparison with explosion magnitudes seems to be connected with the fact that, magnitude determination for sufficiently large earthquakes being included in World statistics, is based more often on observations from stations in most favorable remote (3rd) zone. Besides, it is possible that when using the data of local stations for explosion magnitude determination, no necessary system of station corrections was elaborated, which could be connected with the consideration of different oscillation frequencies; the effects of instrumentation parameters also would not perhaps be considered with necessary accuracy.

Let us discuss in more detail some peculiarities of magnitude m spread in the 1st zone. Deviations of individual values of magnitude from average values for explosions of Logan and Blanca remain the same. It is possible to see this from the comparison of Figs. 47 and 48, and also from mutual positions of open and solid circles in Figs. 49 and 50. Comparison of data for all three explosions, Reinier including, leads to the same conclusion. The latter comparison is given in Table 15.

The comparison of standard deviations $\sigma_{\Delta m}$ for magnitude differences Δm (Table 15) with standard deviations of magnitudes σ_m (see Table 14) shows that the first magnitudes are three to six times less than the second. This means that relative magnitudes are determined with an accuracy three to six times higher than the absolute ones. This testifies that the main part of observed deviations is caused by permanent station conditions (local geology, conditions of instrument installations, specificity of instrumentation etc.) and not by random defects in magnitude determinations which can vary from one experiment to the other (for instance, influence of microseismic disturbances).

These deviations are especially big for stations Barret and Woody: the corresponding points *1* and *2* in Figs. 47 to 50 are shifted far below the denser cluster of points for other stations. It is interesting to estimate quantitatively the level of this deviation. The least square processing of initial data, that is the magnitudes M_L, leads to the following figures. The average local magnitude of Blanca explosion for 10 stations from the 1st zone, including Barret and Woody, is $M_{L10} = 4.79 \pm 0.12$, and for 8 stations, excluding Barret and Woody,

Table 15. Differences Δm in Magnitudes Between Blanca and Logan, and Logan and Reinier Explosions

Serial, No.	Station	Δm	
		Blanca-Logan	Logan-Reinier
1	Tinemaha	0.2	0.4
2	Woody	0.2	0.3
3	Riverside	0.2	0.4
4	Pasadena	0.3	0.3
5	Mt. Hamilton	0.3	0.2
6	Barret	0.2	0.3
7	Palo Alto	0.3	0.3
8	Berkeley	0.3	0.3
9	San Francisco	0.3	0.2
10	Mineral	0.3	0.0
Average differences		0.26	0.27
Standard deviations of single determination		± 0.05	± 0.12
of average value		± 0.02	± 0.04

is $M_{L_8} = 4.98 \pm 0.09$. The standard deviations of average values are given here. We can see, that by excluding the two aberrant stations, the average magnitude error decreases from 0.12 to 0.09. Decreasing of station number from 10 to 8, would lead, without such a big shift of these two points, to an increase of the average value error.

Let us calculate now, using the Blanca explosion, the magnitude deviation for Barret and Woody stations from the average value $M_{L_8} = 4.98$ obtained for the other stations. To compare these deviations with the standard deviation of a single measurement for stations of this group $\sigma_{M_{L_8}} = \pm 0.24$.

We obtain the following deviations for stations:

Barret $3.9 - M_{L_8} = -1.08$
Woody $4.2 - M_{L_8} = -0.78$

The triple standard deviation $3\sigma_{M_{L_8}} = \pm 0.72$, i.e. the deviation of both points exceeds the triple standard deviation for the group of other eight points.

Let us, however, apply the other, softer and, perhaps, juster estimation. We assume, in a similar calculation process, that stations Barret and Woody are included beforehand into the station sample for the 1st zone used for the calculation of average magnitude. This results in the approach of the average value $M_{L_{10}}$ to individual values M_L at these stations and to the increasing of standard deviations for the whole considered group of points. As a result, we obtain $M_{L_{10}} = 4.79$, $\sigma_{M_{L_{10}}} = \pm 0.38$ and then find the following deviations for stations:

Barret $3.9 - M_{L_{10}} = -0.89$
Woody $4.2 - M_{L_{10}} = -0.59$.

The triple standard deviation is $3\sigma_{M_{L_{10}}} = \pm 1.14$.

Deviations for Barret and Woody stations are, in the final calculations, within the limits of triple standard deviation; they are close to double deviation $2\sigma_{M_{L_{10}}} = \pm 0.76$; this, however, forces us to use the data of these stations with a certain caution.

Professor Frank Press, the director of the Seismological Observatory in Pasadena and supervisor of stations Barret and Woody, informed us that these stations are placed in bedrocks that have a low level of seismic noise and, generally, are completely trustworthy ones. Considering this authoritative opinion and assuming, at the same time, that some unfavorable uncontrolled abnormalities could have existed at these stations, we applied, when determining the average magnitude $\overline{M}_L$ in the 1st zone, the following procedure recommended by Professor Hans Boete. When calculating $\overline{M}_L$, we exclude points for Barret and Woody with the smallest values but exclude simultaneously two points with the biggest magnitudes, namely Mt. Hamilton and Palo Alto (points 3 and 4 in Figs. 47-50). Professor Boete accompanied his advice with the following general and very didactic discussion. One can suppose that all deviations of M_L, observed in the given zone, are mainly connected with local details of geological structure close to individual stations crossed by seismic waves which propagate from the Nevada test area. The positive and negative deviations of seismic energy—and of magnitude correspondingly—from some average regularity are equally probable at each station. Severe dissipation of energy, however, is generally much more usual than its concentration, as the latter can rarely be reached, in a limited area only. Thus, one has more reason to expect an essential deviation toward underestimation of energy (as, for example, for Barret and Woody stations) than toward overestimation of energy (which is not observed here for this reason).

Following Professor Boete and excluding two points from above and two points from below we obtain the average local magnitude of Blanca explosion in this zone for six stations as $M_{L_6} = 4.86 \pm 0.06$.

This estimate is slightly higher and twice as precise as the average magnitude for all ten stations $M_{L_{10}} = 4.79 \pm 0.14$. There is, however, no reason to search for a better accuracy. Let us, admitting the possibility of a slight decrease of estimate $M_{L_{10}}$, take it as a basis for further calculations.

The process performed for Blanca explosion can be equally related to two other explosions, Logan and Reinier, which generally have the same distribution of magnitudes M_L in the 1st zone. The conclusion concerning the possibility of use in the 1st zone of the average magnitudes estimated from data of all ten stations in this zone can be applied with certainty to both local and unified magnitudes.

Let us now make some remarks concerning the spread of magnitudes in the 2nd zone, the diffraction zone. We have seen (Fig. 50) that this spread is similar to that in the 1st zone. The epicentral distances, however, are bigger in the 2nd zone, frequencies are lower, and when proceeding from spectral aspects alone, a greater averaging of propagation properties for seismic energy and, correspondingly, smaller random spread of magnitudes could be expected. There are, however, other, more powerful causes that change this statement. The reason is that seismic wave intensity in the 2nd zone can vary remarkably following only a slight variation in the structure of these screening layers and their underlying waveguide (after Gutenberg) layers defining the existence of the zone itself. Nonstability of seismic signal for explosions observed in the 2nd zone does not surprise seismologists as they know the same phenomenon well in earthquake seismology. Just this circumstance hampers the elaboration of sufficiently precise or sufficiently stable it might be better to say magnitude scales in the 2nd earthquake zone. Seismological data show that nonstability increases entering the 2nd zone from outside, and that is quite clearly reflected in the explosion magnitude spread (see Fig. 50): when entering the zone from its boundary with the 3rd zone the cloud of points (x, m) spreads.

Determination of average magnitudes. It is clear from the previous discussion that the determination of average magnitude of underground nuclear explosions, as made in [Romney], means that the total averaging of heterogeneous measurements of local M_L and teleseismic m magnitudes is meaningless. Some difficulties also appear in determining average magnitudes on the basis of values expressed using the Gutenberg scale: unified magnitude scale or any other earthquake magnitude scale. These difficulties are caused, basically, by the systematic trend of explosion magnitudes depending on the epicentral distance.

Magnitude elaborated for earthquakes cannot, in fact, serve for explosions, as a parameter of the seismic effect of the source only, i.e., it cannot have the same function as it usually does in the case of earthquake. The use of magnitudes for explosions has a certain meaning under limited conditions. It is possible to compare numerically definite aspects of seismic effects of an explosion with the corresponding effects of earthquakes within certain epicentral distances, oscillation frequencies and other conditions.

Despite such difficulties, one can use the term "magnitude" to mean "average" constant for some intervals of epicentral distance and for frequencies of about 1 Hz. This is realistic for 1st and 3rd zones separately, at least within the segments covered by magnitude determination. Where no high degree of accuracy is needed, one can neglect the small systematic difference (about 0.2) between magnitudes in 1st and 3rd zones and determine the average explosion magnitude conventionally by integrating these two zones. It is necessary to remember that variation of m for 0.2 unit corresponds with the variation of the trotyl equivalent of the explosion by two to three times.

The 2nd zone where a large systematic trend of magnitudes depends on epicentral distances, should be, naturally excluded. This does not, however, mean that observations in this zone cannot be used for magnitude calculation in other zones; it is important that the systematic magnitude trend should be considered in such calculations.

Average magnitudes m of Blanca, Logan and Reinier explosions were listed earlier in Tables 13 and 14 for the 1st and 3rd zones separately. Now we give two versions of the calculation of conventional "average" magnitudes for Blanca and Logan explosions observed in the 1st and 3rd zones, taken together here.

Let us assume in the first version that weights of individual magnitude determinations at the stations, placed in the 1st and 3rd zones, are equal. In such a case we obtain for both explosions, in accordance with figures from Tables 13 and 14:

Blanca $\quad \bar{m}_{B1} = 5.23 \pm 0.06$
Logan $\quad \bar{m}_{L1} = 4.97 \pm 0.08$

The standard deviation $\sigma_{\bar{m}}$ of average values is given here. The corresponding standard deviations of individual determinations σ_m are ± 0.26 and ± 0.28.

In the second version we calculate for the 1st and 3rd zones the weighted average from partial averages for 1st and 3rd zones; the weights p_i ($i = 1, 3$) of these determinations are taken in inverse proportion to standard deviation squares of partial averages $\sigma_{m_1}^2$ and $\sigma_{m_2}^2$. We have for the Blanca explosion $\sigma_{m_1} = \pm 0.10$; $\sigma_{m_3} = \pm 0.03$; the same for the Logan explosion $\sigma_{m_1} = \pm 0.09$; $\sigma_{m_3} = \pm 0.13$.

As a result, we obtain for both explosions:

Blanca $\bar{m}_{B/2} = 5.14 \pm 0.04$
Logan $\bar{m}_{L/2} = 4.95 \pm 0.10$

The standard deviations of general averages are calculated here by the formula:

$$\sigma_m = \sqrt{\Sigma_{p_i} \sigma_{\bar{m}_i}^2 / \Sigma p_i}.$$

We can see when comparing the results of the two calculation versions that they are very close to each other. Considering that because of a systematic difference between averaging values, the nature of the general average calculation is highly conventional, we can accept as a final result the following average magnitudes for both 1st and 3rd zones together rounded to 0.05 unit:

Blanca $m_{1,3} = 5.2 \pm 0.1$ (6.4)
Logan $m_{1,3} = 4.95 \pm 0.1$ (6.5)

The difference between these two values is equal to 0.25, this coincides exactly with that more precise value of 0.26 ± 0.02 determined earlier from the magnitude differencies at stations in the 1st zone when all three explosions were observed (Table 15). Accepting $m_{1,3} = 4.95$ for Logan explosion and substracting the average difference Logan-Reinier $= 0.27 \pm 0.04$ (see Table 15) we obtain the following magnitudes for 1st and 3rd zones for Reinier explosion (rounded, as before, to 0.05):

$$m_{1,3} = 4.7 \pm 0.1 \qquad\qquad\qquad\qquad\qquad\qquad (6.6)$$

Let us note that average magnitudes for all three explosions taken for the 1st zone separately are ca 0.1 greater, and for the 3rd zone ca 0.1 smaller than in the expressions (6.4) to (6.6).

6.3. Relation Between Explosion Power and Magnitude

In order to determine the number of earthquakes exceeding by magnitude the underground nuclear explosion of given power it is necessary to establish in advance the general relation $m(Y)$ between the explosion power Y in kilotons of trotyl equivalent and explosion magnitude m [Gutenberg, Richter].

The history of the problem. There was no possibility of establishing this general function from data of nuclear explosions at the Geneva conference (1958), because experts possessed observed data of one underground nuclear explosion, Reinier, only. Experimental data on chemical explosions were used for this purpose together with some theoretical considerations.

Earlier observations over many chemical explosions showed, according to Carder that amplitude A of seismic oscillations of the ground varies proportionally to the charge Y in power 0.5-1.0. In such a case, the magnitude m at the same oscillation frequency is proportional to $\log A$ (see, for example, [Richter]), the relation between m and Y could be written in the following general form:

$$m = C + n \log Y, \tag{6.7}$$

where, according to Carder, $n = 0.5$-1.0 for chemical explosions certainly.

Two underground chemical explosions of 10 and 50 tons were made in 1957 during the preparation of Plumb bob operation in the area of the future Reinier explosion [Carder et al]. The comparison of amplitudes A of ground motion by these explosions led to the conclusion, that $A \sim Y^{3/4}$, which corresponds to the coefficient $n = 0.75$ in formula (6.7). This value does not exceed the limits 0.5-1.0 mentioned above.

By establishing the values of parameters n and C in the general formula (6.7) at the Expert Conference in Geneva, 1958, the value n was accepted as equal to 2/3. This was the basis for some theoretical considerations and with respect to experimental data for chemical explosions. This figure also lies within the mentioned limits 0.5-1.0. The Reinier explosion data enables the determination of parameter C only in (6.7).

The Hardtack II operation, with its large nuclear explosions Blanca and Logan and small Tamalpais and other explosions, together with the data on Reinier explosions, presented an opportunity to make the dependence (6.7) more precisely on the basis of explosion observations themselves.

It was mentioned in the "Working Document" concerning the "New Seismic Data" that in accordance with Hardtack experiments "... the amplitude varies as the explosion power in power one, within the scale of power from 0.1 to 23 kT, that is in (6.7) the value n seems to equal 1. The clarification of the processing which led to this result was in Romney's paper [Romney, Fig. 3, p. 1492]. It appears that the mutually concerned determinations of explosion magnitudes were not used to establish this relation but relative recorded amplitudes from several stations only. These were temporary stations and only one permanent station Tinemaha was used (although 10 permanent stations participated in observation, as indicated in Table 13). Only one Tinemaha point is shown at the figure from Romney's paper [Romney] for the Reinier explosion. The averaging straight line with the slope $n = 1.0$ was drawn from a group of points for large explosions (Blanca, Logan, Reinier) and also for two small explosions (Tamalpais and Neptun). The Neptun explosion is not known as, essentially, underground because it produced an atmospheric release. Thus the seismic effect of this explosion was reduced. Romney, in order to confirm the slope $n = 1.0$, refers also to the theoretical discussion of the problem presented in [Zatter, et al]. He finally found for his magnitude M ("indifferently to the type of scale") the relation: $M = 3.65 \pm \log y$. The same relation, namely $M = 3.65 \pm 0.3 + \log Y$, was given also in the "Working Document".

This relation, as already seen from the description of working in [Romney], is doubtful. We shall analyze the relation of magnitude on explosion charge as the basis of more complete data on nuclear explosions. Let us consider this problem first for the large explosions of Reinier, Logan and Blanca separately. Much more reliable observed data are collected for these explosions. And additionally, these explosions are of greater interest for the problem of international control. Some consideration will be given later for explosions of lesser power.

Relationship of magnitude on power for large explosions. The equation (6.7) with $C = $ const, $n = $ const is not a unique form of the general relation between explosion power Y and its magnitude m at all. Moreover, there is some evidence that the slope $n = \Delta m / \Delta \log Y$

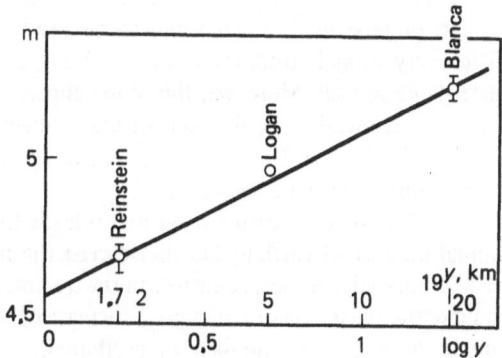

Fig. 51. Relation between explosion power Y (in kilotons) and explosion magnitude for large explosions (Reinier, Logan, Blanca), averaged for 1st and 3rd zones together

of the graph $m(Y)$ can vary for different ranges of power Y. One theoretical consideration leads to the conclusion that with small Y the slope should be close to $n = 1$, and with greater Y it should approach to $n = 2/3$. This theory does not indicate yet at what exact value it is possible to expect the change in the angular coefficient n from 1 to 2/3. It should be considered a possibility that experimental data on chemical explosions [Carder, Cloud] show that the slope could be even smaller than 2/3, decreasing in some cases to 0.5.

It is all the more important, in such an imperfect theoretical situation, to establish first the corresponding empirical relations and trust them more.

As the principal topic that is unclear in the problem is the value of parameter n, it is expedient to use the magnitude *differencies* (Table 15). One should remember that these differencies were determined more precisely (± 0.02-0.04) than the absolute magnitudes (± 0.1). When compiling the graph $m = f(\log Y)$ in a limited range of power—from Reinier, 1.7 kT to Blanca, 19 kT—the averaging curve could be obviously replaced by a straight line.

The graph of the unknown function is presented in Fig. 51. The standard deviations of corresponding magnitude differences (see Table 15) are shown here by vertical segments crossing the points for Reinier and Blanca explosions. In accordance with Fig. 51 and above-mentioned estimates we found:

$$m = 4.6 \pm 0.1 + (0.50 + 0.50 \pm 0.06) \log Y. \tag{6.8}$$

This equation is exactly the solution of the problem on magnitude dependence m for the underground nuclear explosions from its power Y (in kilotons), within the power range from Reinier to Blanca explosions. The regularity found could obviously, within a wide margin of error, be extrapolated close to this range, namely from ca 1 to ca 25 kT.

It is interesting, perhaps, for some comparisons to give here one more function, analogous to formula (6.8), which can be found for local explosion magnitude (it is given without estimations of accuracy):

$$M_L = 3.9 + 0.7 \log Y. \tag{6.8'}$$

Thus, we can see that the considered parameter of slope determined from the experimental data on large nuclear explosions seemed to be equal to $n = 0.5$ for magnitudes, and that differs very strongly from the value $n = 1.0$ given in the "Working Document" and in Romney's paper as well. Moreover, this slope appeared to be only 0.7 even for local magnitudes M_L, to which tend to be the "out of scale" magnitudes of both mentioned documents; the value 0.7 is close to 2/3 or 3/4, that is, to the values which were apparently accepted for calculations in Geneva, 1958.

The fact that increasing n up to 1.0 leads to the essential increasing of the theoretical annual number of earthquakes that exceed the magnitude of the nuclear explosions up to ca 20 kT, played, perhaps, some role in the treatment of "New Seismic Data". Such an increase could create the impression of a great increase of difficulties in recognition of such explosions among "tremendous" numbers of earthquakes.

We can mention that the relatively small value $n = 0.5$ which we obtained for nuclear explosions within the given range of power Y (approximately from 1 to 20 kT) corresponds to the minimum value of n reported by Carder for chemical explosions ($n = 0.5\text{-}1.0$). The cause is, perhaps, the fact that the chemical explosions used by Carder were on average of less seismic* power than the Reinier, Logan and Blanca explosions considered here. The value n, indeed, tends to decrease as power increases and to increase with decrease of power. The result obtained for large underground nuclear explosions is, thus, in complete agreement with existing experimental data for chemical explosions (of average, of smaller seismic power) as well.

Comparison with small nuclear explosions. Romney indicates in his paper that local magnitudes M_L for the small Tamalpais explosion of power 0.072 kT and Neptun explosion of 0.090 kT could be determined at a single station Tinemaha only; magnitudes of all three large explosions were also determined there. The data are placed in columns M_L, the first line in Table 16. The magnitude data of third small Evans explosions are absent in the American documents perhaps because of its "mystery".

Romney believes that one should not trust the magnitudes of two small explosions mentioned at the Tinemaha station. He proposed smaller magnitudes, namely 2.6 and 2.4 correspondingly for Tamalpais and Neptun explosions (see the second line in Table 16). According to [Romney] these data were obtained by comparison of amplitudes and oscillation period for these small explosions and for the Logan explosion. Similar magnitudes for these explosions, namely 2.65 for Tamalpais and 2.45 for Neptun, were indicated in the American list dated Dec. 14, 1959 (see the third line in Table 16). Although there was, traditionally, no indication in the documents produced by American experts concerning the type of magnitude scale to which these figures should be related, it is easy to suspect that they are obtained from comparison of observation in the 1st zone only.

We proceed with the analysis of these data. First of all, we reduce all mentioned magnitudes M_L to the unified scale according to (6.1). These figures are given in Table 16. Further, we shall also determine the differences in our case, taking into account that the magnitude *differences* Δm, seen for the example of Reinier, Logan and Blanca explosions, appear to

---

* The seismic power Y of a nuclear explosion, where Y is in kilotons of trotyl equivalent appears to be essentially less than for chemical explosion with a charge of Y kilotons.—*Author's remark.*

Table 16. Calculation of Magnitudes M_L and m for Tamalpais and Neptun Explosions

Data	Blanca		Logan		Reinier		Tamalpais		Neptun	
	M_L	m	M_L	m	M_L	m	M_L	m	M_L	m
Tinemaha	5.1	5.52	4.8	5.32	4.2	4.89	3.1	4.09	2.9	3.93
Romney							2.6	3.71	2.4	3.55
Report of Dec. 14, 1959							2.65	3.75	2.45	3.59
Average for the 1st zone (see Tables 13 and 14)	4.79	5.30	4.44	5.04	4.06	4.77	2.82	3.88	2.62	3.72

be more exact than the absolute values m. Let us indicate the differences using the first two letters of the explosion names.

The magnitude difference Tamalpais-Neptun is the same and equal to $\Delta m_{TN} = 0.16$. The rigid relation between magnitudes of these two explosions is thus established. Now we determine, from the first line in Table 16, the differences between magnitudes of three large explosions and of Tamalpais explosion. We obtain $\Delta m_{BT} = 1.43$; $\Delta m_{LT} = 1.23$; $\Delta m_{RT} = 0.80$ (corresponding differences for local magnitudes are 2.0; 1.7; 1.1; our estimate 1.7 for the Logan-Tamalpais magnitude difference is close to the value 1.8 reported for this difference in the Document of Dec. 14, 1959). Finally, extracting the obtained differencies Δm from the average magnitudes of Blanca, Logan and Reinier explosions in the 1st zone (see Table 13) we obtain the following values for Tamalpais explosion: 3.87; 3.81; 3.97; and the average magnitude for this explosion is $m_T = 3.88$. Finally, considering, for the Neptun explosion, the difference $\Delta m_{TN} = 0.16$ and rounding it to 0.1, we obtain for the first zone:

Tamalpais $m = 3.9$
Neptun $m = 3.7$

The unrounded magnitudes m for these explosions together with calculated M_L values are given in the last line of Table 16.

The graphical comparison of magnitudes m in function of nuclear explosion power Y for both large (Reinier, Logan and Blanca) and small (Tamalpais and Neptun) explosions, as observed in the 1st zone, is given in Fig. 52. The averaging straight line with the slope $n = \Delta m/\Delta \log Y = 0.5$ determined above is drawn through the points for large explosions (R, L, B), and the other straight line with the theoretical slope $n = 1.0$ for "sufficiently small" charges Y is drawn through the point T for Tamalpais explosion. The point N for the Neptun explosion is left aside, as this explosion according to above reasons, could not be considered as belonging to the united family of other, real *underground* explosions.

One can see from Fig. 52, that both mentioned straight lines cross in the area of power $Y < 1$ kT (by the way, this is valid also where magnitudes M_L are plotted on the graph). It is quite natural to believe that the change in the slope of the function $m(\log Y)$ graph occurs not suddenly but by a monotonously smooth change within the limits of Y-values.

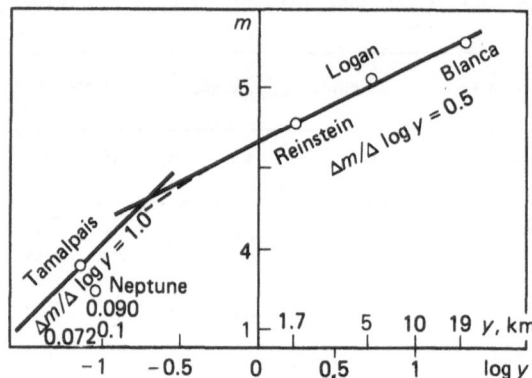

Fig. 52. Comparison of magnitudes of large (Reinier, Logan, Blanca) and small (Tamalpais, Neptun) nuclear explosions for the first zone.

The Neptun explosion does not belong to underground explosions

This is represented by the dashed transition curve. One can see that even if this transition extends to the zone $Y > 1$ kT (not excluded in the present consideration) this cannot disturb essentially the trend of relation $m(Y)$ within the main power range from 1 to 20 kT.

We must stress again that the curve in Fig. 52 corresponds to an average magnitude m for the 1st, local zone. The conventional "average" magnitude m for the 1st and 3rd zones taken together will be ca 0.1 unit less. It does not matter, however, for an explosion like Tamalpais, because it is unlikely that the systematic observations of such small explosions would be carried out at the control points of the 3rd zone.

6.4. Number of Explosion-corresponding Earthquakes

The most reliable statistical data on the average annual number of earthquakes of given magnitude occurring worldwide are published in the Gutenberg's paper. These statistics are based on 50-year long instrumental observations of earthquakes. Their magnitudes are expressed in the unified system of values m, the same system accepted in all previous calculations for underground nuclear explosions.

The seismic effect of an underground explosion is to be compared with that of shallow earthquakes with foci where the depths h do not exceed 60 km. Earthquakes with deeper foci have some known seismological peculiarities which permit recognition and thus allows us to reject the suspicion that these seismic events could be produced by an artificial explosion. It is obvious that the present technical level does not permit us to perform explosions at depths of more than 10 to 15 km, to say nothing of 60 km.

Analysis of Gutenberg's earthquake statistics. The Gutenberg data about the number N of shallow ($h \leqslant 60$ km) earthquakes are given in the first and second columns of Table 17. In the third and fourth columns of the same Table the number of earthquakes N_Σ of

Table 17. Average Annual Number of Shallow ($h \leqslant 60$ km) Earthquakes on the Whole Earth (N—Number of Earthquakes Related to the Given Magnitude Interval $\Delta m = m_i - m_k$; M_Σ—Number of Earthquakes with Magnitude Exceeding m)

m_i-m_k	N	m	N_Σ
	3	7.35	3
7.0-7.3	11	6.95	14
6.2-6.9	80	6.15	94
5.5-6.1	400	5.45	494
4.9-5.4	1300	4.85	1794
4.3-4.8	4500	4.25	6294
3.5-4.2	30000?	3.45	36294?
2.0-3.4	800000?	1.95	836294?

magnitude m exceeding the given value is shown. Those earthquakes obtained by direct counting of the observations of worldwide seismic networks are given with a question-mark. These numbers relate to sufficiently severe earthquakes with magnitudes $m \geqslant 4.2$. Numbers of weaker earthquakes, which are not completely covered by present observations, are marked by a question-mark. The latter figures are obtained by extrapolation of those observations performed on limited territories (California and others) with more dense seismic networks.

We can note immediately that the average magnitudes of our main sequence of underground nuclear explosions, viz. Reinier ($m = 4.7$), Logan ($m = 4.95$) and Blanca ($m = 5.2$), are essentially higher than those magnitudes from Gutenberg's statistics after which the data on earthquake numbers became uncertain.

The graph $N_\Sigma(m)$ compiled by figures from the third and fourth columns of Table 17 is shown in Fig. 53. The smooth curve can be drawn confidently through all observed points and that confirms generally the high quality of Gutenberg's statistics.

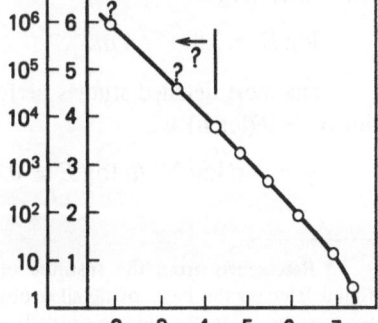

Fig. 53. The annual number N_Σ of shallow ($h \leqslant 60$ km) earthquakes on the Earth exceeding magnitude m, its given value

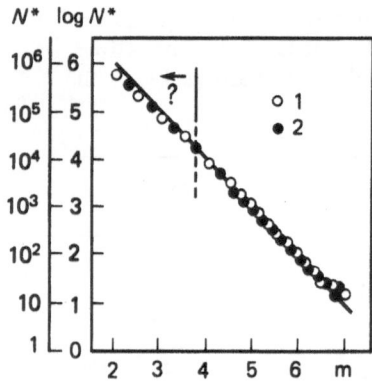

Fig. 54. Distribution of the annual number N_Σ of shallow earthquakes on the Earth at their magnitudes m, $M' = -dN/dm$:

1—graphical differentiation; 2—numerical differentiation

The relation of distribution density of earthquake number on magnitude $N' = -dN_\Sigma/dm$ was calculated (by numerical and by graphical differentiation independently) and presented in a graphical form (Fig. 54). This was to check numerically, whether the obtained function $N_\Sigma(m)$ is smooth enough and to study interesting features of this function. The distribution of calculated points within Gutenberg's interval of reliable estimations of $m > 4.25$ fits perfectly the straight line with the slope:

$$\beta = -d\log N'/dm = 0.99 \pm 0.01. \tag{6.9}$$

There is one exception of a very small, extreme right-hand segment of curve for $m > 7$ which is, as seen from Fig. 53, based on a small number of the largest earthquakes, ca 10 events per year. The value obtained in (6.9) is markably higher (over error limit) than the value given by Gutenberg, $\beta = 0.92$; Gutenberg obtained the latter for $m < 7.1$ by operating direct with earthquake numbers falling into magnitude intervals m of 0.1 unit.

According to Gutenberg, the observed points deviate downward, to fewer numbers of earthquakes from our straight line on the left-hand segment in Fig. 54, i.e. in the area where the statistical data are less reliable. This is probably connected with the uncomplete record of weak earthquakes. The detailed studies of earthquakes performed in some regions of the USSR recently testify, that the graph $\log N' = F(\log E)$ keeps the character of straight line in very large interval of energies E and, consequently, of magnitudes m; here E is seismic energy of the source and it is log-linearly connected with magnitude m (see [Gutenberg] and also [Richter]):

$$\log E = 5.8 + 2.4\,m. \tag{6.10}$$

The most detailed studies performed in Tadjik SSR show that the slope of the graph $\log N' = F(\log E)$ is[*]

$$\gamma = -d\log N'/d\log E = 0.43 \pm 0.05. \tag{6.11}$$

[*] Researchers from the Institute of Earthquake Engineering and Seismology, Tadjik Acad.Sci. found later on the basis of detailed observations of seismicity over the territory of Tadjik SSR that the average value for this territory is $\gamma = 0,48\text{-}0.50$.—(*Editor's note to the Russian edition*).

Taking into account the relation (6.10) we obtain from (6.11) for the value β:

$$\beta = -d \log N'/dm = 1.03 \pm 0.12 \qquad (6.12)$$

which coincide, within the error limits, with the value (6.9) obtained from the reliable part of Gutenberg's statistics ($\beta = 0.99$ corresponds to $\gamma = 0.41$). We note that our estimates (6.11), (6.12) were obtained when studying earthquakes with a range of energy variations from $E = 10^2$ to 10^{13} Joules $= 10^9$ to 10^{20} ergs, i.e., according to (6.10) (energies given in ergs) for the magnitude range from $m = 1.3$ to $m = 5.9$. This interval overlaps essentially the area of uncertain data in Gutenberg's statistics.

The existence of statistical regularity, means the stability of value γ or β over different territories and in wide range of variation of seismic energies E or magnitudes m is a remarkable fact; this permits the extrapolation of the data on earthquake numbers in the direction of small energies or magnitudes with great variability. This seems a good background for accurately estimate the relative numbers of weak earthquakes which may be directly observed and counted in future only following progress in continuously expanding networks of seismic stations.

A suggestion was made in 1958 in the USSR to determine quantitatively the seismic activity A as an annual number of earthquake of any definite or fixed class of energy ($E = 10^{10 \pm 0.5}$ J for example) occurring in specific areas (1000 km^2 for example; see Chapter 2). The activity maps allow us to count easily the average number of earthquakes for any territory. This is not only for the fixed energy class but for all other classes (excluding very rare catastrophic earthquakes) as far as the general relative law of recurrence frequency distribution in energy is presumed.

The author reported this problem [Riznichenko] as far as 1958 at a session of the European Seismological Commission in Utrecht (Netherlands). He appealed to seismologists of other countries to apply the above-mentioned technique and to accumulate the appropriate observations, to process them and to construct seismic activity maps for large territories and finally, for the whole Earth. This suggestion is still valid even now, in connection with the problem of detection of underground nuclear explosions.

The number of earthquakes exceeding by magnitude an explosion of a given power. The annual number of shallow earthquakes N_Σ on the Earth exceeding by their magnitudes m the underground nuclear explosion of given power Y can be found by a simple comparison of magnitude dependence on power $m(Y)$ with the statistics of earthquake numbers $N_\Sigma(m)$.

A comparison of earthquake numbers N_Σ determined in different ways and at various time for several round values of underground explosion power Y is given in Table 18. The four columns of the Table show: (1) earthquake numbers according to data of Expert Conference in Geneva, 1958; (2) the same after "New Seismic data" interpreted in the "Working Document" of January 5, 1959 (see p. 00); (3) earthquake numbers obtained using dependence $M_L(Y)$ for local magnitudes (6.8) with the incorrect theory, following the official position of the American expert delegation in Geneva, 1959, that local magnitudes M_L coincide with those unified magnitudes m used to compile Gutenberg's earthquake statistics [Gutenberg]; (4) earthquake numbers N_Σ determined by combination of our average functions $m(Y)$ for the 1st and 3rd zones according to formula (6.8) or from Fig. 51 and the curve from Fig. 53 for $N_\Sigma(m)$ compiled using Gutenberg's data.

Table 18. Annual Number N_Σ of Shallow Earthquakes on the Earth with Magnitudes Exceeding the Magnitude of Underground Atomic Explosion of Given Power Y

	Number of earthquakes			
Y, kT	Geneva, 1958	"Working Document", Jan. 5, 1959	Data of 1st zone, with erroneous assumption $M_L = m$ (1)	Average for 1st and 3rd zones according to our calculation
1	10 000	26 000	13 000	3200 ± 600
5	3 800	5 800	4 000	1500 ± 300
10	2 400	3 000	3 000	1100 ± 200
20	1 500	1 600	2 000	800 ± 200

Figures with the sign "±" in the last column of Table 18 express the errors in determining number N_Σ of earthquakes caused by inaccuracy of ±0.1 unit in determining average magnitudes m. The same figures permit us to evaluate the number of earthquakes corresponding to observations made separately in the 1st zone (with the sign "−") and in the 3rd zone (with the sign "+").

It was stated in the "Working Document" that, referring to new data, "the annual number of earthquakes equivalent to (explosions of) a given power is two times larger than was previously supposed" (i.e., at the Expert Conference in Geneva, 1959). This is when comparing the figures in the second and third columns of Table 18, at least for explosions of power between 1 to 5 kT. One can see, however, from the same Table that the figures N_Σ of the "Working Document" (third column) for explosions of 1 to 5 kT exceed essentially, by 1.5 to 2 times, those figures (fourth column) which follow from the official position of American delegation at the Expert Conference in Geneva, 1959. The cause is mainly the incorrect determination of magnitude dependence on explosion power Y in the document as it was mentioned above (see Section 6.3).

The figures in the fourth column differ slightly from those which are indicated in the document of Expert Conference in Geneva 1958 (second column). The difference is caused mainly by the small decrease of average magnitudes M_L which occurred because the data from three new stations with lower explosion magnitudes caused by their local peculiarities (mainly stations Barret and Woody) were included into the data processing. This fact played no essential role in changing the results.

The earthquake numbers N_Σ given in the last column of Table 18 were obtained as a result of complete analysis described above; they appear much smaller than all previous esti- (mainly stations Barret and Woody) were included in the data processing. This fact played previously ignored because of unknown reasons. These two circumstances are: (1) a systematic trend in explosion magnitudes in the 2nd zone, the zone of diffraction; this trend forces us to exclude the 2nd zone from the calculation of average values; (2) the difference between local M_L and teleseismic (or unified) magnitudes m established by Gutenberg first for earthquakes [Gutenberg] and confirmed by our work.

The geophysical motivation was needed to consider these two important factors. It was possible to avoid those geophysical contradictions which appeared because of neglect of these two factors only through their analysis and consideration.

One can see from comparison of second, third and fourth columns of Table 18 that despite the statement of "Working Document" it is necessary, in view of new seismic data, to increase markedly the earthquake number N_Σ. A more careful analysis of the same new seismic data leads to the opposite conclusion. It is necessary to decrease approximately by 2 or 3 times the earlier Geneva estimates of 1958 in the 1st and 3rd zones for the explosions of power from 1 to 20 kT. An even larger decrease is necessary for the 1st zone taken separately, by 2.5 to 4 times. In this zone the earthquake number $N_\Sigma = 1200$ events for the explosion of $Y = 5$ kT appears (according to the new more reliable estimations) fewer than that number (1500 events), ascribed earlier to the explosion of 20 kT. We noted the 1st zone here because the observations in this precise zone are mostly used to judge the effectiveness of the first criterium in explosion recognition (in the 2nd zone application of this criterium is practically excluded). This criterium was understood, at both Expert Meetings in Geneva in 1958 and 1959, as the unique criterium recognized by all delegations. Other criteria more effective for the 1st zone and for other zones as well would be elaborated clearly in future.

Graphs for dependence of earthquake numbers N_Σ on explosion power Y corresponding to the second, third and fifth columns of Table 18, are shown in Fig. 55. The slope of our curves $N_\Sigma(Y)$ (one, in the middle, for the combined 1st and 3rd zones and two others on each side, for each of these zones separately) is close to the slope of the straight line drawn for previous Geneva data of 1958. The straight line of the "Working Document" of January 5, 1959 differs from all other straight lines by a clearly exaggerated slope and higher level over the whole main range of explosion power: from 1 to 20 kT.

Remember, that all figures given here relate to the earthquake numbers worldwide. The number of continental earthquakes is approximately two times smaller.

It is necessary to emphasize that, all estimates of seismic effect of underground nuclear explosions made here, should be related to those conditions of their performance which did actually exist at the nuclear test area in Nevada. The technique of analysis and interpretation

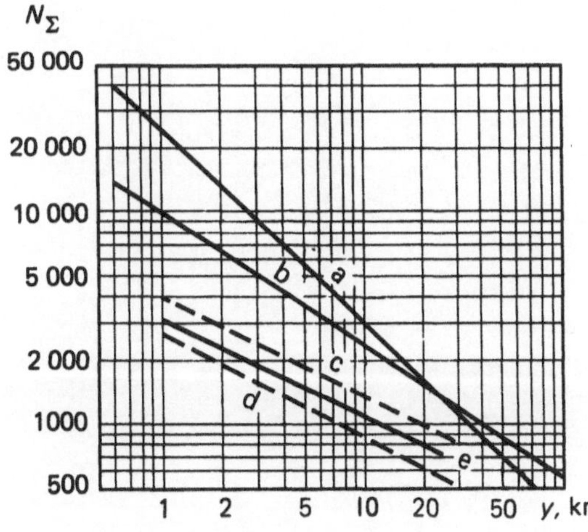

Fig. 55. The annual number N_Σ of shallow earthquakes on the Earth exceeding magnitude m, the underground nuclear explosion of given power Y:

a—Working Document of Jan. 5, 1959; b—Geneva, 1958; c—3rd zone; d—1st zone; e—1st and 3rd zones together of the magnitudes of concentrated underground explosions (regardless whether nuclear or chemical), and the peculiarities indicated in comparison of seismic effects of explosions and earthquakes as well, could be retained generally also for other continental conditions.

The data analysis of American and Soviet observations over the seismic effect, especially for magnitudes of American underground nuclear explosions Reinier, Logan and Blanca of power in order 1 to 20 kT and other explosions of Hardtack II series of much less power showed that, in spite of the statements of American "Working Document" of January 5, 1959 about "New Seismic Data" the previous Geneva, 1958, estimates of numbers of earthquakes corresponding to explosions of similar magnitudes were not underestimated.

Moreover, the precise analysis of new seismic data resulted in earthquake numbers two or three times fewer on average, than was received from previous Geneva estimates for explosions in the indicated range of power.

The problem of control of overground nuclear explosions and search for them among earthquakes is simpler.

II

Seismic Hazard

Maximum possible earthquake and seismic hazard: are they synonyms, at least in a practical sense?

Some people believe that they are. A building that is constructed strongly enough to withstand the maximum possible, for a given place, earthquake, will not collapse during weaker events. However, is it reasonable to make it so strong, if the probability of the maximum possible earthquake is so small that it equals the probability of the impact of a Tungussa or Arizona meteorite? We think not.

The system of seismic zoning now adopted in the USSR is based on the maximum possible earthquake intensity, but not considering the probability of the latter[*]. The rationality of seismic resistance standards is thus still under question. The cost of construction can not be neglected. There is a tendency to diminish the effective "maximum intensity" prior to an earthquake and to exaggerate it after wards. This seems unwise.

The possibility of an alternative background for seismic zoning has been discussed. A general trend can be established for the relation between average frequency of seismic shaking on the intensity of this shaking: severe shaking is rare, weak frequent. One can find the level of such an earthquake, or better its shaking intensity, which corresponds to the *a priori* selected frequency of shaking during the life time of the given building; this level of shaking intensity can be assumed as a design intensity. The problem of the maximum possible earthquake can be then left

---

[1] The new seismic zoning map of the USSR, the GZM-78, includes an indication of the probability of different seismic intensity ranges, in line with proposals made by the author — *Editors note.*

unsolved. The latter may even be infinitely great, for an infinitely small probability. The adherents of such a view appeal to Gumbel's theory of extreme values in statistics, which permits improbable maximum events to be of unlimited magnitude.

However, it is not necessary, in seismology, to adopt such a nihilistic stand concerning the maximum possible event. One can show that an infinitely large maximum possible earthquake would, given the observed law for the distribution of the number of earthquakes according to magnitude, lead to a physically absurd conclusion. This is that the flux density (to be exact, the average power density) of the seismic energy of all earthquake sources should be infinitely great.

We propose something different, namely to define the seismic hazard as the set, at a given place, of all possible shakings of different levels, up to the maximum possible. To be exact, we propose to describe the seismological aspects of seismic hazard by the distribution in the intensity I of shaking of long-term average recurrence B of all seismic shakings possible at any given place. We call the function $B = B(I)$ the long-term average seismic shakeability. The intensity I of shaking can be presented here, either in the traditional, and still not obsolete, term "macroseismic degree of earthquake intensity". Or we can use terms of modern physical or technical indices like displacement, particle velocity, acceleration, energy of seismic oscillations, "seismograms" or frequency spectra. Our spectrum/temporal shakeability belongs to the latter category.

We must emphasize that the problem of the maximum possible earthquake can not be solved by a seismic hazard estimation with the use of shakeability. The maximum possible earthquake should be established for each place at the Earth's surface together with a long-term average recurrence (and, accordingly, with the probability of occurrence) for seismic events of each level up to the maximum one. The proposed new system of seismic hazard estimation is thus more complete than the others mentioned above. It includes values used in both previous systems and enables us to synthesize them with modern physical and technical intensity indices of seismic shaking[*].

---

[*] [Riznichenko, 1971].

7. Quantitative Estimate of Seismic Hazard

7.1. From Activity of Earthquake Sources to the Shakeability of the Earth's Surface[*]

This section is devoted to the background of a quantitative methodology for a transition from source activity presented as a recurrence frequency for earthquakes of different values K to the shakeability of the sites at the Earth's surface obtained as a recurrence frequency of shakings with various intensities I. The intensity itself can be estimated both according to a conventional scale of intensity (officially adopted at the moment) or in terms of other physical values like maximum particle velocity or acceleration at various frequencies, energy of seismic oscillations etc. This which seems to be more suitable for modern geophysical and engineering purposes.

7.1.1. Principal Ideas and Definitions

We shall try to construct a system of definitions and values oriented to the study of the superficial effects of seismicity and its links with underground source activity; the system should be analogous to that used for the study of source activity itself.

We introduce, in parallel to the known parameters describing the activity, like N, K, A, γ and K_{max} [Riznichenko], parameters describing the superficial effect of all earthquakes like I, B, δ, and I_{max}.

Here $n = n(I)$, and B is the recurrence of shakings with intensity I at a given point at the Earth's surface. We shall by tradition, call I the intensity, although we give its value according to a seismic intensity scale as a particular case. We call the graph of the function $n = n(I)$ the shaking recurrence graph or, more concisely, the shakeability graph. In order to discriminate it from previous earthquake recurrence graphs for a source area, it is more convenient to call the latter a (source) activity graph. We refer correspondingly to the activity law as the earthquake recurrence law.

The slope $\delta = -\Delta \log n / \Delta f(I)$ of the shakeability graph is, generally speaking, a variable depending, in particular, on I so $f(I)$ may either be simply I or $\log I$, etc. However, δ may under some conditions be less variable with $f(I)$ and in such areas it can be considered a constant, similar to γ, the slope of the source activity graph. That enables us to simplify the calculations and present the results more compactly.

We shall call B the shakeability. This is the shaking frequency at a point on the Earth's surface produced by shocks of fixed intensity $I = I_i$. This value (we clarify its sense later) can be used to map seismic hazard. It is the quantity which determines the average probability that, at least one case of earthquake impact, of a given I_i or more, will affect the building during its existence T_t. This probability increases with increasing T_i/T_B, where $T_B = 1/B$ is the average period of shakeability. Only for $T_i/T_B \rightarrow \infty$, has this probability a tendency to equal 1.

---

[1] [*] [Riznichenko, 1971].
[**] [Riznichenko, 1965].

A set of shakeability maps B_i for various fixed energy classes or for various finite intensities I_i, $i = 0, 1, 2, 3, ...$ is necessary to characterize seismic hazard in the sense of an average frequency at which points on the Earth's surface undergo seismic shocks of various intensities. However, only one map for shakeability is satisfactory, say a map of B_0, if we approve of $\delta = $ const within an effective range of I; in this case the constant should be indicated as an appendix to the map together with the distribution of I_{max} over the Earth's surface (i.e., the map).

This situation is similar to what we have when describing an average focal seismicity. Again the main problem with shakeability remains the same: it is difficult to determine the maximum possible shaking intensity I_{max}, which essentially depends on K_{max}, the maximum possible earthquake in this area. The problem of K_{max} is important (see Chap. 2) but we will not discuss it here.

One more analogy can be mentioned: this is the possibility of applying the distribution and summation function to shakeability as we did for the source seismicity [Riznichenko]. Just as the earthquake recurrence N and seismic activity A are considered in relation to the source energy classes $K \pm 0.5$ of an earthquake, we can categorize the shaking frequency n or shakeability B in relation to intensity classes $I \pm \Delta I$. The distribution densities n^* and B^* in I will then correspond to the distribution densities N^* and A^* in K. The shakeabilities n and B (the number of events counted from class to class) or n^* and B^* (with upwards summation of number of events starting from some fixed value I) correspond to the earthquake recurrency for a given class K (N_Σ) or starting from a given K(N_Σ^*).

There is still a remarkable difference. All the N, A, etc. values are considered in their usual probability forms of source seismicity, like earthquake recurrence and seismic activity, and so should be related to some spatial area, e.g., per square unit of the zone where sources are localized for a fixed depth, or per volume unit where sources are localized. There is no sense in referring to the earthquake number or earthquake recurrence at one point: these values are simply zero. We may only consider the spatial (volume or, conventionally, superficial) distribution density of these parameters at a point, viz. the area of the focus location. For shakeability the parameters n, B, etc. are evaluated for a point on the Earth's surface. This difference holds for the densities (in K and I), viz. N^* and A_Σ^* for activity, and n^* and B_Σ^* for shakeability.

Here, there is a complete analogy with gravity and other potentials for distributed sources. The "activity" parameters N, A, etc. are analogous to the source density within a volume, but the "shakeability" parameters n, B, etc. are analogous to the potentials of these sources at some point outside the volume, usually at the Earth's surface.

The relation between earthquake focus activity below ground and the shakeability of sites on the Earth's surface caused by this activity generally acts in this same way for long-term average seismicity as the potential theory acts for other geophysical fields.

7.1.2. Formulating the Problem and a General Solution

Let us suppose that both the seismic activity A and distribution of earthquake recurrence $N = N(K)$ with respect to value K ($= \log E$) are specified within the earthquake focal area as point functions $M(x, y, z)$. Furthermore, the intensity I also depends on K and is specified as a function of the points M and $M_0(x_0, y_0, z_0)$, the latter being the point on the Earth's surface for, $z_0 = 0$. We have to find the shakeability B and distribution of shaking recurrence $n = n(I)$ at any specified point M_0 on the Earth's surface. Such a formulation will be more detailed.

Fig. 56. Mutual arrangement of point M on the Earth's surface where shakeability is observed and point M_0 in the vicinity where the source activity is taken place

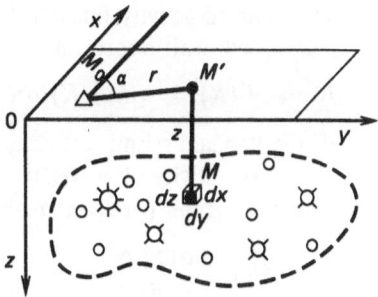

We consider the summation of the earthquake recurrence not the earthquake intensities. We suppose that individual earthquakes occur at different times so that neither earthquake magnitudes K nor their shakeabilities I interfere with each other. This assumption results from the fact that duration of earthquakes both at their foci and on the Earth's surface is negligibly small in comparison with the time intervals between them, so that probability of their coinciding can be neglected. When shocks follow a severe earthquake, their grouping should be treated as a single act.

The seismic regime is presumed to be steady state. We consider the long-term averages so that in case of a cyclic regime, the averages should be for the whole cycle, or for many cycles.

Let us formulate the problem. The volume energy density (by K) of the summed recurrence N_{Σ}^{*} for earthquakes of value K or more is to be determined within the volume element $dx\,dy\,dz$ about the point M, which is within the source area (Fig. 56). N_{Σ}^{*} is a given function of K (Fig. 57):

$$N_{\Sigma}^{*} = N_{\Sigma}^{*}(K). \tag{7.1}$$

The function relating one source activity to another is even more often utilized in seismology, namely the function $N(K)$, the activity distribution in classes K or, the same, the earthquake recurrence function, is given by the expression:

$$N(K) = N_{\Sigma}^{*}(K - 0.5) - N_{\Sigma}^{*}(K + 0.5). \tag{7.2}$$

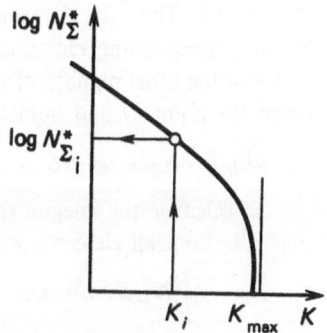

Fig. 57. A diagram of total source activity—the summation of earthquake frequency:

$N_{\Sigma_i}^{*}$—earthquake frequency of the K_i value and more

The summed activity function (7.1) determines all the parameters of the usual activity distribution law with respect to K:

$$N^* = N^*(K) = -dN^*_\Sigma(K)/dK \cong N(K),\qquad(7.3)$$

namely the standard seismic activities, the density activity (in K), $A^* = N^*(K)|_{K=K_0}$ and the class by class activity $A = N(K)|_{K=K_0}$ or the slope γ of an ordinary semi-logarithmic source activity graph (i.e., earthquake recurrence graph):

$$\gamma = \gamma(K) = \frac{d\log N^*}{dK} = -\frac{1}{N^* \ln 10}\frac{dN^*}{dK} =$$

$$= \frac{1}{N^* \ln 10}\frac{d^2 N^*_\Sigma}{dK^2} \cong -\frac{\Delta \log N(K)}{\Delta K}.\qquad(7.4)$$

The value of K_{max}, viz. the maximum possible earthquake in the given area is such that when $K > K_{max}$ in (7.1), we have $N^*_\Sigma = 0$ and thus in (7.3) we also have $N^*(K) = 0$, or, approximately, $N(K) = 0$.

We also consider the function:

$$I = I(M_0, M, K) = I(M_0, r_x, \alpha, z, K),\qquad(7.5)$$

as specified at every point M_0 of the Earth's surface. The function expresses how the shaking intensity I at the given point M_0 depends on the epicentral distance r_x the azimuth α to the epicenter, and the focal depth z and value K of the earthquake (Fig. 56). According to the reciprocity principle, we may say:

$$I(M_0, r_x, \alpha, z, K) = I(M', r_x, \alpha + \pi, r, K).$$

The soil conditions should not be considered when estimating I, since this is a topic of regional seismic zoning, whereas the concrete engineering geological setting should be considered at the stage and at the scale of seismic microzoning.

In general, we fix an arbitrary intensity value I_i so that the construction of a shakeability map of B_{I_i} and the B_{I_i} —value is determined at every point M_0 on the Earth's surface within the map limits.

The first step is to find $K = K_i$ from (7.5) for each point M_0: for each element $dx\,dy\,dz$ about the point M within the source area (see Fig. 56) and for values r_x, α, z, and I_i taken from the mutual positions of points M_0 and M (see Fig. 58).

The second step is to find the corresponding activity $N^*_{\Sigma_i}$ by substituting this value of K_i into the summation function of (7.1) for the point M (see Fig. 57).

Each of the earthquakes composing the sum $N^*_{\Sigma_i}$ causes a shaking of intensity I_i or more at point M_0. Thus, the elementary summed activity $N^*_{\Sigma_i}\,dx\,dy\,dz$ about the point M is equal to the corresponding elementary summed shakeability dB_{I_i} at the point M_0: both values represent the same number of mutually connected events, namely the number of earthquakes about the point M and number of shakings caused by them at point M_0,

$$dB_{I_i} = N^*_{\Sigma_i}dx\,dy\,dz.\qquad(7.6)$$

To calculate the integral shakeability B_{I_i} at point M_0 on the Earth's surface, we have to sum dB_{I_i} from all elements within volume V of the focal area. That is the third step:

$$B_{I_i} = \iiint_V N^*_{\Sigma_i}dx\,dy\,dz.\qquad(7.7)$$

Fig. 58. Dependence of intensity I of shakeability on the epicenter distance r_i at the
given source depth z and the a direction to
an epicenter

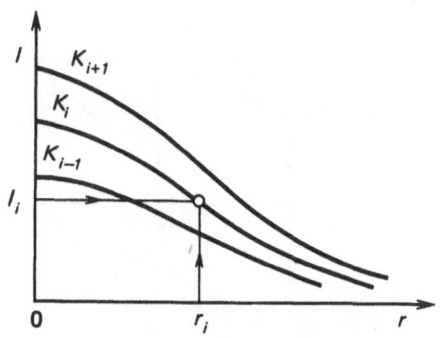

Expression (7.7) is (because of the procedure by which the integrand $N_{\Sigma i}^*$ is connected), the general solution of the problem. The shakeability B_{I_i}, is the average recurrence of shaking with the given intensity I_i or more and is determined by this expression for any point M_0 on the Earth's surface.

The limits of the integration domain V are so defined that the integral in (7.7) is zero when outside it. The boundary of the domain is the locus of the points M, the seismic effect of which at the point M_0 equals I_i when the greatest possible earthquakes K_{max} occur at points M. The domain V need not be singly connected for a heterogeneous source activity field.

When processing the data observed for a region, (7.7) is numerically integrated when the initial data are in a numerical, tabulated, or graphical form. Computer techniques or nomographs like those in gravimetry can then be used. If the initial data can be presented analytically, the solution may also be analytical. Calculation examples are given in Section 7.1.3.

When varying L_i in (7.7) with discrete ($i = 0, 1, 2, 3, ...$) or a continuous sequence of values, we obtain a discrete or continuous function:

$$B_I = B_\Sigma^*(M_0, I). \tag{7.8}$$

It is evident from its construction that this is the summation function, as is reflected in the notation of (7.8). One can also transform it, if necessary, to a distribution function, namely B^*, a density function, or B, a class by class function (within the range $I + \Delta I$). We can also determine the parameters B_0, δ, I_{max} of the "shakeability law", viz. the function $n = n(I)$ i.e., the recurrence of shaking at a shaking intensity I at a point M_0. This transformation is analogous to those which relate different values of the focal activity, (7.1)-(7.4).

7.1.3. Calculation Examples

In order to clarify certain features of the shakeability functions (7.7) and (7.8) with respect to the source activity (7.1) and (7.3), we consider the following simple theoretical examples.

Example 1. The effect of homogeneous activity distributed along an infinite plane. Let the focal activity occur along an infinite horizontal plane at a depth $z > 0$ below the Earth's surface ($z = 0$) (Fig. 59). Let the surface density of the activity at each point M of this plane

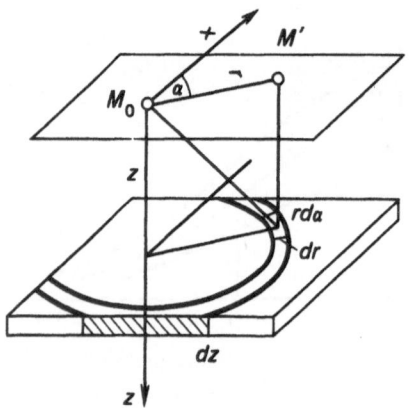

Fig. 59. For the calculation of the M_0 point of the Earth's surface under the action of source activity distributed in a horizontal layer at z depth from the surface

be determined by the usual formulae, viz.:

$$N^* = A^* \times 10^{-\gamma(K-K_0)} \quad \text{by } K \leqslant K_{max} \tag{7.9}$$
$$N^* = 0 \qquad\qquad\qquad \text{by } K > K_{max},$$

where K_{max} is the energy of the maximum possible earthquake for this area. This is an idealization when the source activity is within a layer of limited thickness, say, in the Earth's crust, and when all the sources can be condensed, in order to simplify the calculations, within a layer of negligible thickness dz, which can be simply ignored.

When specifying the function in (7.5) we assume that the medium is homogeneous and non-absorbing, and that the shaking intensity I at the Earth's surface is equal to the surface density of seismic energy flow at the element at point M_0 perpendicular to the ray MM_0. Another way would be to assume that the energy flux through the horizontal surface element at the point M_0 is to be measured. We shall use the first method. In this case we consider I to be given by:

$$I = E/4\pi\overline{MM_0^2} = 10^K/4\pi \, (z^2 + r_x^2). \tag{7.10}$$

Here $E = 10^K$ is the seismic energy of the source. We could assume, according to (7.10), that in (7.5) the intensity I does not depend on the azimuth α.

The shakeability function of (7.7) and (7.8) is found under such conditions.

Prior to searching for a solution, we should transform the focal activity from a distribution function (7.9) to the summation form (7.1). We can rewrite the law using the previous calculations [Riznichenko] in the form:

$$N_\Sigma^* = A^*(10^{\gamma K}/\gamma \ln 10) \, (10^{-\gamma K} - 10^{-\gamma K_{max}}) \quad \text{by } K \leqslant K_{max} \tag{7.11}$$
$$N_\Sigma^* = 0 \qquad\qquad\qquad\qquad\qquad\qquad\qquad \text{by } K > K_{max} \, .$$

Now we start constructing the solution by following the procedure mentioned in Sec. 7.1.2. We are interested in the shakeability B_{I_i} at point M_0 (Fig. 59), which corresponds to the intensity $I \geqslant I_i$, where I_i is an arbitrary fixed intensity, that obviously cannot exceed:

$$I_{max} = 10^{K_{max}}/4\pi z^2. \tag{7.12}$$

This corresponds to the point M_0 just over the focus. In fact, I_{max} is the maximum possible shaking intensity at the epicenter.

The first step. We substitute $I = I_i$ into (7.10) and find $K = K_i$:

$$K_i = \log [4\pi I_i(z^2 + r_x^2)]. \tag{7.13}$$

The second step. We substitute $K = K_i$ into (7.11) and find $N_{\Sigma}^* = N_{\Sigma_i}^*$:

$$N_{\Sigma_i}^* = (A^* \cdot 10^{\gamma K_0}/\gamma \ln 10) \, ([4\pi I_i(z^2 + r_x^2)]^{-\gamma} - 10^{-\gamma K_{max}}. \tag{7.14}$$

It is necessary to limit the variable r_x in (7.14) rather than limit $K \leqslant K_{max}$ in (7.11), the first limitation being related to the second. The limitation can be found from (7.10) by substituting the values $I = I_i$, $K = K_{max}$, and $r_x = r_{x_{max}}$ into (7.10). We find:

$$r_{x_{max}}^2 = 10^{K_{max}}/4\pi I_i - z^2. \tag{7.15}$$

The third step. We construct a surface integral instead of the volume one in (7.7):

$$B_{\Sigma}^*|I_i = \iint_S N_{\Sigma_i}^* dS = \int_0^{r_{x\,max}} N_{\Sigma_i}^* \cdot 2\pi r_x \, dr_x. \tag{7.16}$$

Here $dS = r_x \, dr_x \, d\alpha$ (see Fig. 59). Since we assumed the intensity I is independent and, consequently, the value of N_{Σ_i} at the azimuth α is also independent, the integration with respect to α is computed immediately.

We now substitute $N_{\Sigma_i}^*$ from (7.14) into (7.16):

$$N_{\Sigma}^*|I_i = \frac{2\pi A^* 10^{\gamma K_0}}{\gamma \ln 10} \int_0^{r_{x\,max}} [4\pi]_i(z^2$$

$$+ r_x^2)^{-\gamma} - 10^{-\gamma K_{max}})r_x \, dr_x \tag{7.17}$$

and finally after integration we have:

$$B_{\Sigma}^*|I_i = \frac{\pi A^* \, 10^{\gamma K_0}}{\gamma \ln 10} \left\{ \frac{(4\pi I_i)^{-\gamma}}{1 - \gamma} \left[\left(\frac{10^{K_{max}}}{4\pi I_i} \right)^{1-\gamma} \right. \right.$$

$$\left. - z^{2(1-\gamma)} \right] - 10^{-\gamma K_{max}} \left(\frac{10^{K_{max}}}{4\pi I_i} - z^2 \right) \right\}. \tag{7.18}$$

Here $r_{x_{max}}$ is substituted by its expression in (7.15) as a function of z. The equation (7.18) can be used for calculating the effect of the volume sources distributed in z; the assumption $z = $ var should be made in such a case (see Example 2). We shall nevertheless analyze (7.18) for $z = $ const. In this case we obtain (7.18) in a clearer form by substituting the expression for z^2 from (7.12) into (7.18):

$$B_{\Sigma}^*|I_i = \frac{A^* 10^{\gamma K_0 + (1-\gamma) K_{max}}}{4\gamma \ln 10} \left[\frac{I_i^{-\gamma}}{1 - \gamma} \left(\frac{1}{I_i^{1-\gamma}} \right. \right.$$

$$\left. - \frac{1}{I_{max}^{1-\gamma}} \right) - \left(\frac{1}{I_i} - \frac{1}{I_{max}} \right) \right]. \tag{7.19}$$

We obtain $B_{i_{\max}} = 0$ from (7.19) for the special case where $I_i = I_{\max}$, as was to be expected. This means that maximum shakings may occur very rarely at any fixed point M_0. This can be ignored when working out probability for seismic zoning, but when $I_i \neq I_{\max}$ the shakeability is finite $B_{i_i} > 0$. Let us check this for the particular case of $\gamma = 0.5$. The expression in the square brackets in (7.19) can be simplified to:

$$(1/\sqrt{I_i} - 1/\sqrt{I_{\max}})^2,$$

which proves the statement above.

The formula (7.19) is a summation shakeability function like (7.7) and (7.8). Let us assume that $I_i = I$ and is variable, and let us find the function $B_I^* = - dB_\Sigma^*/dI$ which is the shaking distribution in intensity I at point M_0. This function is the analogue of the distribution function (7.3) of the source activity in K:

$$B_I^* = \frac{A^* 10^{\gamma K_0 + (1-\gamma)K_{\max}}}{4\gamma \ln 10} \frac{d}{dI} \left[\frac{I^{-\gamma}}{I-\gamma} \left(\frac{1}{I^{1-\gamma}} - \frac{1}{I_{\max}^{1-\gamma}} \right) - \left(\frac{1}{I} - \frac{1}{I_{\max}} \right) \right]$$

$$= - \frac{A^* 10^{\gamma K_0 + (1-\gamma)K_{\max}}}{4(1-\gamma) \ln 10} \frac{1}{I_{\max}^{1-\gamma}} - = \frac{1}{I^2} \right). \tag{7.20}$$

When $I = I_{\max}$ we obtain from (7.20) that $B^*|_{I_{\max}} = 0$. At the same time for an area of weak shaking $I \leqslant I_{\max}$ we obtain, clearly from (7.20),

$$B_I^* \cong \frac{A^* 10^{\gamma K_0 + (1-\gamma)K_{\max}}}{4(1-\gamma) \ln 10} \frac{1}{I^2}. \tag{7.21}$$

This was the distribution density of shakeability B^* in I used in (7.20)-(7.21). Because of the very large range of variation in I (several orders of magnitude) it is more convenient to consider the shakeability distribution with respect to $\log I$. This is similar to the usual way of considering the distribution of focal activity with respect to $K = \log E$ rather than energy E. Since $dB_\Sigma^*/d \log I = I \ln 10\, dB_\Sigma^*/dI$, the logarithmic distribution density of shakeability n^* in (7.20) will be:

$$n^* = - dB_\Sigma^*/d \log I = - \frac{A^* 10^{\gamma K_0 + (1-\gamma)K_{\max}}}{4(1-\gamma)} \left(\frac{1}{I_{\max}^{1-\gamma} - I^\gamma} - \frac{1}{I} \right). \tag{7.20'}$$

As for (7.21) we have:

$$n^* = - dB_\Sigma^*/d \log I \big|_{I \ll I_{\max}} = \frac{A^* 10^{\gamma K_0 + (1-\gamma)K_{\max}}}{4(1-\gamma)} \cdot \frac{1}{I}. \tag{7.21'}$$

We can see that for a homogeneous distribution of focal activity along a horizontal plane under the Earth's surface, the logarithimic distribution density (in $\log I$) of shakeability at the point M_0 of the Earth's surface is a curve with the slope δ in the coordinates $\log I$, $\log n^*$. The slope asymptotically approaches the value:

$$\delta|I \ll I_{\max} = - \Delta \log n^*/\Delta \log I = 1. \tag{7.22}$$

The slope of the shakeability graph $n^* = n^*(I)$ in this asymptotic approach and in this coordinate system is constant and does not depend on the slope γ of the focal activity graph (viz. the earthquake recurrence graph). Severe shaking $I \to I_{max}$ occurs less often than might be expected from (7.22); for such shaking the recurrence curve bends downwards, and this bend depends on γ (see 7.20) and (7.20′)).

Example 2. Effect of homogeneous activity distributed within the layer and within the half-space. Let the focal activity be specified within a layer limited by two horizontal planes at depths z_1 and z_2 under the Earth's surface, $z = 0$. The shakeability at a point M_0 on the Earth's surface is to be determined. Let the volume density of activity be specified by (7.9), like the superficial density in the previous case. The value I is the intensity of shaking at the point M_0, and is given by (7.10).

We fix the intensity $I = I_i$ for which the summed shakeability B_Σ^* (i.e., the recurrence frequency of shaking of intensity I_i or more) should be found. In this case the summed shakeability dB_Σ^* at point M_0 caused by sources within a layer of the thickness dz at a depth z is determined by (7.18) multiplied by dz. The summed shakeability $B_\Sigma^*|_{z_1}^{z_2}$ caused at the same point by a source layer of finite thickness at a depth from z_1 to z_2 can be expressed, in accordance with (7.7), by the integral:

$$B_\Sigma^*\Big|_{z_1}^{z_2} = \iiint_V N_{\Sigma_i}^* \, dx \, dy \, dz = \frac{\pi A^* 10^{\gamma K_0}}{\gamma \ln 10} \int_{z_1}^{z_2} \left\{ \frac{(4\pi I_i)^{-\gamma}}{1 - \gamma} \left[\left(\frac{10^{K_{max}}}{4\pi I_i} \right) - z^{2(1-\gamma)} \right] \right.$$

$$\left. - 10^{-\gamma K_{max}} \left(\frac{-10^{-\gamma K_{max}}}{4\pi I_i} - z^2 \right) \right\} dx, \tag{7.23}$$

over the whole volume V of the layer. The integration should only be with respect to z since the integration with respect to two other coordinates was completed before (7.18).

We should consider (7.23) that $0 < z_1 < z_2 < z_{max}$, where z_{max} is the maximum depth at which an epicentral shaking of intensity $I_i = 10^{K_{max}}/4\pi z_{max}^2$ is still produced by a source of maximum value K_{max}, whence:

$$z_{max} = (10^{K_{max}}/4\pi I_i)^{1/2} \tag{7.24}$$

Shaking of intensity I_i or more occurs at point M_0 due to the part of the layer limited by the sphere of radius z_{max} centered at M_0.

By integration of (7.23) we get:

$$B_\Sigma^*\Big|_{z_1}^{z_2} = \frac{\pi A^* 10^{\gamma K_0}}{\gamma \ln 10} \left\{ \frac{\gamma 10^{(1-\gamma)K_{max}}}{4\pi(1-\gamma)I_i}(z_2 - z_1) \right.$$

$$\left. - \frac{(4\pi I_i)^{-\gamma}}{(1-\gamma)(3-2\gamma)} \times (z_2^{3-2\gamma} + z_1^{3-2\gamma}) + \frac{10^{-\gamma K_{max}}}{3}(z_2^3 - z_1^3) \right\}. \tag{7.25}$$

This expression is the summed shakeability at M_0 on the Earth's surface due to source activity distributed homogeneously within the layer at depths from z_1 to z_2 under the Earth's surface.

We will now consider a special case when $z_1 = 0$ and $z_2 = z_{max}$. This corresponds to the homogeneous distribution of source activity over the whole lower half-space under the Earth's surface. In this case we obtain the following by substituting $z_1 = 0$ and $z_2 = z_{max}$

from (7.24) to (7.25):

$$B_\Sigma^* = B_\Sigma\big|_0^{z_{\max}} = \frac{A^* 10^{\gamma K_0 + (3/2 - \gamma)K_{\max}}}{6(3 - 2\gamma)\sqrt{\pi}\ln 10} \cdot \frac{1}{I_i^{3/2}} \tag{7.26}$$

If M_0, where the shakeability is to be determined, is inside a medium with homogeneous activity, (7.26) should be duplicated.

Let us find the density function for the shakeability distribution in value I for activity distributed over the half-space. To do this we assume in (7.26) that $I_i = I$ is a variable and we determine $B^* = - dB_\Sigma^*/dI$. We find the following for the density in terms of I:

$$B^* = \frac{A^* 10^{\gamma K_0 + (3/2 - \gamma)K_{\max}}}{4(3 - 2\gamma)\sqrt{\pi}\ln 10} \cdot \frac{1}{I^{5/2}}, \tag{7.27}$$

and in terms of logarithmic density (by log I):

$$n^* = \frac{A^* 10^{\gamma K_0 + (3/2 - \gamma)K_{\max}}}{4(3 - 2\gamma)\sqrt{\pi}} \cdot \frac{1}{I^{3/2}} . \tag{7.27'}$$

We can state that shakeability graph for point M_0 on the Earth's surface constructed for the logarithmic shakeability density (by log I) in the coordinate system log n^*, log I for a homogeneously distributed source activity within the half-space under the Earth's surface is a straight line with the slope:

$$\delta = - \Delta \log n^* / \Delta \log I = 1.5. \tag{7.28}$$

This slope is greater than the asymptotic value $\delta = 1$ found when the activity was distributed along the horizontal plane. The slope in (7.28) does not completely depend on γ.

We note that shaking at M_0 may, in the spatial case of (7.26) and (7.27), be any value, but as $I = \infty$ the shakeability becomes the recurrence frequency when the shaking is zero, so that there is no reason to consider them from a probability viewpoint.

Finally, we must remember that the intensity I was taken, in all these examples, as equal to the seismic energy flux at M_0 through the surface element, perpendicular to the ray, in a homogeneous medium. Different definition of I, for instance, if I is treated as a macroseismic intensity, depends on the epicentral distance for a fixed focal depth, the shakeability calculations at M_0 will be different.

7.2. Maps of Probable Earthquake Intensity[*]

We propose here a technique for comparing seismic danger maps constructed using two different techniques, namely maximum intensity of seismic shaking $I_{\max}$ [Medvedev] and seismic shakeability $B = B(I)$, where $B(I)$ is the average recurrence frequency for shaking of intensity I [Riznichenko, 1965].

---

[*] [Riznichenko, 1973.]

Fig. 60. Average T periods of shakeability frequencies with the official "maximal" intensity I_{max} at points for Eastern Uzbekistan

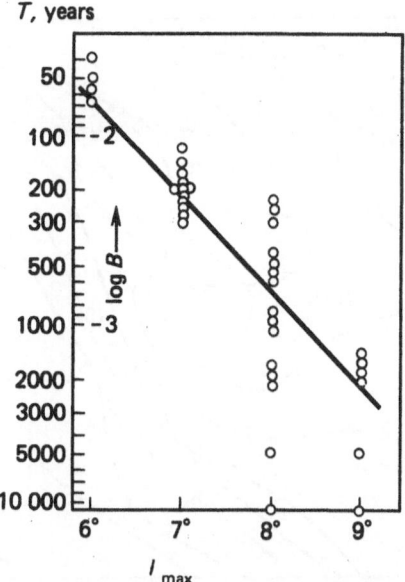

The shakeability can be expressed in terms of intensity either in the form of a macroseismic intensity scale or a physical spectral and temporal index [Riznichenko, 1966]. When comparing the maps we shall use, naturally, as a common index, a macroseismic intensity scale.

The seismic danger for each point (x, y) on the Earth's surface is given as a system of maximum intensity I_{max} in the form of a single value, agreed to be an integer. The map of I_{max} (x, y) is compiled in zones, which are discontinuous horizontal steps in the $I(x, y)$ surface.

The isolines on such map are boundaries of the I_{max} zones, which differ by a single unit of I. To compare shakeability maps which are constructed as continuous smooth maps in isolines and I_{max} (x, y) maps, the latter must be smoothed. Boundaries between zones of different shaking intensity should in any case be treated as the isolines of a smooth surface I_{max} (x, y) with I_{max} equal to the arithmetic average between their values in the zones they had formerly subdivided.

The seismic danger at each point on the Earth's surface is determined in the shakeability system B by the function $B(I)$, viz. a one-dimensional sequence of points (B_i, I_i). The seismic danger for a territory is, in turn, a one-dimensional sequence of maps, either of B_i (x, y)-maps with $I_i = $ const for each individual map, or of $I_i(x, y)$-maps with the parameter $B_i = $ const for each map. The recurrence period $T = 1/B$ can also be used instead of the recurrence frequency B.

The standard seismic zoning map for Eastern Uzbekistan [S. Medvedeva, 1968] was compared with the sequence of shakeability maps B_i of the same territory [Zakharova, et al] compiled in isolines $T_I = 1/B_I$ for $I = 6°$, $7°$, $8°$, and $9°$ in the macroseismic intensity scale. Both maps were covered by the same coordinate grid, and I_{max} values were compared with the T_I, $I \geqslant I_{max}$ from shakeability maps at each grid node.

The results are given in Fig. 60. All the observed points $(I_{max}, \log B)$ are clustered, with a large scatter, around a straight line that rises in the direction of greater I_{max}.

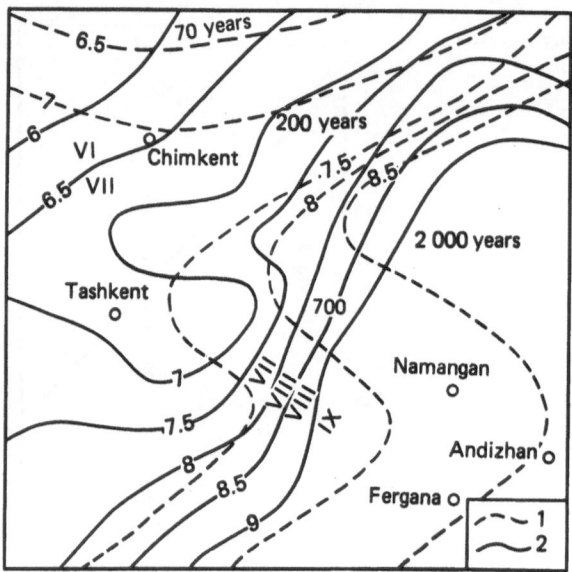

Fig. 61. The map of official "maximal" intensity I_{max} at points (*1*) when comparing a map with a probable intensity I_{max}; (*2*) with the indicated average period of shakeability frequency for Eastern Uzbekistan

The average periods of shaking recurrence $T = 1/B$ for a given intensity $\geq I_{max}$ are:

I_{max}, degrees	6°	7°	8°	9°
T, years	70	200	700	2000
T_M, years	—	290	800	1200

The last line is a rough estimate of the average recurrence periods T_M for the "maximum possible earthquake" (after S.V. Medvedev 93-95). Our T seem to be close to T_M. However, shaking greater than I_{max} within the I_{max} zones on a standard map can occur according to the shakeability maps [Medvedev, 1968]. The average recurrence periods of shaking of intensity $I_{max} + 1$ are, however, several times greater than for the intensity I_{max}, and this is, perhaps, a good reason not to consider such intensities, in the first approximation.

The average function $B = B(I_{max})$ shown in Fig. 60 can be used directly to compile the probable intensity (macroseismic intensity) map that lies at the base of the shakeability maps. This is very interesting since it is the first formal technique for compiling seismic zoning maps for a design intensity I'_{max}, with a meaning close to that of the official "maximum" intensity I_{max}. The latter is, in fact, not a "maximum" but is instead of a value established from qualitative, mostly intuitive suppositions.

An example of a probable intensity map I'_{max} is shown in Fig. 61 in comparison with the standard (official) maximum intensity map I_{max}. When compiling a map of I'_{max} the values of T were taken from the averaging line (Fig. 60) for the intensities $I_{max} = I$. Then the isolines $T = $ const were redrawn to the same map from the shakeability maps for different I. The boundaries of zones with rounded, integer intensities I'_{max} are drawn on the map with continuous lines and labelled by fractional numbers so the isoline $I'_{max} = 6.5$ divides a zone of intensity $I'_{max} = 6 \pm 0.5$ from a zone of intensity $I'_{max} = 7 \pm 0.5$.

We can conclude that the probable intensity map in Fig. 61 is only an illustration of the technique and should not be used for seismic zoning of the territory. For practical applications, analogous comparisons should be made for a set of the main seismic regions of the USSR, and average functions like those in Fig. 61 should be found for the whole country. This is the only way to create a unified technique for estimating seismic danger in equivalent indexes of probable intensity that are valid everywhere. Besides, the final map of probable intensity should include corrections for local soil and other conditions (seismic microzoning).

7.3. Spectral-temporal Seismic Shakeability[*]

We use seismic shakeability B [Riznichenko] as a measure of seismic danger so that seismic zonation should be orientated towards it. The shakeability $B = B(I)$ (which is the long-term average recurrence frequency in terms of the seismic shaking of each intensity possible at a point, site, or construction) allows us to estimate the seismic risk $P = Bt = t/T_B$, where t is the expectation time of a seismic event (the amortization time for a construction etc.), and T_B is the average time between occurrences. The measure of danger formerly used in seismic zonation, i.e. the maximum expected shaking intensity I_{max} in terms of an arbitrary descriptive macroseismic scale, was poorly related to quantitative oscillation indexes for ground and structures and did not contain a probability element. The technique did not yield data for engineering or structural design versus seismic loading or give a value for the background for optimum seismic resistance design in general. The shakeability method provides such data and backgrounds in principle.

7.3.1. Presentation and Calculation of Shakeability

When determining the shakeability, the intensity I can either be used as a simple scalar, the intensity according to the MMS, MSK, or other scale, the maximum velocity, acceleration, energy density etc. or as a function, e.g. Fourier frequency spectrum of ground point oscillations, response spectrum for the modeling system of a structure, oscillation time history, viz. a seismogram, accelerogram etc. Most attention has so far been paid to the first kind of presentation. The second kind is now becoming the main one [Riznichenko].

The spectral shakeability $B = B(S, T)$ is a function of oscillation intensity S for each period T. The same information on seismic danger at a point could be presented as a spectrum set with a synthetic probability shakeability spectra $S = S(T, B)$. Here where the current parameter is T, the period of oscillations is S and the curve parameter is $B = 1/T_B$, the average frequency, T_B being the average period (in years) between oscillation occurrences at a given point, oscillations being measured by intensity S for each given period T. These intensities could be Fourier spectra of ground point oscillation for displacements, velocities, accelera-

---

[*] Riznichenko and Seiduzova as co-author.

tions, or energy density etc. (convenient for seismologists). Or, they could be response spectra, that is spectra of maximum relative displacements, velocities or accelerations of a model or a mechanism imitating the construction, by various oscillation periods T (favorable for engineers).

The temporal shakeability $B = B(f, t)$ is the temporal analogue of spectral shakeability. Here f is a time oscillation expressed by seismogram, or accelerogram, etc., t is the current oscillation time in S. The same information could, in principle, be presented by probability shakeability seismograms, accelerograms etc. of $f(t, B)$ which correspond to a set of spectra $S(T, B)$ in the case of spectral shakeability. They are synthetic shaking seismograms which are expected to be repeated on average at the given point with a frequency $B \cdot \text{year}^{-1}$.

The initial information used to calculate the shakeability B in any version (according to a macroseismic intensity scale, energy density, or in a spectral-temporal form) are data obtained from earthquake catalogues and seismograms, and from some other sources. These data are reduced to the following, which also are of independent interest:

(1) the function $N = N(K)$, viz. the dependence of earthquake recurrence frequency N on earthquake energy $K = \log E$, E is the seismic energy, or on magnitude M;

(2) seismic activity maps [Riznichenko] compiled for different focal depths h;

(3) maps of maximum possible earthquakes K_{max} or M_{max} [Riznichenko] (K_{max} can be determined from seismological data or other geophysical, geological and geodetical data) [Riznichenko, Shebalin, Savage];

(4) dependence of seismic oscillation intensity I on earthquake source value K and source distance r (here the intensity I can be given in macroseismic, energetic or spectral form [Riznichenko, Shebalin, Savage]).

The methodology and techniques for shakeability calculation are described in many papers. Shakeability maps with intensity I presented in terms of seismic energy density ε at the observed point see [Riznichenko, et al, 1967], maps in terms of macroseismic intensity are described in [Riznichenko, 1969]. Computer softwave for calculating shakeability with intensity presented in terms of a macroseismic intensity scale according to the Blake-Shebalin formulae [Shebalin, Sponheuer], or Kövesligethy formulae [Kövesligethy] are described in [Seiduzova, et al]. Softwave to prepare the initial information and compile $N(K)$ functions as A and K_{max} maps are described in [Zakharova]. The accuracy of shakeability determination is studied in [Drumya, et al]. Shakeability in terms of macroseismic intensity is calculated for many seismic territories of the USSR from Moldova [Drumya] to Kamchatka [Vvedenskaya] (for a complete list of references see [Dzibladze, et al,]). Shakeability maps have also been constructed for several territories abroad, e.g., Italy [Riznichenko, at al,] and the Carpathian-Balkan region [Riznichenko, et al].

We shall consider details of the shakeability calculation by presenting intensity I as different spectra (Fourier and response) for seismic oscillations. The compilation of spectral shakeability maps is a matter for the immediate future.

7.3.2. System of Average Spectra of Seismic Oscillations

The dependence of the seismic oscillation spectrum on various main factors $S = S(K, h, r, T)$ has been touched on before using a general mathematical model which also contains some constants. These constants are determined by computer so that the set of observed spectra

Fig. 62. A system of average energetic spectra
of Tashkent earthquakes, 1966.

T, (s)—oscillation period; r, km—a focal distance;
q—energetic spectral density, $\tilde{q} =$
$\log (d\varepsilon/d \log T)$; ε—energy density at the point of
observation, J/m^2

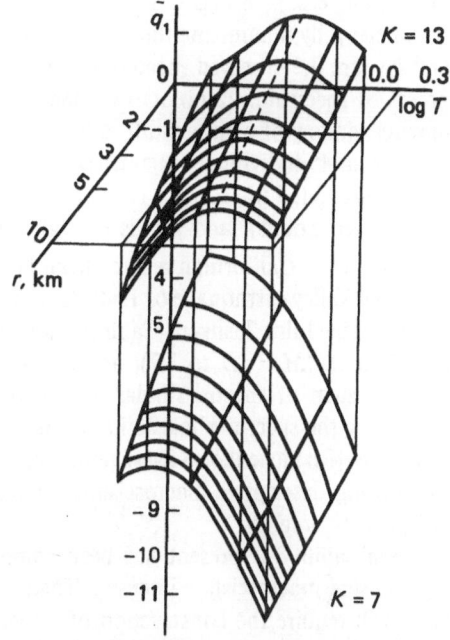

is fitted best (here "best" is used in the sense of the least square or similar methods). Such
a general approach is, for various reasons, more effective than the analysis of partial samples
by fixing all the arguments but the one under study, the latter approach being typical of
manual calculations.

An example of a system of average spectra computed for the sequence of crustal Tashkent
earthquakes of 1966 is shown in Fig. 62. The axes are T, the oscillation period (0.03 to 2.0 s);
r, the focal distance (3 to 10 km), $\tilde{q} = \log (d\varepsilon/d \log T)$, the spectral density of seismic energy
ε, ε in J/m^2. The latter ($\tilde{q}$) was calculated from Fourier spectra of the ground displacement
records. Each individual curve in Fig. 62 is the average oscillation spectrum $S(T)$ of an earth-
quake for a given value K(or M), the oscillation being recorded at a certain distance r from
the focus. The lower part is the sequence of such spectra from a small earthquake of $K = 7$
($M = 1.7$), the upper part is the same for a bigger one, $K = 13$ ($M = 5$).

A mathematical model of a system of spectra was used, in this case in a general form,
with the following equation:

$$\tilde{q} = \sum_{i=0}^{l} \sum_{j=0}^{m} a_{ij}(K - K_0)^j(\log T - \log T_0)^i +$$

$$+ n(\log r_0 - \log r) + \frac{\beta}{T} (r_0 - r) \log e. \qquad (7.29)$$

Here n is the divergence power; $\beta = \alpha T = \nu/c = \pi/cQ$; α is the coefficient and ν is the absorption decrement; Q is a factor of medium quality; c is the wave velocity (the transverse

one, in our case). n and Q were constants to be determined and were calculated using the parameter β by minimizing the function $R = \Sigma[\tilde{q}_{obs} - \tilde{q}]^2$, where $\tilde{q}$ is calculated from (7.29), and $\tilde{q}_{obs}$ are the observed spectral densities. The summation is carried out at all the observed points of spectrum. 22 points were measured along each spectrum after smoothing it by a moving window of octave width, $\Delta \log T = \log 2$. There are about 1000 individual measures, taken from 45 individual spectra. If, in (7.29) $l = 4$, $m = 2$, the total number of parameters to be determined in (7.29) is 17. The number of conditional equations is 1000 (the number of observed points), and this is enough for a firm determination of these parameters.

The set of Californian records from [Trifunac, 1972] the famous El Centro accelerogram (Imperial Valley earthquake of 1940, $M = 6.3$) to 1976 is processed in addition to the Tashkent records. The Joint Tashkent-Californian system of average spectra is obtained in a range of $K = 7$ to 18 ($M = 1.7$ to 7.7), and $r = 2$ to 130 km.

A system of spectra similar to that described above should play the same role when computing the spectral shakeability as the Kövesligethy or Blake-Shebalin macroseismic field equations [Riznichenko, Kövesligethy, or Blake-Sponheuer] do in shakeability calculations when using descriptive macroseismic intensity I. Systems of spectra are, naturally, also of independent interest.

Shakeability at present has been computed for Italy and the Balkan and Carpathian region using macroseismic intensity. The spectral shakeability calculations planned for these areas will require the construction of a regional system of average spectra in advance. The fitness of existing systems, e.g. the Tashkent and Californian one, should be studied and, if necessary, corrected.

7.3.3. Relation Between Fourier Spectra and Response Spectra

A system of average response spectra could be constructed using the same technique in principle, as is used for Fourier spectra. Another simpler way to do this is to establish approximate correlations between Fourier spectra and response spectra.

The general rule [Trifunac] is that the response spectrum RV for oscillation velocity with zero damping $\zeta = 0$ of the system (the structure) approximates to the amplitudes of the Fourier spectrum for acceleration if the absolute oscillation at the system input (i.e., the structure's basement or ground) is the same. To be more exact, when $\zeta = 0$ an inequality like $RV \geqslant FA$ does exist, and but if $\zeta \neq 0$ it may not. This is because the response spectrum contains maximum values of the spectral function which appear within the system while the driving force is acting, while the Fourier spectrum contains those values which appear after the force hads ceased to act but the latter cannot be greater.

To clarify a situation typical in seismology, the correlation between two kinds of spectra were studied experimentally. The data from [Trifunac] were used, in total 20 cases of a vertical component and 40 cases of a horizontal component for Californian accelerograms of earthquakes with $K = 14$ to 18, $M = 5.3$ to 7.7. The response spectrum RV after simple smoothing with a moving octave window was, in each case, compared with the corresponding Fourier spectrum FA. The data were processed by T.G. Kondratieva, and the author expresses his thanks. The results are presented in Table 19, in which $\Delta = \log (RV/FA)$ is the relative difference between the spectral functions RV and FA, $\pm \delta_\Delta$ is the standard deviation of each value from the average Δ. We emphasize that the data in Table 19 are not of deterministic but only correlative in character.

Fig. 63. *RV* reaction spectra in comparison with the Fourier spectra *FA*.

Attenuation ξ for reaction spectra in percent from boundary-aperiodic. The 68° band ±Δδ is shown in a reaction spectrum for ξ = 0

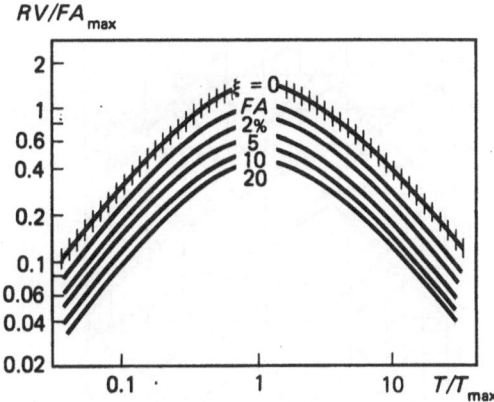

The mutual positions of the response spectra RV and the Fourier spectra *FA* are given in Fig. 63 with a 68%-confidence limit shown as a dashed stripe for *RV*, the spectrum with damping $\xi = 0$. The values of $\pm \delta_\Delta$ for other values of the damping factor ($\xi = 2$ to 20% from an aperiodic damping boundary) are given in Table 19. We ignored, at this stage, the small systematic variations of Δ with changes of T and these were grouped as random deviations $\pm \delta_\Delta$.

Table 19. Comparison of Response Spectra *RV* with Various Dampings ξ of the System and the Fourier Spectra *FA*

%	Vertical			Horizontal		
	RV/FA	Δ	$\pm \sigma_\Delta$	*RV/FA*	Δ	$\pm \sigma_\Delta$
0	1.4	0.15	0.05	1.4	0.15	0.05
2	0.8	−0.10	0.13	0.9	−0.05	0.09
5	0.6	−0.22	0.13	0.7	−0.15	0.13
10	0.5	−0.30	0.18	0.6	−0.22	0.17
20	0.4	−0.40	0.19	0.45	−0.35	0.20

We see from Table 19 and Fig. 63 that the variance $\pm \delta_\Delta$ for the smoothed spectra is rather small. This means that an approximate determination of a response spectrum from a Fourier spectrum is possible, both for $\xi = 0$, and other values of ξ. The inverse is also possible. Similar possibilities exist for other systems of average spectra.

7.3.4. Transformation of Spectra

The consideration of different spectra, e.g., the Fourier spectra F and response spectra R for displacement D, velocities V, accelerations A, and energies $\tilde{q}$, and the transformation of one spectrum to another requires a procedure that can be reduced to simple multiplication

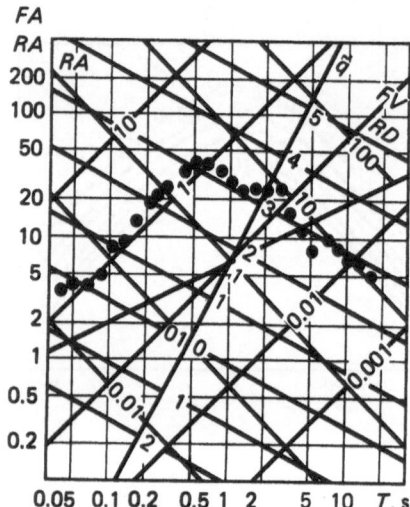

Fig. 64. Transformation net for the calculation of Fourier spectral functions and reaction R

or division by a factor ω, energy spectra require experimentation. These simple operations can be further simplified, and comparisons can be unified and made clearer when using the graphical presentations and operations upon spectra using multiscale logarithmic transformation graphs (Fig. 64). An example is shown in Fig. 64, where a unified smoothed spectrum of the bigger horizontal component of the El Centro accelerogram is plotted on the grid.

The grid (Fig. 64) is a generalization of the tripartite grid used in [Trifunac] and other similar papers from the EERL. The grid is used to demonstrate a unified set of curves with various parameters ξ for various response spectra for displacement RD, velocity RV, and acceleration RA. We have plotted, in addition to the axes used for these spectral functions, additional axes for the Fourier spectra of the ground (basement structure) oscillations, displacement FD and energy density $\tilde{q}$ in particular.

If we take the moduli of the logarithmic scales for RV and T to be 1, the moduli M_{ik} of the scales for other values, angles φ_{ik} of their axes being measured from the vertical axis are given in Table 20, the T axis being horizontal.

Here $\tan \varphi_{FD} = 2$; $M_{FD} = \cos \varphi_{FD}$; $\tan \varphi_{\tilde{q}} = 0.5$; $M_{\tilde{q}} = 0.5 \cos \varphi_{\tilde{q}}$. At the same time $M_{FD} = M_{\tilde{q}}$, and $\varphi_{FD} + \varphi_{\tilde{q}} = 90°$ (but the angle between FD and $\tilde{q}$ axes is not a right-angle, being instead of $\varphi_{FD} - \varphi_{\tilde{q}}$).

The use of this transformation grid can be clarified by an example (see Fig. 64). If $T = 1$, the reaction acceleration is equal to $RA = 0.6 \, g$ (g is gravity acceleration), and the function of the energy spectrum with the same T is $\tilde{q} = 3.0$. If $T = 0.1$, $RA = 1.2 \, g$ and $\tilde{q} = 1.0$.

We can see from Fig. 64 and Table 20 that the axis of the velocity reaction RV coincides with the axis of the Fourier acceleration FA; the axis of the displacement reaction RD coincides with the axis of Fourier velocity FV; the axis of acceleration reaction RA is at the same time the axis of velocity acceleration, that is of third derivative of the displacement; the axis of Fourier displacement FD is the axis of integral in T from displacement response.

The transformation grid (Fig. 64) transforms between displacement, velocity, acceleration and oscillation energy in principle. This transformation between the spectral response functions R and the Fourier functions F retains its approximate and correlative character.

Table 20. Moduli M_{ik} of Logarithmic Scales and Axes Measured from the Vertical Axis of the Transform Grid (Fig. 64) for Fourier Spectra F of Ground (Structure Foundation) Oscillations and Structure Response Spectra R for Displacements D, Velocity V, and Acceleration A, and the Energy $\tilde{q}$ of Seismic Oscillations

Parameter	ω^0	ω^1	$\omega^{1.5}$	ω^2	ω^3
Fourier spectra	FD	FV	$\tilde{q}$	FA	
Response spectra		RD		RV	RA
Modulus M_{ik}	0.447	$1/\sqrt{2}$	0.447	1	$1/\sqrt{2}$
Angle φ_{ik}	63°25'	45°	26°35'	0	$-45°$

7.3.5. Spectral Shakeability at a Point and over an Area

The general principles for calculating and possible ways for presenting seismic danger in terms of spectral shakeability were discussed above. We mention here only those forms which are most convenient.

The spectral shakeability of a point (e.g., a town or construction site) is presented as a set of synthetic probable shakeability Fourier spectra for the energy density $\tilde{q}$ of the ground oscillations. They should be compiled for the oscillation periods $T = 0.1, 0.2, 0.5, 1.0$, and 2.0 s, i.e., for the frequencies 10, 5, 2, 1 and 0.5 Hz, five points in total for each curve. The parameter for each spectrum is the average recurrence period T of the spectral components $T_B = 20, 50, 100, 200$, and 500 years, i.e., the average shaking frequency is $B = 5, 2, 1, 0.5$, and 0.2 for 100 years, five curves in total. The seismic danger is expressed under these conditions, by $5 \times 5 = 25$ numbers.

We can see that the same spectra will also yield after certain recalculations (or given a transformation grid) other spectral Fourier functions for ground oscillations (or the response function of the structures, which are correlated with the latter) for displacement, velocity, and acceleration. The quantity of local initial information does not increase.

Twenty-five maps in isolines of the above point characteristics are required to present the seismic danger for a territory (viz a two-dimensional zone) in the same spectral and temporal terms. The map parameters could be either the oscillation period T (in s) and recurrence times T_B (in years) for seismic shocks, or the oscillation frequency $f = 1/T$ (in Hz) and seismic shakeability $B = 1/T_B$ (in years^{-1}).

The mapping in isolines could represent a spectral intensity (primarily the energy value $\tilde{q}$), and there are already systems of average spectra for this value which are necessary for shakeability computation.

The mapping could, at the same time, be for any other value related with $\tilde{q}$ either functionally or by correlation, e.g., the relative velocity RV, or acceleration RA (in percents of gravity acceleration), plotted on the response spectra.

The set of spectral shakeability maps described above permits us to reconstruct, for each point of the territory, its own system of Fourier spectra or response spectra for the ground level of a structure for given recurrence period T_B (in years), viz. for any given probability or risk.

This is the general scheme for the final solution of how to establish seismic shakeability. As for temporal shakeability, the methodology for determining it has not been completed. Pending the appearance of a rational technique for the synthesis of shakeability seismograms, we propose, for practical purposes, the combination of synthetic shakeability spectra with observed seismograms between close spectral characteristics.

8. Macroseismics

8.1. The Energetics of Macroseismology*

There are two basic systems for assessing the intensity I of seismic shocks. The first is a descriptive system in which I is expressed in conventional degrees of some macroseismic scale (Medvedev and others); the main criteria for some scales are human feelings and expert assessments of the damage to buildings and changes on the Earth's surface. The second is a quantitative system in which the oscillations in seismic tremors are expressed in objective instrumental readings: spectra/temporal, energy readings etc. [Medvedev, Riznichenko]. The two systems have so far been used independently, though the possibility of matching them has been discussed for a long time. It has been attempted by establishing probability correlations between the systems [Dzhibladze and others). Although the importance of this approach cannot be denied, we will take another route. On the basis of hypotheses formulated as a deterministic mathematical model with a clear physical meaning, we attempt to establish deterministic functional relationships between the two systems.

8.1.1. Initial Assumptions

Assume that the focus of an earthquake is characterized by a spherical direction pattern of the emission of elastic energy by body waves. The energy density is expressed as:

$$\varepsilon = \frac{E}{4\pi r_0^2} \left(\frac{r_0}{r}\right)^n e^{-\alpha(r - r_0)} \tag{8.1}$$

Here $r = [(x - x_0)^2 + (y - y_0)^2 + h^2]^{1/2}$ denotes the focal distance; x_0, y_0 are the coordinates of the epicenter; x and y are the coordinates of the point of observation; h is the depth of the focus (hypocenter); $E = 10^K$ is the seismic focal energy, which is given on a reference sphere of radius r_0; K is the earthquake energetic value; n is the exponent of divergence; and α is the coefficient of absorption and scattering of seismic energy in the medium in a particular wave for an observed oscillation frequency. In the simplest case of an unbounded homogeneous, perfectly elastic medium, we have $n = 2$ and $\alpha = 0$. The meaning of equations such as (8.1) was discussed [Berzon, Kogan, Riznichenko] for several other cases.

Furthermore, assume that there exists a logarithmic-linear relationship

$$I/I_0 = 1 + c_0 \log (\varepsilon/\varepsilon_0)$$

or

$$I = c_1 + c_2 \log \varepsilon \tag{8.2}$$

between the macroseismic intensity I in the interval considered (usually $4 \leqslant I \leqslant 10$ on the MMS, MSK, etc. scales) and the seismic energy density ε at the point of observation, where

---

* See Riznichenko, 1974

$c_1 = I_0(1 - c_0 \log \varepsilon_0)$ and $c_2 = c_0 I_0 \log \varepsilon_0$ are constants. A comparison of macroseismic data and instrument observations indicates that a relation of approximately this type exists. The destructive effect I is usually more closely related to the energy density $\varepsilon = w\tau$ than to the power w and the duration τ of the seismic shocks or to certain maximum parameters.

Equations (8.1) and (8.2) define our model. Naturally, the two assumptions expressed by (8.1) and (8.2) are to some extent arbitrary. The two assumptions provide only a crude scheme compared with a more complex real situation. We believe that this scheme is justified because all the basic traditional empirical relationships, i.e., the "macroseismic interpolation formulae" of Köveslighety, Blake, Shebalin, and others have been successfully used for many years by seismologists all over the world when interpreting observational data. These traditional relationships are only particular conclusions drawn from (8.1) and (8.2). The traditional formulae can be deduced from these equations.

It may appear strange that this simple idea has not yet been clearly formulated before or discussed in literature on seismology, though has been touched on several times [Shebalin, Köveslighety, Sponheuer].

One technicality must be considered. In macroseismic formulas [193, 289], the earthquake value is often expressed in magnitudes M rather than in energy units K. Therefore, we have to use a relation between K and M like the Gutenberg-Richter-Rautian relation [Vvedenskaya].

$$K = a + bM, \tag{8.3}$$

where a and b are constants. Equation (8.3) does not lend itself to a smooth incorporation into our model: in principle, macroseismic formulae could be expressed directly through the energetic value K.

All subsequent relations (except the spectral aspect) will be simple consequences of these three equations.

8.1.2. Meaning (in terms of energy) of a Köveslighety formula

By substituting (8.1) into (8.2) we obtain

$$I = c_1 + c_2[-\log 4\pi r_0^2 + K - n \log (r/r_0) - \alpha(r - r_0) \log e], \tag{8.4}$$

or, in a different notation,

$$I = C_{K1} + C_{K2}K - C_n \log (r/r_0) - C_\alpha(r - r_0), \tag{8.5}$$

where

$$c_1 - c_2 \log 4\pi r_0^2 = C_{K1}; \quad c_2 = C_{K2}; \quad c_2 n = C_n; \quad c_2 \alpha \log e = C_\alpha; \tag{8.6}$$

from this we calculate:

$$c_2 = C_{K2}; \quad c_1 = C_{K1} + C_{K2} \log 4\pi r_0^2; \quad n = C_n/C_{K2}; \quad \alpha = C_\alpha/(C_{K2} \log e). \tag{8.7}$$

Equation (8.4) or (8.5) is a macroseismic formula like Köveslighety's equation [Kövesligethy, Sponheuer] in which the earthquake value is expressed in units of K. With (3), the same formula has the following form when the earthquake magnitude M is used:

$$I = C_{M1} + C_{M2}M - C_n \log (r/r_0) - C_\alpha(r - r_0), \qquad (8.8)$$

where

$$C_{M1} = C_{K1} + C_{K2}a; \quad C_{M2} = C_{K2}b, \qquad (8.9)$$

so that:

$$C_{K2} = C_{M2}/b; \quad C_{K1} = C_{M1} - (a/b)C_{M2}. \qquad (8.10)$$

Thus, when macroseismic data are processed with a Kövesligethy formula (8.5) or (8.8), and when the macroseismic parameters C_{K1}, C_{K2} or C_{M1}, C_{M2} and C_n and C_α are obtained, (8.7) and (8.10) can be used to obtain the parameters c_1, c_2, n, and α, which have a clear physical meaning in the initial energy model defined by (8.1) and (8.2). We are therefore able to interpret the results of macroseismic observations in terms of physics.

8.1.3. Meaning (in terms of energy) of the Blake-Shebalin formulas

In (8.1) the attenuation of the energy density ε with increasing distance r was given by $(r_0/r)^n e^{-\alpha(r-r_0)}$. This expression contains two constants n and α, which are determined from a set of observations. Let us assume that $\alpha = 0$ and that the entire attenuation results from a single divergence (an effective divergence in this context) n_e, i.e., we make the formal substitution:

$$(r_0/r)^n e^{-\alpha(r-r_0)} = (r_0/r)^{n_e}. \qquad (8.11)$$

Accordingly, (8.1) becomes:

$$\varepsilon = \frac{E}{4\pi r_0^2} \left(\frac{r_0}{r}\right)^{n_e} \qquad (8.12)$$

We obtain with (8.2) from (8.12):

$$I = c_1 + c_2[-\log 4\pi r_0^2 + K - n_e \log (r/r_0)], \qquad (8.13)$$

or, in another notation:

$$I = C_{K1} + C_{K2}K - C_{n_e} \log (r/r_0), \qquad (8.14)$$

where:

$$c_1 - c_2 \log 4\pi r_0^2 = C_{K1}; \quad c_2 = C_{K2}; \quad c_2 n_e = C_{n_e} \qquad (8.15)$$

and thus:

$$c_2 = C_{K2}; \quad c_1 = C_{K1} + C_{K2} \log 4\pi r_0^2; \quad n_e = C_{n_e}/C_{K2}. \qquad (8.16)$$

By switching from K to M, we obtain the following from (8.14) in accordance with (8.13):

$$I = C_{M1} + C_{M2}M - C_{n_e} \log (r/r_0), \qquad (8.17)$$

or:

$$I = (C_{M1} + C_{n_e} \log r_0) + C_{M2}M - C_{n_e} \log r$$

$$(I = C_{Sh} + b_{Sh}M - S_{Sh} \log r),$$

where, as above, the relations between C_{M1}, C_{M2} and C_{K1}, C_{K2} are (8.9) and (8.10). Equation (8.17) is the macroseismic formula of Blake-Shebalin. The formula in the form proposed by Shebalin is shown in parentheses. The Shebalin formula coincides with (8.17) if r is measured in units of r_0. When this is done (as usual) on the length (km) scale, the free term c_{Sh} in the Shebalin formula becomes dependent upon C_{n_e}, i.e., an effective divergence, which obfuscates its physical meaning.

Once we have established the parameters $c_{Sh} = c_{M1}$ and $b_{Sh} = C_{M2}$ by processing macroseismic data according to (8.17), we can use (8.10) and (8.7) to switch to the parameters c_1 and c_2 of (8.2) of our energetic model. In order to obtain the parameters n and α, a further step must be made. Let us determine the meaning of the effective divergence exponent n_e. We obtain from (8.11):

$$n_e = n + \alpha \frac{(r - r_0)}{\log (r/r_0)} \log e. \tag{8.18}$$

This means that when an approximate meaning in terms of physics is ascribed to (8.1) and when (8.12) is assumed to be a formal equation, the parameter n_e not only depends upon the properties of the medium but also upon the distance r. There are other reasons for this behavior. A great dependence of n_e on the distance to the focus was repeatedly noted by several researchers. I.V. Anan'in, who has made great efforts to map n_e as a "characteristic quantity of a medium" reflecting both absorption and scattering in the medium, tried to eliminate the dependence of n_e on r by determining n_e from data which had been obtained mainly in a fixed, bounded distance interval of r (private communication). In discussions with Anan'in the author pointed out that the absorption and scattering within the medium cannot be exposed as the parameter n_e in Blake-Shebalin formulae like (8.12), (8.17). It is more logical to express absorption and scattering by α, which is derived for this purpose and appears in the macroseismic Kövesligethy formulae like (8.5) and (8.8) and in our energy model (8.1).

Nevertheless, let us assume that a Blake-Shebalin formula (8.17) was used to process observation data. How can we switch from the parameters of this formula, particularly from the parameter n_e (for attenuation) to the energy parameters n and α? Naturally, an independent determination of the two quantities from a single n_e is impossible. However, we can determine α once we have, in some way, fixed n from geophysical concepts such as the average structure of the medium, the travel time curve, the wave type, and the frequency spectrum of the pulse. For example, we can either assume $n = 2$ or we can use some other fixed value or a value expressed as a function of r. Thus, we obtain from (8.18):

$$\alpha = (n_e - n) \frac{\log (r/r_0)}{(r - r_0) \log e}. \tag{8.19}$$

In this formula r must be assumed as the average distance for which n_e was determined.

According to (8.16) and (8.10), n_e is expressed through the parameters $C_{M2} = b_{Sh}$ and $C_{n_e} = S_{Sh}$ of the Shebalin equation (8.17) (when r is expressed in units of r_0) and also by the parameter b of the Gutenberg-Richter-Rautian equation (8.3) in the following fashion:

$$n_e = C_{n_e} b/C_{K2} = C_{n_e}/C_{M2} = S_{Sh} b/b_{Sh}. \tag{8.20}$$

Thus, in this case the interpretation of macroseismic data in terms of energy is possible under certain additional conditions.

8.1.4. The Spectral Aspect of the Problem

So far we have tacitly assumed that the observed frequency of the seismic oscillations is fixed. In this case, the coefficient of adsorption (and scattering), α, can be assumed as a property of the medium. However, in reality seismic pulses are not monochromatic and α depends upon the frequency $f = 1/T$ (T denotes the oscillation period) or wavelength $\lambda = VT$ (V denotes the wave propagation velocity); high-frequency short-wave oscillations are more strongly attenuated. Since the seismic pulse travels in an absorbing medium, the pulse changes its spectral composition and tends to lower frequencies, which means that the attenuation rate decreases. The spectrum of the pulse depends also upon the magnitude of the earthquake at the focus: the low-frequency components of the spectrum increase with increasing K or M. The buildings, from whose damage the macroseismic intensity I is assessed, are affected by different, selective, frequency-dependent properties: low rigid buildings are destroyed by high-frequency oscillations, whereas high, flexible buildings are destroyed by oscillations of lower frequency. Moreover, nonlinear effects are important. All this implies substantial complications in the interpretation of macroseismic observations in terms of physics. Is it possible to take all these details into consideration at least approximately, i.e., a linear approximation? If this is possible, what does the approximation look like?

We will outline in general terms how this could be done. It is not sufficient to consider only the macroseismic intensity I or data on the energy value K of the earthquake or on its magnitude M. Spectral data which are obtained from instrument observations must be considered. A system of average energy spectra of earthquakes must be the basis for the evaluation of the spectral distribution of the seismic oscillation energy as it depends on the magnitudes K or M of the earthquake or the focal distance r. In the system of the loss θ of energy absorption divided by the wavelength λ or the Q factor of the rocks rather than the coefficient α of absorption is assumed to be constant:

$$\alpha = \theta/\lambda = \pi/\lambda Q \tag{8.21}$$

There is no particular doubt that the parameters θ and Q are more or less independent of the wavelength λ, at least under the conditions which apply to the consolidated crust and the upper mantle, where the majority of sources of destructive earthquakes are located. If necessary, one can introduce a certain dependence of θ or Q upon λ, and, hence, upon f, though this is rarely done int he evaluation of macroseismic information which is so uncertain.

Information on the spectral "window" in which various types of buildings are destroyed can be obtained from engineering seismology (Medvedev and others). This filtering window transmits at various frequencies which depend upon the transmissivity of partial windows for different types of buildings and upon the weights determined from the relative number of the cost, etc., of the construction of a particular site. The window must overlap our physical system of spectra. The result is a filtered macroseismic system of spectra. The effective macroseismic energy $\varepsilon_M = \varepsilon_M(K, r)$ could be calculated from the macroseismic system of spectra for each particular earthquake magnitude K or M and for each given distance r. In addition, one must determine the corresponding dominating frequency $f_M = f_M(K, r)$. From the viewpoint of the energy spectrum, the quantity ε which appears in (8.1) and (8.2) of our initial model is also ε_M: this quantity accounts for the macroseismic effect I. Local conditions, which are a matter of seismic microzoning, may be included as spectral considerations [Plotnikova, Ratnikova].

This spectral approach is essentially linear. It is beyond the scope of our spectral method to take into consideration nonlinear phenomena.

Let us state the main problems encountered in the processing of macroseismic data for our subsequent discussion. The first problem is the construction of functions of the average macroseismic field $I = I(K, h, r)$ and the determination of the depth h of the foci of individual earthquakes from the functions. The second problem is the mapping of the absorption properties of the medium on the basis of a set of isoseismals of individual earthquakes in the region.

8.1.5. The Average Macroseismic Field and the Depth of the Focus

Until recently, the function of the macroseismic field was usually described in the USSR in the form given by Blake- Shebalin (8.17), i.e. on the basis of a model of the type defined by (8.12) with only one attenuation parameter (the effective divergence exponent n_e). This approach was dictated by the simplicity of the calculations, required when calculations are made manually. Moreover, a single parameter n_e can be determined with greater accuracy than the two parameters n and α. This is particularly important when a relatively small number of statistical results are simultaneously processed, i.e., when the calculations are manual.

Since computers were introduced, these technical advantages have diminished in view of the flexibility, and better approximations of real conditions provided by the more complex but clearer (in terms of physics) model defined by (8.1) and (8.2) with the two attenuation parameters n and α, and the corresponding macroseismic formulae Kövesligethy's type [Dzhibladze, Riznichenko, Kövesligethy].

The general idea of using the models in computer calculations is as follows. The parameters c_1, c_2, n, and α of the model defined by (8.1) and (8.2) (i.e. the average values of the parameters) are assumed to be constant for the set of simultaneously processed macroseismic data $I(K, h, r_x)$ in the region (r_x denotes the epicentral distance with $r^2 = h^2 + r_x^2$). By minimizing deviations of the observed I from those generated by the mathematical model, the optimal parameter values c_1, c_2, n, and α are determined for the region, along with the depths h_i of the earthquake foci considered in the data processing.

A special communication could outline these ideas in detail. However, let us note that this form of our model disregards the spectral aspect of the problem. It might seem that this is immaterial when only the depth h of the foci is determined. However, this is not the case.

In view of the selective attenuation of high-frequency oscillation components in the medium, the effective absorption coefficient α which corresponds to the frequency f_M (frequency of the oscillations leading to the destruction of buildings) varies with the focal distance r: with small r, the macroseismic frequency f_M and, hence, the effective value α_{obs} are larger, but they decrease for large r. This shortens the distances between close isoseismals and increases them for distant isoseismals, by comparison with the distances predicted by a theoretical model with a constant α. If this is disregarded, excessive depths are obtained in calculations involving remote isoseismals of low frequencies rather than nearby, high-frequency isoseismals.

Since the spectrum of oscillations originating from large earthquakes is richer in low frequencies, the details described must lead to an overall increase in the distances between all isoseismals of severe earthquakes in comparison with the averaged model and, hence, must imply a systematic overrating of the focal depths of large earthquakes. Further modifications of our basic model as defined by (8.1) and (8.2) are required for these details.

The same spectral dependences must affect focal depth determinations based on a Blake-Shebalin formula with constant n_e. The only difference is that the true reasons for possible errors are less clear. Disregarding the spectral aspect is possibly the main reason for the systematic overestimates of the depths of large earthquakes relative to the focal depths of weak earthquakes. This overestimate occurs when the well-known method of [Shebalin] is employed to process macroseismic data. Excessive depths were obtained in cases where the macroseismic source depth data of large earthquakes could be compared with the more accurate depth values derived from instrumental observations.

A method for including the frequency dependence in determinations of the depth h can be based upon our macroseismic system of spectra and the resulting "theoretical" functions $\varepsilon_M(K, r) = \varepsilon_M(K, h, r_x)$, where r_x denotes the epicentral distance and $r^2 = h^2 + r_x^2$. After transforming from ε_M to $I = I(K, h, r_x)$, the problem is solved, as above, by minimizing the deviations of the observed I from the theoretical I values.

8.1.6. Mapping the Absorption and the Q Factor of the Medium

Mapping can be made to establish and plot zones in the crust and the upper mantle with increased absorption, scattering, fissure concentration, reduced strength, etc. with which seismic motion may be associated. The maps are a possible indicator of potential seismicity. They may be used, in combination with other indicators, to determine the long-term average seismic activity A and the greatest possible earthquake magnitudes K_{max} from a combination of data obtained from seismology, recent motion data, neotectonics, geophysical fields, and structural details of the crust and the upper mantle [Riznichenko et al]. The determination of A and K_{max} is an important part of the determination of the seismic hazard in terms of the seismic shakeability [Riznichenko] (see Chapter 7).

The general outline of the mapping of the absorption loss θ or of the Q factor on the basis of macroseismic data is basically as follows. Spectral instrument observations are used, as indicated above, to construct an average system of functions $\varepsilon_M(K, h, r_a)$ and $f_M(K, h, r_x)$, wherein the macroseismic energy ε_M is converted into the "theoretical" intensity $I(K, h, r_x)$. On the other hand, a set of macroseismic fields obtained in a particular region is employed, i.e., maps of the isoseismal lines $I_{obs}(K, h, r_x)$ of earthquakes, their epicenters, and the parameters K and h are determined.

Local characteristics of the theoretical macroseismic field $I(K, h, r_x)$ and the observed macroseismic field $I_{obs}(K, h, r_x)$ are compared at each point of the map, i.e., the absolute values of the horizontal gradients $|\text{grad } I| = \partial I/\partial r_x$ and $|\text{grad } I_{obs}| = \partial I_{obs}/\partial l$ are considered, where l denotes the distance reckoned in the direction orthogonal to the isoseismal lines. The corresponding distances between isoseismal lines can be considered in place of the gradients, i.e. the distances between the theoretical Δr_x values and the observed Δl values. When we assume that the difference between the model which reflects the average conditions and the local natural conditions is given by the difference in absorption, i.e., when the divergence of the waves at a particular point is the same in the model and in reality, we obtain:

$$|\text{grad } I_{obs}|/|\text{grad } I| = \Delta r_x/\Delta l = \alpha_{obs} - \alpha = \Delta \alpha. \tag{8.22}$$

Furthermore, assuming that the dominating macroseismic frequency at that point is equal to the theoretical frequency f_M, we can use (8.21) to switch from the increment of the absorption coefficient α to the increment of the absorption loss θ:

$$\Delta\theta = \Delta\alpha\lambda = V(\Delta\alpha/f_M) \tag{8.23}$$

Since the value of the absorption decrease θ is known in the spectral model, we can determine the decrease θ_i at each point of the map:

$$\theta_i = \theta + \Delta\theta_i \tag{8.24}$$

and, hence, we can determine the Q factor $Q_i = \pi/\theta_i$ at each point.

Average values of θ and Q can be determined where the macroseismic fields originating from various earthquakes overlap. The point values of θ and Q are generalized by the isolines of a continuous field. If desired, zonal maps of discontinuous θ or Q fields can be constructed.

We can see that the basic formulae of the Kövesligethy and Blake-Shebalin formulae for the intensity I of the macroseismic field can be obtained by deduction from a known expression for the density ε of the seismic energy emitted from a focus with spherical symmetry of emission, for the dissipation parameter n and for the absorption parameter α, assuming that a linear relation between $\log \varepsilon$ and I exists.

Based on this model, we have shown that it is possible to determine the energy parameters n and α from macroseismic observations, provided that either the energy value K of the earthquake in the source or the earthquake magnitude M is known in advance.

8.2. Macroseismic Master Curve*

The macroseismic intensity I expressed in terms of descriptive scales has recently been reestablished during quantitative studies of seismicity and seismic danger, to become probability indexes of seismic shakeability, the base of the new seismic zoning of the USSR. In spite of the imperfections of macroseismic determinations, improving them is important as they are connected with establishing better correlations between empirical macroseismic parameters and the physical characteristics of the process.

Many methods have been proposed to express the dependence of macroseismic intensity I on the focal depth h, epi- or hypocentral distance r and magnitude M, or the energy value of an earthquake $K = \log E$. The principal method and its mathematical model remains in Kövesligethy's. The attenuation of intensity I (or energy density ε) is considered as a function of two parameters, namely divergence s (or n) and absorption (together with scattering) α of seismic energy. The Blake and Shebalin approach is more widely used in the USSR and describes the attenuation by a single parameter s_l (or n) of effective divergence, which includes the divergence proper and also the energy absorption; this approach is a particular case using Kövesligethy theoretical model.

The physical or energetic sense of a Kövesligethy equation was discussed under Sec. 8.1. We continue this discussion to some practical conclusions. But first we discuss the determination of the parameters of mathematical model of the macroseismic field used for processing observational data.

---

* See [Riznichenko].

8.2.1. Current Position

Dozens of methods and technical approaches for determining such macroseismic field parameters as focal depth h, and absorption index are known. The majority are described in detail, commented on, and illustrated by calculation in the "classical" book by Sponheuer*. The approach (which has become classic in the USSR) is described in papers by Shebalin etc. New ideas do not cease to appear. Inspiration by seismologists on this simple problem develops because of its practical importance (especially to interpret past earthquakes with their poor instrumental records) and not because of some kind of sporting interest (although its influence can not be excluded). However, because all the approaches proposed before have some essential demerits and were, in particular deprived of necessary physical sense, clarity and universality.

Some approaches described in [Sponheuer, 1960] can be used only with fixed n (or s) and α. This is also true for the technique proposed by Sponheuer himself where according to Köveslighethy the constant $s = 3$ is used. The epicentral intensity I_0 is presumed, in most approaches, to be already known although it would be natural to establish it together with other parameters from a set of observational data. In some methods the whole observed data set is not used and only samples are needed and this leads to inconsistencies in the results and to less stable solutions. For some cases, it is recommended to start from the third isoseismals (counting from the epicenter), and ignore the rest. Such a recommendation eliminates from the calculation numerous earthquakes with only a limited number of isoseismals. Some approaches use the area of the isoseismal zones instead of intensities I and distances r as the initial information. This leads to complications when the observed isoseismals are not closed and does not permit the calculation of damping parameters (individually or in general) various directions from the epicenter. Moreover, it would be better to use, direct, primary values I_i and r_{x_i}, which are fairly objective rather than the isoseismals themselves.

Isoseismals are, at best, drawn by different researches and may be inconsistent.

We shall see that the technique proposed here is free from most of these defects.

8.2.2. Formulation of the Problem

A Köveslighethy equation has the general form:

$$I = c_0 + b_0 M - s \log r - c_\alpha r, \quad r = (h^2 + r_x^2)^{1/2} \tag{8.25}$$

Here I is the macroseismic intensity; r is the hypocentral and r_x is the epicentral distance; h is the focal depth, M is the magnitude; c_0, b_0, s, c_α are constants, where s describes the divergence, and c_α describes absorption. The magnitude M and the earthquake epicenter are presumed known in the simplest formulation of the problem.

The problem is, generally, to select values of the parameters c_0, b_0, s, c_α within the area and h_k for the observed earthquakes of magnitudes M_k that make the mathematical model

---

* Sponheuer, W., Maaz, R. Ein Beitrag zur makroseismischen Berechnung der Erdtiefe. – Communs Observ. roy Belg. Sér. géophys., 1971, N 101, p. 85-87.

of (8.25) fit the set of observed pairs I_{ik}, $r_{x_{ik}}$ as well as possible (in the sense of least squared deviation $I_{obs} - I$).

In the particular case, where $c_\alpha = 0$ we obtain from a Köveslighethy equation (8.25) a Blake-Shebalin equation, i.e.

$$I = c_0 + b_0 M - s_0 \log r, \quad r = (h^2 + r_x^2)^{1/2} \tag{8.26}$$

The parameter s_0 of the intensity damping with distance r determines the "effective divergence" as it describes the divergence and absorption together. The general character of the problem does not change.

It is necessary to discriminate between methods for determining one or two damping parameters when searching for a concrete rational processing technique for macroseismic data. The first approach is more physics-oriented but when simultaneously processing rather limited amounts of data, it is much less stable than the second method. The "manual" solution of this problem, as we will see, is easy to realize in both methods. The second method, is better, as noted in [Shebalin, 1968], since it is usually done with differentiation in limited portions of the whole information. The idea of simultaneously calculating a mutually conformed solution of the problem from all the data was described for both methods in the previous section. We consider here how both methods yield "manual" solutions, and especially for the second method because of its efficiency.

In order to simplify the "manual" solution of an inverse macroseismic problem of geophysics, that is the determination of the parameters in this general model of the process from the field observations associated with this process, we try to construct the general solution of the "direct problem" in a clear form, and to get the set of theoretical fields connected with this model in a convenient graphical form for all possible values of the model parameters. Since the parameters used in the calculation of the theoretical fields are known, the inverse problem could be solved by comparing the observed field with all the possible theoretical ones and selecting the field which best fits the observed one.

This general approach is widely used in geophysical prospecting to find a solution, given a selected general model, and to eliminate incorrect, unstable inverse problems of potential theory and similar problems, among which we find other macroseismic problems as well. The approach was realized in seismic prospecting by constructing the theoretical travel time of the master curve used to determine the position of the reflecting boundary and the seismic velocity in the covering layer. In seismology the focal depth and wave velocities from observations of the arrival times of near-field seismic oscillations were determined [Riznichenko]. This approach can easily be applied in this macroseismic problem to determine those macroseismic field parameters which vary most often and most strongly. These are the hypocentral depth h and damping parameters s and α, or s_{eff}.

8.2.3. Determination of Depth and Attenuation Using a Set of Master Curves

For a compact graphic representation of the set of macroseismic fields for all possible values of the parameters let us represent (8.25) as:

$$\log s + \log \left(\log r + \frac{c_\alpha}{s} r \right) = \log (c_0 + b_0 M - I), \quad r = (h^2 + r_x^2)^{1/2} \tag{8.25'}$$

Fig. 65. Diagram of a macroseismic chart for determining h depth source and the parameters of intensity attenuation s, c_α or s_e.

The chart axes are *continuous lines*, the axes of the observed diagram (*circles*) are *broken lines*

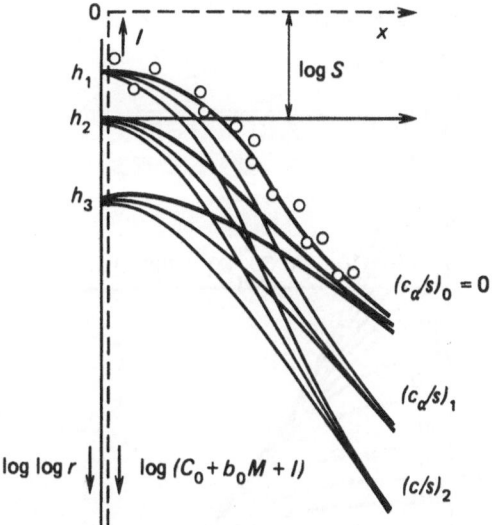

and in the particular case where $c_\alpha = 0$ let us represent (8.26) in the form

$$\log s_0 + \log \log r = \log (c_0 + b_0 M - I), \quad r = (h^2 + r_x^2)^{1/2} \qquad (8.26')$$

Equation (8.25') for the two attenuation parameters s and c_α and (8.26') for the single parameter s provide the key to the construction of a set of theoretical curves $I(r)$ with parameters h and c_α/s or h for $c_\alpha/s = 0$, which we call a set of macroseismic master curves.

A diagram of the set of master curves to explain how to use them is shown in Fig. 65. These master curves depict the function $\log [\log r + c_\alpha/sr]$. Their argument is r_x, and the parameters are h and c_α/s. The coordinate axes of the set of master curves are shown in Fig. 65 by solid lines, while the axes of the theoretical graph being compared with it are shown by the broken lines (see Fig. 65).

The problem may be solved as follows. The values of $\log (c_0 + b_0 M - I_i)$ are calculated from the observed values of I_i for the c_0 and b_0 chosen in the zero approximation. They are taken as the ordinates of the observed points, the abscissas of which are r_{x_i}. This graph is plotted on a separate sheet of paper from the points: they should not be connected beforehand by an "observed curve". The observed graph is superimposed on the set of master curves in such a way that the y-axes of the graph and the set of master curves coincide. The graph and the set of master curves are shifted relative to each other along the y-axes until an optimum coincidence of the observed sequence of points (with allowance for the natural dispersion) with one of the theoretical curves of the set is achieved. The parameters h and c_α/s determine the hypocenter depth and the ratio between the attenuation parameters. The attenuation parameter s (or s_e) is determined by the length of the segment $\log s$ of the y-axis between the x-axes of the observed graph and the set of master curves. Since the ratio c_α/s is already known, the second attenuation parameter c_α is thereby determined. Naturally, for such operations it is necessary that the observed graph or the set of master curves be plotted on transparent paper.

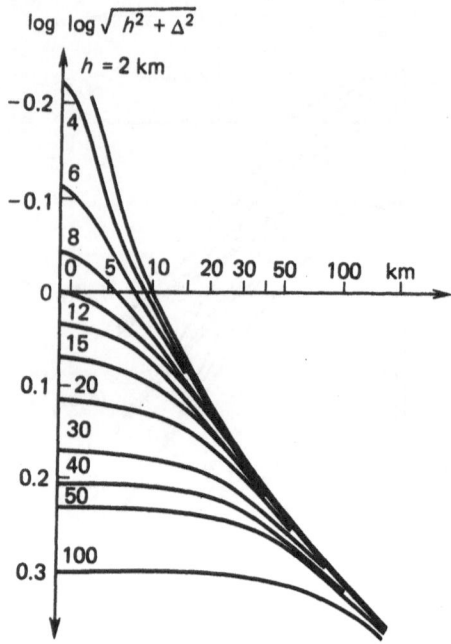

$\log \log \sqrt{h^2 + \Delta^2}$

In the particular case where $c_\alpha = 0$, i.e. for (8.26′) only a single-parameter family of curves with the parameter $s = s_e$ for $c_\alpha / s = 0$ is used. A diagram of a working set of curves is given in Fig. 66, for simplicity's sake, just for this case. Now we can be guided by it.

The intensity I_0 at the epicenter is not determined from some individual observed point, but from the ordinate of the initial point of an averaging theoretical curve plotted in the coordinate system of the observed graph. This value can be calculated from the formula

$$I_0 = c_0 + b_0 M - s_0 \log h. \tag{8.27}$$

This value of I_0 follows from the entire set of observed points I_i, r_{x_i} and is automatically consistent with all the other parameters. This eliminates the need for a special study of the "I_0 problem", which has sometimes been necessary when using other more particular methods of treating macroseismic data.

Let us cite a few other advantages and possibilities of using a set of macroseismic master curves.

Generality of the solution. A set of master curves makes it possible to make a comparison with the general mathematical model used simultaneously for an entire set of observed data I_i, r_{x_i} for each individual k-th earthquake without prior treatment of the data: without isoseismals, determining their areas, sampling, etc. This reduces the possibility of additional arbitrariness in the processing. On the other hand, if the primary determinations of I_i, γr_{x_i} at individual points are difficult, then pairs of $I_i \gamma r_{x_i}$ may be taken from an isoseismal map, for example, from eight rays drawn from the macroseismic epicenter.

The dependence of the parameters on direction. The parameters h, s, c_α, and s_e may be determined with the aid of a set of master curves both from the entire set of observed points and from groups of points corresponding, for example, to the directions along ∥ and across ⊥ of the main extension of the isoseismals or the strike of geological structures. Appropriate sectors are chosen on the map, and the points pertaining to them on the observed graph are denoted appropriately. Then the points may be processed either by groups or all together.

When determining separately to the parameters in different directions it is necessary to note both: $h_\parallel$ and $h_\perp$, $s_\parallel$ and $s_\perp$, etc. and to establish the mean values of h and s with ± deviations in the ∥ and ⊥ directions. This gives us an estimate of the stability of the solution due to systematic variations depending on the direction.

Estimation of the accuracy related to data scatter. In the simplest case this can be done by visually establishing the limits of the parameters h, s_e, etc. of the various theoretical curves in the set with which the set of observed points coincides satisfactorily.

The accuracy, as it depends on the data scatter, may be analyzed in more detail by calculating the rms deviation $I_i - I = 1, 2, \ldots, n$ of the observed points from the averaging curve: for a single measurement we have

$$m_i = \sqrt{\frac{\Sigma(I_i - I)^2}{n - 1}},$$

and for a set of points we get:

$$M_i = \frac{m_i}{\sqrt{n}}.$$

An estimate of the corresponding rms variance (at about the 70% probability level) $\pm M_h$, $\pm M_{se}$ or h and s_e is given by

$$M_h = \left|\frac{\partial h}{\partial I}\right| M_I = \left(\frac{h^2 + r_x^2}{h s_e} \ln 10\right) M_I \tag{8.28}$$

$$M_e = \left|\frac{\partial s_e}{\partial I}\right| M_I = \frac{1}{\log \sqrt{h^2 + r_x^2}} M_I. \tag{8.29}$$

Here the partial derivatives are obtained from (8.26). The epicentral distance r_x in (8.28) and (8.29) may be regarded as a mean value in the operating range of observations.

Determination of the epicenter. Using a two-sided set of curves plotted according to the same principle in a uniform linear scale of distance r_x the location of the macroseismic epicenter can be refined with respect to various cross sections of the field. The master curve shown in Fig. 66 does not enable us to do this. Firstly, this set of master curves is only one-sided. Secondly, for the sake of compactness, its scale along the r_x-axis is not uniform: in the range $r_x = 0$ to 10 km it is linear, while further on, for $r_x \geqslant 10$ km, it is logarithmic. In this form, the set of curves is only suited for determining the parameters h and s_e, which

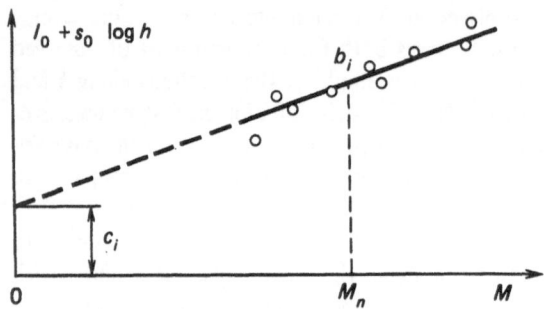

Fig. 67. Determination of the improved meaning of c_i and b_i parameters of the macroseismic field

is its main purpose. The location of the macroseismic epicenter in various cross sections of the field, on the other hand, may be found from symmetry considerations even without using a set of master curves, for example, by visually determining the position of vertical line to get the best bilaterally symmetrical halves of I_i, r_x, graphs plotted on transparent paper overlays.

8.2.4. Refinement of the Parameters c_0 and b_0 of a Macroseismic Field

The parameters c_0 and b_0 cannot, in principle, be obtained by processing macroseismic data for only a single earthquake. For this purpose, we need at least two earthquakes with different M's, and, in practice, a series of earthquakes of different magnitudes. The approximate values of these parameters are usually known, and they can be taken as the initial values. A refinement of these parameters may be carried out by the following graphic or analytic method.

In the graphic solution for each k-th earthquake we determine from (8.27) the value of:

$$I_{0k} + s_{0k} \log h_k = c_0 + b_0 M_k, \tag{8.27'}$$

taken as the ordinate of the point for this earthquake on the graph in Fig. 67. The abscissa of this point is equal to M_k. The set of such points for various earthquakes is averaged by a straight line (this may also be done by the method of least squares). The initial ordinate of this line is equal to the refined value c_1, while the slope is b_1. If these first-approximation values differ significantly from the previous, zero-approximation c_0, b_0, then the determinations of the depths and attenuations with the aid of a set of master curves must be repeated for the new c_1, b_1, etc. In other words, we must resort to successive approximations.

In practice, given a proper choice of the initial values c_0, b_0, it is sufficient to repeat the process with a set of master curves and only 2 or 3 earthquakes to estimate the variations which occur in h and s_e and to verify that they are small.

The corrections to h and s_e may also be estimated from the following analytic representations. Let us assume that the variations in these quantities are small. Then from (8.26) we obtain:

$$\frac{\partial h}{\partial c_0} = \frac{h^2 + r_x^2}{h s_e} \ln 10,$$

so approximately:

$$\Delta h_c = \left(\frac{h^2 + r_x^2}{hs_e} \ln 10\right) \Delta c_0,$$

and similarly:

$$\Delta h_b = M \left(\frac{h^2 + r_x^2}{hs_e} \ln 10\right) \Delta b_0.$$

The total correction Δh due to the variation of c_0 and b_0 is equal to:

$$\Delta h = \left(\frac{h^2 + r_x^2}{hs_e} \ln 10\right)(\Delta c_0 + M\Delta b_0). \tag{8.30}$$

In a similar manner we find that

$$\Delta s_e = \frac{1}{\log \sqrt{h^2 + r_x^2}} (\Delta c_0 + M\Delta b_0). \tag{8.31}$$

In (8.30) and (8.31), as previously in (8.28) and (8.29) $\bar{r}_x$ may be regarded as the mean r_x epicentral distance in the operating range. Let us cite a numerical example. Let $h = 10$ km, $r_x = 20$ km, and $s_0 = 3.5$; and note that $\ln 10 = 1/0.4343$. Then from (8.30), (8.31) we obtain $\Delta h = 33(\Delta c_0 + M\Delta b_0)$ km, $\Delta s_0 = 0.74(\Delta c_0 + M\Delta b_0)$. In determining $\Delta c_0 = c_1 - c_0$, and $\Delta b_0 = b_1 - b_0$, we must take the signs into account: the corrections may partly offset each other. Note that an increase in b_0 often corresponds to a decrease in c_0. This is because c_0 is measured from $M = 0$, usually far to the left of the operating range of M_k. Such a consistent change in b_0 and c_0 has little effect on the results for all values in the operating range.

Of course, in a manual calculation successive approximations and corrections, even potentially, are undesirable. But the proposed solution method does not provide any other way. Successive approximations could not be avoided by other methods. Iterations must be employed even in computer programs, but then it is not inconvenient. In addition, on a computer ascertaining whether an iteration converges is easily solved.

8.2.5. What do the Macroseismic Field Parameters Mean in Energy Terms?

We have shown earlier that a macroseismic formula like (8.25) may be obtained in a logical manner from (8.1) for the seismic-wave energy density ε if we assume that there is a log-linear relation between the energy density ε and the macroseismic intensity I (8.2) and consider that the energy value $K = \log E$ of an earthquake is related to the magnitude M by the expression (8.3).

A macroseismic formula such as (8.26) corresponds to that as in (8.1) where α is taken to equal zero. Then the energy density attenuation is described only by a single effective divergence parameter (8.12).

Using the formulae in Section 8.1, we derive the following dependences of the parameters of macroseismic formulae (8.25), (8.26) on the energy quantities in (8.1)-(8.3), (8.12). In Köveslighethy formulae (8.25) we have:

$$c_0 = c_1 + c_2 \left[a + \log \frac{r_0^{n-2}}{4\pi} + \alpha r_0 \log e \right],$$

$$b_0 = c_2 b, \quad s = c_2 n, \quad c_\alpha = c_2 \alpha \log e, \tag{8.32}$$

In Blake-Shebalin formulae like (8.26) we have:

$$c_0 = c_1 + c_2 \left[a + \log \frac{n_0^{n_e-2}}{4\pi} \right],$$

$$b_0 = c_2 b, \quad s_e = c_2 n_e, \tag{8.33}$$

Hence it becomes clear why the parameter b_0 in the macroseismic formulae (8.25) and (8.26) show surprising stability in various regions, while the remaining parameters do not. Indeed, clearly from (8.32) and (8.33), b_0 is entirely determined by the relations which are independent of local conditions between the scales of the macroseismic intensity I and the energetic intensity ε (8.2), on one hand, and the earthquake energy class K and the magnitude M (8.3), on the other. The parameter c_0 in (8.25) and (8.26), however, depends on the constants c_1, c_2, a, and r_0, and also on the attenuation parameters, which, generally, depend greatly on local conditions.

All the quantities on the right-hand sides of (8.32) have a clear physical meaning, which follows from (8.1)-(8.3). In (8.33), however, the quantity n_e remains conditional and effective. Its meaning is explained by (8.18), where it is expressed in terms of the seismic energy divergence and absorption parameters n and α. Here, as in (8.28)-(8.31), the quantity $r = (h^2 + r_x^2)^{1/2}$ should be taken as the mean epicentral distance $\bar{r}_x$ in the operating range.

Formulae (8.18), (8.32), and (8.33) make it possible to pass from the conditional macroseismic described field parameters, established with the set of master curves (or by some other similar method) to the parameters of (8.1)-(8.3), which have a definite physical and energy meaning. This may be used, in particular, to determine (and map) the elastic energy absorption index α in the medium or the Q factor on the basis of data obtained from macroseismic studies (see Section 8.1).

intensity I and the seismic energy density ε in terms of the macroseismic indices and the parameters of (8.3). From 8.33 we obtain:

$$c_1 = c_0 - \frac{b_0}{b} \left[a + \log \frac{r_0^{\frac{b}{b_0} - s_e^{-2}}}{4\pi} \right], \tag{8.34}$$

$$c_2 = \frac{b_0}{b}. \tag{8.35}$$

A numerical example. Let us assume that in (8.3) $a = 4$, $b = 1.8$, and $r_0 = 10$ km; in (8.26) we assume, according to [Shebalin] that $c_0 = 3$, $b_0 = 1.5$, and $s_e = 3.5$. Then from (8.34) and (8.35) we obtain $c_1 = -1.25$ and $c_2 = 0.834$. For these values (8.12) takes the form:

$$I = c_1 + c_2 \log \varepsilon = -1.25 + 0.834 \log \varepsilon.$$

Here ε is given in J/km^2. It should be emphasized that the values obtained for c_1 and c_2 must be regarded as very rough. Reliable values must be established from a careful simultaneous treatment of extensive macroseismic data.

Now, let us estimate the attenuation energy indices n and α, starting from the same rough macroseismic indices. For the numerical values adopted, in accordance with (8.33), we obtain:

$$n_e = s_e/c_2 = 3.5/0.834 = 4.2.$$

In (8.18) we must take $r_0 = 10$ km. Here we tentatively assume that the divergence energy index is $n = 2$. Let us assume, in addition, that the mean hypocentral distance in the operating range is $\bar{r} = 20$ km, i.e., that we are dealing with observations close to a shallow source ($h = 8$ km). Then from (8.18) we obtain:

$$n_e = n + \alpha \frac{r - r_0}{\log (r/r_0)} \log e = 2 + 14.4\alpha,$$

when the attenuation energy coefficient is:

$$\alpha = \frac{n_e - 2}{14.4} = 0.15 \text{ km}^{-1}.$$

To go from the attenuation coefficients α to the Q factor, let us assume that the propagation velocity V_S of the transverse waves, which carry most of the destructive energy, is 2.5 km/s, while the period T of the destructive oscillations ranges from 0.1 to 0.5 s. This approximates local conditions of the Tashkent earthquake of 1966. Then for larger and smaller values of T we obtain, correspondingly:

$$Q = \frac{\pi}{VT\,\alpha} \cong 16 - 80.$$

This reasonable order of magnitude for Q under the conditions even of rough calculations indicates the concepts are realistic.

The set of macroseismic master curves makes it possible to determine directly from observations of the macroseismic intensity (scale rating) of seismic tremors at a number of points the focal depth (hypocenter depth) and the intensity attenuation parameters of the tremors with increasing distance from the source. These values are established simultaneously from the entire set of observation data being processed. In this respect the proposed method differs from other better known methods, where the solution involves sampling. For this reason the method makes it possible, in principle, to obtain results which are more stable and internally consistent. It may also be used in a number of cases where former methods were inapplicable, owing to the difficulty or impossibility of obtaining the required samples.

8.3. Macroseismic Program*

In the generalization of the macroseismic, energy and other data on the intensity I of seismic shakings due to a number of earthquakes observed at many points of the Earth's surface, we come across the problem of constructing theoretical models of the intensity fields, optimal-

---

* See [Riznichenko, 1977] Co-authors are S. S. Seiduzova and L. M. Matasova.

ly conforming to the entire observed data complex. With this problem, the model parameters which are to be determined are (i) the common parameters for the entire observed data set, describing the dependence of the intensity I on the energy value K of the earthquake at the focus or on its magnitude M and the attenuation of the intensity I in the function of the focal distance r and the individual parameters, such as the hypocentral coordinates x, y, h, for each earthquake.

The hypocentral coordinates determined from the macroseismic or energy data on the one hand, and from the instrumental observations on the arrival time of seismic waves on the other, need not necessarily agree for physical reasons. This is because the energy parameters relate to the source center considered as a volume source of seismic energy, whose density shows itself at the points of the Earth's surface [Riznichenko]. The instrumental observations, however, relate to that proportion of the source where the rupture started and gave rise to the seismic waves observable on the seismograms. Instrumental data are generalized by wave kinematic methods, which usually involve the use of travel time curves. A comparison of macroseismic data, in essence the energy data of the earthquakes, with the kinematic data concerning them yields additional inferences about the foci and the properties of the surrounding medium.

In processing macroseismic data on the intensity I of the shakings or the energy data connected with them, we usually employ two simple mathematical models of the I-field using a Köveslighethy "point" focus (8.4) or a Blake-Shebalin [Riznichenko, Shebalin] (8.13) focus.

In (8.4) and (8.13), n, α, and n_e are the conventional constants characterizing the attenuation of the intensity I of the shakings with distance from the focus; c_1 and c_2 are other parameters which must be determined along with the n, α, n_e and x_i, y_i ($i = 1, 2, \ldots, n$) for all the foci, from the set of I-values observed at a number of points with known coordinates on the Earth's surface.

This is a traditional problem of classical seismology. Many have investigated the problem. A good survey of the research in the field is given in [Sponheuer, et al.]. Some subsequent research is listed in Section 8.2. The same reference contains a discussion of the earlier methods of solution. Until recently, the problem was not solved in the most rational way: the solution was sought piecemeal, not simultaneously from all the available original data and not in an optimal way using the whole data set. Most often, not all the parameters of the models which could be determined, in principle were calculated. This is partly true for our method of solving the problem using macroseismic master curves (see Section 8.2). We have combined macroseismic data for each individual earthquake, but not for the entire series of earthquakes.

It has been demonstrated in Section 8.2 that it is theoretically possible to formulate and solve the problem in a more complete form and under a single natural condition of optimizing of the solution using the entire observational data set. Given a large volume of initial material, such a solution is only realizable in practice by computer. This possibility is discussed below by means of concrete examples.

While discussing the formulation and solution of the problem, we shall at first and for the sake of brevity, deal only in macroseismic empirical terms. However, (8.1) and (8.13), as shown in Section 8.1, possess a physical, energetic sense too, under certain conditions. Later, we shall return to the physical aspect of the problem.

8.3.1. General Formulation of the Computation Problem

For each of $i = 1, 2, \ldots, m$ earthquakes, the energy value K or the magnitude M assumed to be known (from instrumental observations or by other methods) at a series of $j = 1, 2, \ldots, l$ points on the Earth's surface, with coordinates x_{ij}, y_{ij}, the shaking intensity values I_{ij} in terms of macroseismic scale degrees, are given. From these initial data, the parameters c_1, c_2, n and α or c_1, c_2 and n_e of the model (8.4) or the model (8.13) of the macroseismic field and such value of the hypocentral coordinates x_i, y_i and h_i of the earthquakes in question, which conform to the total macroseismic data set in the best way can be determined.

Other versions of the formulation are also possible. These are discussed below. A common feature of all these alternatives is that some of the parameters (under solution) of the models are common to the whole data set (series of earthquakes); the specific for some subset of the earthquakes (each of the earthquakes individually); we get a common and optimized solution in respect to all the data.

8.3.2. Algorithm and Computer Program

To develop an algorithm for a computer program, (8.4) and (8.13) were rewritten as:

$$I = c_1' + c_2'(K - \log 4\pi r_0^2) - c_3' \log \frac{\sqrt{r_x^2 + h^2}}{r_0} - c_4'(\sqrt{r_x^2 + h^2} - r_0), \qquad (8.4')$$

$$I = c_1' + c_2'(K - \log 4\pi r_0^2) - c_3 \log \frac{\sqrt{r_x^2 + h^2}}{r_0} \qquad (8.13')$$

In expression $(8.4')c_1 = c_1'$, $c_2 = c_2'$, $c_2 n = c_3'$, $c_2 \alpha \log e = c_4'$, and hence $n = c_3'/c_2$, $\alpha = c_4'/c_2 \log e$; in $(8.13')c_1 = c_1'$, $c_2 = c_2'$, $c_2 n_e = c_3'$, and hence $n_e = c_3'/c_2$; r_x^2 is the epicentral distance; h is the focal depth. The program was written to conform these two models.

The initial data for a computer are the macroseismic data presented for each earthquake separately. The following information is included in the code for punching: K is the energy value of the earthquake, determined directly or calculated from the magnitude M; l is the number of points at which this earthquake is registered and the intensity I of the shakings is determined; x_0, y_0, h_0 are the instrumental coordinates of the hypocenter or, if they are not available, the approximate macroseismic coordinates; I_1 is the macroseismic intensity fixed at the first point; r_{x_1} is the epicentral distance up to it, x_1, y_1 are its coordinates; I_2 is the intensity at the second point, r_{x_2} is its epicentral distance and x_2, y_2 are its coordinates, etc; finally I_l is the intensity at the last point where the earthquake was observed, and x_l, y_l are the epicentral distance and the coordinates of this point.

The coordinates of the observation points (x_k, y_k) and the hypocenter of each earthquake (x_0, y_0, h_0) are given in a Cartesian fame. The origin of the system is chosen arbitrarily. For each earthquake, a convenient origin can be chosen. As a matter of fact, this is preferable, since by a free choice of the origin we can reduce the computation error caused by replacing the spherical system of coordinates by Cartesian ones.

The program is executed in stages as the solution develops. Therefore, the algorithm is composed of several, fairly independent parts.

At the *first stage*, the coefficients c_i' in the expressions (8.4') and (8.13') are sought. The linear equations, whose solutions are c_i', are set up by the method of minimizing the sum of squares of the deviations:

$$R = \sum_{}^{N} [I(Kr, r, c_i) - I_{obs}]^2 = \min \tag{8.36}$$

As a result of the minimization in the first technique we have a system of four linear equations, and three in the second:

$$\partial R/\partial c_i' = 0, \quad i = \overline{1, 4} \quad \text{or} \quad i = \overline{1, 3} \tag{8.37}$$

In explicit form the system of four equations is

$$Nc_1' + \sum_{}^{N} (K - \log 4\pi r_0^2)c_2' - \sum_{}^{N} \log \frac{\sqrt{r_{x_k}^2 + h^2}}{r_0} c_3'$$

$$- \sum_{}^{N} (\sqrt{r_{x_k}^2 + h^2} - \gamma_0^2)c_4' = \sum_{}^{N} I_{obs},$$

$$\sum_{}^{N} (K - \log 4\pi r_0^2)c_1' + \sum_{}^{N} (K - \log 4\pi r_0^2)^2 c_2'$$

$$- \sum_{}^{N} \left[\log \frac{\sqrt{r_{x_k}^2 + h^2}}{r_0} (K - \log 4\pi r_0^2)\right] c_3' - \sum_{}^{N} [(\sqrt{r_{x_k}^2 + h^2}$$

$$- r_0)(K - \log 4\pi r_0^2)] c_4' = \sum_{}^{N} I_{K obs}(K - \log 4\pi r_0^2);$$

$$\sum_{}^{N} \left[\log \frac{\sqrt{r_{x_k}^2 + h^2}}{r_0}\right] c_1' + \sum_{}^{N} \left[(K - \log 4\pi r_0^2) \log \frac{\sqrt{r_{x_k}^2 + h^2}}{r_0}\right] c_2'$$

$$- \sum_{}^{N} \left[\log \frac{\sqrt{r_{x_k}^2 + h^2}}{r_0}\right]^2 c_1' - \sum_{}^{N} \left[(\sqrt{r_{x_k}^2 + h^2} - r_0) \log \frac{\sqrt{r_{x_k}^2 + h^2}}{r_0}\right] c_4'$$

$$= \sum_{}^{N} I_{K obs} \log \frac{\sqrt{r_{x_k}^2 + h^2}}{r_0};$$

$$\sum_{}^{N} (\sqrt{r_{x_k}^2 + h^2} - r_0)c_1' + \sum_{}^{N} \left[(K - \log 4\pi r_0^2)(\sqrt{r_{x_k}^2 + h^2} - r_0)\right] c_2'$$

$$- \sum_{}^{N} \left[\log \frac{\sqrt{r_{x_k}^2 + h^2}}{r_0} (\sqrt{r_{x_k}^2 + h^2} - r_0)c_3' - \sum_{}^{N} (\sqrt{r_{x_k}^2 + h^2} - r_0)^2 c_4'\right]$$

$$= \sum_{}^{N} I_{K obs}(\sqrt{r_{x_k}^2 + h^2} - r_0). \tag{8.37'}$$

The terms on the left-hand sides of the first three equations, omitting the parameter c_1' and the right-hand sides corresponding to these, constitute a system of equations for the Blake-Shebalin model. The index N of the sums of the coefficients of the equations, is equal to $N = l_1 + l_2 + \ldots + l_m$, where l is the number of observation points where I-observations were taken for each of the earthquakes, and m is the number of earthquakes in the calculations. The first stage ends with the printout of the solutions namely, the c_i' and n and α in the first, or n_e in the second, variant.

Now commences the second stage namely, the determination of the macroseismic coordinates of the hypocenter (x_{0m}, y_{0m}, h_{0m}) of each earthquakes. The corresponding algorithm is based on (8.4) and (8.13) or (8.4') and (8.13'). The hypocentral coordinates implicit into these formulae through the expression for the hypocentral distances r of the points k:

$$r = \sqrt{(x_k - x_0)^2 + (y_k - y_0)^2 + h_0^2}. \tag{8.38}$$

It is not possible to apply least squares directly to (8.4') and (8.13') to get the most probable values of x_{0m}, y_{0m}, h_{0m} since the function I is nonlinear in these parameters. Therefore, the function I was expanded as a Taylor series about the point (x_0, y_0, h_0) in powers of $(x_{0m} - x_0)$, $(y_{0m} - y_0)$, $(h_{0m} - h_0)$ and only the first power of these (presumably small differencies) was retained. Thus, we have:

$$I = I(x_0, y_0, h_0) + I_{x0}'\delta x + I_{y0}'\delta y + I_{h0}'\delta h, \tag{8.39}$$

where:

$$\delta x = x_{0m} - x_0, \quad \delta y = y_{0m} - y_0, \quad \delta h = h_{0m} - h_0.$$

The 'minimum' condition was applied to the functional:

$$F = \sum^{l} [I(x_{0m}, y_{0m}, h_{0m}) - I_{\text{obs}}]^2 = \min \tag{8.40}$$

From (8.40) we obtain a system of three linear equations

$$\partial F/\partial x_{0m} = 0, \quad \partial F/\partial y_{0m} = 0, \quad \partial F/\partial h_{0m} = 0 \tag{8.41}$$

In (8.41) the coefficients of the unknowns δx, δy, δh are formed from combinations of the derivatives I_r', I_y', I_h'. The functional expression of these derivatives at the point (x_0, y_0, h_0) for the first variant (Köveslighethy's formula) is:

$$I_{x0}' = c_3' \frac{x_k - x_0}{r^2 \ln 10} + c_4' \frac{x_k - x_0}{r}$$

$$I_{y0}' = c_3' \frac{y_k - y_0}{r^2 \ln 10} + c_4' \frac{y_k - y_0}{r} \tag{8.42}$$

$$I_{h0}' = - \left[c_3' \frac{h_0}{r \ln 10} + c_4' \frac{h_0}{r} \right],$$

where r is represented by (8.38). For the second variant (Blake-Shebalin formula) the derivatives of the function I with respect to the same variables are equal to the first terms on the right-hand sides of the equation (8.42).

The second stage begins with the computation of the derivative functions of I and of this function itself at the points (x_k, y_k) of the macroseismic observations. The derivatives are computed from (8.42), while that of the function I (depending on the variant) is computed according to (8.4') or (8.13'). In these formulae, we use the stage I solutions as coefficients c_i', the initial conditions (x_0, y_0, h_0) being the instrumental coordinates of hypocenters in our coordinate system. If these are not available, the initial conditions can be the estimated values, for instance, the coordinates of the point at which the impact of the earthquake was maximum and the assumed mean focal depth in the region.

After forming the matrix and right-hand sides of the system of equations (8.41) in the computer, it solves the system. The computed differences δ are compared with the ε, i.e., the accuracy of the solutions we established. If a δ-value exceeds, in absolute value, the corresponding ε-value, then the new focal coordinates $x_{01} = x_0 + \delta x$, $y_{01} = y_0 + \delta y$, $h_{01} = h_0 + \delta h$ become the initial conditions for the succeeding iteration. After obtaining the next approximation of the solution of the systen (8.41) should be checked again for the accuracy ε. The iteration continues until the approximation of the solution of the system in absolute value becomes less than the corresponding ε. Towards the end of the iteration the macroseismic coordinates of the hypocenter are printed out. If the iteration diverges, a further search for the macroseismic coordinates of the processed earthquake stops; the initial conditions remain as before. These are also printed out so that order in the system of output of all the stage by stage results, is not broken.

The *second stage* ends with the determination of the hypocentral coordinates of the last earthquake.

At the *third stage* the coefficients of (8.4') and (8.13') are recalculated. The difference between this and the first stage is that in place of measured epicentral distances in solving (8.37'), we use the calculated distances in conformity with the observed macroseismic hypocentral coordinates. Otherwise, there is no major difference between this and the first part of the program. The program ends with the printout of the calculated c_i' again.

The *fourth stage* involves sorting out the initial data by the 3σ method. The standard deviation σ of the individual measurement and of the result σ_1 are calculated as

$$\sigma = \sqrt{\frac{\sum\limits_{}^{m}(I - I_{obs})^2}{m - 1}}, \quad \sigma_1 = \frac{\sigma}{\sqrt{m}}. \tag{8.43}$$

The values of σ and σ_1 are printed out and then the differences $|I - I_{obs}|$ are checked for individual earthquakes having three times the value σ. If, at any observation point, the difference $|I - I_{obs}|$ is larger than 3σ, this point is rejected.

After analyzing the macroseismic information of all the earthquakes included and rejecting spurious points, the *fifth stage* begins, i.e., the recalculation of the c_i' with the selected data, i.e., the third stage is repeated completely. After repeated execution of the third stage, i.e., the coefficients c_i' are selected, once again the second stage is repeated. As a result of this new macroseismic hypocentral coordinates are determined.

The final step is the refinement of the c_i' using the new macroseismic coordinates. In order to do this, the third stage is repeated. The program ends with the printout of the final results: c_i' and σ, σ_1 and the theoretical curves determined from (8.4') or (8.13'). The theoretical curves are printed out in the form of a point set (log r, I) having a given step in r and K.

8.4. Analysis and Synthesis of Isoseismals*

Models of intensity fields of seismic shakings based on Köveslighethy and Blake-Shebalin formulae were considered in Sections 8.1-8.3. These models are widely used in seismological practice in the USSR and elsewhere. These models like similar ones [Shebalin, Sponheuer] are based on conventional assumptions that the tremor source (earthquake focus) is concentrated or spherical with spherical (axisymmetric in some cases) directional radiation, and that the medium is homogeneous and isotropic. This was assumed in our simplest version of the energy model of the intensity field, both in the macroseismic and spectral presentations [Riznichenko, 1974-75]. These assumptions result in circular theoretical isoseismals.

Observed isoseismals, however, are not circles for many generally known reasons. This difference does not always need to be considered depending on the type of problem to be solved and the accuracy required. Spherical homogeneous models when determining, for instance, the focal (hypocentral) depth from macroseismic data are still quite acceptable. One reason is the general absence of any independent control over the macroseismic solution; the other reason is the absence of distinct accuracy requirements as the results (focal depths) are of little interest in practice. However, when seismic hazard must be estimated, in equations of seismic shakeability [Riznichenko, et al.] in particular, accuracy is more important, and a different kind of control is needed. Often here a rough calculation based on a spherical model is found to be unacceptable.

The present section is directed mainly to the satisfaction of requirements in the field of seismic hazard estimation in seismic zonation, although the results could have wider applications in seismological problems. The effects of focal extension to isoseismal shape were considered in [Riznichenko]. Here we examine the influence of the surrounding medium, its anisotropy and heterogeneity. We present some techniques for constructing theoretical intensity fields of seismic shaking given the arbitrary shape of isoseismals close (an average sense) to the observed isoseismals of a given area. This technique is orientated to the processing of vast quantities of data that are common now in seismic observations, both in terms of energies (magnitudes) and of macroseismics. It is also presumed that spectral data may need to be processed. There is not a great quantity of such data yet, but it may increase soon, and it is necessary to be prepared methodologically and theoretically in advance.

8.4.1. Possible Approaches

Let us discuss the various approaches to the formulation of the general problem and to the construction of its solutions.

1. The most general physical approach would be the consecutive consideration of the chain: geodynamics—recent tectonics—seismicity—individual earthquakes with their field of shaking intensity. This should be based on a physical analysis of the causes and effects and on a theoretical synthesis of these events. Such an approach is undoubtedly promising, but

---

* Section 8.4 presents unpublished manuscript which Riznichenko considered completed series on macroseismics. The section is devoted to the simultaneous consideration of heterogeneity, anisotropy and non-ideal elasticity of the medium when constructing the theoretical intensity fields of seismic shakings—(*Editor*).

in practice it is hopeless given the need for acceptable results. A detailed coordination of several physical theories in geology and geophysics is needed to realize it, and this has only just begun. Besides, its practical realization, under natural conditions, would require the consideration of tremendous volumes of information on the medium's structure, its properties and its internal processes.

We do not yet possess such information.

2. The next approach also has a physical orientation, but it is narrower and more schematic. It is a combination of two widely developed approaches in seismology, namely the fault plan solution technique and the theory of wave propagation. The earthquake focus is represented by a known model, namely a center of extension, concentrated force, moment couple, double couple without resulting moment, dislocation, and fracture. The motion near to the focus is described completely, usually assuming homogeneity and isotropy in the surrounding medium. The asymptotic part of the solution is added for large distances from the focus, comparing size and wavelength. This part is considered as a seismic impulse radiated by the focus treated as a generator of seismic oscillation. The propagation of this seismic wave over long distances is studied by wave seismology considering the heterogeneity and, perhaps, anisotropy of the medium along the route from the focus to the observation point. The ray approximation is usually used to process the data. This approach is widely used in structural seismology. Local heterogeneities in the medium in the vicinity of the reception site are not considered in the regional aspect (general or detailed seismic zonation). A consideration of the situation at the reception site is realized by local corrections to the regional solution (seismic microzonation).

The main reason for such a schematic approach is the abstraction from geodynamic processes at the earthquake focus and the use of an asymptotics solution to analyze the motion far from the focus.

3. There is a more formal way to consider the problem. This is to reduce it to smoothing and interpolating the observed intensity fields. A mathematical model of a family of possible intensity fields is constructed containing a series of values describing the location of the sources, their parameters, and the location of the observation point and its conditions. It also contains a series of constants with values found by an optimization process that should guarantee agreement between the model and the family of observed intensity fields in the area. The formulation of such a mathematical model should permit its application given the non-spherical directional characteristics of the focal radiation, and also anisotropy and heterogeneity of the medium. Thus, some physical ideas should be inserted. Some formal elements could also be included to improve the approximation of the theoretical model to the observational data.

The formulations and solutions of the problems on how to analyze and synthesize the intensity field includes statements from the second and third approaches.

8.4.2. Energetic Models of Earthquake Sources

We express the intensity I of seismic shakings at the observation point in terms of a density ε of energy flow in the seismic wave (usually a transverse, viz. the most intense and destructive, wave) incident to the element of surface perpendicular to the seismic ray at the observation point. This has to be calculated for the whole time the wave passes through the point.

An energetic model of the intensity field for the point focus for a spherical focus in the outer zone [Riznichenko] can be described assuming central symmetry of radiation by a general formula (8.1).

In the case of an extended focus, considered as a combination of (8.1)-type foci distributed along a line, or part of a surface, or in volume, we have:

$$\varepsilon = \int\limits_V \frac{E^*}{4\pi r_0} \left(\frac{r_0}{r}\right)^n e^{-\alpha(r-r_0)} dV, \tag{8.1'}$$

where the energy density E^* is generally a point function, and the integral is taken over the distribution of focal—domain V.

The relation between the earthquake energy value K, $K = \log E$ (in joules) and magnitude M is usually given as an (8.3)-type formula. The relation between the energy flux ε and the macroseismic intensity I is given by an (8.2)-type formula, if the intensity I is expressed in terms of MSK or similar (MMS etc.), and ε is given in J/km^2. Very approximate values of the constants in this formula are: $c_1 = -1.25$ and $c_2 = 0.834$.

We can use (8.2) and (8.3) to proceed from (8.1) to Köveslighethy macroseismic formula like (8.8) [Köveslighethy]. Assuming that $\alpha = 0$ and thus that the whole attenuation of ε or I, with distance, is due to an effective divergence n_e, we can use a Blake-Shebalin formula like (8.26) [193]. The expressions for the parameters in (8.8) and (8.26) in terms of the parameters of (8.1) can be found in Sections 8.1 and 8.2.

The value α, the coefficient of absorption (and scattering) of seismic energy in the medium should be considered in the models as depending on K and r. In fact, $\alpha = \vartheta/\lambda = \pi/QVT$ where ϑ is the absorption reduction for each wavelength λ, V is the wave velocity (usually for S-waves), T is the predominant period of the oscillations in the pulse, $f = 1/T$ is the frequency, and Q is the quality factor of the medium. It is usually supposed that for shallow earthquakes the parameter of the medium Q (or ϑ) does not depend on T or f. It is possible, however, to include, if desired, the relation of Q on T, for example, as $Q = Q_0(T/T_0)^\nu$, where T_0 is fixed, say $T_0 = 1$ s, Q_0 and ν are the parameters of the medium independent on T determined from all observations.

The greater the earthquake's energy value K and the distance r, the greater the predominant period T of the oscillations in the pulse will be. The dominant period T in the ε-spectrum as a function of K and r could be obtained from the average earthquake energetic spectra [Riznichenko, et al.]. To be more exact, we should consider for macroseismic purposes the natural frequency distribution for structures, which (when damaged) are used to estimate the intensity.

Let us keep n free for the present. The dependence $n = n(K, r)$ could be established by fitting the mathematical model to all the intensity fields. We return to this problem later.

Considering the directionality of the focal radiation. The central symmetry of the focus is generally absent and in order to account for it, a directionality parameter ϱ for the seismic energy radiated by the focus must be introduced. This parameter could, in principle, be calculated from data on the focal mechanism [Vvedenskaya, Kostrov, etc.], now being accumulated from routine observations, for the double couple source model (with zero resultant moment) only. The parameters of this model come from observations on the direction of first arrivals. A theory describing this focal model assumes a four-lobe directional characteristic for longitudinal and transverse waves. However, the actual energy distribution, certainly for crustal foci, does not usually conform to the theory. The actual characteristics are circular or elliptical in shape. This might be associated with several circumstances. In particular, the shape may be due to the existence, in addition to relatively low-frequency radiation caused by the main rupture (already noted by the theory) of relatively high-frequency radiation, produced by numerous partial ruptures. These, in turn, might be distributed randomly throughout the focal volume and could possibly be of the same intensity and causes ruptures.

Let us consider this and approximate the observed distribution of seismic energy of the focus by an ellipsoid with a center at the earthquake hypocenter and a length ϱ of the radius vector proportional to the energy density ε for a reference sphere in the given direction. Let us express this length in terms of the lengths of the main ellipsoid half-axes (big a_1, middle a_2, small a_3) and in terms of the angles between the direction of the radius vector and these axes φ_1, φ_2 and φ_3. We present the ellipsoid equation in a coordinate system related to the main axes:

$$x_1^2/a_1^2 + x_2^2/a_2^2 + x_3^2/a_3^2 = 1. \tag{8.44}$$

In (8.46) we have $x_1 = \varrho \cos \varphi_1$; $x_2 = \varrho \cos \varphi_2$; $x_3 = \varrho \cos \varphi_3$; where $\cos^2 \varphi_1 + \cos^2 \varphi_2 + \cos^2 \varphi_3 = 1$. After substituting these expressions into (8.44) we obtain the unknown from the equation, i.e.:

$$\varrho^2 = (\cos^2 \varphi_1/a_1^2 + \cos^2 \varphi_2/a_2^2 + \cos^2 \varphi_3/a_3^2)^{-1}. \tag{8.45}$$

To make ϱ equal the energy density ε for the corresponding element dS of the reference sphere's surface, the condition of energy equality should be applied, i.e.,

$$E = \iint_S \varrho \, dS. \tag{8.46}$$

where the integral should be taken over the whole surface of the reference sphere $S = 4\pi r_0^2$, r being the radius of the sphere.

8.4.3. Energetic Models of Isoseismals

In our consideration of the intensity fields of seismic shakings presented in the form of an isoseismal we proceed from an energetic model such as (8.1), with two attenuation parameters, a divergence n and an absorption α. We consider the medium as anisotropic and heterogeneous in relation to these parameters. We take the quality factor Q as a main medium property which reflects its anisotropy and heterogeneity, as the absorption parameter α is closed in relation to the quality factor Q. We assume for simplicity that the Q-factor does not depend, in first approximation, on the frequency f or period T. This cannot be done for α.

Considering the anisotropy. Let us assume, when studying the anisotropy of the medium, that at each point of the medium the polar diagram of the Q-factor is, in the three-dimensional case, a triaxial ellipsoid, and in the two-dimensional case, an ellipse. Then the Q-factor is a tensor. The quality tensor $\bar{Q} = Q_{ik}|q_{ik}|$ is symmetric ($q_{ik} = q_{ki}$). It can be expressed by six scalars in the three-dimensional case, and by three scalars in the two-dimensional case.

The simplest way to construct the isoseismals in an anisotropic heterogeneous medium should, in principle, be to transform the space as in geometrical seismology [Riznichenko]. Basically the technique is to consider the (Euclidean) space Π_1 as being anisotropic and heterogeneous in relation to Q. Then the tensor or ellipsoid Q is assumed known for each point in the space. We deform the space Π_1 in order to transform the ellipsoids at all points of the space into spheres of the same radius. The space Π_2 obtained will be isotropic and homogeneous in relation to Q. The complexity of the medium's properties we replace by the complexity of the space measurements.

Any isoseismals of interest can be compiled for the isotropic homogeneous medium in the space Π_2. These isoseismals will be circular. The space Π_2 is now inversely transformed back into the initial space Π_1, which is anisotropic and heterogeneous in relation to Q, and the isoseismals will take whatever shape is defined by the spatial distribution function $\tilde{Q}(x, y, z)$. This is the general solution for the problem.

This technique could easily be realized "for small areas", viz. in the neighborhood of each point. "On the whole" and for large areas its application is more complicated, as the space might be non-Euclidean or Riemannian. In such a case all the difficulties in compiling the isoseismals remain and the change is only in the form of their presentation. Some particular distribution functions $\tilde{Q}(x, y, z)$ exist which lead to a Euclidean space Π_2. Such functions for transforming an initially isotropic heterogeneous medium Π_1 were considered in [Riznichenko]. In this case the transformations $\Pi_1 \rightleftarrows \Pi_2$ are conformal. If the medium Π_1 is anisotropic and homogeneous, the transformations $\Pi_1 \rightleftarrows \Pi_2$ are affine. The second particular case of identical and identically orientated quality ellipsoids $\tilde{Q}$ should lead to an Euclidean space Π_2, in which case the solution should be easy.

Let us now turn to the more general case of anisotropic and heterogeneous medium in which the $\tilde{Q}$ ellipsoids differ in space by axis lengths and axis directions. The transformation of this space is not possible. There is another way. We limit ourselves here, to the two-dimensional case. We select three values for a convenient definition of the tensor or ellipse of the quality Q. When processing intensity field observations, in particular using macroseismic master curves [Riznichenko], we usually determine the parameters h and S, and then n and α; this should be done for the two directions separately, i.e. parallel to and perpendicular to the elongation of isoseismals. Thus it is convenient to describe the anisotropy ellipse $\tilde{Q}$ in terms of the parameters: the quality factors $Q_{\parallel}$ and $Q_{\perp}$ along and across the isoseismals and the azimuth $Az_{\parallel}$ of the large axis of the ellipse $Q_{\parallel}$.

In order to calculate the Q value along each given direction we fix the azimuth Az of this direction. Then the angle β between Q and $Q_{\parallel}$ will be $\beta = Az - Az_{\parallel}$. Thus, the Q-value, in any direction, at each point of the observational plane, could be determined for anisotropy by four parameters, $Q_{\parallel}$, $Q_{\perp}$, $Az_{\parallel}$ and β.

If the coordinate system is associated with the direction of the isoseismals' elongation, i.e. with the main axes of the ellipse $\tilde{Q}$, the Q-value in this direction should be determined by three parameters only, namely $Q_{\parallel}$, $Q_{\perp}$ and β. The function $Q = Q(Q_{\parallel}, Q_{\perp}, \beta)$ could easily be obtained from equations analogous to (8.44) and (8.45). In our two-dimensional case we find:

$$Q = (\cos^2 \beta_{\parallel}/Q_{\parallel}^2 + \cos^2 \beta_{\perp}/Q_{\perp}^2)^{-1}$$

$$= (\cos^2 \beta/Q_{\parallel}^2 + \sin^2 \beta/Q_{\perp}^2)^{-1}. \tag{8.47}$$

Here $\beta_\| = \beta$, $\beta_\perp = 90° - \beta_\|$. If $\beta = 0$, we obtain from (8.47), as we would expect, $Q = Q_\|$; when $\beta = 90°$, $Q = Q_\perp$ when $\beta = 45°$, then:

$$1/Q_{45}° = 1/2(1/Q_\|^2 + 1/Q_\perp^2).\tag{8.48}$$

If, in the last case, $Q_\| = 2$, $Q_\perp = 1$, then $Q_{45}° = 1.6$. The linear interpolation by angle β with the result $Q = 1.5$ would be erroneous.

When each of the scalars $Q_\|$, $Q_\perp$ and $Az_\|$ is determined by the technique described in [Riznichenko] at various points $[x, y]$, i.e., the earthquake epicenters, we can determine by interpolation and plot, in isolines, the continuous field of these values in the x, y plane. In this way we obtain three maps, $Q_\|(x, y)$, $Q_\perp(x, y)$, $Az(x, y)$, which together describe the field of the quality tensor $Q = \tilde{Q}(x, y)$ for the region. This field could also, for a clearer presentation, be drawn as a family of ellipses, of sufficient scale, scattered over the map.

The field of quality tensor Q could be used in future to synthesize theoretical isoseismals for the intensity field of seismic shakings assuming the elliptical isoseismals, the anisotropy and the heterogeneity considered in the zero approximation. The calculation of non-elliptical isoseismals still requires more detailed analysis and more detailed consideration of heterogeneity.

III

Seismotectonic Flow of Rock Masses

Seismic flow is part of the general seismotectonic motion of large space-time regions of the Earth's crust and upper mantle. It also concerns mountain building associated with residual displacements in a group of earthquake foci.

In a macroscopic treatment of the tectonic flow of rock masses, we can distinguish between plastic flow, continuous in the broad meaning of the word, and intermittent-continuous seismic flow [Riznichenko]. The latter includes elastic deformations preceding individual earthquakes and the earthquakes themselves, from a multitude of weak ones to very rare, catastrophic ones. These partial processes are discrete in time and space and form. They are the microstructure of a macroscopically quasi-continuous seismic flow.

Seismic methods enable us to assess the proportion of total deformation of rock mass due to earthquakes. To a certain extent, these assessments may also relate to the parameters of tectonic flow, as a whole, because the level of seismicity is an indicator of its intensity. In general, however, tectonic flow as a whole, is better covered by methods of geodesy, geology and geomorphology, rather than by seismology.

The present study is an attempt to consider theoretically the seismic flow of rock mass due to displacements in numerous earthquakes as a part of tectonic flow over large regions. The macroscopic parameters of the flow of rock mass—the gradient of the flow velocity, the average stress, and the effective (apparent) viscosity—are compared with the properties of the material of the rock and with the quantitative characteristics of seismicity.

In modern geology and geophysics, the development of comprehensive quantitative methods for studying crustal or deeper movements using derived data from various sciences is an important topic. For each Earth sciences, e.g., seismology, a condition for successful joint projects is to find a way of presenting data that are compatible with data from other sciences.

We shall only discuss problems that fall within the competence of seismology; viz. the seismic part of tectonic flow. We discuss how the permanent average values of the velocity gradient of a flow and other comparable quantities derived, by other methods from an investigation of slow deforming Earth movements, can be calculated. Rock properties and deformation characteristics of individual earthquakes, together with general statistical laws, that require groups of earthquakes (the law of earthquake repetition), are related to the effective macroscopic parameters of the Earth's slow movements.

This general trend of studies is not new [Grudeva et al., Gurevitch, Benioff]. It seems to us, however, that the detailed quantitative methods of studying seismicity developing in the USSR [Riznichenko, etc.], and the results they obtain will help these studies to progress.

9. Seismic Flow of Rock Masses

9.1. Relation Between Rock Mass Flow and Seismicity*

9.1.1. The General Nature of Process and Medium Model

Consider the following simple scheme of a possible tectonic flow including a seismic flow. Let A and B in Fig. 68 be large, comparatively strong and only slightly deformable regions of the Earth's crust or the upper mantle. Factors that are external, relative to the volume process interesting us (subcrustal currents, forces of gravity adjustment resulting from volume change associated with heating and cooling, etc.) cause A and B to shift relative to each other. C is a zone intermediate between A and B consisting of weaker and easily deformable material. Our tectonic flow process occurs in zone C, this allows regions A and B to shift.

Zone C is like a layer of lubricant in engineering between solid components A and B. In hydrodynamics or aerodynamics it is like the boundary layer of fluid (air) between the undisturbed medium and a solid moving in it. In geology it may be like a deep fracture zone including the main fracture and its wake, with corresponding crush zones, or like part of an active geosyncline or activated platform region between two consolidated blocks or other tectonically quieter regions.

The arrows in Fig. 68 show the relative velocities. A point in zone A has been adopted arbitrary as zero. The thin lines are stream lines. Where the lines converge the movement is most intense; since the velocity gradients are highest, the highest seismicity can also be expected.

It is natural to begin a quantitative analysis with a consideration of a steady state quasi-plastic flow of a macroscopically homogeneous, but structurally inhomogeneous, solid-fluid material. The latter is internally destroyed and restored in the plane-parallel layer C between two solid half-spaces A and B joined to the material of layer C along its boundaries (Fig. 69).

The material in zone C as a whole is under uniformly simple shear and over long time-intervals flows like a viscous Newtonian liquid. This flow is turbulent, with rotation of the particles as a whole (clockwise in Fig. 69). Microscopically, on the other hand, the flow may either be turbulent or laminar. We begin our treatment with the simpler laminar case, leaving the no less important turbulent flow for further studies.

The model of the microstructure of the material is approximately as follows. "Seismic elements" are continuous with the viscous (or visco-elastic) continuously deforming "plastic" substance of zone C like crystals in an amorphous rock mass. These elements, in the simplest case, are elastically deformable with brittle failure. When failing, they liberate accumulated elastic energy in the form of seismic waves. After this, the seismic elements consolidate again. It is the deformation and failure of these elements that ensure the seismic part of the flow. The embedding plastic substance enables them to move relative to one another without disturbing the quasi-continuity of the medium as a whole.

Material may be exchanged between the undestroyed plastic and destroyed seismic parts of the medium, but the relative amounts of material of the two kinds at steady state must, naturally, remain constant.

---

* See [Riznichenko].

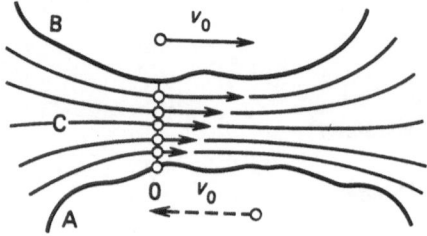

Fig. 68. Tectonic flow in the yielding zone C between two consolidated, mutually shifting blocks A and B

The size of the seismic elements determines the energy they accumulate and liberate. This energy and the frequency of the ruptures of elements of various size are related by the earthquake recurrence.

The structure, properties, and action of the individual seismic elements can be described as follows. In the initial undeformed state, each element is a cube with edges of length l_i oriented along the coordinate axes of Fig. 69. In the course of flow (Fig. 70), the cube linearly and elastically deforms (according to Hooke's law). It becomes a parallelepiped and accumulates potential elastic energy with a maximum of

$$E_i = (\tau_{max}\varepsilon_{max}/2)\,V_i. \tag{9.1}$$

Here $V_i = l_i^3$ is the volume of the cube. When the ultimate strain ε_{max} is reached, at the ultimate shear stress $\tau_{max} = \mu\varepsilon_{max}$ (μ is the shear modulus), the seismic element fails by shear and separates into parts that straighten out completely and emit all the elastic energy (9.1) they have stored, as seismic waves. After this, the seismic element ceases to exist and merges with the surrounding continuous medium. Such seismic elements of various sizes appear chaotically at different times and at different places in a medium.

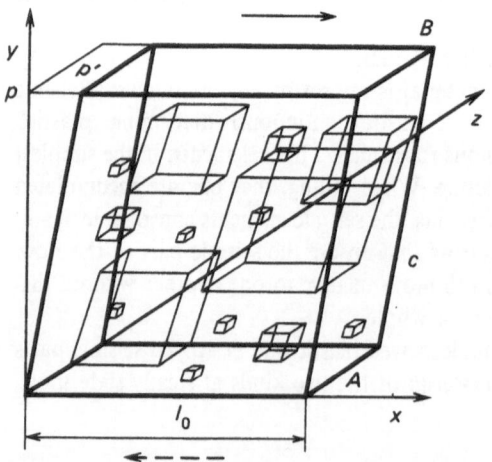

Fig. 69. Quasiplastic flow of the material in the zone C.

The structural "seismic elements" that are elastically stressed fail. They are restored during the flow and are shown in a macroscopically continuously sheared volume

Fig. 70. A seismic element:

ABCD—in the initial unstressed state: *AB′C′D*—
in the ultimate stressed state before fracture; and
AD/B′C′—in the final unstressed state after fracture (along plane *OO′*) and the release of elastic
energy

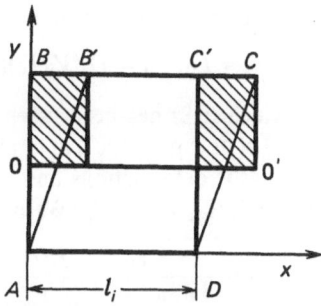

Every seismic element is thus a model of a Reid-Benioff seismic focus [Benioff] with all its known drawbacks [Gurevitch et al.]. There is no need to demand more from it than is required for initial rough appraisals.

The main feature of our scheme, unlike that of Reid-Benioff, is that we immediately consider a set of such seismic elements distributed by the earthquake recurrence law.

The specific properties of the seismic elements adopted (Hookean elasticity, brittle failure, etc.) are not an important feature of the approach we are developing. These elements may and must subsequently be assigned with general theoretical rheological and strength properties, which are better substantiated experimentally. Something has been already done in this direction [Gurevitch et al., Kuznetsova]. The local mechanisms for these elements have also been clarified by Soviet and foreign investigators [O. Gotsadze, H. Honda, K. Kasahara, Benioff].

9.1.2. Calculation of Seismic Flow

Consider the transfer of masses occurring during the failure of seismic elements. Assuming that the density ϱ of the material is the same everywhere, we shall only consider the transfer of volume.

Assume that the failure of a seismic element results in the transfer of material as shown in Fig. 70: half of the volume of the cube moves from position *BC* to *B′C′* so that the volume of the hatched parallelepiped with edge *BB′* is transferred, to position *CC′*. This transfer of volume also occurs during the preceding maximal elastic deformation of the seismic element, namely wedge *ABB′* is transferred to position *DCC′*. If the elastic shear strain is $\varepsilon_{max} = BB′/AB$, the volume ΔV_i transferred as a result of this strain or failure is

$$\Delta V_i = (l_i \varepsilon_{max} l_i/2) l_i = (\varepsilon_{max}/2) l_i^3, \tag{9.2}$$

and the magnitude of its transfer is l_i.

Let us calculate the increment v_0 in the average velocity of the centers of mass of the macrovolume $V_0 = l_0^3$ in the form of a cube with an edge of $l_0 = 1$ (see Fig. 69) due to the transfers of seismic microvolumes ΔV_i during time T_0. The displacement L_0 of the center of mass (volume) related to these transfers is $L_0 = \Sigma \Delta V_i l_i / V_0$. During time T_0, this corresponds to an increment in the average velocity of

$$v_0 = L_0/T_0 = (1/V_0 T_0)(\varepsilon_{max}/2) \Sigma l_i^4 \qquad (9.3)$$

in which (9.2) has been taken into consideration.

Let us express the l_i in terms of the energy E_i accumulated by the seismic elements during the elastic deformations and released as seismic waves in earthquakes. From (9.1), we have $V_i = 2E_i/\tau_{max}\varepsilon_{max} = l_i^3$. We obtain from (9.3):

$$v_0 = \frac{1}{V_0 T_0} \frac{\varepsilon_{max}}{2} \frac{2^{4/3}}{\tau_{max}^{4/3} \varepsilon_{max}^{4/3}} \Sigma E_i^{4/3}$$

$$= \frac{2^{1/3}}{\tau_{max}^{4/3} \varepsilon_{max}^{4/3}} \frac{1}{V_0 T_0} \Sigma E_i^{4/3},$$

or in integral form:

$$v_0 = \frac{2^{1/3}}{\tau_{max}^{4/3} \varepsilon_{max}^{1/3}} \int_{K_1}^{K_2} N E^{4/3} dK, \qquad (9.4)$$

where

$$N = A \times 10^{-\gamma(K - K_0)}, \qquad (9.5)$$

when $K \leqslant K_{max}$ we have $N = 0$; when $K > K_{max}$, we have the earthquake recurrence in its simplest form. Here N is the frequency of earthquake recurrence with a seismic energy of $E = 10^K$, K is the energy class of earthquakes, A is the volume seismic activity, γ is the slope of the recurrence plot, and K_0 is the class used to calculate the activity $A = N|_{K=K_0}$. All the quantities in (9.4) and (9.5) are assumed to relate to a continuous distribution relative to K. The relation, between these "density" quantities for N and A and the quantities related to the decimal energy classes, is established elsewhere [Riznichenko]. Under ordinary conditions, the transition from one set of quantities to another does not change their order.

It has been shown [Riznichenko] that in an integral such as the one in (9.4) at the usually observed values of $\gamma < 1$, we can confidently assume that $K_1 = -\infty$ and $K_2 = K_{max}$, where $K_{max} = \log E_{max}$. Here E_{max} is the maximum possible seismic energy in the earthquake region.

We emphasize that because the main result of these calculations—(9.4)—was derived: (1) assuming that the law of earthquake recurrence (9.5) is observed, and (2) from the calculation of the velocity of center of mass transfer by (9.3). The latter can be interpreted as the conservation of the additional momentum associated with transfers of the masses of the seismic elements in the course of seismic flow. Indeed, (9.3) can be written in the form $V_0 v_0 = \Sigma \Delta V_i (l_i/T_0)$ or, multiplying both sides by the density ϱ and adopting the notation $\varrho \Delta V_i = m_i$—the mass of each seismic element being transferred, $l_i/T_0 = v_i$—the average velocity of its transfer during the time T_0, and $\varrho V_0 = m_0$—the total mass of the macrovolume, we obtain the conventional form of the law of conservation of momentum $m_0 v_0 = \Sigma m_i v_0$.

We shall stress once more that (9.4) only defines the increment in the velocity of the center of gravity of the macrovolume due to the transfer within it of the microvolumes ΔV_i associated with the seismic flow. Moreover, there may be another velocity increment associated with flow in the enclosing plastic material. If we tentatively assume, however, that all the movements in the macrovolume are only associated with seismic flow, (9.4) will give the total velocity increment, underlying the tectonic flow as a whole. It is possible that this assumption is close to the actual conditions: the plastic flow of rock itself in the solid state may consist of the rupture of particles and their merging in other positions, as we adopted for our seismic elements.

9.1.3. Relation to the Macroscopic Parameters of a Tectonic Flow

We consider tectonic flow as a whole as the flow of an incompressible Newtonian viscous liquid. It obeys Newton's formula:

$$\tau_0 = \eta_0 (\partial v_0'/\partial y), \tag{9.6}$$

where v_0' is the velocity of the flow along the x-axis (see Fig. 69); τ_0 is the shear stress along this axis, and η_0 is the viscosity. The energy flux dissipated by this flow per unit volume of unit time is

$$w_0 = \tau_0 (\partial v_0'/\partial y). \tag{9.7}$$

Adhering to the condition (Subsec. 9.1.2) that the entire tectonic flow is due only to the seismic flow, the velocity gradient $\partial v_0'/\partial y$ in (9.6) can be compared with the velocity increment v_0 in the displacement of the center of gravity of a uniformly sheared unit cube. Since the displacement of the center of gravity is half that of the extreme point $P \to P'$ in Fig. 69 used to determine Δv_0 in the expression $\Delta v_0/\Delta y \sim \partial v_0'/\partial y$, we have the equation $\partial v_0/\partial y = 2v_0$. Now equating $\partial v_0/\partial y = \partial v_0'/\partial y$ and assuming in (9.4) that $K_1 = -\infty$ and $K_2 = K_{max}$, the validity of which has already been mentioned, we obtain the required formula:

$$\partial v_0'/\partial y = (2^{4/3}/\tau_{max}^{4/3} \varepsilon_{max}^{1/3}) \int_{-\infty}^{K_{max}} NE^{4/3} dK. \tag{9.8}$$

It is this formula that relates the velocity gradient of tectonic flow or, more exactly, its seismic part to the properties of the medium and the characteristics of seismicity.

In order to determine the tectonic flow in (9.6), provided that all of it is due to seismicity, we must determine the stress τ_0 at which it occurs. To do this, we introduce a significant new assumption, viz. energy conservation. We assume that the energy flux supplied to the tectonic flow from outside, accumulated by the seismic elements in the form of potential elastic energy, and, finally, dissipated upon their destruction as seismic waves is equal to the energy flux (9.7) dissipated by the tectonic flow.

In this case we can write that $w_0 = \tau_0 (\partial v_0'/\partial y) = \int_{-\infty}^{K_{max}} NE \, dK$, where the integral is the expression of the flux of seismic energy from foci from unit volume in unit time (see [116, 120]). Hence we obtain the required quantity τ_0:

$$\tau_0 = \frac{1}{\partial v_0'/\partial y} \int_{-\infty}^{K_{max}} NE \, dK. \tag{9.9}$$

Since $\partial v_0'/\partial y$ has been determined from (9.8), equation (9.9) gives the stress in a tectonic flow in terms of the properties of the medium and the seismicity parameter.

The last step is to determine the effective (apparent) viscosity η_0 in Newton's formula (9.6) for a tectonic flow. From (9.6), we have:

$$\eta_0 = \tau_0 / \partial v_0' / \partial y), \tag{9.10}$$

where the numerator is determined by (9.9), and the denominator by (9.8).

Hence, the problem is solved in a general form by (9.8) and (9.10) with the assumption that the entire tectonic flow is due to seismicity.

We could formally eliminate this restriction by introducing quantitative indices of the presumed relative role of the seismic flow in the total deformation and total energy of the tectonic flow. This would require at least another two constants in (9.8)-(9.10). Since they would remain indefinite until additional information is included (which could be extracted from geodetic data), we shall not do this here.

To make the calculations more specific, we integrate (9.8)-(9.10) for the case when the recurrence law is given in its conventional simple form of (9.5). After the relevant substitutions, we obtain:

$$\frac{\partial v_0'}{\partial y} = \frac{2^{4/3}}{\tau_{max}^{4/3}\, \varepsilon_{max}^{1/3}} \frac{A \cdot 10^{\gamma K_0}}{(4/3 - \gamma) \log 10}\, 10^{K_{max}(4/3 - \gamma)} \tag{9.8'}$$

$$\tau^0 = \frac{\tau_{max}^{4/3}\, \varepsilon_{max}^{1/3}}{2^{4/3}} \frac{(4/3 - \gamma)}{(1 - \gamma)} \frac{1}{10^{K_{max}/3}} \tag{9.9'}$$

$$\eta_0 = \left(\frac{\tau_{max}^{4/3}\, \varepsilon_{max}^{1/3}}{2^{4/3}}\right)^2 \frac{(4/3 - \gamma)^2}{(1 - \gamma)} \frac{\ln 10}{A \cdot 10^{\gamma K_0}} \frac{1}{10^{K_{max}(5/3 - \gamma)}}\,. \tag{9.10'}$$

Examination of these formulae reveals that given all these assumptions the velocity gradient of tectonic (seismic) flow $\partial v_0'/\partial y$ is directly proportional to the seismic activity A and grows with increasing energy $E_{max} = 10^{K_{max}}$ of the maximum possible earthquake in the given region in proportion to $E_{max}^{1.3-\gamma}$, i.e., to $E_{max}^{0.8}$ when $\gamma = 0.5$ or to $E_{max}^{0.3}$ when $\gamma = 1$.

The stress τ_0 at which this flow occurs does not depend on the seismic activity. With an increase in E_{max}, however, the stress τ_0 diminishes in proportion to $E_{max}^{-1.3}$ regardless of the value of γ.

The effective viscosity η_0 of the material of rock masses diminishes with a growth in seismicity, the material becomes more pliant to shear forces: with a growth in A it decreases in proportion to A^{-1}, and with a growth in E_{max}, to $E_{max}^{-0.5}$ when $\gamma = 0.5$ or to $E_{max}^{-0.7}$ when $\gamma = 1$.

Equations like (9.8)-(9.10) include the energy E_{max} of a maximum earthquake. Thus, we may assume that this approach could be useful, in particular, when seeking a quantitative solution of the problem of the possible value of E_{max} in a region from combined seismic, geological, and geodetic data.

Hence, from comparatively simple assumption on the structure and properties of the structurally inhomogeneous material of rock masses, we have compared the macroscopic parameters that part of the tectonic flow defined by seismicity only with the properties of the structural elements and quantitative parameters of seismicity in the flow zone.

The calculations were based on the following considerations: (1) the identical movement of the center of mass of a macrovolume and its constituent microvolumes, "seismic elements", which took the form of the law of conservation of momentum; (2) the conservation of the amount of energy participating in the process, when it transforms from the acting external forces to the potential elastic energy of the seismic elements, and further, to the energy of the seismic waves appearing in earthquakes. All the energy dissipated by the process of macroscopic flow of the rock masses is also equated to this energy; and (3) observance of the law of earthquake recurrence in the course of flow.

The assumptions on the rheological and strength properties of the medium components adopted above for specific calculations are extremely schematic. They must naturally be clarified in further studies.

We deem that such studies will help develop a comprehensive quantitative approach to the investigation of the deformation motions of the Earth's crust and upper mantle in the joint use of the data of seismology, geodesy, geology, geomorphology, and other Earth sciences.

9.2. Calculation of the Strain Rates in the Seismic Flow of Rock Mass*

9.2.1. General

Seismic flow is only associated with the elastic deformations (strains) and stresses that drop, are withdrawn, or released in the focal region in an earthquake [Aki, Brune, M. Randal]. In the mass of studies of the focal dynamic parameters the directions of principal strain axes are generally indicated as tension T, compression C, and the axis O of zero deformation. The scalar value of the seismic moment M_0, related to the average withdrawn shear strain γ and the relevant stress τ over the area of the main fracture [Aki, Brune, Randal] or over the volume of the source [Riznichenko], has been established. In contrast to this, we refer here to average removed shear strains and stresses over the volume of a focus of a certain size [Riznichenko], being in mind their tensor nature. In other words, together with their maximum scalar values, we shall consider the directions of the principal deformation axes.

In the schematic treatment of each separate source instead of the involved non-uniform stress-strain state of the medium (rock masses) in the focal region we refer only to the average value of the drop in the deformation on the fracture area or in the volume of a focus. In the same way when considering the rates of strain of a seismic flow, we use the average indices in each space-time volume element of the medium and give a set of foci. From the energy viewpoint, we characterize this set by the principal parameters of the permanent average seismic conditions, i.e., the slope of the plot of the recurrence versus the distribution of earthquakes by energy $\gamma_K = -d \log N/dK$ (N is the number of earthquakes, and K is the energy value of an earthquake at its source, or we can consider the magnitude M), seismic activity A, and the quantity K_{max} of the maximum possible earthquake [Riznichenko].

The drop in stress at each focus is given in a local system of coordinates related to the principal axes of the removed strain. On the other hand, it is useful to give the parameters

---

* See [Riznichenko].

of the seismic flow and the set of variously oriented foci a single common system of coordinates. These should be related, for example, at each point to the geographical system. Our primary task will be first, to establish a relation between the scalar quantities determining the directions of the principal axes of the removed strain at a focus, the value of the seismic moment M_0, and the main values of the removed strain ε and stress σ. Secondly, we determine the components of the tensor of the released strain, expressed in a local system of coordinates of a focus. We shall also have to change from a local system of coordinates to a common one.

Further, our task will be to integrate (in the common coordinate system) the strains and rates of strain at the included foci in any given space-time volume element of the medium. This is to establish the components of the tensors of the strain and rate of strain of the seismic flow within the element. If necessary, this should include not only the laminar, but also the turbulent motions of the medium [Riznichenko].

All these tasks involve calculations and contain no fundamentally new ideas or approaches. They had been advanced earlier [Riznichenko, Kostrov, Aki, Benioff, Brune, Kendal]. Their calculations were too general at times, so they were difficult to absorb into the practice of interpreting field seismological observations. Our aim, by presenting the material more specifically to open to seismologists, the real possibilities of mass data on the mechanism of foci and the permanent average seismicity in quantitative geodynamic images certainly for simple cases.

9.2.2. Velocities of Vertical Motions in the Seismic Flow of Rock Mass*

An elementary theory of vertical motion in the seismic flow of rock masses was set out briefly in [Riznichenko]. Here we substantiate it, especially concerning the unambiguity of the general tensile and compressive strain of the volume region of a focus, regardless which of the two possible principal shear planes occur. For additional information on the mechanism of the foci see Vvedenskaya. When calculating the seismic flow parameters we use data on the permanent average seismicity, i.e., the parameter $\gamma = -d \log N/dK$ of the distribution of $N(K)$ earthquakes with respect to energy $K = \log E$ (E is the seismic energy), the seismic activity A, and the magnitude of the maximum possible earthquakes K_{max} [Riznichenko]. The velocities of the vertical motions due to seismic flow are compared with the velocities of present vertical motions obtained by geodesy (repeated leveling) and geomorphology (in places poorly accessible for work with instruments). They can be correlated with the velocities derived from the latest motions by geology.

Previously [Riznichenko] we used arbitrary average maximum stresses and strains in the focal region (after Benioff) in the calculation of the seismic flow. Instead we use the average dimensions of a focus as a rupture shear dislocation. This can be found from spectral and other observations, viz., the area $S = \pi R^2$ of the rupture surface (R is its average radius) and the average displacement D over it. We can also use Aki's seismic moment [Aki] $M_0 = \mu SD$, where μ is the elastic shear modulus of the rock mass material in the source region. By considering a set of dislocations jointly, the quantities S, D, and M_0 can be expressed as a function of K by correlation expressions [Riznichenko].

---

* See [Riznichenko]. E. A. Dzhibladze is the co-author.

Fig. 71. Motions in an earthquake source.

Axes: TT'—tension; CC'—compression; PP'—possible rupture planes; D-displacement along rupture

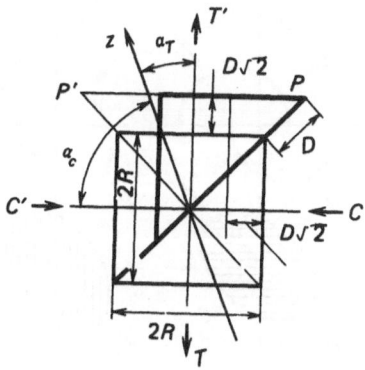

The motions at the focus of a core earthquake in our assessments can be represented schematically as shown in Fig. 71. The volume of a focus, initially a cube, is split and shifted by D ($\ll R$) along one of the two possible shear planes P or P'. Under conditions of homogeneity and isotropy, these planes are at $45°$ to the principal axes of tensile TT' and compressive CC' strains. Let us assess the vertical motions of the rock masses at the focus due to motion along the TT' and CC' axes. We assume complete freedom of motion only toward or away from the Earth's surface: elevation or subsidence.

Inspection of Fig. 71 reveals that the displacement of the cut-out part of the focal volume along tension axis TT' is $D\sqrt{2}$. If the axis is at an angle α_T with the vertical z, regardless of whether the displacement D occurred along plane P or P' (or the displacement $D/2$ along each of them), the uplift of the rock masses in the source region associated with this displacement will be $(D/\sqrt{2}) \cos \alpha_T$. Similarly for a displacement of the opposite sign along the compression axis CC', if it is at α_C to the vertical z, we obtain subsidence by $(D/\sqrt{2}) \cos \alpha_C$. No movements occur along the third "zero" axis orthogonal to TT' and CC'. Hence the total uplift of the rock mass in the focal region associated with motion along both the TT' and CC' axes will be:

$$\Delta h = (D/\sqrt{2})(\cos \alpha_T - \cos \alpha_C). \tag{9.11}$$

If the cross-sectional area of a focus is S, the volume of the uplifted rock in the focal region will be

$$\Delta V = (SD/\sqrt{2})(\cos \alpha_T - \cos \alpha_C) = (M_0/\mu\sqrt{2})(\cos \alpha_T - \cos \alpha_C). \tag{9.12}$$

This increase in the size of the focal region upward is obtained reducing its size in the horizontal direction so that the volume of the focus as a whole does not change.

Since $D \ll R$, the estimations of Δh, and consequently ΔV, are close if we assume an initially spherical focus that is transformed by an earthquake with the same average density of the moment M_0 into an ellipsoid of equal volume with the same principal axes.

The sequential transition from rupture strains that have different space frequencies to a continuous strain can be seen in Fig. 72. Here the example involves tension T and compression C axes in the plane of the drawing and at $45°$ to the vertical z. In this case, one of the two possible rupture planes (or systems of planes) is horizontal (thrust) and the other

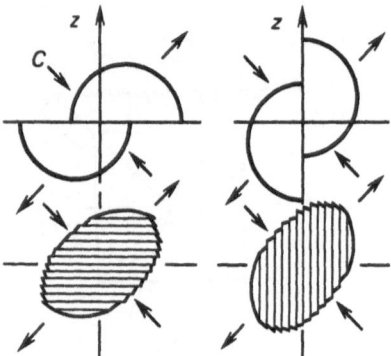

Fig. 72. Total strain of a source region in simple and composite ruptures with fixed tension T and compression C axes.

With growth in the space frequency of ruptures, regardless of their direction, a deformed source transforms into the same ellipsoid

is vertical (fault). At the top of Fig. 72 simple ruptures, each along only one plane, are shown while the lower part shows composite ruptures along a series of parallel planes of the same two principal directions (imbricate thrust or imbricate fault). The transition from composite ruptures to continuous strain is obvious. It is the same regardless of the direction of the ruptures. The total strain is determined by the directions of the principal tension and compression axes, and by the strains and accordingly the displacements along them. It does not depend on the direction of the rupture planes. The total strain of the enveloping surface of a focus also does not depend on the frequency of the ruptures. The space forms of a ruptured focus with any number of ruptures from infinity to unity is, as it were, a single sequence of the same stepwise approximation of the same continuous three-dimensional figure—an ellipsoid of the strain removed by an earthquake.

It is significant that in all rupture strains shown in Fig. 72 and in the corresponding cases of continuous strain, the total displacements D over the opposite edges of a volume focus, and also the total seismic moment, remain the same. We can introduce the concept of the volume density M_0 of the seismic moment M_0^*, and now this will be expressed by the constancy of the integral in the Lebesgue meaning $M_0 = \iiint_V M_0^* \, dxdydz = \text{const}$,

where V is the volume of a focus. If desired, it can also be considered as unlimited (the parts of a focus remote from a rupture only slightly affect the integration).

The physical meaning of the volume density of the seismic moment is the shear stress σ in the elementary volume $dV = dxdydz$ of the focal region. To convince ourselves of this, we choose a system of coordinates such that the planes of rupture or glide (for a continuous focus) are parallel to the x, y-coordinate plane; z is the direction of a perpendicular to it, and y is the direction of displacement. Hence $\sigma = \sigma_{yz} = \mu\varepsilon_{yz}$, where ε_{yz} is the shear strain, and we obtain:

$$\iiint_V M_0^* dxdydz = \iiint_V \sigma_{yz} \, dxdydz = \iint_S dxdy \int_0^h \sigma_{yz} \, dz$$

$$= S \int_0^h \mu\varepsilon_{yz} \, dz = \mu SD = M_0.$$

Here h is the dimension of a volume focus in the z direction at right angles to the rupture or glide plane.

Since for all foci, ruptured or continuous, we have $M_0 = $ const, these foci are indistinguishable with respect to M_0. To distinguish them, additional information is required. But to solve this problem, i.e., determining the velocities of vertical seismotectonic motions, is not necessary as they can be expressed in terms of the seismic moment.

We also note that we have only considered the case when the principal axes of the strains are directed equally at all points in the volume of the focus. If this were not true, instead of a very simple scalar representation of the seismic moment we would have to revert to its tensor representation, e.g., as proposed by B. Kostrov.

Aggregate of foci. The density N^* of distribution per unit area and time of the number of earthquakes with respect to their magnitude K is expressed by:

$$\log N^* = \log A^* - \gamma(K - K_0), \tag{9.13}$$

where A^* is the density (with respect to K) of seismic activity. It is related to the conventional activity by classes within the limits of $K \pm 0.5$, i.e. $A = A_0$ when $K = K_0$ by the formula [Riznichenko]:

$$A = A^*(10^{0.5\gamma} - 10^{-0.5\gamma})/\gamma \ln 10. \tag{9.14}$$

When $\gamma \leqslant 0.5$, the ratio is $A/A^* \leqslant 1.06$, so that the difference between A^* and A is insignificant practically and may often be disregarded.

Let us express the seismic moment M_0 as a function of the magnitude of earthquakes by the correlation [Riznichenko]

$$\log M_0 = C_1 + C_2(K - K_0). \tag{9.15}$$

The total volume of the rock lifted above all earthquake foci per unit surface area in unit time (in these units) equals the mean velocity of the seismotectonic uplift of the Earth's surface in the focal region, i.e.,

$$v = \int_{-\infty}^{K_{max}} \Delta v \, N^* \, dK = \frac{1}{\mu\sqrt{2}} \int_{-\infty}^{K_{max}} M_0 N (\cos \alpha_T - \cos \alpha_C) \, dK$$

$$= \frac{1}{\mu\sqrt{2}} \int_{-\infty}^{K_{max}} \frac{M_0 A^*}{10^{\gamma(K - K_0)}} (\cos \alpha_T - \cos \alpha_C) \, dK$$

$$= \frac{1}{\mu\sqrt{2}} \int_{-\infty}^{K_{max}} \frac{M_0 A^\gamma \ln 10}{10^{\gamma(K - K_0)}(10^{0.5\gamma} - 10^{-0.5\gamma})} (\cos \alpha_T - \cos \alpha_C) \, dK. \tag{9.16}$$

Here $M_0 = M_0(K)$. The value of v is determined for each point of the map using averages over each elementary area surrounding the point. When averaging $\cos \alpha_T$ and $\cos \alpha_C$, they must be weighted with respect to M_0. Let us for the present assume that γ, C_1, and C_2 do not depend on K for $-\infty \leqslant K \leqslant K_{max}$. This is not dangerous because v only depends on earthquakes close to K_{max}. This is defined later.

Inserting (9.14) into (9.15) and integrating, we obtain the following formula for the velocity of vertical seismotectonic motion:

$$v = \frac{A M_0(K_0) \gamma [\cos \alpha_T - \cos \alpha_C]}{\mu (C_2 - \gamma)(10^{0.5\gamma} - 10^{-0.5\gamma})\sqrt{2}} 10^{(C_2 - \gamma)(K_{max} - K_0)}. \tag{9.17}$$

Here $M_0(K_0)$ is the value of M_0 corresponding to a fixed value of K_0 in (9.14). For this value of K_0, we must also determine the seismic activity A. For instance, if $K_0 = 10$, we have $A = A_{10}$, and if $K_0 = 15$, we have $A = A_{15}$.

Equation (9.16) has a physical meaning and may be used when the quantities in it are expressed in an agreed system of units. The dimensions of v and AM_0/μ are the same. The remaining quantities in (9.16) are dimensionless. This formula differs from (9.8) with regard to the difference expressed by (9.13) between A and A^*. In [Riznichenko], this small difference has been disregarded.

To plot a map of v using (9.16), we must generally have four maps, namely, one for each A, K_{max}, $\cos \alpha_T$, and $\cos \alpha_C$. The last two can be combined into $\cos \alpha_T - \cos \alpha_C$. The maps of A and K_{max} are plotted as for the quantitative study of seismicity and seismic danger [Riznichenko, etc.]. The values of α_T and α_C are determined when studying the mechanism of sources [Vvedenskaya etc.].

Numerical calculations. For a numerical assessment of v, let us assume that $\gamma = 0.5$ and $\mu = 3 \times 10^{11}$ dyn/cm^2 (for granite). The quantities A and K_{max} in (9.17) are independent, but to simplify our calculations, we can use their known average correlation in a particular case [Riznichenko], viz.,

$$\log A = C_3 + C_4 (K_{max} - K_0) \tag{9.18}$$

Here $K = \log E_J$. Let $K_0 = 15$, hence $A = A_{15}$. It is customary to determine A_{15} by the number of earthquakes with a magnitude within the range of $K = 15 \pm 0.5$ in one year and an area of $10^{5.5}$ km^2 = $10^{15.5}$ cm^2. On the other hand, the activity $A = A_{10}$ is determined by the number of earthquakes with a magnitude of $K = 10 \pm 0.5$ in one year and an area of 10^3 km^2. If, as we have assumed, $\gamma = 0.5$, then $A_{10} = A_{15}$. For A_{10} and A_{15}, the parameters in (9.18) are: $C_3 = \bar{2}.84$, and $C_4 = 0.39$ for $K \leqslant 15$. When changing (for A) from the indicated areas to parameters per cm^2, the parameters in (9.18) will be $K_0 = 15$, $C_3 = \bar{2}.84 + 15.5 = 14.340$, and $C_4 = 0.39$. When changing from joules to ergs, the parameters in (9.18) do not change. We shall adopt the average correlation dependence of M_0 on K in the form]Riznichenko]:

$$\log M_0 = 11.842 + 0.889K = 25.177 + 0.889(K - 15) \tag{9.19}$$

so that in (9.15) we shall assume that $C_1 = 25.177$, $C_2 = 0.889$, and $K_0 = 15$. Taking all this into account and disregarding the small difference between A and A^*, we find:

$$v = \frac{10^{C_1 - C_3(C_2 - \gamma)/C_4}}{\mu(C_2 - \gamma)\sqrt{2} \ln 10} A^{1 + (C_2 - \gamma)/C_4} (\cos \alpha_T - \cos \alpha_C) \tag{9.20}$$

For the parameters given above this formula yields:

$$v = 0.18 A^2 (\cos \alpha_T - \cos \alpha_C) \text{ (cm/year)}. \tag{9.21}$$

Below we derive formulae for the components of the rate of strain tensor in three-dimensional space and simultaneously refine the elementary formulae for the vertical components of the velocities of seismotectonic motion.

9.2.3. Removed Stress-strain State at a Focus and the Seismic Moment*

We shall model the region of a three-dimensional intermittent-continuous focus before and after an earthquake by a sphere of radius $R = R(K)$ or the relevant ellipsoid filled with a homogeneous isotropic elastic medium. We shall assume that these properties of the medium are only disturbed at the "instant" of an earthquake.

Stresses and strains. An estimate of the order of magnitude of the average elastic strains removed in the region of a source by an earthquake can be obtained from information on the average radii R and displacements D along the fracture in crustal earthquake foci of various energies $K(= 4 + 1.8 \, M)$ or magnitudes M using world data [Riznichenko]. According to the theory of a three-dimensional intermittent-continuous focus [Riznichenko], a displacement D of the same order of magnitude is received by the ends of the focal diameter $2R$:

K	10	15	20
M	3.3	6.1	8.9
R, km	0.54	7.9	120
D, cm	0.19	24	3100
$D/2R$	0.17×10^{-5}	1.5×10^{-5}	13×10^{-5}

Hence, the estimated values of the average strain $D/2R$ at a source grow somewhat with an increase in K or M, but still remain small, viz. less than 10^{-4}. This justifies the use of the linear theory of elasticity.

The tensor of a small strain U_{ik} is in general [Landau, Lifshits]

$$U_{ik} = \frac{1}{2} \left(\frac{\partial U_i}{\partial x_k} + \frac{\partial U_k}{\partial x_i} \right) = \begin{pmatrix} U_{11} U_{12} U_{13} \\ U_{12} U_{22} U_{23} \\ U_{13} U_{23} U_{33} \end{pmatrix}. \tag{9.22}$$

The tensor is symmetric: $U_{ik} = U_{ki}$. Here the subscripts i and k run through the values 1, 2, 3, and U_i is a displacement vector defined by its components U_1, U_2, and U_3 along the axes $x_1 = x$, $x_2 = y$, and $x_3 = z$. When $i = k$, we obtain the normal (elongation—tension, or compression) components of the strain tensor, while when $i \neq k$, we obtain the tangential (shear) components of this tensor. For instance, the elongation along the axis x_1 is $\varepsilon_{x_1} = U_{11} = \partial U_1 / \partial x_1$, while the shear, i.e., skewing of the initially right angle between the axes x_1 and x_2, is $\gamma_{x_1 x_2} = 2U_{12} = \partial U_1 / \partial x_2 + \partial U_2 / \partial x_1$. Elongation is considered to be positive, compression negative, while shear is positive if the original right hand side becomes smaller. We shall assume that (9.22) describes only the part of the strain at the focus removed by an earthquake.

---

* From [Riznichenko].

The stress tensor σ_{ik} for an isotropic linearly elastic (Hookean) body is expressed in terms of the strain tensor U_{ik} by the general equations [Eshelby] $\sigma_{ik} = KU_{ll}\delta_{ik} + 2\mu(U_{ik} - \frac{1}{3}\delta_{ik}U_{ll})$, where K is the modulus of compression, μ is the shear modulus, $U_{ll} = U_{11} + U_{22} + U_{33}$ is the volume expansion, and $\delta_{ik} = \partial x_i/\partial x_k$ is a unit tensor. For pure shear, which we shall consider, volume expansion is absent, $U_{ll} = 0$, and therefore:

$$\sigma_{ik} = 2\mu U_{ik}. \tag{9.23}$$

Hence, for a focus with pure shear strain, the tensor in (9.23) of the removed stresses σ_{ik} differs from the tensor in (9.22) of the removed strains U_{ik} by the scalar 2μ.

Consider a volume focus with pure plane shear strain. We choose a local rectangular system of coordinates related to the principal strain axes at the focus (Fig. 73): x is the axis of tension T, y is the axis of compression C, and z is a zero axis 0, along which there are no strains (not shown in Fig. 73), consequently the strain is planar in the T, C coordinate plane.

In the coordinate system $(T, C, 0)$ of extreme elongations, let us imagine an initial "undeformed" sphere of radius R—the future focus:

$$x^2 + y^2 + z^2 = R^2. \tag{9.24}$$

In Fig. 73, a circle with inscribed and circumscribed squares corresponds to this state. It is assumed for simplicity that $R = 1$. This can be an element of the focal region similar to the focus as a whole (homogeneity assumption). Immediately prior to an earthquake, the focal region is in the elastically deformed state that is removed by the earthquake. In this state, the sphere transforms into an ellipsoid with the semi-axes $R(1 + \varepsilon)$, $R(1 - \varepsilon)$, and R directed along the same axes:

$$x'^2/R^2(1 + \varepsilon)^2 + y'^2/R^2(1 - \varepsilon)^2 + z'^2/R^2 = 1. \tag{9.25}$$

In the plane of Fig. 71, an ellipse circumscribed by a rectangle and inscribed by a rhombus corresponds to this state. It is clear that in the transition (9.24) → (9.25) no change in the volume occurs.

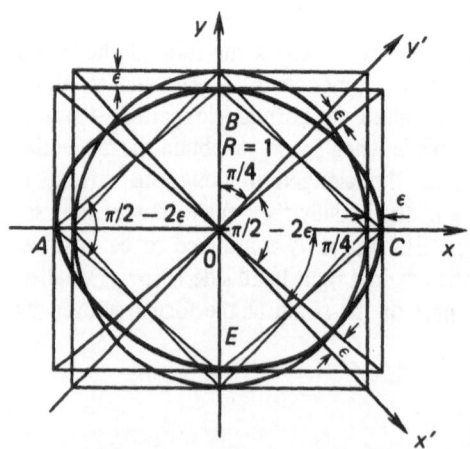

Fig. 73. Strain in the plane x—tension T and y—compression C of a volume source.

The circle transforms into an ellipse, and a square into a rhombus $ABCD$; ε is the tensile strain; ε is the compressive strain, and 2ε is the shear strain, skewness of the initial right angle

At the "instant" of an earthquake in the focal region, i.e., the ellipsoid (9.25), rapid fracturing and plastic motion occur, i.e., an earthquake. The latter transfers the medium within its confines from the elastically deformed to the undeformed state. Recall that we are only speaking of the removed strain. This strain turns from elastic into inelastic, a residual form. In this transition, only the details of the microstructure of the medium in the focal region are disturbed, the general configurations of the region remaining as before, i.e., expressed by the same ellipsoid (9.25). This is explained in greater detail in [Riznichenko].

It should be noted that the idea of an initial "undeformed" sphere (9.24) in the focal region is quite arbitrary. It only concerns that part of the elastic strain that is removed. Actually the medium is not perfectly elastic, and in the focal region it is constantly in an elastic-inelastic state. The reasons for this are (1) hydrostatic stresses associated with the pressure of the heavy overburden, and (2) non-hydrostatic stresses, both elastic—only part of which is removed by an earthquake, and inelastic. The latter are spent (relax) by continuous plastic and creep movements that are slower than an earthquake. Because of residual elastic stress in addition to the removed ones, the elastic energy released from a focus by an earthquake is generally greater than the energy that calculated by the conventional theory of elasticity (for an elastic potential) from (9.22) and (9.23) for the tensors of the removed stress and strain.

To make specific calculations possible, we shall write out the formulas from the linear theory of elasticity for elongation and shear in any direction. The elongation ε_e in the direction of the unit vector $e = e_i = e_1, e_2, e_3$ is:

$$\varepsilon_e = e_1^2 U_{11} + e_2^2 U_{22} + e_3^2 U_{33} + 2(e_1 e_2 U_{12} + e_2 e_3 U_{23} + e_3 e_1 U_{31}), \tag{9.26}$$

where e_i are the direction cosines of the angles between the vector and the coordinate axes, e.g., $e_i' = \cos (e, x_1)$.

The change in the angle between two directions e and e' is expressed by:

$$\cos (e, e') = (e_1 e_1' + e_2 e_2' + e_3 e_3') - (1 - \varepsilon_e - \varepsilon_e')$$

$$+ 2\{U_{11} e_1 e_1' + U_{22} e_2 e_2' + U_{33} e_3 e_3'$$

$$+ U_{12}(e_1 e_2' + e_1' e_2) + U_{23}(e_2 e_3' + e_2' e_3) + U_{31}(e_3 e_1' + e_3' e_1)\}. \tag{9.27}$$

If the directions e and e' are at right angles, we have $e_1 e_1' + e_2 e_2' + e_3 e_3' = 0$, and (9.27) enables us to obtain the shear $\gamma_{ee'}$ between these directions in the form:

$$\gamma_{ee'} = 2\{U_{11} e_1 e_1' + U_{22} e_2 e_2' + U_{33} e_3 e_3'$$

$$+ U_{12}(e_1 e_2' + e_1' e_2) + U_{23}(e_2 e_3' + e_2' e_3) + U_{31}(e_3 e_1' + e_3' e_1)\} \tag{9.28}$$

It can be seen directly from Fig. 73 that in the coordinate system $\{T, C, 0\}$ of extreme elongations, the components of the strain tensor (9.22) for a focus with pure shear are $U_{11} = \varepsilon$ and $U_{22} = -\varepsilon$, when $i \neq k$ all the U_{ik}'s equal zero, whence

$$U_{ik} = \begin{pmatrix} \varepsilon & 0 & 0 \\ 0 & -\varepsilon & 0 \\ 0 & 0 & 0 \end{pmatrix}. \tag{9.29}$$

Using (9.26) and (9.28), we obtain a table of the components U_{ik} of the same tensor in the system of coordinates x', y', z' for the extreme shifts. The x' and y' axes of this system are at $\mp\pi/4$ to the specific axes of the preceding system, while the z axes coincide (see Fig. 73). We obtain

$$
U_{ik} = \begin{pmatrix} 0 & \varepsilon & 0 \\ \varepsilon & 0 & 0 \\ 0 & 0 & 0 \end{pmatrix}.
\tag{9.30}
$$

Let us find, for instance, $U'_{12} = \varepsilon$. In the x, y, z coordinate system the skewness of the right angle between the directions $\mathbf{e} = \mathbf{x}'$ and $\mathbf{e} = \mathbf{y}'$ correspond to component. For $\mathbf{e} = \mathbf{x}'$, the direction cosines, as can be seen from Fig. 73. will be $e_1 = \cos(x', x) = \cos(\pi/4) = \sqrt{2}/2$, $e_2 = \cos(x', y) = -\sqrt{2}/2$, and $e_3 = \cos(x', z) = 0$. For $\mathbf{e} = \mathbf{y}'$, we obtain $e'_1 = \sqrt{2}/2$, $e'_2 = \sqrt{2}/2$, and $e'_3 = 0$. By (9.26), elongations along x' and y' are absent, whence $\varepsilon_e = \varepsilon_e = 0$. Further, from (9.29) we have $U_{11} = \varepsilon$, $U_{22} = -\varepsilon$, and $U_{33} = 0$. When $i \neq k$, all the U_{ik}'s equal zero. Inserting all this into (9.28) for $\gamma_{e,e'}$ we find that in the polynomial in braces, only the first and second terms differ from zero. This yields $\gamma_{e,e'} = 2\varepsilon$. Hence, $U'_{12} = 0.5\gamma_{e,e'} = \varepsilon$, Q.E.D.

This result can also be found purely geometrically by considering Fig. 73. The initial right angle BAE of the relevant square with sides parallel to the axes of extreme shifts is skewed into the rhombus $ABCE$ through the same angle 2ε.

Seismic moment. For a transition to the seismic moment M_0, the rhombus $ABCE$ has been transferred to Fig. 74, which also depicts the classical scheme for determining M_0 for fracture along plane S at right angles to the plane of the drawing:

$$
M_0 = \mu S D.
\tag{9.31}
$$

Here S is the fracture area, D is the displacement over it, and μ is the shear modulus of elasticity. If the length of the square or rhombus is unity, we have $D = 2\varepsilon$.

For a cubic volume focus with side $2R$ and base parallel to the plane of fracture, we assume in (9.31) that $S = (2R)^2$ and $D = (2R)2\varepsilon$. This yields $M_0 = 16\mu\varepsilon R^3$. For a cylindrical focus with a round base of radius R and altitude $2R$, we find $M_0 = 4\pi\mu\varepsilon R^3$. For a spherical

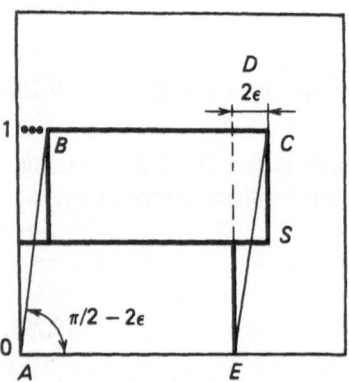

Fig. 74. Determination of the seismic moment $M_0 = \mu S D$.

S is the fracture area, and D is the displacement. Rhombus $ABCE$ is the same as in Fig. 70

focus with radius R:

$$M_0 = \frac{8}{3} \pi \mu \varepsilon R^3. \tag{9.32}$$

Formula (9.32) establishes the relation between the scalar magnitude of the seismic moment M_0 for a spherical focus of pure shear (9.29) and the maximum values (by direction) of the average (over the points of space) strain in the focal region of the elongation ε and shear 2ε.

To change over to a tensor expression of the seismic moment $M_{0_{ik}}$, we shall consider the seismic moment to be an integral of the removed stress over the volume of the focus [Riznichenko]. Proceeding, as previously, from a homogeneous spherical focus, we shall assume that at any point of the focus the removed stress σ_{ik} is the same and equals the average for the focus over its volume. It is expressed in the tensor form in terms of the removed strains U_{ik} by (9.23). Hence we obtain

$$M_{0_{ik}} = \int_v \sigma_{ik} dv = v\sigma_{ik}$$

$$= (4\pi R^3/3) 2\mu u_{ik} = \frac{8}{3} \pi \mu R^3 u_{ik}. \tag{9.33}$$

Inserting into this formula the scalar seismic moment M_0 from (3.32), we also obtain a tensor expression for $M_{0_{ik}}$:

$$M_{0_{ik}} = M_0 u_{ik}/\varepsilon = M_0 \theta_{ik} \tag{9.34}$$

Here θ_{ik} is a tensor which in the coordinate system $\{T, C, 0\}$ of extreme elongations by (9.29) has the following components:

$$\theta_{ik} = \begin{pmatrix} 1 & 0 & 0 \\ 0 & -1 & 0 \\ 0 & 0 & 0 \end{pmatrix}$$

The seismic moment tensor in this coordinate system acquires the form:

$$M_0 = \begin{pmatrix} M_0 & 0 & 0 \\ 0 & -M_0 & 0 \\ 0 & 0 & 0 \end{pmatrix} \tag{9.35}$$

This expression is, in essence, the result obtained earlier by B. V. Kostrov from much more general considerations.

Drop in the stress. If we assume that, in our case, the drop in the shear stress is $\Delta\sigma = \mu 2\varepsilon$, we can obtain the following estimate from the volume model (9.32): $\Delta\sigma = M_0/1.33\pi R^3 = 0.24 M_0/R^3$. This estimate was given earlier in [Riznichenko]. Let us compare it with the dislocation estimate of Brun, viz. $\Delta\sigma = \frac{7}{16} M_0/R^3 = 0.44 M_0/R^3$.

We see that these estimates obtained for various focal models (volume and dislocation) are within $\log (0.44/0.24) = 0.16$, viz. an order of magnitude, error limits for determining M_0 and R^3.

9.2.4. Transition to a General Coordinate System

We shall adopt the geographical system as a general orthogonal coordinate system at any point of space. Its axes x, y, and z are oriented as follows: x along a parallel to the east, y along a meridian to the north, and z upward to the zenith. Using a definition of focal mechanism such as the one proposed by A. V. Vvedenskaya, we can find the angle φ_T between the direction of the tension axis T and a vertical z, and the azimuth α_T of the axis, i.e. the angle between its horizontal projection and a meridian (Fig. 75). Similar angles φ_C and α_C can also be found for the compression axis C and the angles φ_0 and α_0 for the zero strain axis 0.

Proceeding from (9.27) for the components of the tensor of the removed stresses at the focus in the $\{T, C, 0\}$ system of coordinates of the principal elongations and using (9.23) and (9.24) for transforming the tensor components to another system of coordinates, let us find expressions for the components of the same tensor, in our general geographical system x, y, and z. These expressions must include the known value of the tension or compression ε and functions of the angles φ_T, α_T, φ_C, and α_C.

Directions of the principal strain axes in a focus. From Fig. 75, we find the direction cosines l_T, m_T, and n_T of the axis T in the coordinate system x, y, and z:

$$l_T = \sin \varphi_T \sin \alpha_T, \quad m_T = \sin \varphi_T \cos \alpha_T, \quad n_T = \cos \varphi_T. \tag{9.36}$$

Similarly, for the axis C, we obtain

$$l_C = \sin \varphi_C \sin \alpha_C, \quad m_C = \sin \varphi_C \cos \alpha_C, \quad n_C = \cos \varphi_C. \tag{9.37}$$

The direction cosines l_0, m_0, and n_0 of the O axis for zero removed strain can be obtained as a function of the arguments defined in (9.36) and (9.37) from the system of equations $l_T l_0 + m_T m_0 + n_T n_0 = 0$, $l_C l_0 + m_C m_0 + n_C n_0 = 0$, and $l_0^2 + m_0^2 + n_0^2 = 1$.

The first two equations are the conditions that O axis is perpendicular to the T and C axes, and the third equation is the uniqueness of the vector of the O direction. Solving this system for l_0, m_0, and n_0 and considering the conditions of mutual perpendicularity

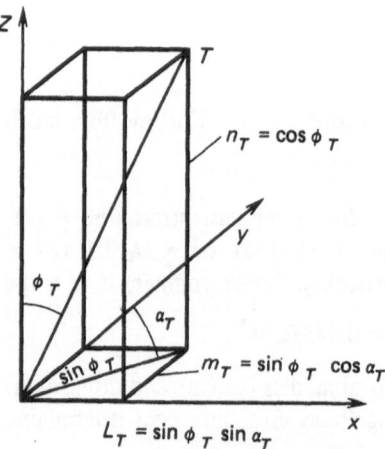

Fig. 75. Directing cosines l_T, m_T, and n_T of the tension axis T in the geographical coordinate system x, y, and z:

α_T is the azimuth, and φ_T is the angle with the vertical

and the uniqueness of the vectors in the T and C directions we find:

$$l_0 = m_T n_C - m_C n_T, \quad m_0 = n_T l_C - n_C l_T, \quad n_0 = l_T n_C - l_C m_T. \tag{9.38}$$

The azimuth α_0 of the O axis and its angle φ_0 with the vertical are determined in terms of the T and C directions by the formulae:

$$\tan \alpha_0 = l_0/m_0 = m_T n_C - m_C n_T/n_T l_C - n_C l_T,$$

$$\cos \varphi_0 = n_0 = l_T n_C - l_C m_T. \tag{9.39}$$

The geophysical meaning of the O directions for a set of earthquake foci in a region is that they indicate the general extension of the fracture surfaces in the earthquake foci in the region [A. Ritsema]. It should be noted that when φ_0 is small, the azimuth α_0 solution is not stable.

Strain tensor components at a focus. In the (T, C, O) coordinate system, the components of the strain tensor are given by the matrix in (9.29). Let us find expressions for the components U_{ik} of the same tensor in the new general geographical coordinate system x, y, and z. We shall use (9.26) for the components $U_{xx} = \varepsilon_x$ of the elongation along the x axis of the new system. In the old $\{T, C, O\}$ system the direction cosines of the x axis will be $l_{T_x} = l_T = \sin \varphi_T \sin \alpha_T$, $m_{C_x} = l_C = \sin \varphi_C \sin \alpha_C$, and $n_{0_x} = l_0 = \sqrt{1 - \sin^2 \varphi_T \sin^2 \alpha_T - \sin^2 \varphi_C \sin^2 \alpha_C}$.

Hence we obtain the following from (9.26) and (9.29) for the elongation along the new x axis, formula: $\varepsilon_c = l_{T_x}^2(\varepsilon) + m_{C_x}^2(-\varepsilon) + n_{0_x}^2(0)$.

We find the final form of the formula by inserting the above l_{T_x}, etc. In a similar way, we also obtain the remaining formulae for the elongations along the axes of the new geographical system, viz.

$$\varepsilon_x = \varepsilon(\sin^2 \varphi_T \sin^2 \alpha_T - \sin^2 \varphi_C \sin^2 \alpha_C) \tag{9.40}$$

$$\varepsilon_y = \varepsilon(\sin^2 \varphi_T \cos^2 \alpha_T - \sin^2 \varphi_C \cos^2 \alpha_C) \tag{9.41}$$

$$\varepsilon_z = \varepsilon(\cos^2 \varphi_T - \cos^2 \varphi_C). \tag{9.42}$$

These equations are used to determine the components of the strain rate tensor during the seismic flow of rock.

In the same way, we can obtain expressions for the shear γ_{ik} components U_{ik} $(i \neq k)$ of the strain tensor at a focus in the new general geographical coordinate system x, y, and z:

$$\gamma_{xy} = 2\varepsilon(\sin^2 \varphi_T \sin \alpha_T \cos \alpha_T - \sin^2 \varphi_C \sin \alpha_C \cos \alpha_C) \tag{9.43}$$

$$\gamma_{yz} = 2\varepsilon(\sin \varphi_T \cos \alpha_T \cos \varphi_T - \sin \varphi_C \cos \alpha_C \cos \varphi_C) \tag{9.44}$$

$$\gamma_{zx} = 2\varepsilon(\sin \varphi_T \sin \alpha_T \cos \varphi_T - \sin \varphi_C \sin \alpha_C \cos \varphi_C). \tag{9.45}$$

9.2.5. Calculations of Seismic Flow

Scalar relations. Let us single out an elementary volume region $\Omega = SL$ of rock masses with a length of L along an axis and with a cross-sectional area of S (Fig. 76). We assume that during time t, some events with maximum tension along the same axis took place in this region $i = 1, 2, 3, \ldots, N$, so that it is the T axis. We find the average elongation along this axis due to foci in the region Ω and the average rate of strain v of the elongation.

We first consider the contribution of the i-th focus. Let its length along the T axis be l_i, its cross-sectional area be s_i, i.e., its volume is $\omega = s_i l_i$, and let its relative strain be ε_i. Hence, the area s_i covers the volume $s_i l_i \varepsilon_i$ (for a pure shear focus it is naturally compensated by compression along the other axes). Let us distribute this volume over the area S of region Ω. Now we obtain the absolute elongation of this region in the same direction, viz. $s_i l_i \varepsilon_i / S$. The relative elongation, i.e. the strain of region along the T axis, is therefore $(s_i l_i / SL)\varepsilon_i = (\omega_i / \Omega)\varepsilon_i$. Assuming now that the volume of a focus is $\omega = (4/3)\pi R^3$, we express it by (9.32) in terms of the seismic moment M_0 of the focus. Dividing the resultant strain of region Ω of the rock masses by the time t, we obtain the average rate of strain v_i (elongation) of the region in the T direction due to the i-th focus, $v_i = M_0 / 2\mu\Omega t$. It is obvious that this is the rate of tensile strain along the T axis, v_i being positive.

If in the region Ω during the time t there were $i = 1, 2, 3, \ldots, N$ foci with the same direction of maximum tension, the total strain rate in this region along the T axis would be:

$$v = \frac{1}{2\mu\Omega t} \sum_{i=1}^{N} M_{0_i}. \tag{9.46}$$

If we had good statistics for the distribution of earthquakes by M_0, the summation in (9.46) could have been performed directly or replaced by integration. At present, the statistics for earthquakes for their energy value K (or magnitude M), are more reliable. New correlations

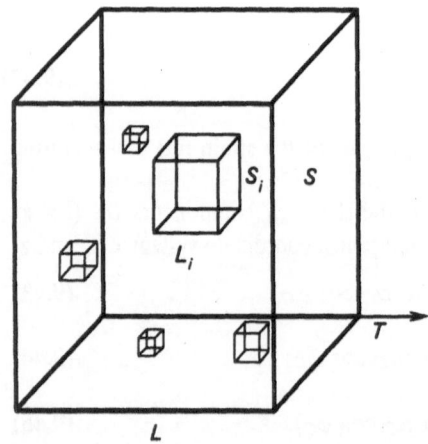

Fig. 76. For the determination of the rate of strain of the volume rock mass region $\Omega = SL$ along the tension axis T:

$\omega_1 = S_i L_i$ is the volume of the i-th source

between K and M_0 are being established. We shall therefore sum (integrate) over K. We write the expressions for N and M_0 versus K [Riznichenko]:

$$\log N^* = \log A^* - \gamma(K - K_0) \quad \text{when} \quad K \leqslant K_{\max} \tag{9.47}$$

$$\log M_0 = C_1 + C_2(K - K_0). \tag{9.48}$$

Here $N^* = dN/dK$, and A^* is the density (with respect to K) of seismic activity. It is related to the more popular class activity A, calculated within the limits of the class $K_0 \pm 0.5$ of the magnitude of an earthquake by [Riznichenko]

$$A = A^*(10^{0.5\gamma} - 10^{-0.5\gamma})/\gamma \ln 10. \tag{9.49}$$

We shall assume that A and A^* are volume activities corresponding to the numbers N of earthquakes per unit volume of Ω and unit time t.

We assume that Ωt is a unit in a space-time region identical to the one used to normalize the seismic activity A. We shall replace the summation in (9.46) with integration over K. This yields the following formula for calculating the rate of strain v of the rock masses along the axis T of maximum tension:

$$v = \frac{1}{2\mu} \int_{-\infty}^{K_{\max}} M_0(K) N^*(K) \, dK = \frac{\gamma A M_0(K_0) \, 10^{(C_2 - \gamma)(K_{\max} - K_0)}}{2\mu(C_2 - \gamma)(10^{0.5\gamma} - 10^{-0.5\gamma})} \cdot \tag{9.50}$$

Formulae (9.47)-(9.49) were considered when integrating in (9.50). Here $(C_2 - \gamma) > 0$, $M_0(K_0) = 10^{C_1}$ is fixed, the seismic moment M_0 (9.43) at the same value $K = K_0$ being adopted when determining the seismic activity A (usually K_0 is taken to be 10 when $A = A_{10}$ or $K_0 = 15$ when $A = A_{15}$).

Tensor relations. In a very simple case, all the foci in a space-time region have identically directed principal axes T, C, and O for the removed strains and stresses and identically directed principal axes (latent vectors) for the tensors of the seismic moments M_{0ik} of the foci. Here the rates of strain tensor of the seismic flow of rock obviously has the same principal axes T, C, and O [Fedorov]. In the (T, C, O) coordinate system, its matrix is similar to (9.29) and has the form:

$$v_{ik} = \begin{pmatrix} v & 0 & 0 \\ 0 & -v & 0 \\ 0 & 0 & 0 \end{pmatrix}.$$

The coordinate system is transformed to the geographical system for the tensor v_{ik} using (9.40)-(9.45), just as for the tensor U_{ik} of the strains in a focus.

For the general case, we can write, as in (9.46):

$$v_{ik} = \frac{1}{2\mu\Omega t} \sum_{t=1}^{N} M_{0ik}. \tag{9.51}$$

Let us derive the formulae for the rates of tensile strain in the seismic flow of rock along the axes of the geographic system: x to the east, y to the north, and z upward:

$$v_{xx} = v(\sin^2 \varphi_T \sin^2 \alpha_T - \sin^2 \varphi_C \sin^2 \alpha_C) \qquad (9.52)$$

$$v_{yy} = v(\sin^2 \varphi_T \cos^2 \alpha_T - \sin^2 \varphi_C \cos^2 \alpha_C) \qquad (9.53)$$

$$v_{zz} = v(\cos^2 \varphi_T - \cos^2 \varphi_C). \qquad (9.54)$$

Here v is defined in (9.50) in terms of γ, A, K_{max} the permanent average seismic conditions, φ_T and φ_C are angles with the vertical, and α_T and α_C are the azimuths of the tension T and compression C axes at the earthquake foci, as determined from their mechanism. All these data are obtained by standard large-scale processing of seismic observations.

Formulae similar to (9.40)-(9.45) can also be derived for the remaining, shear components v_{xy}, v_{yz}, and v_{zz} of the tensor v_{ik} of the strain rates in the seismic flow in the geographic coordinate system:

$$v_{xy} = 2v(\sin^2 \varphi_T \sin \alpha_T \cos \alpha_T) - \sin^2 \varphi_C \sin \alpha_C \cos \alpha_C)$$

$$v_{yz} = 2v(\sin \varphi_T \cos \alpha_T \cos \varphi_T) - \sin \varphi_C \cos \alpha_C \cos \varphi_C) \qquad (9.55)$$

$$v_{zx} = 2v(\sin \varphi_T \sin \alpha_T \cos \varphi_T - \sin \varphi_C \cos \alpha_C \cos \varphi_C).$$

Equation (9.54) refines the equations from the elementary theory of vertical seismotectonic motion [Riznichenko] for the velocities $v_z = v_{z_{appr}}$ of these movements. For $\gamma = 0.5$ and other values of the parameters in the elementary theory resulted in the following working formula for the velocity of vertical motion (more exactly, the velocity of the change in the thickness of the active layer) in the seismic flow of rock masses:

$$v_{z_{appr}} = 0.18A^2(\cos \varphi_T - \cos \varphi_C), \qquad (9.56)$$

where the small difference between A and A^* in (9.49) was disregarded. For the same parameters, but considering the difference between A and A^*, a tensor treatment instead of the preceding expression leads to the following refined formula (see also below):

$$v_{z_{ref}} = 0.12A^2(\cos^2 \varphi_T - \cos^2 \varphi_C) \qquad (9.57)$$

The difference between the squares of the cosines should appear in the parentheses in equations such as (9.54) instead of the difference between their first powers. This was pointed out to me by B. Kostrov, for which I express my profound appreciation. The correction factor for a transition from the approximate formula of the elementary theory [Riznichenko] to the refined formula of a tensor nature is $(\cos \varphi_T + \cos \varphi_C)/\sqrt{2}$ (here the difference between A and A^* is disregarded). This factor is usually of the order of unity, while on average, when $\varphi_T = \varphi_C = 45°$, it simply is unity, so that its introduction should not lead to a qualitative change in the previous results.

Verification of convergence of calculations*. To verify the last thesis, I requested comparative calculations of the vertical component of the velocity of seismic flow, to be performed

---

* From [Riznichenko]. Co-author E. A. Dzhibladze.

Fig. 77. Comparison of the results from the calculation of the vertical velocity components in the seismic flow of rock masees:

$v_{z_{appr}}$—according to the formulas of the elementary theory [Riznichenko]; $v_{z_{ref}}$—in the tensor treatment (the present work); C—denotes for the Vrancha region in the Carpathians; B—for the region of the Baikal rift

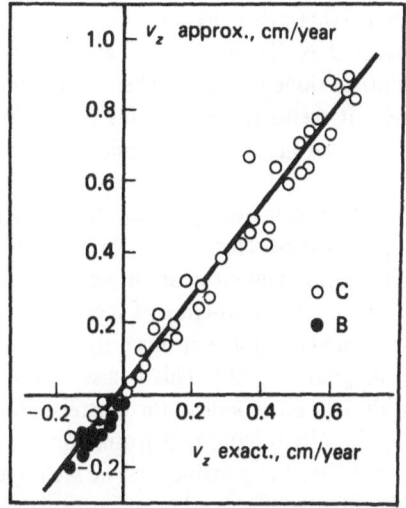

in two ways: (a) using the elementary [Riznichenko] and (b) using the refined (the present work) theory. I also requested that three regions in the USSR be selected so that they differ greatly in their seismotectonic conditions. The relevant calculations were performed for the Vrancha region in the Carpathians by N. A. Simonova, for the region of the Baikal rift by L. A. Misharina, and for the Dzhavakhet highland in the Caucasus by E. A. Dzhibladze. I express my profound gratitude to all of them. The quantities $v_{z_{appr}}$ and $v_{z_{ref}}$ for the Carpathians and Baikal are compared in Fig. 77. We see that the differences are indeed quite small. A similar result was obtained for the Caucasus. It is nevertheless better in the following examples to use the refined relations directly.

It does not matter in (9.56) and (9.57) whether $A = A_{10}$ or A_{15} (when $\gamma = 0.5$). It can be seen from (9.56) that only at very high activity $A > 1$, does the velocity v of seismotectonic motion become of the order of 1 cm/year. This is because in (9.18) $K_{max} \geqslant 18$, i.e., (at $K = 4 + 1.8M$) we have $M_{max} \geqslant 8$. Such a situation is rare under continental conditions: in the USSR it occurs for crustal sources in small regions of Central Asia; for subcrustal ones in the Hindu Kush region in Central Asia adjoining the USSR; and in the Vrancha region in the Carpathians. Under these conditions, our seismic velocities $v = v_s$ are usually small fractions of a centimetre a year. These are much smaller than the total velocities v_g for modern vertical tectonic motion as established by geodesy and geomorphology. Specific data for the Caucasus are given below.

Under the conditions of marginal-continental and oceanic seismicity of the Benioff zone and in part of the mid-oceanic ridges and rifts (here the seismicity is weaker), the values of A and K_{max} observed there indicate that velocities v_s of the order of centimetres a year should not be rare. J. N. Brune et al. used seismological considerations similar to ours [Riznichenko] to calculate the velocities of the type v_s of the relative displacements of asthenospherical platforms under such conditions. They obtained values of the order of several centimetres a year. Our estimates agree with these.

We know from global plate tectonics that approximately the same velocities are the rule in the expansion of ocean floors and in the region of median ridges and rifts. It is also presumed that, when oceanic plates move under continental ones in Benioff zones, such velocities occur. J. N. Brune et al. [G. Davies, Brune], and other later authors, concluded that seismic motion alone (v_s) is sufficient to ensure the principal total motion of global plate tectonics (v_g). It seems to us that this conclusion deserves discussion.

Regardless of geotectonic concepts, total tectonic motion includes the following main components: (1) continuous motion—viscous, plastic, etc. and (2) fracture motion. The latter, in turn, divide into (a) rapid, "noisy" seismic movements, and (b) slower, "quiet" motions— rapid and slow creep. Field observations of fractures in recent years have shown that continuous motion and creep are more important than seismic dislocation. Laboratory experiments involving the destruction of specimens of rock and model materials have shown that the energy of noisy ruptures forms only a small fraction, about 1%, of the total deformation energy [Vinogradov et al.]. This is also true of the relevant v_s and v_g velocities. All this, together with data on v_s and v_g for crustal seismicity, indicates that the contribution of the seismic, rapid, noisy fracturing deformations to the overall fracture-continuous deformation is small. It is difficult to imagine, and we see no reason to do so, that the situation should be appreciably different for border and oceanic tectonics.

Hence we suspect that in Benioff zones and in the median ridges and rifts, as well as in the transform fractures associated with them (the famous San Andreas Fault in California is such a seismogenic structure), seismic motion must only form a modest part of the total tectonic motion.

How then can we explain the closeness between the v_s velocities in Benioff zones and the v_g velocities at which the ocean floor plates move apart? Perhaps the velocity of the total tectonic motion v_g locally in a zone is nevertheless appreciably higher than the average v_g velocity in the large ocean plates. Exactly such a difference between the local "microscopic" (v_g) and the regional "macroscopic" ($\overline{v}_g$) velocities is known in cases of complicated deformation of structurally inhomogeneous media. We can obviously only go from conjecture to fact by directly determining both v_s and v_g velocities, in ocean ridges and rifts and especially in Benioff zones. Such determinations are very necessary.

Let us now assess the convergence of the integral in (9.16) for small K, in other words, let us see what proportion of seismic flow belongs to major earthquakes close to K_{max}. From (9.17), we obtain:

$$\Delta v/v = \left(v \Big|_{-\infty}^{K_{max}} - v \Big|_{-\infty}^{K} \right) \Big/ v \Big|_{-\infty}^{K_{max}} = 1 - 10^{-(c_2-\gamma)(K_{max}-K)}. \tag{9.58}$$

At the same values of $\gamma = 0.5$ and $c_2 = 0.889$ as in the preceding calculations, (9.59) leads to

$K_{max} - K =$	1	2	3	4	5
$\Delta v/v =$	0.59	0.82	0.93	0.98	0.99

Consequently, over 50% of the effect of seismic flow is due to earthquakes varying from the maximum possible ones by, not less than, an order of magnitude of seismic energy. On the other hand, the values the closest to $K_{max} = 3$-5 orders of magnitude of the energy account for virtually all the effect, viz. 93-99%.

Averaging and consideration of turbulence*. When processing data from specific regions, the calculations of the rate of strain v_{ik} in seismic flow in the rock masses can be represented in the form of a number of maps (or cross sections). These are isoline maps for the individual components of the v_{ik} tensor in the geographical system of coordinates: a map of the rates of elongation along the vertical v_{zz} (similar to maps of the velocities of vertical motion [Riznichenko, etc.]); two maps of the rates of horizontal elongation v_{xx} and v_{yy}, and three maps for the shear components v_{xy}, v_{yz}, and v_{zz}. One can also plot more compact and more expressive maps of the vector lines of the rates of the maximum tensions V_T and maximum compressions V_C in a horizontal plane. These can be attended by isolines of the magnitudes of these quasi-vectors.

When the points for determining the mechanism of foci in compiling such maps are dense, the problem is to find a rational procedure for smoothing and averaging the observed data. The choice of region size for averaging is determined by the contradictory requirements of ensuring the highest possible accuracy and the most detail. These have been discussed as applied to maps of the seismic activity A [Riznichenko]. Here we shall only consider the averaging of data on the rates of deformation v_{ik} given that they are tensors. Also, we examine the different contributions to the total deformation of the averaging region containing earthquake foci of various magnitudes.

The above analysis reveals that the contribution of individual earthquakes to the rate of strain in any region (in this case—in averaging region) is proportional to their seismic moments. Consequently, the seismic moments can be adopted as the weights of the strain rates, this holds both in the scalar and tensor treatment of the problems. In the tensor treatment, like components of the tensors are averaged. A component of the strain rate tensor must be included in the averaging with a weight equal to the relevant component of the seismic moment tensor.

If, in a region in which the rates of strain v_{ik} are being determined, the foci are differently directed, their direction tensors differ, $\theta_{ik} = \mathrm{var}$, and the v_{ik} should be determined directly by the general formula for v_{ik} in terms of the sum of the seismic moments of the foci M_{0ik}. If (9.50)-(9.55) are used, on the other hand, the different focal directions can be considered in part, if in some regions the data on M_{0ik} are representative and make it possible to find the coefficient of turbulence of the seismic flow:

$$\Theta = 1 - M_{0\Sigma} \bigg/ \sum_{j=1} M_{0j}. \tag{9.59}$$

Here $M_{0\Sigma}$ is the maximum eigenvalue of the tensor of the sum of the seismic moments ΣM_{0ik} and ΣM_0 is the sum of the scalar seismic moments of the foci, $0 \leqslant \Theta \leqslant 1$. In a region that is more or less homogeneous tectonically, we can find the average $\bar{\Theta}$ for those parts of it where success was achieved in finding the partial values of Θ. These Θ could be ascribed to adjacent parts of the region where only γ, A, and K_{max} are representative. Now it should be possible to substitute $v(1 - \Theta)$ for v in (9.52)-(9.55). This would lead to an approximation of the different directions of the sources for elementary sections of the region.

---

* From [Riznichenko].

9.3. Examples of the Flow Velocity Calculations for Various Regions

9.3.1. Vertical Seismotectonic Motion of the Caucasus*

The following initial data were employed to calculate the velocities v of the vertical seismotectonic motion in the Caucasus.

1. Determinations of the direction of the tension and compression axes for 69 earthquake foci: with magnitudes $K = 8\text{-}16$ ($M = 2\text{-}6.5$) by I. S. Shengeliya using A. V. Vvedenskaya's procedure.

2. Determinations of the permanent average parameters of seismicity: $\gamma = 0.5$ (from [Dzhibladze et al.]; a map of A based on observations for 15 year data of E. A. Dzhibladze); a map of K_{max} based on comprehensive seismological-geophysical and geodetic data from Riznichenko and after E. A. Dzhibladze, I. N. Bolkvadze, and P. O. Dzhidzheishvili.

3. A correlation between M_0 and K found from Caucasian observations [Riznichenko] in the form:

$$\log M_0 = 20.95 + 0.65(K - 10). \tag{9.60}$$

Within the range $K = 8\text{-}16$, this regional relation differs from that for the world within the limits of the usual scatter of observations, i.e., $\log M_0 \pm 0.7$; $K \pm 0.5$.

The velocity v is calculated as:

$$v = \frac{A M_0(K_0)(\cos \alpha_T - \cos \alpha_C)}{\mu(C_2 - \gamma) 10^{13} \sqrt{2} \ln 10} 10^{(C_2 - \gamma)(K_{max} - K)}, \tag{9.61}$$

where it was assumed that $\mu = 3 \times 10^{11}$ dyn/cm, $M_0(K_0) = 10^{20.95}$, $C_2 - \gamma = 0.65 - 0.5 = 0.15$, and $K_0 = 10$. In this formula, the small difference between A and A^* has been disregarded.

Let us appraise the accuracy of determining v by (9.61) depending on the accuracy of the arguments. The errors of the latter are approximately: $\Delta\gamma = \pm0.02$, $\Delta C_2 = \pm0.1$, $\Delta K_{max} = \pm1$, and $\Delta(\cos \alpha_T - \cos \alpha_C) = \pm0.1$. Calculations now reveal that v is determined with a relative accuracy of about 35%.

The calculating of v by (9.61) for the Caucasus are presented in Fig. 78. In addition to the seismological velocity $v = v_s$ isolines (solid), the velocities v_g isolines (dashed) of the vertical present motion from the results of repeated levelings and geomorphology are shown for comparison. The overall configuration of the two systems of isolines, v_s and v_g, can be seen to be similar in general, though there are differences in detail.

The main Caucasian range is characterized by positive values of v_s and v_g. These and other data show the mountains are continuing to rise. In the transition to the Caspian depression, in particular in the vicinity of Makhachkala, and also in the Kurinskaya basin to the southwest of Baku, v_s and v_g are negative. The depression is continuing to deepen.

The relations between v_s and v_g can be characterized quantitatively by a correlation field (Fig. 79). Inspection of the plot shows that between v_s and v_g within the limits of about $-0.01 \leqslant v_s \leqslant 0.03$ and $-5 \leqslant v_g \leqslant 12$ there is on average a linear relation

---

* From [Riznichenko]. Co-author E. A. Dzhibladze.

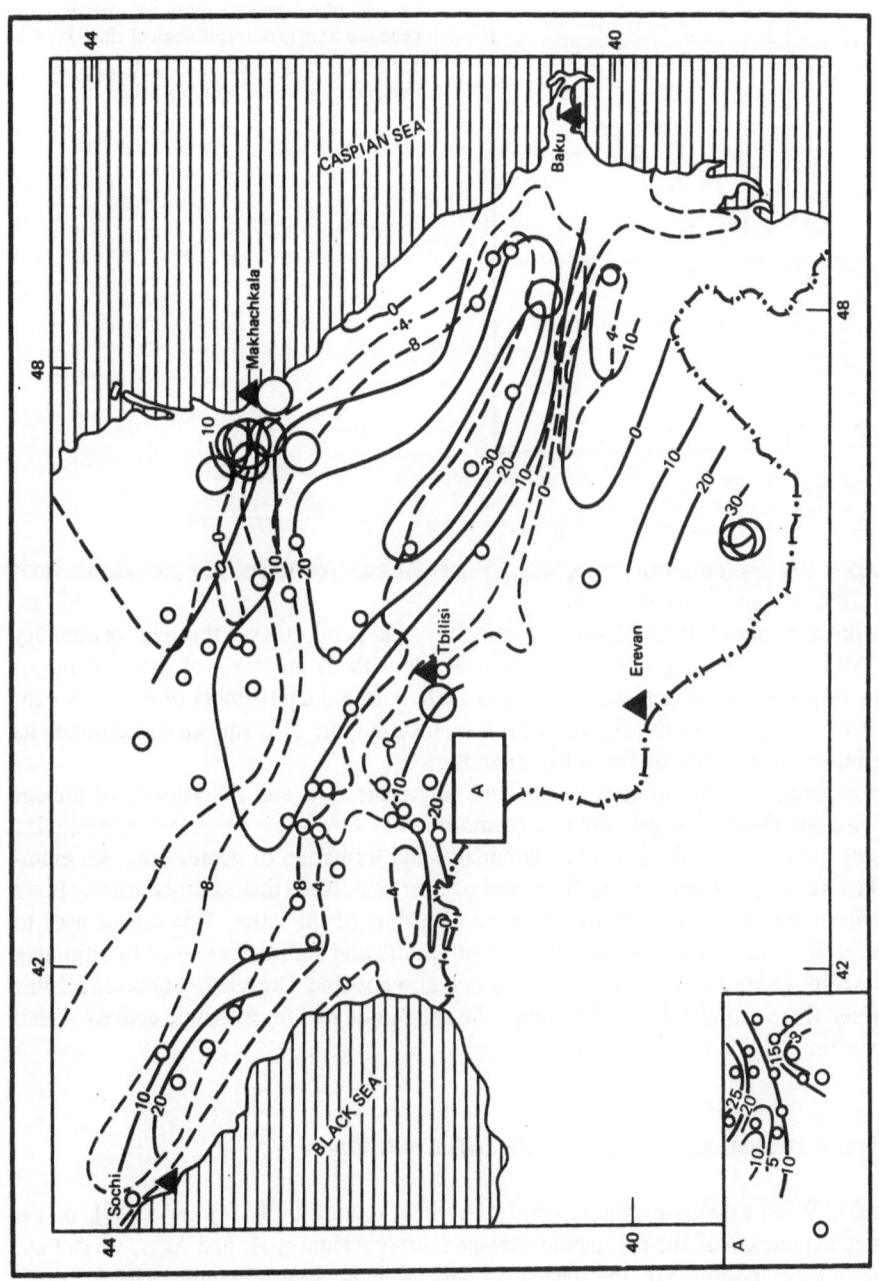

Fig. 78. Map of the vertical seismotectonic motions of the Caucasus.

The *solid isolines* are the velocities of vertical seismotectonic motions v_s, mm/year; the *dashed isolines* are the velocities of the present vertical motions according to geodetic and geomorphological data v_g, mm/year. The insert (A) is the region of the Dzhavakhet highland

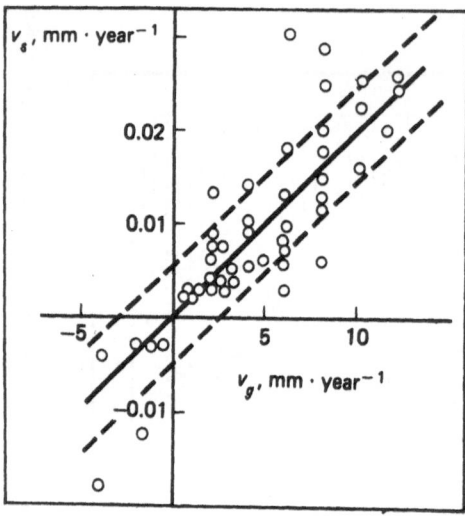

Fig. 79. Correlation of the rates of vertical seismotectonic motions v_s with the velocities of the vertical present motions according to geodetic and geomorphological data v_g.

The *dashed lines* confine the confidence, interval including 70% of the observed points

$v_s \pm 0.005 = 0.002 v_g$ (mm/year) or $v_g \pm 2.5 = 500 \, v_s$ (mm/year). The $\pm \sigma$ confidence band includes 70% of the observed points.

The presence of a clear correlation between the v_s and v_g velocities at this relative accuracy (about 35%) of determining v_s (which is comparable with the accuracy of determining v_g, especially by geomorphological data) allows us to delve into the possibility of mutually controlling both quantities. In the absence of information on one of them, we can estimate its values from measurements of the other quantity.

A comparison of the uplift velocities, due to seismic flow and the velocity of present vertical motion, obtained by geodesy and geomorphology shows that the role of earthquakes is generally quite small in the growth of mountains and formation of depressions. For example, in the Caucasus, it does not usually exceed one per cent. Nevertheless, since seismic flows are correlated with tectonic ones, they may be indicators of the latter. This can be used to continue geodetic data, generally profile data on uplifts and subsidences, over to adjoining territories using indirect seismic data. The reverse is also possible. The relation between seismic and tectonic flows can also be used to judge the magnitude of future seismic activity A and the maximum possible earthquake K_{max}.

9.3.2. Vertical Seismotectonic Motion in the Baikal Rift Zone*

Initial data. When calculating the velocity of vertical motion by (9.56) and (9.57), of the three basic parameters of the permanent average seismic regime γ, A, and K_{max}, we can use only one in explicit form, viz. the seismic activity A. It is given as a map. The following is assumed with respect to the other two parameters.

---

* From [Riznichenko]. V. M. Kochetkov, L. A. Misharina, and N. A. Gileva are co-authors.

It is implied in (9.56) that $\gamma = 0.5$. This is the usual average of γ for most regions in the world. It is also an average for the zone of the Baikal rift. We have adopted a single common value of γ here to simplify the calculation of v_z. This was not done when mapping activity A.

K_{max} is absent in (9.56) because the equation uses the average correlation between K_{max} and A. It was adopted in the form given for major earthquakes ($K \geqslant 15$) on the Eurasian continent and Japan. This has been done because it is with major earthquakes that seismic flow is associated. Naturally, this assumption is also to simplify the calculations.

To improve the calculations of the seismic flow in the region, all three parameters γ, A, and K_{max} if desired may be considered as mutually independent and demarcated geographically over the territory. Here A is the volume seismic activity for the same unit area as the generally adopted surface activity, but in addition it is also related to the unit of thickness h of the seismically active layer. It is generally accepted that $h = 10$ km, and here for A_{15} the unit volume is $10^{6.6}$ km^3, and for A_{10} it is 10^4 km^3. Hence, as before, for $\gamma = 0.5$ we find that A_{15} and A_{10} are equal.

Equation (9.57) differs from the approximate one (9.56) by the correction factor $(A^*/A)^2$ $(\cos \varphi_T + \cos \varphi_C)/\sqrt{2}$. On average, $\varphi_T = \varphi_C = 45°$, and we obtain $(\cos \varphi_T + \cos \varphi_C)/\sqrt{2} = 1$. For $\gamma = 0.5$, we have $(A^*/A)^2 = 0.90$, so that here the approximation does not differ substantially from the refined result. The velocities v_z calculated by (9.56) and (9.57) were directly compared for the South and Middle Baikal regions (as an example). The correlation field includes 30 points and the $v_{z_{appr}}$ and $v_{z_{ref}}$ for this region is shown in Fig. 80. The equation of the averaging straight line has the form $\log v_{z_{ref}} = -0.18 + 0.99 \log v_{z_{appr}}$. The coefficient of linear correlation is close to unity, namely, $\gamma = 0.974$.

We did not recalculate the v_z by formula (9.57) that had previously been calculated by (9.56) when considering the Baikal rift.

Seismic activity. Maps of seismic activity A $(=A_{10})$ have been compiled many times already for the Baikal region, including the Baikal rift [Riznichenko]. Apart from the map in [Riznichenko], these are constant detail maps and constant accuracy maps; the latter are especially important where epicenters are rare. The present map of A [Fig. 81], on the other

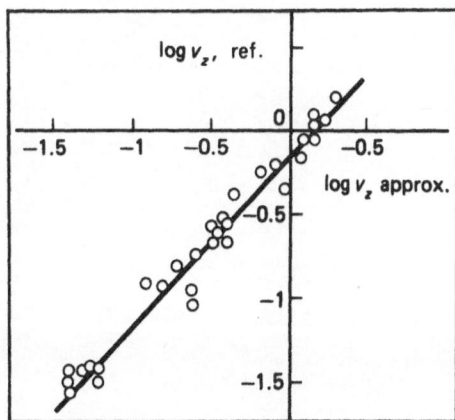

Fig. 80. Comparison of the velocities of the vertical seismotectonic motions in the Baikal region v_z, mm/year.

$v_{z_{approx}}$ has been calculated by the approximate (9.56), and $v_{z_{ref}}$, by the refined (9.57) formulas

Fig. 81. Map of the seismic activity $A = A_{10}$, Baikal region, for 1968-1974

hand, like that in [Riznichenko], is one of the "established quality" [Riznichenko]. Its meaning is explained below. It has been compiled with γ changing from section to section.

The homogeneity of the seismic field in the region is retained only on the largest scale, when the rift zone is treated as a single quasi-homogeneous geological structure of the first order. Actually, as can be seen from the map of the epicenters and seismic activity in Fig. 81, the detailed seismic field is very inhomogeneous. In the following approximation it can be considered to have a block nature [Riznichenko]. From Tunka to Olekma, five large blocks of the Earth's crust can be singled out with reliability. They form the rift zone and are characterized by their individual average parameters of earthquake recurrence. They are the Tunka, Baikal, Central, Muya-Chara, and Olekma region blocks. The angular coefficients γ of the earthquake recurrence plots for the regional blocks were evaluated by the method of least squares with the following results:

Block	Tunka	Baikal	Central	Muya-Chara	Olekma region
γ	0.52 ± 0.01	0.51 ± 0.05	0.60 ± 0.05	0.49 ± 0.08	0.50 ± 0.08

When plotting and calculating the earthquake recurrence graphs, we took into consideration the different weights of the material, namely, for earthquakes K = 8-7 years (1968-1974), for K = 12-20 years (1955-1974), and for K = 13-50 years (1925-1974). Prior to compiling the map of the activity A, we analyzed the earthquake grouping. The groups were appraised by N. S. Borovik's criteria for r = 30 km and τ_2 = 0.35 day. Earthquakes in the vicinity of the epicenter of a major earthquake at a distance up to 10 km were considered to be aftershocks. The latter were excluded from the number of earthquakes for which the activity A was calculated. A swarm was taken as a single event with its total energy.

We excluded the Olekma region from the calculations of the seismic flows and also of the activity A because very few determinations of the mechanism of earthquake foci are available within its confines.

The map of the seismic activity A ($=A_{10}$) was compiled with the usual formula of the summation method [Riznichenko] $A = (N/St) \, [(1 - 10^{\gamma})/10^{-\gamma(K - K_0)}] \, 1000$.

In this formula, $(N/S) \, 1000 = \nu$ is the density of epicenters per 1000 km^2. If $N = N(\nu)$, then $1/S = \nu/N(\nu)$. Thus, we can see that the accuracy, which depends on N, and the detail, which depends on $1/S$, are reciprocals—increasing one diminishes the other. The quality of the map is, in essence, established as a compromise solution of this relation. The details of the field must be determined as far as possible with enough accuracy. This is why it is rational to select a function $N(\nu)$ so that increasing the density ν of the epicenters, improves both the accuracy and the detail of the map within certain limits. At a low epicentral density, the accuracy and accordingly the number of epicenters N_{min} are the limiting factors, while at a high density, the minimum size of the averaged area S_{min} is the limiting one. They must be selected with a view to the accuracy of determining the epicenters and to the map scale. Hence, in a plot of N against ν, we can fix the ends of a section of the curve or straight line $N = f(\nu)$ and use this plot to assess the size of the averaged area.

To compile a map of the activity A for any established quality, we assumed that N_{min} = 10, which corresponds to the mean-square error of determining A of about 33%. For the minimal density ν corresponding to A = 0.02, the maximum size of the area is S_{max} = 4000-5000 km^2. On the other hand, at the virtual maximum density, which corresponds to A = 4.0, we obtain S_{min} = 314 km^2. The relevant averaged area, with a definite number of epicenters in it, was employed to find each isoline of the activity A.

Both elliptical and circular averaged areas were tried in the compilations. Elliptical areas improve the agreement between the outlines of the obtained zones of the activity A and outlines of the tectonic structures along which the major axes of the ellipses are directed. At some places (the central section), the activity zones do not follow the general regional structures. In such cases, it would not logically be justified to make the activity zones follow the structures by elongating the areas. Less extended ellipses with an axis ratio of about 1.5 only slightly affect the picture obtained. We adopted, at least at this stage, the simpler and more objective circular averaging areas. Their radii varied from $R = 10$ to 40 km.

Mechanism of foci. The directions of the principal axes of the measured stressed-strained state at the earthquake foci were determined from data on the displacements of the first entries of longitudinal waves by one of two ways.

1. The focal mechanism was established separately for each earthquake. This was done by the generally adopted procedure of A. V. Vvedenskaya. In most cases, these were comparatively large earthquakes with a magnitude M of at least four. But with a successful location of the epicenter relative to the observing stations, the focal mechanism was also determined for smaller earthquakes. A total of 40 earthquakes were considered individually.

2. A single focal mechanism was established for a group of observed earthquakes. This has to be done to process small earthquakes with magnitudes M usually below four. This procedure is based on the assumption that in restricted volumes of the Earth's crust, the processes at the foci are approximately similar. In practice, the earthquakes are combined into groups. The foci of each group in calculations are arbitrarily considered to be at one point, viz. at the center of the epicentral region, and the signs of their displacements are plotted onto the same stereographic projection—the Wulff net. If the observations for a group of earthquakes with nearby epicenters satisfy the same system of nodal lines, the focal mechanism found for one is ascribed to all. If, on the other hand, the observations do not lend themselves to joint processing, a group is divided into subgroups in which the signs of the displacements agree. The epicentral regions of these subgroups either adjoin or partly overlap. The overlapping in the plane of the epicentral fields of the groups of earthquakes with a different focal mechanism is most naturally explained by their separation either in space or in time. In the first case, they may relate to different levels. But to verify this assumption, one must have better information on the depths of the Baikal region earthquakes than that available at present. Additional studies are also required for time analysis.

The legitimacy of including observations of small earthquakes for study-sets of earthquake foci in the Baikal rift zone is demostrated by the customary coincidence of the focal mechanism of earthquakes of various magnitudes relating to a specific epicentral district. Such agreement is also usually observed when the focal mechanisms for a relatively large individual earthquake and for a group of small earthquakes are compared. This enables us to ascribe the orientation of the axes found for an individually considered earthquake to a group of earthquakes territorially close by. This can be done when the data for such a group do not make it possible to determine unambiguously the position of the nodal planes in the generalized group focus.

The number of events in a group vary significantly, from several units to two or three hundred earthquakes. The possibility of determining the focal mechanism by collective observations is not established by the number of events considered jointly. It is determined by how successfully the conditional points of observation are arranged on the Wulff net relative to the nodal planes of the waves characterizing the given group of foci. The experience gained in processing observations has shown that the average diameter of an epicentral zone for a group with a single focal mechanism is about 25 km.

This group method has been used to determine the source mechanism for more than 1500 earthquakes divided into 93 groups.

These materials were mainly obtained by observations during 1961-1974 and were published, except for data relating to the latest period, by Misharina et al. The distribution over the area of the values of $\xi = (\cos \varphi_T - \cos \varphi_C)$ was calculated from information on the orientation of the tension T and compression C axes at the foci and is shown in Fig. 82. The hatched sections combine both separately located and overlapping epicentral fields of the groups of earthquakes differing in focal mechanism. Where the fields are superimposed, the average quantity being mapped was ascribed to the shared region. The same figure also shows this quantity for individual foci. The velocities v_z of vertical motion in the seismic flow of rock masses in the Baikal rift zone at this stage, as mentioned, were calculated by (9.56) of the elementary theory [Landau].

Mapping procedure. The velocities were generally determined at points of the isolines on the map of the seismic activity A (see Fig. 81). The ξ values were either adopted as equal to the ones obtained for separately located groups of earthquakes or, when the epicentral zones overlap, to the arithmetical means for the overlapping fields. Outside of the hatched section, i.e., where there is no direct information on the focal mechanism, the mean values of ξ obtained for groups of earthquakes and individual events on adjoining territories were used. To facilitate plotting the isolines of v_z, in a number of cases it was necessary to calculate the values for auxiliary points not on the isolines of the activity A. Everything is explained in the captions to Fig. 83, where the initial data used in the calculations and the calculations of v_z are given in detail, for two regions in the Baikal rift.

Map of velocities v_z. A map of the vertical component of the velocity of seismic flow of the rock masses for the Baikal rift zone as a whole is presented in the main part of Fig. 83. It can be seen from the figure that the main part of the rift zone from the southern end of Lake Baikal to Vitim is characterized by negative v_z—subsidence. Only at the flanks of the zone, in the Tunka and Muya-Chara regions, do positive v_z appear, i.e. uplifts. Consequently, calculations of the velocities of seismic flow confirm the generally known geological fact that subsidence motion prevails in the rift zone and impart a new quantitative appraisal to them.

Attention is attracted by the circumstance in which the maximum v_z gravitate in space to regions with increased seismic activity A.

In the field of the velocities v_z in Fig. 83, we again see the same blocks that were singled out on the activity maps A. Whereas in the Tunka, Baikal, and Muya-Chara regions, the isolines of the velocity v_z follow the general course of the rift structures, this is not true in the Central section. Here an extensive area from the Barguzin region to the Upper Angara and North Baikal depressions is involved in subsidence, and it is oriented as a whole across the rift zone.

The most complicated picture of differently directed motion is in the Tunka and Muya-Chara sections (see Fig. 83), where "non-rift" mechanisms of earthquake sources reveal themselves. At Tunka, the positive motion is connected with the uplift of the Large Sayan and

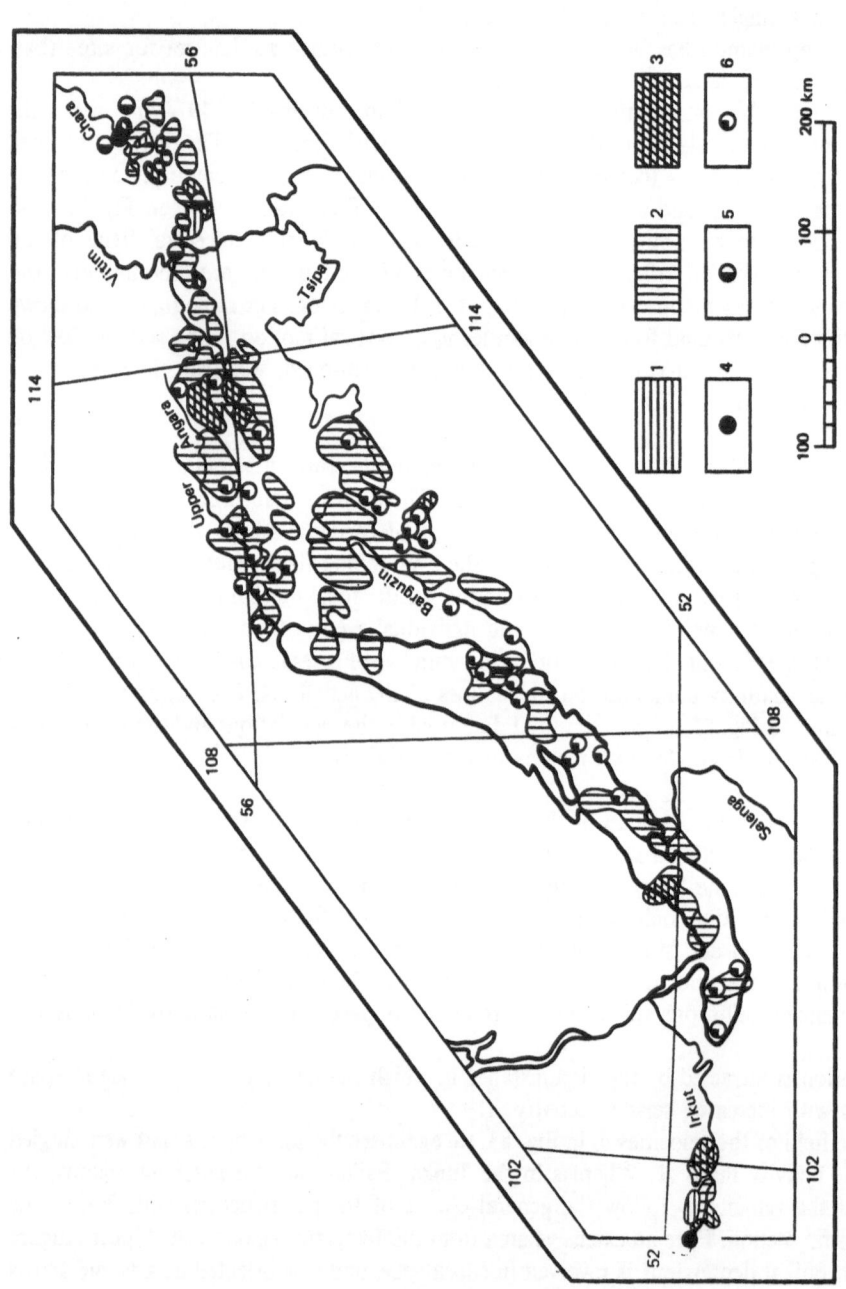

Fig. 82. Map of (cos φ_T − cos φ_C) according to data on the mechanism of earthquake sources combined into groups (1-3) and processed individually (4-6). The values of (cos φ_T − cos φ_C) are: 1, 4—≥0; 2, 5—from 0.5 to 0; 3, 6—from −1 to −0.5

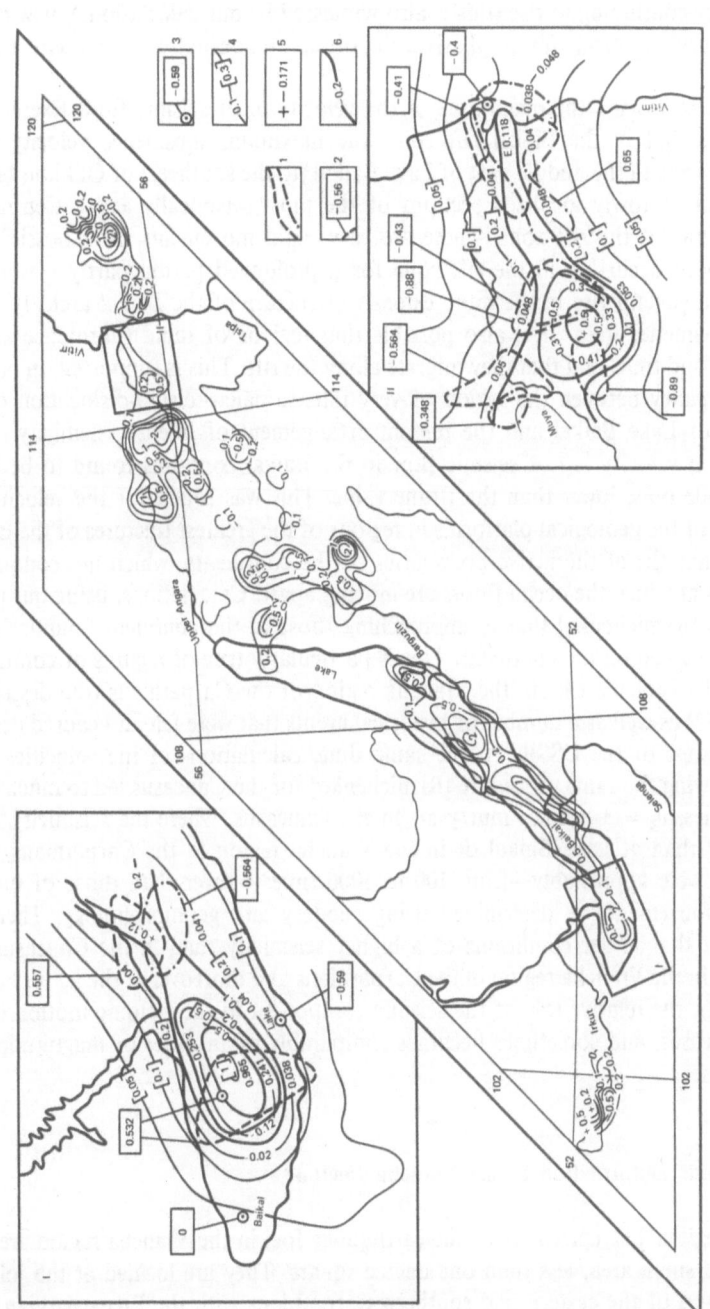

Fig. 83. Map of the velocities v_z of the vertical seismotectonic motions in the Baikal rift region. Inserts *I* and *II* show the initial data and the results of calculating the velocity values for the southern Baikal (*I*) and Muya (*II*) sections of the region:

1—contours of areas with identical values of the differences (cos φτ − cos φc); values of the differences: *2*—within the confines of an area; *3*—in the epicentral region of individually considered earthquakes; *4*—isolines of the seismic activity (the *numbers in brackets* are the values of seismic activity, *without brackets* are the values where points are lying on the isolines of activity); *5*—values calculated for auxiliary points; *6*—isolines of v_z, mm/year

the Tunka bald peaks. In the Muya-Chara block, the signs of the vertical seismotectonic motion—uplift and subsidence—alternate within the major, intersystem, mountain commissure involved in the riftogenic process. Here against the background of high mountain ridges, which are possibly continuing to rise (this is also witnessed by our calculations), new depressions and fault troughs form; this emphasises the differentiated nature of modern tectonic motion.

Let us pause to discuss the magnitude of the velocity v_z of seismic flow. Examination of Fig. 83 reveals that in the Baikal rift zone, the maximum subsidence velocity $|v_z|$ is 2 mm/year and occurs in the middle part of Lake Baikal (to the southeast of Ol'khon Island), and on individual narrowly localized sections of the most seismically active Central and Muya-Chara regions of the rift zone. These are very rapid movements. It is possible that subsidence has been occurring in the rift zone for a prolonged period partly isostatically compensated by a general rise in the more extensive structure of the Baikal arch. It is also witnessed by gravimetric data. It is also possible that regions of rapid subsidence are not constant in space and time, and that they migrate along the rift. This is supported, in particular, by the discrepancy between the region of very intense paleo-seismo-dislocation on the northwest shore of Lake Baikal and the present arrangement of intense seismicity.

The maximum velocity $|v_z|$ of seismic flow in the Baikal zone was found to be by an order of magnitude only, lower than the Brune value. This was found for the velocities of "seismic gliding" of the geological platforms in regions of the greatest fractures of the Earth's crust and upper mantle, at the active boundaries of the continents, which are comparable with the velocities at which the ocean floors are moving apart. Calculations, using our procedure, have shown [Riznichenko] that v_z approaching those at the continent boundaries are also sometimes obtained on the continent. This is particularly true of regions of continental intense subcrustal seismicity, e.g., in the Vrancha region of the Carpathians (the destructive earthquake of 1977) as well as a number of historical events that were felt and caused destruction in the southwest of the USSR. At the same time, calculations of the velocities v_z of seismic flow following the same procedure [Riznichenko] for the Caucasus led to much lower maximum velocities, $v_z = 3 \times 10^{-2}$ mm/year. In the Caucasus, where the seismicity is undoubtedly weaker than at Lake Baikal or in the Vrancha region in the Carpathians, these seismic velocities were appreciably—from 100 to 1000 times—lower than those of modern vertical tectonic movements as determined using geodesy and geomorphology. There are grounds to expect that under conditions of a higher seismicity than in the Caucasus, say at Lake Baikal or in the Vrancha region in the Carpathians and moreover at the active boundaries of continents, the relative role of the seismic component in the tectonic motion of the Earth's crust increases, and sometimes becomes comparable to an order of magnitude with their total velocities.

9.3.3. Seismotectonic Deformation in the Vrancha Region*

Vrancha source region. The epicenters of the earthquake foci in the Vrancha region are concentrated within a small area, less than one degree square. They are located at the joint of the folded structures of the eastern and southern Carpathians with the Precarpathian fore-

---

* From [Riznichenko]. A. V. Drumea, N. Ya. Stepanenko, et al. are co-authors.

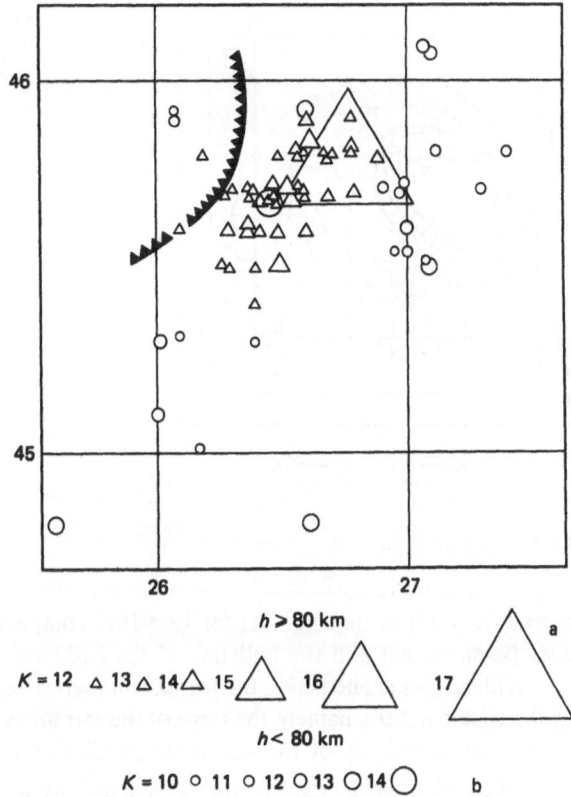

Fig. 84. Map of earthquake epicenters in the Vrancha region for the years 1965-1977:
a—subcrustal sources; b—crustal sources

deep. In depth, the foci are traced from the upper layers of the crust down to 180 km into the mantle, the depth of the foci diminishing to the east and south-east.

This region is a zone of contrasting motion that appeared at the boundary of the rising system of mountain structures of the Carpathians and the sinking Precarpathian foredeep. During the Neogene-Quaternary period, i.e. for 40 million years, the eastern Carpathians rose, discounting erosion, 2-2.5 km, while the Precarpathian foredeep sank 5-7 km; the Odobesti depression sank 13 km [Dobrev]. Information about modern vertical motion indicates that here the uplift and subsidence velocities reach 4-5 and 0.5 mm/year, respectively [V. Somov]. This indicates that, at present, intense tectonic rearrangement is continuing in this region. The consequences are frequent, sometimes strong earthquakes.

Permanent average seismicity. The seismic conditions in the Vrancha region have been discussed from the quantitative aspect by a number of authors [Drumea]. The subcrustal foci are the most active and dangerous in this region. Figure 84 presents a map of the epicenters of subcrustal ($M \geqslant 4, 2$, i.e. at $K = 4 + 1.8M$—an energy value of $K \geqslant 12$) and

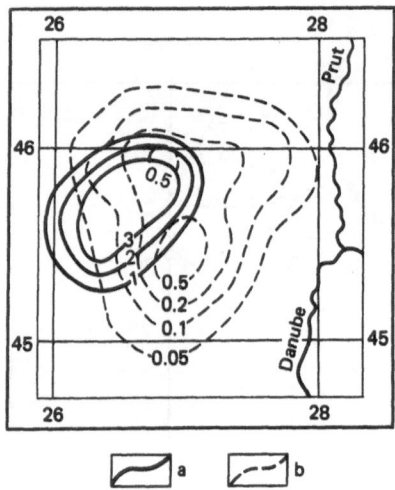

Fig. 85. Map of seismic activity A ($= A_{10}$) in the Vrancha region:

a—subcrustal activity; b—crustal activity

crustal ($K \geqslant 10$) earthquake foci for 1965-1979 compiled from data by Romanian seismologists [Radu, et al.] and the bulletins of the Kishinev seismic station.

With respect to energetics, the permanent average seismicity is known to be characterized by three main indices, namely, the slope of the earthquake recurrence plot, the seismic activity A, and the magnitude of the maximum possible earthquake K_{max}.

Figure 85 presents maps of the seismic activity $A = A_{10}$ in isolines for subcrustal and crustal sources. They have been compiled by the summation method with a constant accuracy. The observations for 1965-1977 were employed to calculate the activity for the subcrustal sources. It was assumed that $\gamma = 0.47$. The level of earthquake representation was $K_{min} = 12$. For crustal sources, the activity was calculated from data for 1952-1977, $\gamma = 0.45$ and $K_{min} = 11$. The subcrustal activity $A_{max} = 3$ in the Vrancha region is very high for a continental region. The crustal activity is much weaker and only reaches $A_{max} = 0.5$.

Maps of the maximum possible earthquakes K_{max} for subcrustal and crustal earthquakes in the Vrancha region are given in Fig. 86. When compiling these maps, we used the method of correlating K_{max} with the average seismic activity A for the area where the earthquakes prepare. The standard correlations were used [Riznichenko], viz.

$$\log A = \overline{2}.84 + 0.21(K_{max} - 15) \quad \text{at} \quad K < 15 \tag{9.62}$$

$$\log A = \overline{2}.84 + 0.39(K_{max} - 15) \quad \text{at} \quad K \geqslant 15.$$

For subcrustal earthquakes, the isolines of $K_{max} = 13$-17 coincide [Drumea], so that within the limits of the entire contour in Fig. 86, the maximum possible calculated earthquakes are those with $K = 17$, i.e. with a magnitude $M = 7.2$. The observed maximum earthquakes are close to this, namely, $M = 7.3$ (1940), and $M = 7.2$ (1977). The difference does not exceed the possible errors for calculation and observation.

The maximum crustal earthquakes with $K_{max} = 16$ ($M_{max} = 6.7$) are possible by calculations in the zone surrounding the town of Focsani. But during the time from 1965 within the entire region being considered, only crustal earthquakes with $K \leqslant 14$ ($M \leqslant 5.6$) occurred.

Fig. 86. Map of the maximum possible earthquakes in the Vrancha region:

a—for subcrustal sources; b—for crustal sources

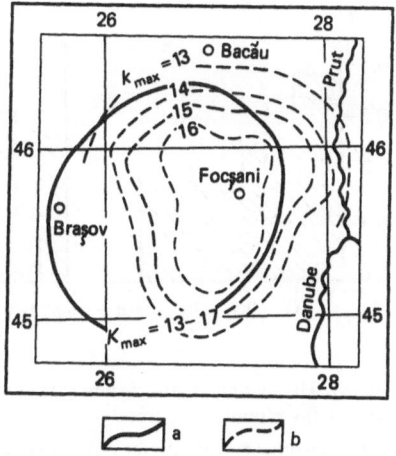

Mechanism of earthquake foci. Data on the mechanism of the foci of earthquakes with $M \geqslant 5.0$ from 1929 to 1965 and with $M \geqslant 4.0$ from 1966 to 1972 were used as the initial information. This information is given in Simonova.

The mechanisms were appraised for reliability using the parameters of accuracy, stability, and representation proposed by V. N. Averyanova. The accuracy of an individual mechanism δAz is defined to be the limits within which the orientation of the principal axes may vary with a given field of the signs of the displacements without violating the relation between consistent and inconsistent signs. The stability is defined to be the relative weight in comparison with other possible solutions: if one solution is possible, the stability is 1, if two are possible, it is 0.5, and so on. The ratio between the consistent N and inconsistent $n_{(-)}$ signs $\delta n = n_{(-)}/N$ yields the stability parameter. The accuracy and stability of a set of mechanisms are described by statistical and probability methods. It is necessary to obtain standard distributions of the parameters being studied to estimate their means and variances for the given confidence level. The representation characterizes the number of sources with a mechanism with respect to the total number of sources, counting from a certain level, e.g., from $K \geqslant 11$ $(M \geqslant 4)$.

An analysis of the accuracy with which the source mechanism is determined has shown that for 80% of all earthquakes, the directions of the principal stress axes may vary by 30°. For 40% of them, the variation did not exceed 10°. The less accurate determinations relate to smaller earthquakes. In 14% of cases, two variants are possible for all solutions. Histograms of the accuracy and stability parameters are given in Fig. 87. Their means for a 99% confidence level are $\delta Az = 22°$, $\sigma\delta Az = 4°$, $\delta n = 0.09$, and $\sigma\delta n = 0.08$ for the accepted solutions; $\delta n = 0.17$ and $\sigma\delta n = 0.05$ for the discarded solutions. If in an individual case two or more variants are possible for which the stability parameter ranges from 0 to 0.17, it is difficult to choose one. Consequently, if δn exceeds 0.17, such a decision was not made.

We have already indicated that from 1966 the mechanism of almost all sources with $M \geqslant 4$ has been studied, so that the data are representative for this period. Prior to 1966, only severe earthquakes with $M \geqslant 5$ were studied. We shall adopt the quantity $v = \Sigma M_{0,\text{stud}}/\Sigma M_0$ as a representative measure for our purposes. The numerator is the sum of the seismic moments of earthquakes with a known focal mechanism, while the denomina-

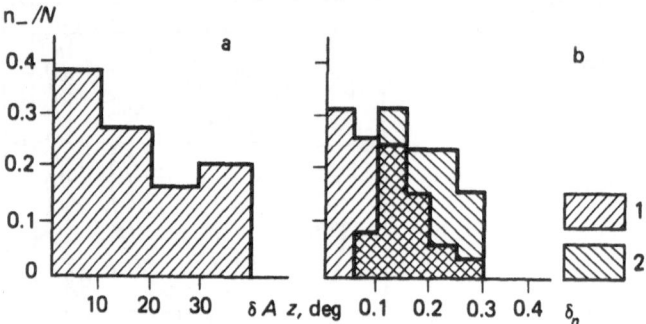

Fig. 87. Histograms of the accuracy (a) and stability (b) of determining the directions of the principal axes of the strains at sources:

1—for subcrustal sources; *2*—for crustal sources

tor is the same for the entire set of earthquakes beginning from a certain energy level, e.g., from $K = 11$ ($M = 4$) during the period. The representation parameter ν determined by this formula for earthquakes in the Vrancha region during the period from 1929 to 1977 is 0.88. The seismic moment M_0 was calculated on the basis of K (or M) from a correlation expression such as (9.15), in which the coefficients C_1 and C_2 were obtained from globular data for crustal foci [Riznichenko]. The possibility of doing this for subcrustal sources also is discussed below.

Studying the mechanism of the earthquake foci revealed two main directions for the rupture planes, viz. northeast-southwest with an inclination to the northwest, and northwest-southeast with an inclination to the southwest. The 0 axis of the revealed deformation is oriented along these directions. For the subcrustal sources, the compression axes are close to horizontal, tension axes close to vertical. The compression axes are predominately at right angles to the Carpathian range. The type of motion in the sources is upthrust and shift-upthrust. The number of crustal earthquakes with a known mechanism is not great, and it is difficult to assess the prevailing orientation of compression and tension. It should be noted that the compression axes here are generally steeper than the tension ones. Upthrusts and shift-upthrusts prevail accordingly.

Summing and averaging the strains at foci. In the region being studied, points with known parameters of the focal mechanism are not uniformly located. They are bunched in the centre. To determine the average rates of strain $\bar{v}_{ik}$ of seimsic flow at individual, arbitrarily selected, points in the region, at the nodes of the geographic degree network in practice, it becomes necessary to single out elementary areas around each of these points. Within the confines of these areas or, more exactly, the volume regions under them, the effects due to the earthquake foci whose epicenters are on each area were summed and averaged.

The effects were summed and averaged by the components of the tensors v_{ik} of the direction of the focal seismic moments M_{0ik} with the scalar values of M_0 as weights. The components of the average direction tensor θ_{ik} for each elementary area for the relevant index of seismic flow—the rate of strain or the velocity of displacement in a vertical direction—are determined from

$$\bar{\theta}_{ik} = \sum_{j=1}^{n} M_{0j} \theta_{ik, j} / \sum_{j=1}^{n} M_{0j}. \tag{9.63}$$

Hence for each area, or more exactly, for the volume region under it, we have:

$$\bar{\theta}_{ik} = v \, \theta_{ik}. \tag{9.64}$$

The components of the tensors θ_{ik} are determined by the trigonometric expressions in parentheses in (9.52)-(9.56), and for shear components, also by a factor 2. The scalar multiplier is determined from (9.50).

When choosing the size of the elementary areas for summing and averaging, we were guided, as when determining the seismic activity A, etc. by two contradictory requirements, namely, (1) detail, and (2) accuracy. If we consider that the accuracy of determining the coordinates of an epicenter is about $\pm 0.1°$, i.e., about 10 km, while the size (average radius) of a focus of the maximum earthquake in this region $M \approx 7.5$ is about 30 km [Riznichenko], calculations over the network, every $0.1°$ with areas $0.2° \times 0.2°$ in size seem reasonable. It is obvious that the areas partly overlap.

To determine M_0, we shall use the correlation of the seismic moment and magnitude in the following form:

$$M_0 (M) \pm 0.6 = 15.4 + 1.6M. \tag{9.65}$$

This relation was obtained by the author for crustal foci, and its applicability for the subcrustal foci of the Vrancha region is doubtful. For verification, we also correlated K or M with M_0 for the subcrustal foci. The straight line in Fig. 88 depicts the correlation expression of (9.63), while the circles depict the seismic moment and magnitude M for deep-focus earthquakes in other regions [H. Berckhemer], as well as for the subcrustal foci of Vrancha. The crustal relation of (9.63) is consistent with both the subcrustal and deep-focus data.

Summing and averaging yeild a symmetric tensor with six independent components in the geographical system of coordinates: three elongations in the two horizontal directions (to the east and north) and in a vertical direction (to the zenith) three shift components. To determine the directions of the maximum lengths, in particular the maximum tensions and maximum compressions, we must go from an arbitrary (geographical) system of coordinates to an intrinsic system $(T, C, 0)$ of the strain tensor. This is a classical problem of the theory of elasticity or tensor algebra (see [341, pp. 31-36], the EIGEN program). In practice, we used the computer programme placed at our disposal by S. L. Yunga.

Numerical calculations of the strain rate modulus. The value of v is calculated from (9.50). To do this, we shall take $\gamma = 0.47$, $\mu = 0.7 \times 10^{12}$ dyn/cm^2 for the subcrustal foci, and $\gamma = 0.45$, $\mu = 0.3 \times 10^{12}$ dyn/cm^2 for the crustal foci [Riznichenko] : $K_0 = 10$, $M_0 (K_0) = 10^{C_1}$, $C_1 = 20.732$, and $C_2 = 0.889$ from (9.15). Insertion of these values yields the following expression for the strain rate modulus for subcrustal foci:

$$v = 2.46A \times 10^{(0.419K_{max} + 4)} \tag{9.66}$$

and for crustal ones

$$v = 3.46A \times 10^{(0.439K_{max} + 4)}. \tag{9.67}$$

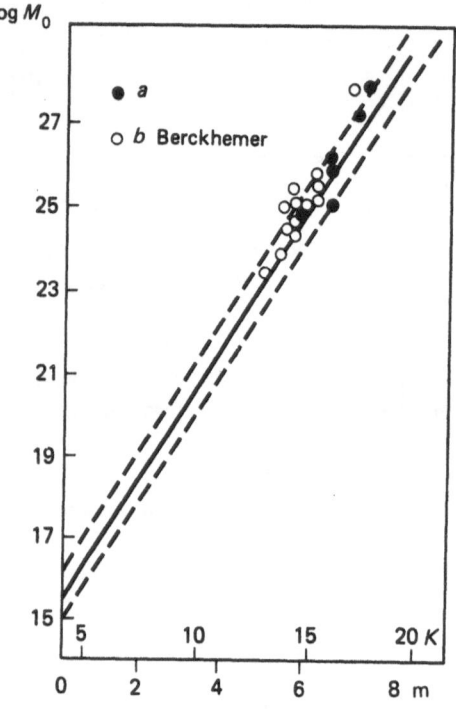

Fig. 88. Dependence of the seismic moment
M_0 on the magnitude M by correlation formula (9.65):

a—for crustal sources (*straight line*) in comparison
with the results of observations of subcrustal Vrancha earthquakes; b—for deep sources [Berckhemer
et al]

The calculations were performed for the entire active region. The seismic activity A was related to an area of $S = 1000$ km^2 $= 10^{13}$ cm^2 and to a layer with a thickness of $h = 10$ km, and so $Sh = 10^{19}$ cm^3, over $t = 1$ year. Since for the subcrustal regions $K_{max} = 17$, these quantities yeild $v = 3.26A \times 10^{-8}$ year^{-1}. At the average activity for the subcrustal foci of $A = 2$, we obtain a rate of strain v of 0.6×10^{-8} year^{-1}.

Results of determining the seismic flow indices. Using (9.50), (9.52)-(9.58), we calculated the components of the tensor of the average rate of seismic flow strain in the geographical coordinate system for the Vrancha region as a whole and for elementary areas. The latter were $0.2° \times 0.2°$ in size with centers at the nodes of the geographic network spaced $0.1°$ apart. The total number of areas was 60. Moreover, the principal directions and main values of the tensor $\overline{v_{ik}}$ were calculated for all these summation and averaging regions. In the follow-ing, we shall omit the bar over v_{ik} for averaged quantities. The tensors of the rate of strain v_{ik} are given by three mutually perpendicular vectors v_1, v_2, and v_3 of the maximum rates of tension and compression along the three principal axes of the tensor v_{ik}. The values of v_1, v_2, and v_3 can also be used, if desired, to find the directions and values of the extreme shifts. The same data can also be used to restore the initial values of all the components of the tensor v_{ik} in the geographical coordinate system.

In all cases, the tensor of the average strain rates v_{ik}, as expected, was found to be a deviator—the dilatation $\Sigma v_{ii} = 0$ ($i = 1, 2, 3$), so the volume of the medium in the summation and averaging regions, as in the initial foci, does not change. This is only a result of the previous formal assumption about the pure shear mechanism of the foci.

But if the strain was pure plane shear in each individual focal, when only two of the three principal elongations are nonzero (and are equal in magnitude and opposite in sign), the tensors of the strain and their rates in the summation and averaging regions were found to have, generally speaking, three nonzero principal elongations.

If we single out a sphere in a conditionally homogeneous and isotropic underformed medium, it will become an ellipsoid with the axes $1 + \varepsilon_1$, $1 + \varepsilon_2$, and $1 + \varepsilon_3$, as a result of a small strain. The strain rate will have the axes $1 + v_1$, $1 + v_2$, and $1 + v_3$. Figure 89 shows the horizontal and vertical projections of such an ellipsoid of the average strain for the entire Vrancha region. The strain values ε_x, ε_y, and ε_z shown in Fig. 89 correspond to a million years of the seismic flow strain rates v_x, v_y, and v_z observed here. It can be seen from Fig. 89 that the Vrancha region experiences almost bulk compression in a horizontal plane $\varepsilon_x < 0$ and $\varepsilon_y < 0$. The maximum compression is directed at an angle of $\alpha = 6°$ to the horizon and is in the azimuth $Az = 293°$. Tension close to vertical $\varepsilon_z > 0$ is observed in the vertical plane. Its axis is inclined to the horizon at an angle of $\alpha = 73°$ in the azimuth $Az = 44°$.

The individual elementary areas in the Vrancha region (their horizontal size is $0.2° \times 0.2°$ for an active layer with thickness $h = 10$ km) mostly showed that one of the principal quantities of the strain is greater than zero (tension), while the other two are less than zero (compression), i.e., $\varepsilon_1 > 0$, ε_2 and $\varepsilon_3 < 0$ so uniaxial tension prevails. But there are cases when ε_1 and $\varepsilon_2 > 0$, while $\varepsilon_3 < 0$, and so uniaxial compression prevails. Finally, there are sometimes cases when $\varepsilon_1 = -\varepsilon_2$ and $\varepsilon_3 = 0$, i.e., pure shear in one plane is observed. In all cases here ε_3 is at least an order of magnitude smaller than ε_1 and ε_2, though generally ε_3 is nevertheless no zero.

Figure 90 summarises the calculated directions of the average strains ε_{ik} (or v_{ik}) for the elementary areas of the Vrancha region. The vector lines of the azimuths of the axes of maximum tension (dashed lines) and maximum compression (solid lines) are shown in the figure.

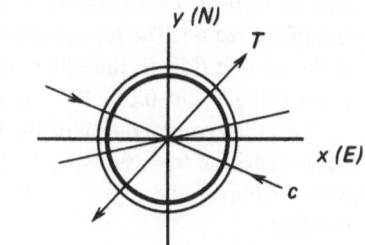

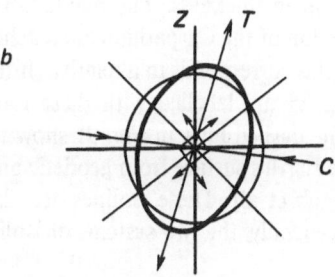

Fig. 89. Projections of an averaged strain ellipsoid for the Vrancha region:

a—horizontal projection; *b*—vertical projection

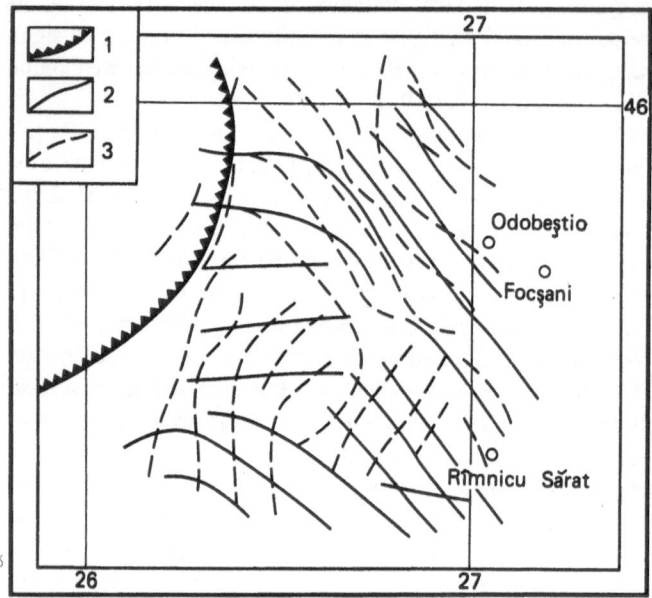

Fig. 90. Vector lines of the azimuths for the directions of the maximum elongations and contractions: *1*—Vrancha arc; *2*—contractions; *3*—elongations

In the center of the region, the vector lines of compression can be seen to be almost perpendicular to the Carpathian arc, and those of tension paralled to it. Their orientation changes at the edges. The compressive strains are close to horizontal ones, and the tensile ones are close to vertical. Reorientation of the axes occurs to the east and south-east of the central part of the region. The tension axes become close to horizontal. The maximum rate of strain of the seismic flow in the direction of the maximum elongation over the entire region was found to be about 0.1×10^{-6} year^{-1}.

A detailed consideration of the horizonal strain components in the Vrancha region shows that a transition from sublatitude tension to bulk compression is observed from east to west. Deformation of the region in the latitudinal direction prevails over deformation along a meridian.

The solid lines in Fig. 91, show the average rates of strain of the active layer in a vertical direction. Its thickness is taken to be $h = 10$ km $= 10^6$ cm, and the strain is taken to be constant in thickness. The maximum rate of strain of the seismic flow along a vertical in the region of the Carpathian arc reaches 0.9×10^{-7} year^{-1}. When reduced to the layer thickness, this corresponds to a relative shift rate of the roof and floor of 0.9 mm/year. The isolines in Fig. 91 are labelled with these rates (im mm/year).

The dashed lines in Fig. 91 show the isolines of the velocities of present vertical motion of the Earth's surface from geodetic and geomorphological data from map of Eastern Europe [Sologub et al]. These isolines are also labelled in mm/year. Inspection of Fig. 91 reveals that generally, the two systems of isolines are consistent, at least in the region of the highest positive velocities. The regions of negative values are not as consistent. The magnitudes of the seismic velocities are somewhat lower than the geodetic ones.

It should be borne in mind that for vertical motion the two sets of velocities are not fully compatible. Seismic velocities characterize the relative displacements of the active layer boundaries, which geodesy does not recognise, whereas the geodetic velocities characterize the absolute displacement of only the Earth's surface, which is not a concern for seismology. A comparison would be possible if we assumed that the lower boundary of the seismically active layer remains fixed. This assumption seems justified for small regions and not too long time intervals. For large space-time regions, on the other hand, it becomes incorrect owing to the presence of compensation motions of an isostatic nature in the elasto-fluid lithosphere. When the latter thickens, its free surface rises, while its base, together with both

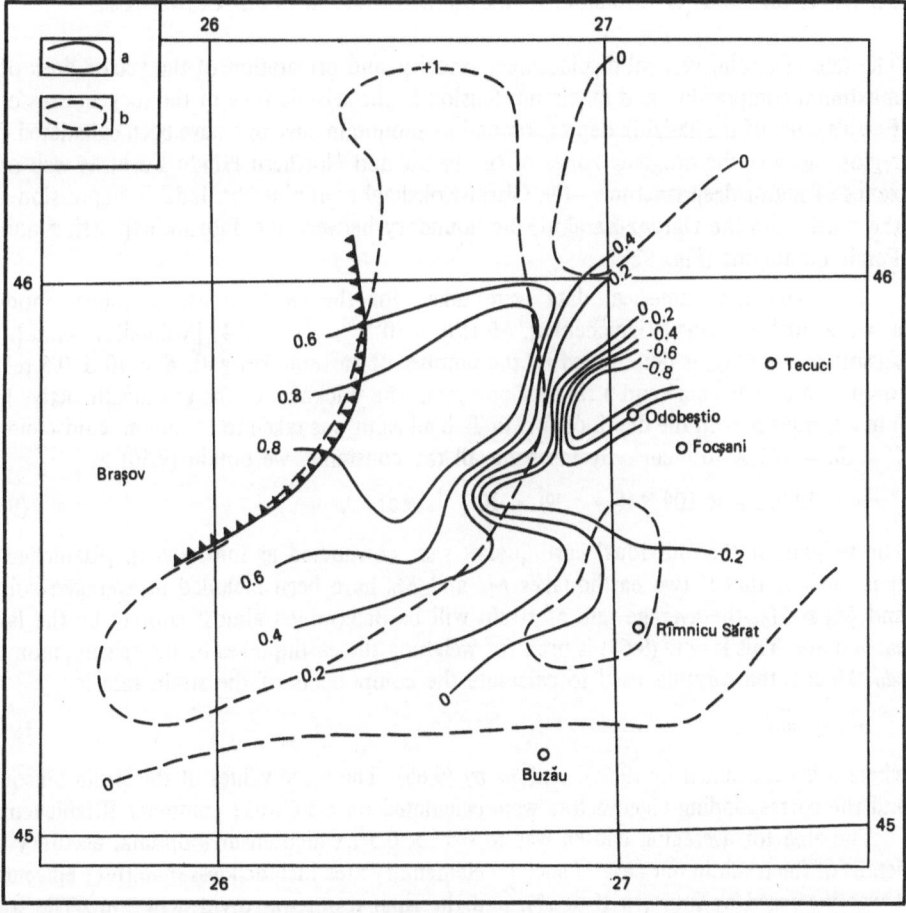

Fig. 91. Vertical component of the strain rate (mm/year):
a—for seismic flow according to our calculations; b—for present vertical motions [Riznichenko et al]

boundaries of the seismoactive layer bends downward. For a quantitative analysis of these phenomena, the theory of seismic flow needs to be merged with the dynamics of the lithosphere with respect to general tectonic motion.

The lack of agreement between seismological and geodetic data may be affected by other factors, including inaccuracies in both sets of data.

It should be noted that the seismological data on deformation in a region have greater detail in this given stage than geodetic data (see Fig. 91). For instance, seismological data give a clear picture of the region of maximum uplift in the Vrancha region and the region of maximum subsidence in the vicinity of Odobesti, known from geological observations including boring. On the other hand, the geodetic (and geomorphological) information describes the uplift in the Vrancha region very well, but remains poor for the Odobesti subsidence region.

9.3.4. Seismotectonic Deformation of the Earth's Crust in South Central Asia[*]

The rate of strain, vertical displacement velocity, and orientation of the vector lines of the maximum compression and maximum tension in the seismic flow of the rock masses of the Earth's crust of the Tadzhik depression and its mountain envelope have been calculated. This region includes the orogenic zones of the Pamir and Northern Hindu Kush, as well as the zones of major deep fractures—the Gissar-Kokshaal confining the Tadzhik depression from the north, and the Darvaz-Karakul, the boundary between the Tadzhik depression and the Pamir mountains (Fig. 92).

The following numerical data were taken for the calculations: the shear modulus $\mu = 3 \times 10^{11}$ dyn/cm^2 [Riznichenko], $M_0 (10) = 10^{20.73}$, $\gamma = -0.45$ [Mikhailova et al.]. The seismic activity A_{10} is determined by the number of earthquakes with $K = 10 \pm 0.5$ related to an area $S = 10^3$ cm^2 and a time of one year. The thickness of the seismically active layer h in accordance with the depth of foci in Tadzhikistan was taken to be 30 km. Consequently, $V = Sh = 1/3 \times 10^{19}$ cm^3. By inserting all the constants, we obtain (9.50) as

$$v = 2.83 A_{10} \times 10^{0.44 (K_{max} - 10)} \times 10^{-11} \text{ year}^{-1}, \qquad (9.68)$$

The weights of the individual earthquakes were considered as indicated in [Riznichenko]. It is obvious that if two earthquakes M_0' and M_0'' have been included in averaged volume and $M_0' \gg M_0''$, the average rate of strain will be determined almost entirely by the larger earthquake. This is why (9.63) is used for weighing the earthquakes by the seismic moments M_0. Hence, the formula used to calculate the components of the strain rate is

$$v_{ik} = v\bar{\theta}_{ik} \qquad (9.69)$$

where v is determined by (9.68), and θ_{ik} by (9.63). The basic values of the strain tensor $\bar{\theta}_{ik}$ and the corresponding eigenvectors were calculated on a EC-1033 computer [Riznichenko].

The area for averaging chosen was to 0.5° × 0.5°, which ensures optimal accuracy and details of the result in our case. If such an elementary area included less than three epicenters, it was increased by four (to 1° × 1°), and the final result was divided by four. The strain

---

[*] See [Riznichenko]. O. V. Soboleva, O. A. Kuchai, et al. are co-authors.

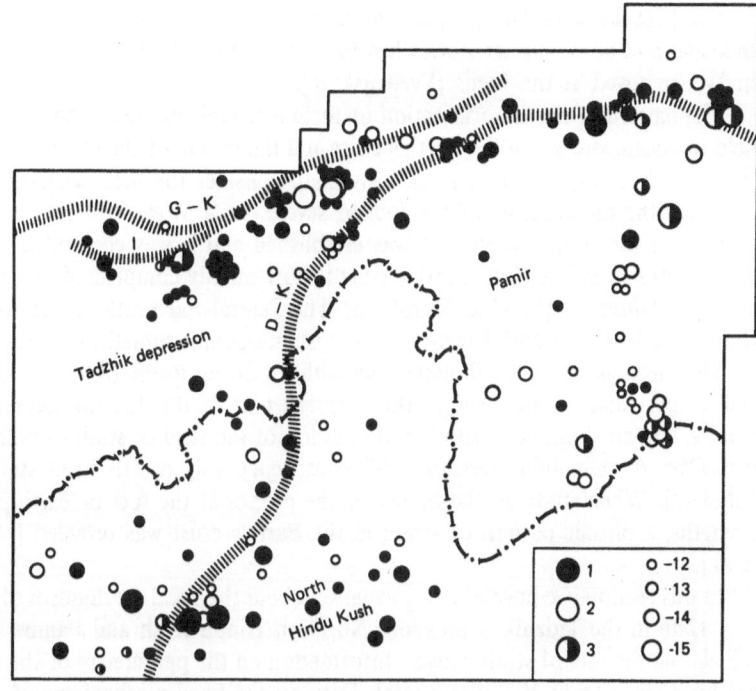

Fig. 92. Map of the earthquake epicenters used in this work.

Classification of the earthquakes according to the type of shift at the sources: *1*—thrusts and shift thrusts; *2*—fault-shifts and faults; *3*—pure shifts; *12-15*—classification according to energy class. The figures correspond to the value of $K = \log E_J$. The *letters* denote geological fractures: G-K—Gissar-Kokshaal, I—Ilyak, D-K—Darvaz-Karakul

rate components were calculated over the entire map (see Fig. 92) with the averaging areas overlapping by half their linear dimension (a total of 260 points).

Maps of A_{10} and K_{max}. The values of A_{10} and K_{max} were taken in numerical form from the relevant maps. The seismic activity was evaluated by summing earthquakes [Riznichenko] with a constant averaging area of $0.2° \times 0.2°$ (constant detail) for sections with a large epicentral density, and with a varying averaging area (constant accuracy) where there was low activity for the period 1955-1970. Only the eathquakes themselves (without aftershocks and groups) were considered. The minimum number of earthquakes in an average area was three, and consequently the relative error in determining the activity did not exceed 58% [Riznichenko]. A map of the activity in A_{10} isolines was published by R. S. Mikhailova and D. G. Bibarsova.

The K_{max} maps were evaluated by correlating K_{max} with the seismic activity A as developed in [Riznichenko]. It was assumed that for each major earthquake of class K there is a preparation region of radium R, namely, $\log R$ km $= 1.505 + 0.111 (K - 15)$ within which relation (9.62) between the average seismic activity and K_{max} is satisfied. The K_{max} were calculated with a pitch of $0.1°$ in latitude and longitude; a map in isolines was published by R. S. Mikhailova.

Focal mechanism. The extensive material on the mechanism of earthquake foci in Tadzhikistan and adjoining territories has been studied on the basis of the traditional ideas of stresses removed at the focus [Vvedenskaya].

The pattern of spatial orientation of focal mechanisms yields some indication about the field of elastic stresses in the Earth's crust and the nature of the strains in individual section [Kuchai, Nersesov, Soboleva et al.]. We shall consider the most general of the results.

When the mechanisms of the foci of severe ($M \geqslant 5$) earthquakes were considered, the stability of the compression axes was established and it was concluded that, over most of the territory considered, the Earth's crust is horizontally compressed in a north-west or submeridional direction [Landau, Shirokova]. The Central and Southern Pamirs are an exception. Here there is a substantial region of almost horizontal sublatitudal tension [Landau].

The mechanisms of earthquake foci with moderate forces ($K = 12\text{-}13$) are more diverse and in individual regions, e.g. in the northern part of the Tadzhik depression and Central Pamirs, and are consistent with the orientation of the field of strains obtained for the severe foci. Over the remaining territory, this consistency only reveals itself statistically [Landau, Soboleva]. When studying the nature of the pushes at the foci of earthquakes of different strengths, a mosaic pattern of strain in the Earth's crust was revealed [Nersesov, Soboleva et al.].

In this section, we have used information about the focal mechanism of earthquakes with $K = 12\text{-}16$ in the Tadzhik depression, Northern Hindu Kush and Pamirs as initial data for calculating the rate of strain tensor. Information on the parameters of the focal mechanisms has been given by O. V. Sobolev et al. Data on the focal mechanisms of 23 earthquakes in the Garm region was provided by V. G. Leonova and A. A. Lukk. A total of 182 earthquakes were used (see Fig. 92). A glance at the figures shows that the epicenters are non-uniformly distributed over the area. This makes it impossible to achieve similar accuracies and detail over the entire region.

The period of observations of the focal mechanisms is short, from 1960 to 1975, i.e. the period of observations is not long in comparison with that of recurrence of powerful earthquakes. We have therefore to assume that the focal mechanisms are stable in time.

Strain rate distribution over an area. Figure 93 presents maps of the components of the strain rate characterizing the relative elongation (positive values) or contraction (negative values) of the elementary volume in the relevant direction. Figure 94 presents maps of the rate of strain characterizing the relative angular shear strain between the orthogonal directions. The positive values indicate an acute, and the negative ones, an obtuse angle between the relevant coordinate axes.

Let us first consider the territorial distribution of the strain rates in magnitude. The highest rates belong to the contact zone between the Pamirs and Southern Tien Shan, and also to the epicentral zone of the deep Hindu Kush earthquakes (Northern Hindu Kush). In individual sections, the strain rate exceeds 5×10^{-8} year^{-1}. The same zones also have the highest seismic activity, which is a multiplier in the formula for calculating the strain rate. This is why the maps of the strain rates, as a whole, are similar to the map of seismic activity, while the distribution of the strain rate over the area may be a consequence of the calculation method. On the other hand, seismic activity is nothing but the manifestation of a high level of accumulated strains, and hence the coincidence of large A_{10} with maximum strain rate areas is quite natural.

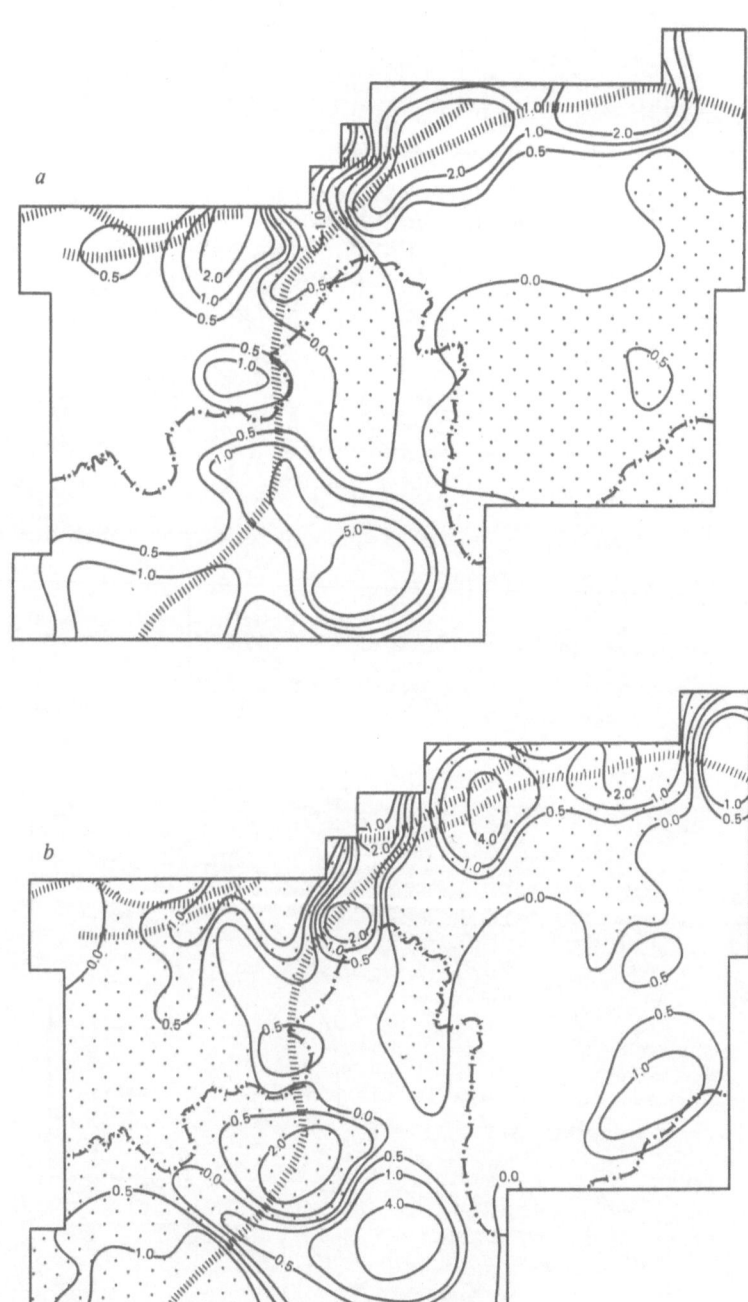

Fig. 93. Maps of the isolines for the normal components of the strain rate in the seismic flow of rock masses (in units of 10^{-8} year^{-1}).

The *dotted areas* show the regions of negative rates (contractions); *a*—vertical strain component v_{zz}; *b*—latitude component v_{xx}; *c*—meridian component v_{yy}

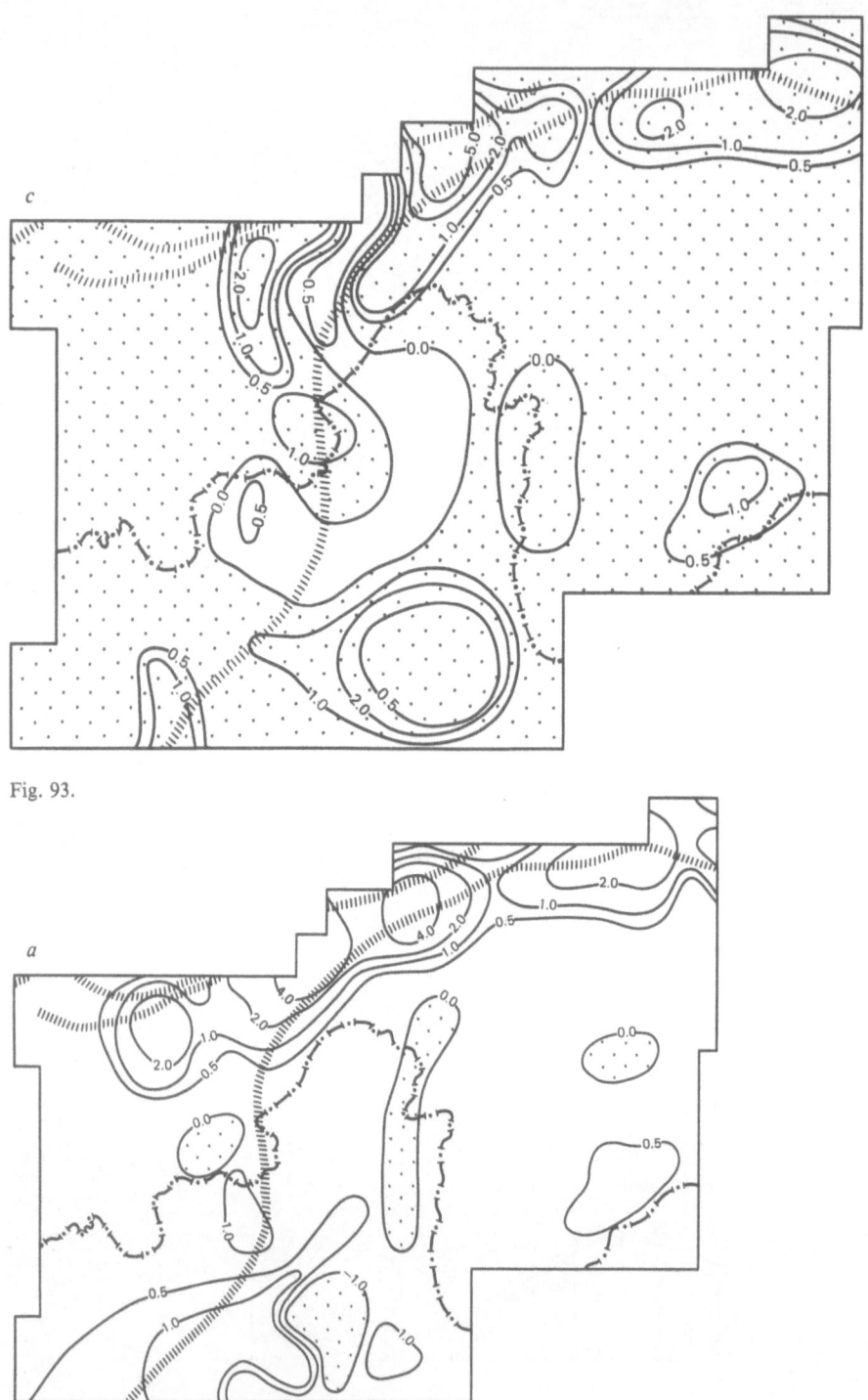

Fig. 93.

Fig. 94. Maps of the isolines for the tangential components of the strain rate in the seismic flow of rock masses (in units of 10^{-8} year^{-1}).

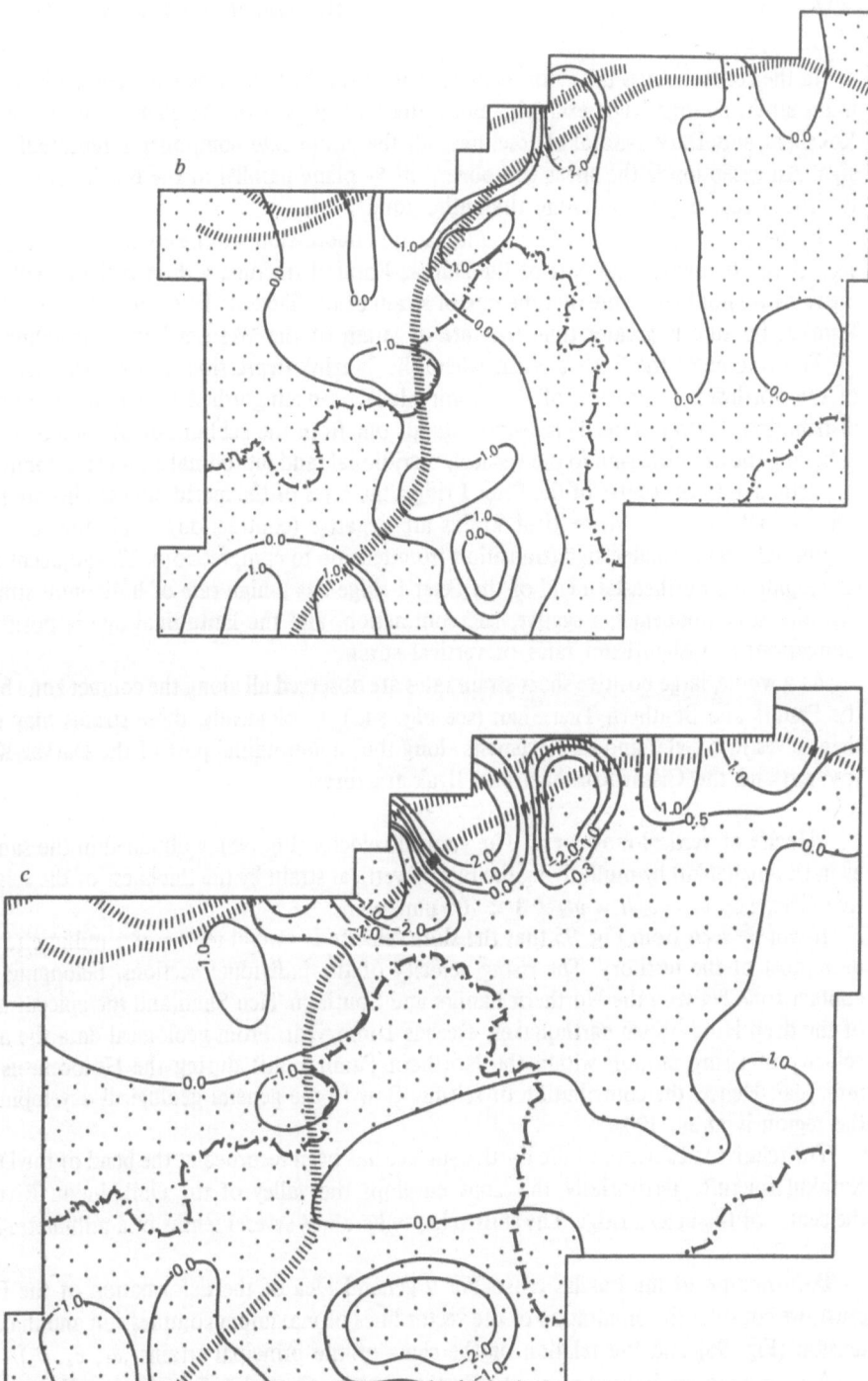

Fig. 94 (continued)

The *dotted areas* show the regions of negative rates: *a*—component in horizontal plane v_{xy}; *b*—component in the meridional plane v_{yz}; *c*—component in the latitudinal plane v_{zx}

In the contact zones of major tectonic structures, the high rates are attended by the maximum strain contrast. For example, along the frontal part of the Pamir arc in the Gissar-Kokshaal and Darvaz-Karakul fractures, all the strain rate components repeatedly change sign. An exception is the shear component in xy plane parallel to the Earth's surface—here the strain rate is positive along the entire zone.

In the contact zone of the Pamir and Tadzhik depression, all the components also change sign along the meridional part of the Darvaz-Karakul fracture, although the magnitudes of the rates are not high. Note that the rate of strain of the Tadzhik depression along the Darvaz-Karakul fracture is greater than the rate of strain of the Western Pamir structures.

The volume of the Earth's crust, where the Tadzhik depression narrows sharply in front of the northwest protrusion of the Pamir block, is distinguished by the very complicated distribution of strain rate. This section stands out from the sublatitudinal zone of high rate in having the opposite sign to the vertical, meridional, and latitudinal strain rate components. In particular, in the centre of the Peter I ridge, the rates of the meridional strains are positive (elongation), those of the vertical strains are negative (contraction), while the rates of the latitudinal strains change sign (transition from tension to compression). The adjacent section enveloping the northeastern end of the Peter I ridge has a high rate of horizontal strain (the meridional component is negative, i.e., contraction, and the latitudinal one is positive, i.e., elongation) at insignificant rates of vertical strain.

As a whole, large positive shear strain rates are observed all along the contact zone between the Pamirs and Southern Tien Shan (see Fig. 94a). Geologically, these strains may express themselves in a righthand displacement along the sublatitudinal part of the Darvaz-Karakul fracture and the Gissar-Kokshaal and Ilyak fractures.

Velocity of vertical movement. The vertical velocity (Fig. 95) is obtained in the same way as in [Riznichenko] by multiplying the rate of vertical strain by the thickness of the seismoactive layer, i.e., $v_z = v_{zz}h = v_{zz} \times 3 \times 10^7$ mm/year.

It can be seen from Fig. 95 that the shift velocity is several tenths of a millimetre a year over most of the territory. The rising velocity of the individual sections, belonging to the contact zone between the Northern Pamirs and Southern Tien Shan and the epicentral zone of the deep Hindu Kush earthquakes, exceeds 1 mm/year. From geological data the average velocity of rising motion within the Northern Pamir uplift during the Holocene is 15-20 mm/year. Hence, the contribution of seismic flow to the general geological development of the region is about 10%.

The relative subsidence of the Earth's surface has been recorded at the bend of the Darvaz-Karakul fracture. Territorially, this zone envelops the valley of the Obikhingou River and the center of the Peter I ridge. The shift velocity here was several tenths of a millimetre a year.

Deformation of the Earth's crust. For a general idea of the deformation of the Earth's crust, we consider the orientation of the vector lines of maximum compression and maximum tension (Fig. 96) and the relation of the signs of the principal strains (θ_1, θ_2, θ_3).

According to establihsed ideas, the Earth's crust in Central Asia in the late Cenozoic was horizontally compressed [Gubin, Kuchai, Shirokova, Topopnier, P. Molnar]. The substantial role of background compression in seismotectonic deformation has also been uncovered in our studies: the vector lines of maximum contraction are directed from southeast to northwest throughout the region (see Fig. 96), which coincides approximately with the direction of the

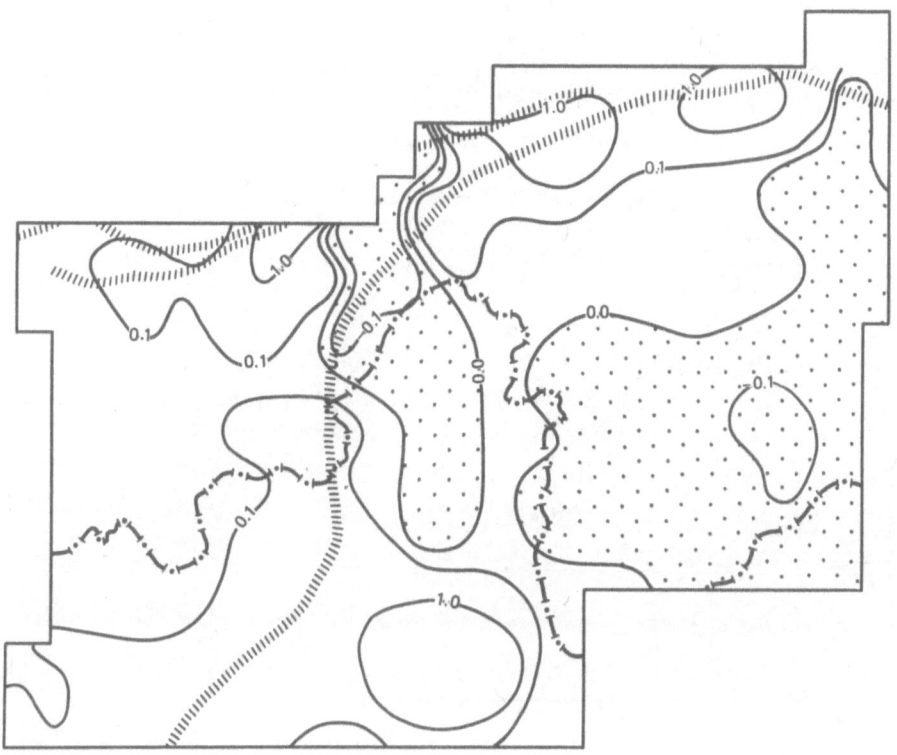

Fig. 95. Map of the isolines for the velocity of vertical shifts of the Earth's surface (mm/year).
The *dotted areas* show the subsidence regions

background compression [Kuchai, Shirokova] and with that of the Pamir block shift relative to the rock masses of the Tadzhik depression [Kuchai]. The trajectories of the maximum elongation are more involved. Their general direction is northeast, but they often repeat the configuration of the surface lines of geological fractures, and sometimes become close to vertical. Judging from the ratio of the signs of the principal strains, the deformation of the Earth's crust is sharply differentiated—only an insignificant part of the territory is characterized by simple compression ($\theta_1, \theta_2 > 0$; $\theta_3 < 0$). Most of the crust is under tension ($\theta_1 > 0$; $\theta_2, \theta_3 < 0$) or pure shear ($\theta_1 = -\theta_3$; $\theta_2 = 0$).

In particular, the crust in the Central Pamirs is under plain horizontal tension. This was noted by O. A. Kuchai and was explained from the standpoint of the gravitational spreading of the crust where it is thick and there is background compression [Guseva, E. Artuashkov, Kuchai].

The crust in the Tadzhik depression, in front of the northwestern projection of the Pamir block, is also deformed by simple tension. The maximum elongation is in the horizontal direction, and the vector lines of the elongation repeat the direction of the Darvaz-Karakul fracture (see Fig. 96). The Pamir block when moving northwest [Kuchai] probably encounters

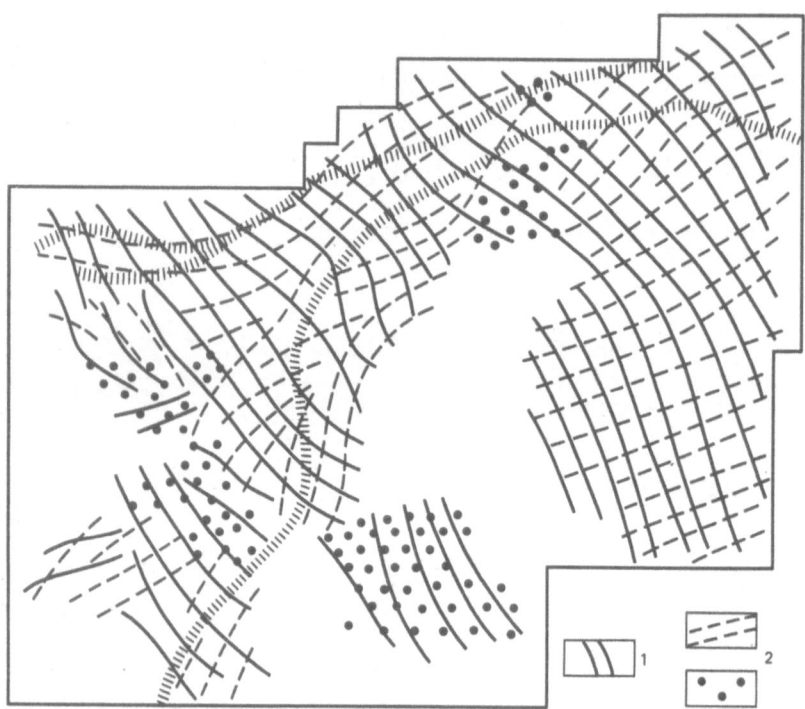

Fig. 96. Diagram of the seismotectonic deformation of the Earth's crust for the southern Central Asia: *1*—vector lines of maximum elongation; *2*—vector lines of maximum shortening (*broken lines* are lines of near-horizontal elongation, *points* are near-vertical orientation)

resistance in the form of the rigid mass of Paleozoic rock of the southern Tien Shan and creates compression in this region [Lukk, Nersesov]. As a result, the soft rock of the Tadzhik depression is confined between the two rigid blocks and spreads out along the fracture line. This region includes the southeastern part of the Garm geophysical test ground on which extensive geodetic studies are in hand. These have revealed the presence of horizontal tension in the subsurface of the crust in some sections of the test site [Guseva, Kuchai, Lukk, Pevnev]. The results here are consistent with the geodetic data and indicate that horizontal tension at a rate of about 4×10^{-8} year^{-1} is characteristic of both the subsurface, and deeper layers of the crust.

A section of the crust including the zone of the Darvaz-Karakul fracture, where its direction changes from southwestern to meridional, has been deformed under pure shear strain. The orientation of the vector lines of maximum contraction (close to horizontal, SE-NW) and of maximum elongation (close to horizontal, SW-NE) indicates that the rock in the Pamir and Tadzhik depression along this part of the fracture is relatively displaced to the left. Prevalence of a lefthand shift along the meridional part of the fracture was also deduced from seismological data by O. A. Kuchai et al. This conclusion is consistent with geological field work, during which the left displacements were discovered in this region.

An extensive section of the crust including the southern bend of the Darvaz-Karakul fracture was also deformed by pure shear at an average rate of over 5×10^{-8} year^{-1}. The vector lines of maximum elongation here have a subvertical direction. The difference, between the orientation of the lines of maximum elongation on this section and on the previously considered ones of pure shear, is apparently explained by the fact that the Pamir block, in its relative movement northwest, does not encounter as strong frontal resistance as in the northern region. The spreading of the rock parallel to the fracture line is prevented by its configuration.

In conclusion, we note once more that the determining factor in the seismotectonic deformation of the crust in this region is the background SE-NW compression, whereas the observed mosaic nature of the deformation may be due to differences in the mechanical properties of the rock, the geological conditions of its occurrence, and the presence and orientation of the fracturing faults.

10. Extended Focus and Seismotectonic Flow of Rock Masses[*]

Proceeding from an energy model of an earthquake focus, this present chapter deals with the intensity of seismic tremors, in an external and internal region, relative to the focus. Cases of point, linear, surface, and volume focus are considered.

We also discuss the relations between the seismic moment, stress and strain drops, the size of a fracture-continuous focus, and the seismotectonic flow of rock masses.

Depending on the aims and conditions of the investigation, an earthquake focus can be represented as a point, a segment of a line, a part of a surface, or a three-dimensional body. For the very simple representation as a point, a focus is characterized by the five known parameters x, y, z, t, and E, where the magnitude M is often employed instead of the seismic energy E. Lately a sixth parameter, the seismic moment M_0, has begun to be added. The directions of the tension and compression axes at the focus, the directions of the fracture plane, and the components of the stress and strain tensors are other parameters of a point focus. These parameters must also involve the directions at which the various waves propagate away from the focus, and so on. Hence, even a very simple representation of an earthquake focus as a point is very involved. When models of a linear, surface, and volume focus are employed, other parameters have to be introduced.

The effect of an earthquake focus on the Earth's surface is called the *intensity* of seismic tremors. The intensity can be characterized by various quantities or sets of quantities, or by functions or sets of functions. In the simplest scalar case, this is the macroseismic intensity expressed according to a descriptive macroseismic scale. It may also be the density of the energy of a seismic pulse during the time τ takes to pass through a unit surface area at right angles to a ray at the point of observation. It can also be the maximum shift, velocity, acceleration in this pulse, or its duration, etc. In a more complicated case, the intensity can be determined by a set of scalar quantities and considered to be a "multicomponent vector", e.g., three-dimensional with average amplitude A of the oscillations (of the shifts, velocity, or acceleration), the prevailing period T of the oscillations, and their duration τ as the components. In a more complicated case, it can be characterized by functions, namely, the frequency spectrum $\Phi(\omega)$ of the oscillations and the time function $f(t)$ of these oscillations—a seismogram. In principle, a complete description of the intensity of tremors at a point in a linear approximation can be obtained by a set of six seismograms: three components of translational motion and three of rotation. The intensity of tremors on a segment of a line or on an area is now determined by a set of such sets for the numerous points contained in the object. Consideration of the non-linear and recurrent effects of seismic action makes matters still more involved.

Consequently, the task of expressing the intensity of seismic tremors in a general way is not readily visualized. It is inexhaustible for an earthquake focus.

---

[*] See [Riznichenko].

In this investigation, the first part of which is aimed at the intensity field of tremors for an extended focus as compared to a point focus, we limit ourselves to the simplest models of both the focus and tremor intensity, namely, energy models, so as to obtain answers to specific questions. The choice of energy as the basis for all the calculation is first its scalar nature, additivity, and chiefly, its conservation. The second motive is its practical nature and its correlation with the destructive effect of earthquakes, which is important for mankind. This correlation is better than those for other scalar quantities.

The main aim of the second part of our investigation is to establish a pattern linking fracturing motion at a focus and the continuous elastic motion in the region before a fracture occurred. We also wanted to establish a correlation between the fracturing motion at each focus and the quasi-continuous motion of the medium in large space-time regions, viz. the seismic and tectonic flows of the rock. All these relations are expressed in a natural way in terms of the seismic moment of an extended volume focus in tensor form.

10.1. Energy Models of a Focus

We decided above that the seismic energy E of a focus as a whole or as elementary parts will be the basic characteristic of a focus in different models. We shall express the intensity of a seismic tremor in terms of the density e of the seismic energy at the observation point. From the energy density e (which is in itself of interest) we obtain a macroseismic intensity I. In a simple one-dimensional scalar model, we shall use the idea developed in [Riznichenko], and when considering the spectral aspect, the notation in [Riznichenko].

Point Focus. This focus model can be considered independently or as an elementary unit in a model of an extended focus. For a point, or small concentrated focus, or spherical focus of finite size with central symmetry of radiating seismic energy, the energy density e of a seismic pulse per unit of wavefront for the entire duration of its passage at the observation point is [Riznichenko]:

$$e = (E/4\pi r_0^2)\,(r_0/r)^n\, \exp\left[-\alpha\,(r - r_0)\right], \tag{10.1}$$

where $E = 10^K$, usually in joules, is the seismic energy of the focus;

$$K = a_M + b_M M \tag{10.2}$$

where M is the magnitude (it is generally assumed that $K = 4 + 1.8M$), r_0 is the radius of the reference sphere for the energy E, and r is the distance from the focus (hypocenter) to the observation point:

$$r = \sqrt{h^2 + \xi^2} = \sqrt{h_2 + (x - x_0)^2 + (y - y_0)^2} \tag{10.3}$$

where h is the depth of the source, ξ is the epicentral distance, x_0 and y_0 are the focus coordinates, x and y are the coordinates of the observation point on the Earth's surface, n is a discrepancy exponent, and α is the absorption coefficient. The approximate meaning of (10.1) in general, and also in the particular cases when it is accurate, were discussed above [Riznichenko].

The attenuation of the energy density e with the distance r is described in (10.1), by two parameters: n and α. More formally, but simpler to calculate, this can be done with a single attenuation parameter or effective discrepancy n_e, i.e.,

$$e = (E/4\pi r_0^2)(r_0/r)^{n_e}. \tag{10.4}$$

In the energy model [Riznichenko] the macroseismic intensity I is obtained from the energy density e using the expression:

$$I = C_1 + C_2 \log e. \tag{10.5}$$

If e in this formula is in J/km^2 and I is measured on the GOST, MMS, MSK or other scale, then very approximately we have $C_1 = -1.25$ and $C_2 = 0.854$. Given (10.2), we move from (10.1) to a macroseismic formula like N. Köveslighety type:

$$I = C_{M^1} + C_{M^2}M - C_n \log (r/r_0) - C_a (r - r_0), \tag{10.6}$$

and from (10.4) to a Blake-Shebalin formula:

$$I = c + bM - s \log r. \tag{10.7}$$

The relations between the parameters of (10.1) and (10.6), and also between (10.4) and (10.7) are given in [Riznichenko].

Given that macroseismic data in the USSR is "manually" processed by Blake's and Shebalin's methods [Shebalin, A. Blake], and also using our macroseismic plate, generally (10.7) is employed with one attenuation parameter s and accordingly n_e, which equals [Riznichenko]:

$$n_e = (b_M/b) s = n + \alpha [(r - r_0)/\log r/r_0)] \log e. \tag{10.8}$$

Here we have approximately $b_M/b = 1.8/1.5 = 1.2$. A transition from n_e to n and α is possible, if we assign a definite value to n in (10.8), e.g., assume that $n = 2$, or adopt a definite relation $n = f(r)$ [Riznichenko] or $n = F(K, r)$ as $n \to 2$ and $r \to r_0$ ($r \geqslant r_0$). Here we consider that r is the average distance within the interval of observations.

The experience gained in employing a macroseismic plate [Riznichenko] in the Caucasus (E.A. Dzhibladze), in Tadzhikistan (T.A. Kinyapina) and in Turkmenia (G.L. Golinsky) indicate that macroscopic data are usually described satisfactorily by the empirical relation (10.7) within the entire range of distances r used. We can thus consider in a first approximation that the effective parameter s or n_e does not depend on r. This is a favorable circumstance for employing the effective quantity from (10.7) as the initial one in subsequent calculations by (10.8) of the quantities n and α contained in our energy model and having a more definite physical meaning [Riznichenko].

The quantity α should in general be considered to depend on K and r. Indeed,

$$\alpha = \theta/\lambda = \pi/Q\nu T, \tag{10.9}$$

where θ is the absorption decrement over the wavelength λ; ν is the propagation velocity of a wave, firstly the lateral, more destructive one, T is the prevailing oscillation period of a pulse, $f = 1/T$ is the frequency, and Q is the quality of the medium. During crustal earthquakes, we can generally assume that θ and Q are parameters of the medium independent of T or f. If desired, one may include the possible dependence of θ on T in the calculations, for instance by assuming that $Q = Q_0(T/T_0)^\nu$, where Q_0 is a parameter of the medium independent of T and changing in space, $T_0 = $ const, and ν is a constant parameter to be determined from the set of observations. The prevailing period T of the oscillations in a pulse is generally larger when the K value for an earthquake and the distance r are greater. The peak period T in term of energy density e as a function of K and r can be obtained from systems of the average spectra of earthquakes [Riznichenko]. More exactly, we ought to consider for the purposes of macroseismicity, the frequency characteristics of damaged structures used to assess the intensity I [Riznichenko].

The dependence of n on K and r can be established from a best fit of the mathematical model to the set of observed intensity fields. But this deserves special discussion.

Extended source. As with the procedure used by T. G. Rautian [Riznichenko] for foci distributed along a plane, we shall synthesize an extended focus from elementary energy foci distributed discretely or continuously along a line L, over a surface S, or in a volume V so that, the total flux of their seismic energy through the surface of a sphere of radius r_0 around each element, equals the seismic energy E of the focus as a whole.

If we label the seismic energy density E^*, then for a volume focus we have:

$$\iiint_V E^* dV = E. \tag{10.10}$$

Here the volume density E^* of the energy foci, generally speaking, is a function of the point (x, y, z) in region V (or, accordingly, the surface density on S, or the linear one on L). The surface density e of the seismic energy at the observation point for models of extended foci of type (10.1) is an integral of the form:

$$e = (r_0^{n-2}/4\pi) \iiint_V E^* r^{-n} \exp\left[-\alpha(r - r_0)\right] dV \tag{10.11}$$

and for models line (10.4), an integral similar to a potential:

$$e = (r_0^{n-2}/4\pi) \iiint_V E^* r^{-n} dV. \tag{10.12}$$

In the region outside the foci, these integrals are not unusual. Let us consider their behavior inside the region where the sources of seismic energy E are distributed with volume density E^*. From mathematical physics we know that improper integrals of the type:

$$\iiint_V E^* r^{-n} dV, \tag{10.13}$$

converge continuously when $r \to 0$ in the focus region provided that $n < 3$ and, naturally, if E^* is finite, which is implied. Our expression (10.12) relates directly to (10.13). In (10.11), on the other hand, the integrand also contains the multiplier $\exp\left[-(r - r_0)\right]$ to account for the absorption of the seismic energy along the path r_0, r. It remains finite for any finite r including $r = 0$, and so it does not affect the convergence. We can also generalize: for both the inside and outside of the reference sphere, along any segment of the path r_1, r_2, we must instead of the multiplier write:

$$\exp\left[-\oint_{r_1}^{r_2} \alpha dr\right], \tag{10.14}$$

where α is a function of a point along this path. Absorption only reduces the yield of the seismic energy e. Near the point $r = 0$, where the convergence of the integral in (10.13) is being studied, (10.14) tends to unity as $r_1, r_2 \to 0$. Now the integral in (10.11) is converted into an integral like (10.13), so that its convergence is solved in the same way as (10.12).

Intensity of tremors inside a volume focus. Our energy model of a volume focus is depicted in a general form by an equation such as (10.11) with the refinement of (10.14). At small r, the value of n tends towards $n = 2$, so that the convergence requirement $n < 3$ is observed with a large margin. This enables us to obtain, for this model, a restricted intensity of tremors in the form of e or I and so on, both outside the region, and inside it, viz. in the neighborhood of the focus itself.

This is consistent with seismological observations. For instance, it has been repeatedly noted [Ambrayses] that when major earthquakes occur, with fractures extending to the Earth's surface, observations near the fracture, i.e., undoubtedly in the volume region of the focus itself, show that the intensity of the tremors remains moderate and only depends slightly on the distance to the fracture.

Now let us revert to (10.12) with the effective discrepancy n_e. When n_e is determined by observations outside the region, $n_e > 3$ is quite frequent. Here (10.12) must diverge inside the region, which is not justified physically. This is yet another indication, in addition to those already discussed [Riznichenko, etc.], that this parameter must be carefully treated when assuming it to be constant, especially near a focus. It is clear that when introducing the observation interval in the neighborhood of the focus itself, we cannot assume that $n_e \gg 2$. Here both n_e and n should tend towards 2.

Let us calculate the intensity of tremors e or I inside a volume energy model of a focus at the center of a spherical focal of radius R with a constant focus density as a specific example of the behavior of these quantities:

$$E^* = E/1.33\pi R^3. \tag{10.15}$$

We shall study the convergence of the integral in (10.13). Let ϱ be the radius of a small sphere surrounding the center of our focus. Hence:

$$\iiint\limits_{V-V_\varrho} E^* r^{-n} dv = \int_0^{2\pi} d\phi \int_0^{\pi} \sin\theta \, d\theta \int_\varrho^R \frac{E^*}{r^{n-2}}$$

$$= \begin{cases} 4\pi E^* \left[\dfrac{1}{3-n} r^{3-n} \right]_\varrho^R & \text{if } n \neq 3 \\[3mm] 4\pi E^* \, [\ln r]_\varrho^R & \text{if } n = 3. \end{cases} \tag{10.16}$$

At the limit when $\varrho \to 0$, which holds when $n < 3$, we obtain for the case $n = 2$ (of interest to us here)

$$\iiint\limits_V E^* r^{-n} dV = 4\pi E^* R. \tag{10.17}$$

Inserting this expression into (10.11), where we assume that $n = 2$ and $\alpha = 0$, we find the surface density $e = e_c$ of the seismic energy at the center of the spherical focus as $e_c = E^* R = E/4/3\pi R^2$.

Let us compare it with the surface density of the seismic energy $e = e_c$ on the surface of a sphere of the same radius R for point focus (10.11). It is obvious that the same energy density e_s is also retained on the surface of our spherical focus $e_s = E/4\pi R^2$.

These densities can be seen to differ by $e_c/e_s = 3$ times. Equation (10.5) transforms e into the macroseismic intensity I. The value of C_2 in (10.5) is 0.854. We thus find that the relevant macroscopic intensities I_c and I_s differ only by $\Delta I = I_c - I_s = c_2 \log{(e_c/e_s)} = 0.4$ points.

Hence, in the three-dimensional energy model, the intensity in the neighborhood of the focus itself increases only slightly from the periphery to the center, which is consistent with macroseismic observations.

We presume that this is not accidental, and that a volume focus correctly approximates reality. Indeed, there is no doubt that seismic energy is drawn not from the surface of a fracture (it is only spent here), but from the volume region surrounding the fracture. Perfect elasticity and continuity may also be appreciably altered in this region, which can be assessed by the aftershocks, most of which occur during the main phase of an earthquake and immediately after it, but so densely in space and time that it is difficult to distinguish them separately. The mechanical processes in the volume neighborhood of a focus as a whole, are also shown by the residual phenomena, in particular the echelon and plumage fractures noted in major earthquakes. Such partial fractures may remain concealed under the surface. At the same time, they may cause a large fraction of the relatively high-frequency oscillations with which the destructive effect of an earthquake is associated.

Intensity of tremors in the region of a surface and linear focus. In a two-dimensional case for energy foci distributed on a part of surface S, i.e., geophysically, in the focal fracture plane, the convergence conditions of the integral corresponding to (10.13) is stricter: $n < 2$. In a one-dimensional case on L it is even stricter: $n < 1$. This is why, in these cases, energy models with $n = 2$ cease to converge, or yield plausible results for the focal region. They are suitable for approximations of thephenomena only that are adequate far outside the region where the foci are distributed on S or L.

Let us now compare the external effect of an extended focus with that of a concentrated one. We have already mentioned that for an extended focus with a central, spherical symmetry of the distribution of the density of the sources E^* has the same effect as a point focus. In both cases, circular isoseismal lines are obtained for a homogeneous and isotropic medium. The situation is different for an extended focus. The extreme case, where the elementary sources are located on a segment of a straight line continuously or discretely, is illustrative when studying its external effect.

Effective parameters of an elongated focus. An elongated focus bunches the isoseismal lines close to it in the direction of their extension. For a quantitative analysis of the distribution effects of the tremor intensity e, I in the surrounding region, we consider a discrete uniform distribution of three point foci of energy on a segment of straight line L. We assume that this focus is at a depth h under the observation surface—the Earth's surface.

Using (10.4) with one effective discrepancy n_e we have the following at the observation point M_1 (Fig. 97) in the direction of focal extension:

$$e_{\parallel} = (10^K r_0^{n_e-2}/3.4\pi)\{1/[h^2 + x^2]^{n_e/2} + 1/[h^2 + (x + L/2)^2]^{n_e/2}$$
$$+ 1/[h^2 + (x - L/2^2]^{n_e/2}\} \tag{10.18}$$

and at point M_2 across the direction of extension:

$$e_{\perp} = (10^K r_0^{n_e-2}/3.4\pi)\{1/[h^2 + x^2]^{n_e/2} + 2/[h_2 + (L/2)^2 + x^2]^{n_e/2}. \tag{10.19}$$

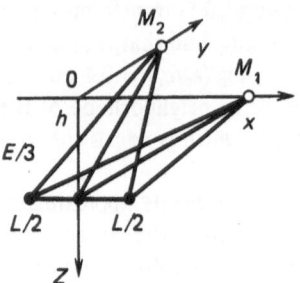

Fig. 97. Discrete model of a source:

$E/3$—seismic energy of elements of a triplex source; L—the length of source; h—the source depth; x—the direction along the strike; y—transverse to strike; M_1 and M_2—observation points

We calculate for $M = 5$, i.e., $K = 4 + 1.8M = 13$, $r_0 = 10$ km, and $h = 10$ km. Let the focal length be $L = 20$ km. This is much larger than the average length $L = 8$ km for a crustal earthquake with such an energy value K or magnitude M [Riznichenko]. In (10.7) we assume that $c = 3.0$, $b = 1.5$, and $s = 3.6$, so that by (10.8) $n_e = 1.2$ and $s = 4.2$.

The $e_\parallel$ and $e_\perp$ calculated by (10.18) and (10.19) for an extended focus are compared to e calculated by (10.4) for a concentrated (or spherical) source. These e, $e_\parallel$, and $e_\perp$ are transformed to the corresponding macroseismic intensities I, $I_\parallel$, and $I_\perp$ using (10.5) for the C_1 and C_2 indicated theorem.

Figure 98 shows the arrangement of the (I, x) and (I, y) points for a concentrated I and extended $I_\parallel$ and $I_\perp$ foci with $K = 13$ $(M = 5)$ in the coordinate system of a macroseismic plate [Riznichenko]. It can be seen that for an extended focus the intensities $I_\parallel$ and $I_\perp$ at the epicenter are somewhat smaller than I for a concentrated focus of an earthquake with the same K (or magnitude M). But at $x \cong L/2$, i.e. 10 km, we find that $I_\parallel > I$, but $I_\perp < I$. For greater distances $x, y \gg L/2$, the effects of concentrated and extended sources tend asymptotically in all directions, to the same value $I_\parallel \rightarrow I \leftarrow I_\perp$.

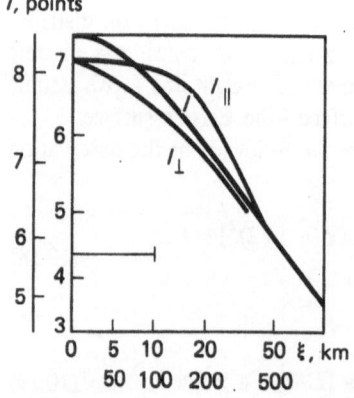

I, points

Fig. 98. I intensity of shakings from the source of L length and depth $h = L/2$ at an epicentral distance ξ:

$I_\parallel$—along the strike; $I_\perp$—transverse to source strike; I—for concentrated source. ξ—axis = 0-100 km corresponds to $I = 3$-7 points; $\xi = 0$-1000 km; $I = 5$-8 points

The same curves of I, $I_{\parallel}$, and $I_{\perp}$, but at a different scale (in Fig. 98) show the macroseismic effect of an earthquake with $K = 18.8$ ($M = 8.0$) with a focus at a depth of $h = 10$ km, concentrated (I) or extended ($I_{\parallel}$, $I_{\perp}$) with a length of $I = 200$ km. In this case, the value of K (or M) is chosen so that $L = 200$ km is the average focal length for an earthquake of such a magnitude [Riznichenko]. The outer scales $\xi = 0\text{-}500$ km and $I = 5\text{-}8$ scale units correspond to this in Fig. 98.

It is interesting to see how the length of a focus affects the determinations of the effective parameters h and s (or n_e), which are evaluated by methods based on a Blake-Shebalin point model (10.4), (10.7). Figure 98, when compared with our macroseismic plate [Riznichenko] readily allows us to study this. For an earthquake of $K = 13$, the curve of I for a concentrated focus corresponds to the curve of a plate with $h = 10$ km, and as expected, $s = 3.5$. The curve of $I_{\parallel}$ for an extended focus with the "observation" points along its length does not coincide with any plate curve, but is located, on average, with some scatter of observation points near the theoretical curve of I and $h = 10$ km and at $s = 3.5$. On the other hand, the "observed" curve of I for an extended focus across the length of a source is approximated well by the curve of a plate with the increased depth of $h = 12$ km, but the same attenuation $s = 3.5$. Similarly, for an earthquake $K = 18.8$ ($M = 8.0$), $h = 100$ km, an extended focus with a length of $L = 200$ km with observations across the source leads to a proportionally overstated depth of $h = 120$ km and to the same attenuation s.

Hence, when $L/h = 2$, the depth h in observations across the source direction is overstated by about 20% which the actual attenuation parameter s (or n_e) remains approximately the same. With a growth in the relative length of a source L/h, the relative overstatement of the depth h naturally increases. It is especially great for shallow and very long sources.

The effects of the length of a source as shown by an energy model naturally do not cover all the possible cases. Perhaps the main factor we succeeded in showing here is the effectiveness of an energy model in such studies, in particular for a volume source with its internal region.

10.2. Seismic Moment of a Fractured-continuous Focus

Here we shall consider two aspects of the relation between fracturing and continuous deformation during earthquakes. The first related to one source, and chiefly to elastic continuous deformations before an earthquake and fracturing inelastic deformation after it. In general it was explained by Reid's "theory of elastic recoil". Here we additionally give some considerations associated with the concept of a seismic moment and its tensor, which in Reid's time were not yet known.

The second aspect concerns a set of earthquake sources and relates to establishment of relations between fracturing motions in foci and quasi-continuous motion in the large space-time regions where these earthquakes occur. The problem of relating the two kinds of motion was probably first stated in [Riznichenko] on the "seismic flow of rock". The starting point for our calculations, now over 10 years old, included the concept that the ultimate stresses and strains are constant for a given material, as indicated by Benioff. This was the conventional theory in seismology of that time. These publications were soon followed by the less known Aki work on the seismic moment M_0. It subsequently replaced to a certain extent, the previous initial extreme values in later publications on a focus and the seismic flow.

Our initial calculations [Riznichenko] were for the general three-dimensional case. Brune reasoned as we did, but used the seismic moment which had been established then and calculated the average velocities of the relative displacements of the ends of large fractures for the two-dimensional case. Later, B.V. Kostrov used the seismic moment for a theoretical discussion of a three-dimensional case—the seismic flow of rock. Randall and then Kostrov expanded the concept of the seismic moment and introduced its tensor.

In developing the general idea, we recently proposed an elementary theory for estimating the velocities of the vertical seismotectonic motion of the Earth's surface [Riznichenko]. It was based on data on seismic moment and directions of compression and tension axes at earthquake foci as postulated by A.V. Vvedenskaya. It also included the parameters γ, A, and K_{max} of the permanent average seismic regime. These are established from seismological [Riznichenko] or combined seismogeological-geophysical and geodetic data [Riznichenko]. Together with E.A. Dzhibladze we applied the theory to observation data from the Caucasus. We showed there was a satisfactory correlation between the seismic velocities and the velocities of modern tectonic motion established by methods of geodesy and geomorphology for this region. The approach made it possible to supplement the mapping of the directions of the tension and compression axes of the seismotectonic motions practised earlier [Vvedenskaya] with the mapping of the velocities of these motions. True, we mapped only their vertical components in [Riznichenko].

Here we discuss the possibility of further progress. Proceeding from the concept of volume fracture and continuous earthquake foci, the tensors of removed stress and strain, the tensor of seismic moment, and the idea of the seismic flow of rock, we outline the use seismic observation data to determine quantitative vertical and horizontal seismotectonic motions.

Fractured and continuous focus. The classical dislocation model of a shear focus depicts it as a surface, usually a fracture plane along which shear occurs [Vvedenskaya, Kostrov]. It is assumed that otherwise the medium remains continuous and perfectly elastic. This case is shown schematically for an initially spherical volume focus in Fig. 72 (top). In the zero approximation, generally employed to process the data of seismic observations, the dual nature of the possible orientation of the fracture plane with the same arrangement of the tension TT' and compression CC' axes must be recognized.

In [Riznichenko] one fracture in a bulk focus was considered as the limiting case of a multitude of parallel fractures (see Fig. 72). The total seismic moment M_0 of a complicated focus can be required to remain the same as for a simple one.

With an infinite increase when the number of fractures tends to infinity, we progress from a complicated fracture focus to the other limit of a continuous volume focus with the same moment M_0. Figure 72 shows that a deformed spherical focus turns into the same ellipsoid regardless of the orientation of the fracture planes. Whatever the directions of the planes (Fig. 72, right and left), different fracturing foci can be regulated as different stepped approximations of the general form of the ellipsoid. It is an ellipsoid of the fraction of the elastic deformation that is removed at the focus during an earthquake. These general considerations can quide us in obtaining the following quantitative conclusions.

Seismic moment and reduced stresses. The expression $M_0 = \mu SD$ is the classical definition of seismic moment. Here μ is the shear modulus, S is the area of a fracture, and D

is the dislocation shift over it. The concept of the volume density $M_0^* = M_0^*(x, y, z)$ of the seismic moment M_0 in [Riznichenko] was introduced for a volume V including a focus, such that:

$$M_0 = \iiint\limits_V M_0^* \, dx \, dy \, dz. \tag{10.20}$$

The physical meaning of the volume density M_0^* of the seismic moment M_0 is the shear stress $\Delta\sigma^*$ removed in each elementary volume $dV = dx \, dy \, dz$ of the focal region. To verify this, we choose a coordinate system such that the planes of fracture or, for a continuous source, the glide planes are parallel to the coordinate planes x, y, the z axis is perpendicular to it, and y is the direction of shift. Hence, $\Delta\sigma^* = \sigma_{yz} = \mu e_{yz}$, where e_{yz} is the shear strain. We assume for simplicity that all the displacements in the focus are colinear. We have:

$$\iiint\limits_V M_0^* \, dx \, dy \, dz = \iiint\limits_V \sigma_{yz} \, dx \, dy \, dz = \iint\limits_S dx \, dy \int\limits_0^h \sigma_{yz} \, dz = S \int\limits_0^h \mu e_{yz} \, dz = \mu SD = M_0.$$

The last in this series of identities is the classical definition of seismic moment, which proves that $M_0^* = \Delta\sigma^*$. Here h is the dimension of a volume focus in the z direction at right angles to the fracture planes for a fracturing simple or complicated source, or to the glide planes for a continuous volume source. It should be noted that the definition of the density $M_0^* = \Delta\sigma^*$ of the seismic moment remains true for a volume focus even if it encompasses the entire Earth, or for an unconfined space: the parts of a source remote from a limited fracture, where $M_0^* = \Delta\sigma^*$ are small, affect the integration only slightly (Saint Venan's principle).

If we assume that all the density of the moment M_0^*, or (the same), all the stress $\Delta\sigma^*$, which is important for the integral in (10.20), is localized in a sphere of radius R, the average value of $\Delta\sigma = \Delta\sigma^*$ over the volume inside the sphere will be:

$$\Delta\sigma = M_0/1.33\pi R^3. \tag{10.21}$$

If we adopt the average radius of the fracture surface of a simple dislocation focus as the radius R of this sphere, (10.21) gives a new estimate of the average stress removed on this surface and in its vicinity, i.e., of the quantity generally adopted in seismology and called the "stress drop", in the sense used by Aki, Brune, and Kostrov et al. Our assessment of $\Delta\sigma$ (10.21) differs from Brune's $\Delta\sigma = 7/16 \, M_0/R^3$, which follows from Ashelby's and Keilis-Borok's formulae by only $\Delta\sigma_{Rz}/\Delta\sigma_{Br} = 0.55 \approx 0.5$ (times) [Riznichenko]. This discrepancy is no greater than that associated perhaps with inaccurate knowledge of the shape of the fracture surface. The latter may for example, be round or extended, rectangular [Kostrov, Riznichenko]. We have already used (10.21) when processing spectral data on earthquakes in the Caucasus [Riznichenko].

It is easy to transform the stress drop $\Delta\sigma^*$ at a point or the average drop $\Delta\sigma$ in a volume to the relevant elastic shear strains removed by an earthquake i.e., $\Delta u^* = \Delta\sigma^*/\mu$ and $\Delta u = \Delta\sigma/\mu$.

To go directly from that to the elastic energy released by a focus remains impossible, however, because M_0 and M_0^* are not related to the absolute values of the stresses and strains that determine the energy, but rather to their differences. Indeed, the elastic energy released by an earthquake from each element dV of a focus (we omit the superscripts[*]) is:

$$dE = dV \int_{(2)}^{(1)} U\sigma dU = \frac{dV}{2}\mu U^2 \Big|_{(2)}^{(1)} = \frac{\mu dV}{2}(U_1^2 - U_2^2)$$

$$= \frac{\mu dV}{2}(U_1 + U_2)(U_1 - U_2) = \mu \frac{U_1 + U_2}{2}\delta U dV.$$

If the stresses and strains are released completely, $\sigma_2 = U_2 = 0$, we have:

$$dE = \mu\,[(\Delta U)^2/2]\,dV, \qquad\qquad\qquad\qquad\qquad\qquad (10.22)$$

while if they are not released completely: $dE = \mu\,[(U_1 + U_2)/2]\,\Delta U dV$.

At the same value of $\Delta U = U_1 - U_2$ and, consequently, at the same $\Delta\sigma^* = M_0^*$ in this volume element, the second quantity dE will be greater than the first one, the difference being higher if the stress and strain remain unrelieved and thus greater. This naturally remains true for the focus as a whole with its average released elastic stresses $\Delta\sigma$ and strains ΔU, source energy, and seismic moment M_0.

We can now make some prognostic conclusions for the possibility of major earthquakes, for which sufficient unreleased elastic energy is needed. For the same M_0 and incompletely released strain, the energy of a source and, therefore, generally the seismic energy $E = 10^K$ is greater than when the strain is completely released. Consequently, for the same earthquake energy value K or magnitude M, smaller M_0 should be expected with an incomplete release, compared with the average standard value [Riznichenko], than with complete release. Under the same conditions with incomplete release, as was shown in [Riznichenko], variation considerations for the energy of a set of earthquakes, should lead to a smaller $\gamma = -d\log N/dK$ for the distribution of the number N of earthquakes by K. Hence, smaller K or M than the standard M_0 and γ for earthquakes of a given magnitude are indicators of increased danger of a major earthquake at a given locality.

Total strain of a focal region before and after an earthquake. If we refer only to the fraction of elastic strain removed by an earthquake in the medium in the region of a focus before the earthquake, which is directly related to the moment M_0, then ignoring details, this is the same as the residual ultimate strain appearing as a result of an earthquake.

An earthquake transforms this continuous elastic restorable strain into the fracturing inelastic residual strain approximating it. Both strains (see Fig. 72) are characterized by the same seismic moment M_0; see (10.20), where $M_0^* = \Delta\sigma^*$.

For a number of schematic calculations of the total regional strain, it is much more convenient to use the continuous than the discrete form. We shall do this for calculations of the seismic flow.

In an earthquake, the transition from continuous strain to fracturing strain is attended by the evolution of source energy, including its seismic energy $E = 10^K$. This has only been studied by observation for which data are systematic and numerous. The strain at a source, and accordingly the seismic flow due to all the strains from many sources, is related by a determinant to M_0 rather than E. If we need to express the seismic flow E, however, we must use stochastic correlations of M_0 with E.

Tensors of strain, stress, and seismic moment. We have already mentioned that the seismic moment tensor was introduced by M.J. Randall and B.V. Kostrov. They did this without considering the direct relation between the moment and the stresses and strains in the volume region of the source, however, we shall do this.

It is expressed by (10.20), where $M_0^* = \Delta\sigma^* = \Delta u^*/\mu$. It is only necessary to consider the stresses and strains being removed as a tensor and not the scalar quantities.

Let us determine the strain being removed (the strain drop) at any point of a volume source, as is customary in continuum mechanics [Landau, Lifshits, etc.], by the strain tensor U_{ik}. In the case of elasticity implied here, this strain is small, and therefore:

$$U_{ik} = 0.5\,(\partial U_i/\partial x_k + \partial U_k/\partial x_i). \tag{10.23}$$

Here the subscripts i and k have the values 1, 2, and 3, corresponding to the coordinate axes x, y, and z, and U_i is the component of the displacement vector. We shall avoid, as we did before, the symbol Δ to denote increments and shall retain this symbol for the Laplacion operator.

The tensor of the elastic stress being removed σ_{ik} corresponds to the strain being removed U_{ik}. They are related by Hoode's law, which for the very simple case of isotropy contains two elasticity constants. But when only the shear (and not volume) strains and stresses are significant, as in Fig. 72, for rough estimates, our present interest, it is often sufficient to use the shear modulus μ. Somewhat schematically, we can write the following expression for the seismic moment tensor M_{0ik}:

$$M_{0ik} = \int_G \sigma_{ik}\,dV = \int_G \mu U_{ik}\,dV = V\bar\sigma_{ik} = V\mu\bar U_{ik}. \tag{10.24}$$

Here G is the volume, while $\bar\sigma_{ik}$, $\bar U_{ik}$, and μ are the average stress and strain being removed in this volume, and the shear modulus. With pure shear total stresses and strains, the tensors M_{0ik}, $\bar\sigma_{ik}$, and $\bar U_{ik}$ are deviators. The principal values of these prins correspond to the scalar quantities M_0, $\Delta\sigma$, and ΔU adopted in seismology.

The components of these tensors can be projected as usual to the principal axes, which are local systems of directions of the tension, compression, and zero stress strain axes for each source. Conversely, if the directions of these axes have been determined beforehand by the usual methods of studying the "source mechanism" adopted in seismology [Vvedenskaya], while the principal values of $\Delta\sigma$, ΔU, and M_0 have been established, say, after I. N. Brune or in similar way, we can progress from local coordinate systems to a single common system, e.g., connected to the geographical network. Here the components of all these tensors, chiefly M_0 can be represented for all the earthquake foci in a region being studied in one common coordinate system. It is rational to do this by calculating the seismic flow of rock in tensor form.

We consider the main result of the present section to be the introduction of the volume density of the seismic moment and the showing that, in a simple case, this density equals the stress in the volume removed by an earthquake. Although this was also clear from the formulas for the seismic moment (Brune's formula, etc.), I have not encountered a direct acknowledgement thereof. This may be because recently, after the classical period of seismology (Benioff et al.), the concept of focal volume is ufashionable. I think that this is mistaken.

10.3. Seismotectonic Flow

Seismic flow. As before [Riznichenko], we shall use the term seismic flow of rock to denote the fraction of tectonic deformation—the tectonic flow—of the Earth's crust and upper mantle in large space-time regions due to the set of earthquakes in it. Digressing from the details, which otherwise individual foci become, we shall consider the seismic flow as a whole to be quasi-continuous and try to model using the simple case of a Newtonian incompressible viscous fluid. We shall consider the seismic flow to be steady state, i.e., its parameters only depend on the space coordinated x, y, and z, though generally, they can also be considered to be functions of time t.

We shall consider that the vector v of the flow velocity (or in its components v_i), and also the rate of strain tensor v_{ik} whose components are known combinations of the derivatives of the velocity vector components v_i for the coordinates x, y, and z to be the main indices of the seismic flow. Let us establish a relation between the velocities of seismic flow v_i and the seismicity indices, including the set of seismic moments M_{0ik} of the foci contained in any given elementary space-time volume WT.

Let us first refine our concept of the elementary volume. In principle, in a physical space, the volume W should be much greater than the volume of any of the possible sources in it, while the time T should be much longer than the average recurrence period in W of the maximum possible, rarest earthquakes. At the same time, the volume W must be as small as possible for a detailed description of the space distribution of the seismic flow indices. The requirements simultaneously of the adequate size and small elementary region WT contradict each other, and we have to compromise. We shall return to this problem later. It should be noted that such an approach to the determination of the "total seismic dislocation" was used in [Averyanova] and [Solonenko, et al.].

If we use the concept of a volume source and its average indices, the question of the relation of v_i and v_{ik} to M_{0ik} is exeedingly simply solved (more extensive explanations can be found in [Kostrov or Honda]). The average strain $U_{ik} \mid W$ in the volume W due to the average strains $\overline{U}_{ik} \mid V$ in the volumes V of the earthquake foci that occurred in it during time T and are contained in W is:

$$U_{ik} \mid W = (1/W) \Sigma V \overline{U}_{ik} \mid V = (1/W) \Sigma (1/\mu) M_{0ik}. \tag{10.25}$$

In (10.25), we used (10.24) for the average indices of a volume source. Hence the average rate of strain in an elementary space-time volume WT is:

$$v_{ik} = \partial U_k \mid W/\partial t = (1/WT) \Sigma (1/\mu) M_{0ik}. \tag{10.26}$$

Tensor equation (10.21), i.e., a system of six (according to the number of independent components of the symmetric tensors M_{0ik}) scalar equations gives us the required basic relation between the rates of strain v_{ik} of the seismic flow and the seismic moments M_{0ik} of the sources in WT, from which the v_{ik} are determined completely. Equation (10.26) was obtained earlier by Kostrov in a more general form and in a more involved way. He circumvented the concept of the volume of a source, but in essense he reasoned as here and for a very simple case coinciding with ours.

To go from the strain rate to the velocities of seismic flow at different points of space, we have to integrate (10.26). In a general case, if the data described by (10.26) does not satisfy the other compatibility conditions, this cannot be done exactly without emerging from the physical Euclidean space. This can be done approximately and in an optimal sense. We shall discuss this later.

Hydrodynamics of a seismotectonic flow. For a seismic flow, as for a tectonic flow as a whole, we can formulate a somewhat non-standard hydrodynamic problem. It invloves compiling the equations of motion of the liquid, in the simplest Newtonian case, and finding their solution for known boundary conditions. The solution must correspond to these conditions and the set of observation data (10.26) in an optimal way. The boundary conditions are as follows: on the free boundary (the Earth's surface), the normal stresses are zero; on the other surfaces confining the region of the "seismic liquid" (the boundaries of the platforms between which there is a mobile seismic zone, and the underlying surface) the vectors of the velocities of the points of these surfaces and the points of the liquid are equal. It is thus assumed that the "seismic liquid" flows up to the boundaries of the region.

The conditions of the Earth's surface are universal. On the other boundaries, however, they depend on the environment, assumptions about which are quite often controversial. In this connection, our first calculations of the velocities of the seismic flow [Riznichenko] were for the vertical component of the flow velocities of rock at the Earth's surface, where the boundary conditions are well known. The question of the compatibility of (10.26) was set aside, at that time, because in the other cases the motions of the medium were not determined.

In the optimal solution of our general problem on the seismotectonic flow, the number of equations with observations of data and boundary conditions should be (and usually is) much larger than the number of required functions and constant parameters of the equations that are unknown and are to be determined. When considering seismic flow, the solution of the problem of hydrodynamics must be consistent with observed data (10.26). When considering tectonic flow, on the other hand, the seismic flow indices can be used as indicators for measuring its intensity.

The first step to a joint quantitative study of the two flows was made in [Riznichenko] and led to the conclusion that there is quite a close direct correlation between them, at least in the area being studied. To a first approximation, the velocities of these flows were found to be, on average, linearly related. This enables us to simplify the formulation of the problem of a single seismotectonic flow. Here, if the hydrodynamic approach can be considered for a seismic flow as only a conditional procedure for generalizing observed data, then for a seismotectonic flow, including a seismic one, it can be considered as the solution of a definite physical problem.

A complete system of equations of hydrodynamics, as was shown in [Landau, Lifshits] consists of five scalar equations essential for the quantities completely characterizing the state of a moving liquid at each point, namely, the three components of the velocity v_i and any two thermodynamic quantities, e.g., the pressure p and density ϱ. In general, they include the equation of continuity (scalar):

$$\partial \varrho / \partial t + \operatorname{div} pv = 0. \tag{10.27}$$

Euler's equation (vector, i.e., three scalar ones):

$$\partial v_i / \partial t + (v\nabla) v = (1/\varrho) \operatorname{grad} p + \mathbf{g}, \tag{10.28}$$

and an equation of state or one replacing it (scalar). Here ∇ and Δ are the Hamiltonian and Laplacian operators, and $\mathbf{g}$ is the acceleration of free fall.

When modeling a seismic or tectonic flow by the steady flow of a Newtonian incompressible viscous fluid, the basic equations of hydrodynamics can be simplified. Owing to incompressibility, $\varrho = \mathrm{const}$, and the continuity equation (10.27) becomes div $v = 0$. If we consider, as is usual when processing a great deal of seismological data, that earthquakes have a purely shear nature, the tensors M_{0ik} are deviators (pure shear), and our system of equations with the observed data (10.26) satisfies this condition identically. If, further, we introduce the velocity potential $\Phi = \Phi(x, y, z)$ such that $v = \mathrm{grad}\,\Phi$, then for Φ instead of (10.27) we simply obtain Laplace's equation:

$$\Delta\Phi = 0. \tag{10.29}$$

Euler's equation (10.28), on the other hand, under steady state $\partial v/\partial t = 0$ and given the equation of state for an incompressible viscous liqiud, leads to the Navier-Stokes equation in the form:

$$(v\nabla)\,v = \ -(1/\varrho)\,\mathrm{grad}\,p + \mathbf{g} + (\eta/\varrho)\,\Delta v. \tag{10.30}$$

Here η is the (dynamic) viscosity coefficient, $v = \eta/\varrho$ is the kinematic viscosity.

Consequently, when a seismic flow is modeled with a Newtonian liquid, three flow velocity components v_i and the pressure p—four functions of a point determined by solving four scalar equations (10.29) and (10.30) with the relevant boundary conditions—give the required coordinate functions. The density ϱ and viscosity η of the "seismic liquid" are constant to be determined optimally.

For a tectonic flow, we cannot arbitrarily vary the density ϱ. It must be taken to be that of the rock in the seismic region. The effective tectonic, macroscopic viscosity η of the rock in the region can be obtained, on the other hand, by fitting the solution of the hydrodynamic problem to the observed data. The latter need not only be seismological as in (10.26), but also geological, geophysical, and so on, as is being done in the quantitative solution of other problems of seismology.

Construction of the field of the seismic rate of strain. Locally and for a small WT for each elementary region separately, this problem is formally solved completely by (10.26), where the rate of strain v_{ik} is determined from observations of the seismic moments M_{0ik} in the same region. This does not require a knowledge of the boundary conditions and integration of the equations, so that the fundamental difficulties of solving the problem as a whole associated with this are circumvented. But some methodical difficulties remain, and we shall discuss the way of overcoming them. They are associated with the problem of choosing the size of the elementary space-time regions in which to average the observed ΣM_{0ik}. This is closely related, in turn, to the problem of getting enough detail in the result. For simplicity, we refer to "mapping" the quantities that interest us. The associated vertical and other seismic sections can be plotted like maps. The construction of three-dimensional fields can be reduced to mapping by layers.

When the seismic parameters of a regime fluctuate, e.g., the average volume density of the moment M_{0ik} (10.26), the accuracy with which it is determined is higher the greater the number of determinations of M_{0ik} in the region and, consequenly, the larger the ik region. But increasing the region is equivalent to reducing the details of the result. We encounter a similar problem when mapping other seismic parameters, namely, γ, A, and K_{max} (or M_{max}). This experience [Riznichenko] can also be transferred to the case being considered.

Further, the K (or magnitude M) of an earthquake is determined, at present, on a much larger scale than the seismic moments M_0. The latter correlate with K or M [Riznichenko], and this can be used to determine M_0 indirectly from data on K or M. The difficulty is that M_0 has only been correlated with K or M for scalar M_0, whereas we are interested here in the tensor M_{0ik}. The transition from M_0 to M_{0ik} requires determinations of the directions of the principal axes of the "source mechanism" [Vvedenskaya, etc.]. But there are fewer of these data than for K or M. To circumvent this difficulty, we propose to divide our main problem into two. The first is to establish the scalar densities M_0^* and accordingly the strain rates v on the basis of data on γ, A, and K_{max} and the correlation $M_0(K)$. The second is to establish the fields of the average directions of the principal axes of the moment tensors M_{0ik} and accordingly of the strain rates v_{ik} from data on the source mechanism. The final result is obtained by combining these two parts.

Scalar value of the seismic flow strain rate. This part of the problem [Kostrov, Riznichenko, Brune] consists in transferring from a sum such as ΣM_0 to the integral proceeding from the distribution of $N^*(K)$ earthquakes with respect to K or magnitude M and the correlation $M_0(K)$. This is possible in the scalar form if we assume that in each averaging region the principal axes of the tensors M_{0ik} and correspondingly v_{ik} are directed in the same way. Hence, following (10.26), but in the scalar variant, and assuming that $\mu = \text{const}$, we can write the following for the average rate of the maximum shear strain in the region:

$$v = \frac{1}{\mu WT} \Sigma M_0 \cong \frac{1}{\mu ShT} \int_0^{K_{max}} \int_0^{S} \int_0^{T} M_0(K) N^*(K) \, dK \, dS \, dT, \tag{10.31}$$

where $N^*(K)$ is the K distribution density of the number of earthquakes per unit area S and time T. Here we progress from volumes W to areas S on a map, $W = Sh$, assuming that the seismicity is localized in a horizontal layer of thickness h. For crustal seismicity— the most widespread and practically important type—this calculation is sufficient to a first approximation. In general, however, a layer-by-layer calculation is needed.

We shall adopt the functions $M_0(K)$ and $N^*(K)$ as the average correlations [Riznichenko]

$$\log M_0 = C_1 + C_2(K - K_0) \tag{10.32}$$

$$\log N^* = \log A^* + \gamma(K - K_0) \tag{10.33}$$

The density (with respect to K) seismic activity $A^* = dN^*/dK$ is calculated from the activity in the classes $K = K_0$, $A = A_{10}$ for $K = 10 \pm 0.5$, or $A = A_{15}$ for $K = 15 \pm 0.5$ ($E = 10^K$) by the formula [Riznichenko]

$$A = A^*(10^{0.5\gamma} - 10^{-0.5\gamma}) \gamma \ln 10. \tag{10.34}$$

Inserting (10.32), (10.33), and (10.34) into (10.31) and integrating, we obtain the final expression for the strain rate v in terms of γ, A, and K_{max}:

$$v = \frac{A M_0(K_0) \gamma}{\mu h (C_2 - \gamma)(10^{0.5\gamma} - 10^{-0.5\gamma})} \cdot 10^{(C_2 - \gamma)(K_{max} - K_0)}. \tag{10.35}$$

Here $M_0(K_0) = C_1$, see (10.32) is a fixed value of M_0 corresponding to the K_0 adopted in (10.32) and (10.33). Formula (10.35) has a physical meaning and requires that all the quantities are expressed in a consistent system of units.

Estimates of the strain rates of seismic and tectonic flows. To estimate v as a function of A and h, we can express K_{max} in terms of A, using the average correlation [Riznichenko]

$$A = C_3 + C_4(K_{max} - K_0). \tag{10.36}$$

If we bear in mind that the seismic energy $E = 10^K$ is expressed in joules, the activity $A = A_{15}$ if for an area of $10^{15.5}$ cm^2 and a period of one year, and the seismic moment M_0 is expressed in ergs, then assuming $\mu = 3 \times 10^{11}$ dyn/cm^2 (for granite), $K_0 = 15$, $\gamma = 0.5$, $C_1 = 25.177$, $C_2 = 0.889$, $C_3 = 11.340$, $C_4 = 0.39$, and ignoring the small difference (about 6%) between A^* and A, we obtain, in agreement with [Riznichenko] the approximation:

$$v = A^2/4h, \text{ year}^{-1}. \tag{10.37}$$

Here the thickness h of the seismic layer should be expressed in centimetres, and it does not matter whether $A = A_{10}$ or A_{15}, because they are equal when $\gamma = 0.5$. If we now assume that $A = 1$ (for crustal earthquakes under continental conditions this is a very high activity) and assume that $h = 20$ km $= 2 \times 10^6$ cm, then from (10.37) we find that $v \cong 10^{-7}$ year^{-1}.

In principle, we can go from the strain rate for a seismic flow $v = v_s$ to the strain rate for the general tectonic (or seismotectonic) flow v_t if we succeeded in establishing their sufficiently close correlation. To date, such a relation has been studied and established only for the vertical components of the velocities of the flows V_s and V_t and only for one region, viz. the Caucasus [Riznichenko]. This relation on an average was $V_t = 500 \ V_s$.

If we tentatively assume that the ratio of the strain rates v_t/v_s equals the ratio of the flow velocities V_t/V_s, which in the given situation does not seem unreasonable in principle, under the conditions of the above calculation (10.37) we find that $v_t \cong 500 \ A^2/4h = 100 \ A^2/h$, year^{-1}, where the strain rate v_t of the tectonic flow is expressed in terms of the seismicity A and h that in this case act as "intensity parameters" of the tectonic process. Again using (10.37), but for $A = 1$ and $h = 20$ km, we obtain $v_t = 10^{-5}$ year^{-1} for the strain rate of the tectonic flow.

This calculation must be treated as illustrative of how to do something like this in principle. In order to be confident in the results, it is necessary to accumulate and analyze, first of all statistically, considerable quantities of observations of both flows in a number of regions under various seismological conditions.

Account of turbulence. Recall that all our last calculations were based on the assumption that the principal axes of the tensors summed in (10.26) for the rate of strain are directed identically, and that the flow in the averaging region W is laminar. In general, however, things are different: the flow is only quasi-laminar, and it would be useful to estimate quantitatively and if possible consider the degree of its "turbulence". We shall do this by introducing the coefficient of flow turbulence in the form:

$$\Theta = 1 - \{v_{ik}\}/v, \tag{10.38}$$

where $\{v_{ik}\}$ is the principal value of this quantity calculated from the tensor relation (10.26), while v is calculated from the similar scalar relation on the assumption that the principal axes are parallel. It is obvious that the quantity $\{v_{ik}\}/v \leqslant 1$. When the directions of the axes are the same, we have $\{v_{ik}\} = v$, and therefore the coefficient of turbulence (10.38) is zero, and the flow is laminar. Otherwise $\theta > 0$, and θ grows with the turbulence.

The turbulence coefficient θ can be established from observations in averaging regions W where the number of determinations of M_{0ik} is sufficiently great. The average θ derived from such sampling observations can be used to correct the generally systematically overstated rates of strain v for the entire region as obtained by (10.35) for the general case when the flow is only quasi-laminar.

The chief result of this chapter is the formulation of the problems of seismic flow and tectonic flow as conventional problems of hydrodynamics with boundary conditions. It should be noted that the simple formulation of the problem given here for a steady flow of an isotropic viscous liquid can only give a very rough approximation. The formulation of the hydrodynamic problem of seismotectonic flows could undoubtedly be improved.

We have discussed several ways of how to employ the concept of an extended, in particular a volume, earthquake source of two main regions. The first is to estimate the energy effect of such a focus at the observation point determined at the point by the intensity of the seismic tremors. The second is the expression of the seismic moment of an individual volume source, and for a set of such foci, a description of the seismic flow of rock.

For the energy part, the following are the main results.

1. For a volume source, expressions for the intensity of tremors have been obtained both for the external and internal region relative to a source. It has been shown that for an energy model of a volume source, the intensity of the tremors in the internal region remains a finite quantity, whereas for models of a surface, linear, and point source, the intensity in the source itself tends to infinity. The energy volume model of a source is in better agreement with observations than the other two.

2. For the external field of the intensity from a linearly extended source, the functions and effective parameters of this field have been calculated, namely, the effective depth of the focus and the attenuation parameter (the effective discrepancy) of the tremor intensity. It has been shown, in particular, that the use of the conventional macroseismc formulae for such a source will result in an overstatement of its depth.

As regards the seismic moment and the flow of rock, the main results are:

1. It was shown that the concept of a volume source makes it possible to obtain the expression for the tensor of the seismic moment introduced by Randall and B.V. Kostrov naturally and simply. The volume density of the seismic moment in the simplest case is the average removed elastic stress in a volume element of a source. In other words, the seismic moment is the integral of the stress drop over the volume of the focus. This is its physical meaning.

2. Whether the problem of seismic, and also general tectonic (seismotectonic), flow of rock can be formulated is a problem of integrating hydrodynamics equations for definite boundary conditions. The procedure for constructing the field of the strain rates of a seismic (or seismotectonic) flow was discussed.

IV
Geoacoustics

Sonic and ultrasonic methods are being widely used now in many fields of science and engineering. Ultrasound helps to drill metal, make emulsions, grow crystals; it is used to influence biological processes; for communication and location underwater; as well as searching for fish shoals in oceans or searching for cracks in metals or concrete blocks. Use of sound and ultrasound in geophysics and mining is, figuratively speaking, just one jet in this strong and variegated stream.

However even this single "jet" is very diverse. It embraces the simulation of seismic waves, study of mechanical properties of rocks in samples and in mass, acoustic logging and echo sounding of bottom sediments, engineering research of foundations in construction and studies of rock pressure in mines using natural and artificial sources of sonic and ultrasonic vibration.

In spite of the diversity, all these applications of sound and ultrasound are united by their close connection with geophysics. All this is, in fact, the study of the Earth with acoustic methods and solution of problems related to it. It seems high time to give a proper name to the field of knowledge under consideration. I think it could be called geoacoustics.

What is characteristic of this new field of knowledge—geoacoustics—is its close connection with already known branch of geophysics—seismics, i.e., seismology and seismic prospecting—the science of natural and artificial earthquakes.

It is difficult to make a definite distinction between geoacoustics and seismics. Both sciences study the same elastic waves: longitudinal and transverse, body and surface waves. The only difference is in vibration frequency: seismics, even high frequency seismics, makes use of frequencies no higher than several hundred Herz. While in geoacoustics frequencies of hundreds of Herz, form usually just the lower section of the range employed. Usual geoacoustic frequencies comprise tens and hundreds of kilo-Herz. The upper boundary of geoacoustic frequencies lies at present at about one or several mega-Herz.

The relation of geoacoustics and seismic is determined by two basic circumstances. The first one consists in the unity of physical substance of the process under use (elastic waves) which leads to principally the same physical and technological problems in both the fields: excitation, propagation of elastic waves and reception of vibration. Principally the same questions arise in seismics and geoacoustics: instrument sensitivity and signal selection in the noise; dependence of wave velocity on frequency; vibration damping; frequency spectrum and form of impulses; resolution of methods depending upon observation techniques and structural complexity of the medium. Methods of information processing and interpretation are also similar. Instrumental problems are principally the same, though they have some specific aspects of their own. In particular, the problem of automated observation and information processing is equally acute. In some cases instrumental problems, especially in data processing, are so close in both fields, that they can be solved by the same means.

The second factor determining the close links between geoacoustics and seismics consists of the unity of physical meaning of investigated problems.

One of the basic problems in seismics is use of elastic waves in studies of the properties of media, where these waves propagate; in a more general form, the problem consists in revealing correlation of properties and structure of the medium versus character of the elastic process. The second problem of seismics and seismology, in particular, is to study the very process of earthquake origin, to reveal its nature. It studies the origin of general or large enough faulting of a material by means of observations over elastic waves, generated by natural or artificial sources.

It is easy to see that the same two problems are studied by geoacoustic methods as well. That is why researchers in seismics (seismic prospectors and seismologists) should be deeply interested in results of geoacoustic investigations. Miners are interested in study of structure of the medium and geoacoustic problems no less than seismic researchers.

Let us return to the substance of the process employed—physics of elastic waves. Geophysics has experience of using both stationary and impulsive elastic vibration and waves. Both types of vibration can be used to study properties of a medium. This concerns both seismics and geoacoustics. For example, there are investigations using stationary harmonic oscillations in studing elastic and inelastic properties of material and rock samples, as well as of rock properties in natural conditions—in mines, wells or on the Earth's surface. In most cases, however, impulsive vibration from both artificial and natural sources appeared the most efficient. Although stationary harmonic vibration may be used in some specific cases in geoacoustics, still the pulse method should be recognized as the main one in this field at present, as well as for the foreseable future. It undoubtedly possesses higher resolution as compared to the stationary vibration technique, which results from active use of time as a dividing factor of the vibration process.

As known, the impulsive ultrasonic method was first developed in physics and engineering by S. Ya. Sokolov in 1934. This is the starting point for its application in our field. But the method has greatly progressed during the last 10-20 years with the general development of pulse technology. This was connected with progress in radiologication methods. The first research using impulsive ultrasonic methods in geophysics, more exactly, in the simulation of elastic waves, was performed in the modeling laboratory of the Geophysical Institute of the USSR Academy of Sciences in 1947. It was probably the beginning of its large scale application in geophysics and later in mining.

Use of natural pulses, generated during rock faulting, has more ancient and less definite origin. Hearing rock cracking in mines is probably as old, as mining itself. Tens of years ago miners began improving their "ear" by means of geophones with amplifiers. In the Geophysical Institute of the USSR Academy of Sciences magnetic recordings of mine crackle were first used in 1951 by M. S. Antsyferov.

Let us abstain, however, from going deeper in geo-acoustic history, and concentrate on present development.

1. Simulation of seismic waves. The aim is to help seismology and seismic prospecting in interpreting observed wave patterns to discover structure of the investigated medium. When research in wave simulation was just beginning, there were no good theoretical methods to solve dynamic problems quantitatively about propagation of elastic waves even in heterogeneous media with rather simple structure. That is why modeling had greater significance than now, when many theoretical problems have quantitative solutions. Still, some of them are difficult to solve. For body waves, these include problems, where diffraction is involved, whose effect exceeds the limits of ray approximations of the elastic wave theory, widely used in practice, e.g., the problem of a waveguide with low velocity and indistinct boundaries in the Earth's upper mantle. Concerning surface waves, the specific problems relate to the media, unlike horizontally layered ones.

However, in some cases modeling may also be useful for solution of problems that lend themselves to theoretical analysis only. It seems reasonable to combine methods of theory or modeling of more or less complicated cases, when complete numerical solution can rarely be found for a theoretical problem. The way it can be solved may be clear enough, and some of its elements can be calculated. For rather complicated cases like this, it may be useful to employ reciprocal checking of specific results and to use modeling for research into important factors when deriving a theoretical solution or in geophysical interpetation of numerical results.

Simulation of waves in various non-ideally elastic media and simulation of seismic wave sources remain very important. Concerning the latter problem, it should be noted, however, that simple imitation by simulating a theoretical source as, for example, the centre of expansion or combination of foci (dipoles and multipoles of different types) does not seem to be of any interest. The wave pattern for a dipole or a multipole may be found through superimposing elementary wave patterns obtained from calculation or simulation of corresponding elementary sources. What is much more important is autosimulation, i.e., reproduction of rupturing of the medium at laboratory scale, including non-ideally elastic rheological processes leading to and accompanying the rupture. Simulation of fracture origin and examining its rate of growth, as well as of generated elastic waves remains one of the most important geo-acoustic problems in prospect.

The main directions of development of seismic wave simulation methods are as follows: development of effective simulation methods not only for longitudinal, but also for transverse and surface waves; various types of converted waves; an increase in resolution of simulation methods, in particular, by use of higher vibration frequencies; improving methods of constructing two-dimensional models of gradient media; construction of three-dimensional models of solid media, including gradients; seeking improvement in simulation between the model and nature by various parameters including density and absorption; development of simulation methods for the earthquake source as discontinuity in rocks.

Improving instruments and simulation methods should also be accompanied by further theoretical research to provide the ground range of applications for various simulation techniques. Such theoretical research should be performed in close connection with simulation practice, otherwise it may lead, as has already happened, to false concepts about real possibilities in this domain.

2. Study of mechanical properties of rocks with samples. Data obtained this way, are of importance for seismics (seismology and seismic prospecting), mining and construction. Only the problem concerning investigation of the elastic properties of rocks in laboratory conditions seems to have been considered satisfactory, so far. Thus, to determine the propagation velocities of longitudinal waves in these conditions, the accumulation of data, their systematization, geological and geophysical interpretation and generalization are the main problem, rather than the further improvement of measuring techniques. However, measuring velocities of transverse waves under the same conditions also needs methodological improvements. To still greater extent this applies to the study of elastic wave absorption in different rocks.

The matter is much more complicated when studying the elastic properties of rocks under various types of pressure, not hydrostatic alone.

There are some good results in this research, but we are just beginning the great work. Little has yet been done in the study of elastic, rheological and strength properties of rocks under different pressure and temperature conditions and at different modes of deformation. Meanwhile, these very questions are the focus of the deepest interest in the Earth's physics, including the problem of earthquake origin and prediction, mining and construction, as well as problems of stability and destruction of mines and buildings.

3. Determination of elastic properties of rocks and study of structure of media under natural conditions. Acoustic (impulsive ultrasonic) log is the main method in this direction. After the first experiments using this method were conducted in the USSR (in modeling laboratory of the Geophysical Institute of the USSR Academy of Sciencies in 1954-1955), some geophysical organizations performed a number of interesting instrumental and methodological investigations and experiments. However, very little has been done in the USSR to implement these methods in industry and/or to employ them systematically in geophysical prospecting.

Another important activity in this field consists in the determination of mechanical properties of rocks in borings—natural and artificial—at construction sites, quarries, underground mines. Investigations, such as high-frequency seismic prospecting, acoustic sounding and echo location from mines and boreholes are relevant here. Various researches were carried out in this field in the USSR, which need generalizing. Further development and employment of these methods in construction and mining (geoacoustic engineering) seems to be very promising.

The third important direction is the study of basal sediments to depths of tens or hundreds of meters below the surface by ultrasonic echo sounding. Magnetostrictive or some other device serve here as emitters, and piezo-electric hydrophones—as receivers. Previously these methods were mainly used by oceanologists; now they are being adopted by geophysicists-prospectors. Research employing sound frequency pulses from explosions recorded by piezoelectric or other types of hydrophones, are similar to this type of investigation. These methods allow penetration into basal sediments to the depth of several kilometers, are widely used in oceanology and marine seismic prospecting.

4. Study of the stress state and failure of rocks in mines. Investigations of this type are carried out using both artificial pulses, similar to those in wave simulation and in study of elastic properties of rocks in samples and in mass, and natural pulses, generated as rocks fracture under pressure. Miners may be interested in these investigations in connection with the problem of rock pressure in general (roof control, for example), hazard zoning and prediction of rock bursts, failures and instantaneous coal and gas explosions. Geophysicists-seismologists may become interested in this study due to the similarity between the mentioned

phenomena, rock bursts, in particular, to earthquakes. It is quite possible that geo-acoustic investigations in mines will contribute to better understanding of processes in the potential earthquake area. Thus a solution could be advanced of the most important and difficult seismological problem—prediction of catastrophic earthquakes. Geo-acoustic investigations in this direction alone are a complex of specific problems and solution methods.

It is interesting and important to compare achievements in these basic branches of geoacoustic investigations in the USSR and abroad. Here we shall emphasize problems that seem urgent for improving the general level of geophysical investigations and contributing to the Soviet national economy.

Soviet geoacoustics on average is in line with international achievements. However, this "average" is not distributed evenly among the branches.

As concerns simulation of seismic waves, we do not lag behind the USA and other countries in methodology and theoretical research. However the quality of instruments, techniques of manufacturing individual models, wide scope and high mobility in the solution of some problems, are sometimes a hindrance. Our first goals are—to improve instruments and modeling techniques, to use simulation of transverse and surface waves, to develop three-dimensional models in solid media, to derive models of the earthquake source and its effect on the Earth's surface, for large hydrotechnical installations—earth dams etc.

As far as the study of mechanical properties on rock samples is concerned, our progress in methodology is about the same. But the Americans are better at accumulation and systematization of observational data and in their geophysical interpretation. We aim to catch up and to increase laboratory investigations into the behavior of rocks under conditions of high temperature and pressure, characteristic of the Earth's crust and upper mantle.

Concerning research into the stress and failure of rocks in mines, and into adjacent problems, the USSR seems to be leading the field. But we should not console ourselves with this. We know very well how rapidly individual branches of science and technology may develop, if they are given proper attention. We may easily be left behind, if we slow our advance. But that is not the point. We see that our successes in this field are not sufficient. The main goal—prediction of catastrophic failures in mines—still remains to a certain degree unsolved. Attention should be paid to a further search for its solution.[*]

***

* See [430].

11. Simulation of Seismic Phenomena[*]

Most investigations on the simulation of seismic phenomena use models to study dynamics and seismic waves. These are encountered in "large scale" seismology when dealing with natural earthquakes, and in experimental seismology and seismic prospecting, where elastic waves are generated in the Earth by explosions or other man-made sources of vibration.

The interpretation of observations in seismics (seismology, seismic prospecting) has so far been mostly based on assessments of the time it takes various elastic waves to travel from a source of mechanical vibration—an earthquake, an explosion or other source of shaking—to devices that pick up the ground vibration at the reception points (seismographs). By analyzing the travel times the various waves take over different distances and in different directions from the focus, the structure of the medium through which the waves have passed, their velocity in different media, and underground surfaces which reflect or refract the waves can all be assessed. The surfaces naturally correspond to boundaries between underground layers characterized by different physical properties.

It is these data—the arrival times of the elastic waves—that at present serve in seismics as the main basis for conclusions about the Earth's structure. Depths from several meters to tens of kilometers can be considered, when waves are artificially excited. Tens, hundreds of kilometers or the whole globe can be explained using the waves generated by natural earthquakes. If we also use the dynamic features of waves—their dominant amplitudes and periods—or follow changes in the form of the seismic vibrations on seismograms, then this gives complementary data and a more comprehensive knowledge of the physical properties of the various domains in the medium. As an example, differences in the absorbing ability of the medium of elastic energy, associated with imperfect elasticity can be studied, etc.

The experiment solution of this problem in the field meets with many difficulties. Even in regions with relatively well-known geological structure, e.g., areas with a large number of deep boreholes the physics of the rocks is still not understood enough. The very structure of a natural medium appears too complicated for an investigator to solve a concrete problem. This difficulty can be completely overcome, under laboratory conditions, in experiments carried out with special models of inhomogeneous media.

All these considerations promoted the simulation of seismic phenomena at the Geophysical Institute of the USSR Academy of Sciences and directed the study.

The simulation of seismic waves has never been systematically worked on, which is why it was first necessary to develop the most rational method of simulation in seismics and then use models to study the propagation of elastic waves pertaining to seismology and seismic prospecting.

---

[*] From [Riznichenko].

11.1. Simulation of Seismic Waves

11.1.1. Basic Methods of Wave Simulation

Methods of laboratory simulation of mechanical wave processes to be applied to a study of elastic seismic waves as described in literature, may be divided into four basic groups: (1) grid methods—use of mechanical and electrical "discrete" grid models; (2) hydraulic methods (models using waves on the surface of liquids); (3) optical methods (models for obtaining wave patterns using elastic waves in continuous media with the optical fixing of these patterns); (4) oscillographic methods (models to record elastic vibration of particles in continuous media).

Let us consider these methods as potential simulators of seismic waves applicable to the problems of natural (earthquakes) and experimental (explosions etc.) seismology, and seismic prospecting.

Grid methods (mechanical and electrical models and corresponding computational methods). Mechanical devices capable of representating and studying waves based on the principle of the elastic grid, with inertial nodal elements were described for the one-dimensional case in [Spivak, Riznichenko] and for the two-dimensional case in [B. Ivakin, et al.]. Experience shows that devices of this type provide limited possibilities for comprehensive quantitative study of seismic wave phenomena in inhomogeneous continuous media.

Hydraulic methods [A. Wood]. These methods use gravitational and capillary waves (ripples) on the surface of a liquid (water, mercury and other). Different velocities of wave propagation in a liquid model are obtained at different depths of the liquid layer. Illumination is stroboscopic. The scope of its application in simulating seismic waves is very narrow. This is primarily because the quantitative similarity between gravitational and capillary waves on the surface of liquid, and elastic waves characteristic of seismication is very limited.

Optical methods. This consists of using elastic waves which are actually the same as the simulated seismic waves; these models like those of the previous one, employ continuous media. The wave patterns in space corresponding to certain fixed moments in time, can be observed or photographed.

Gas, liquid or solid elastic media were used in models of this type. The vibration source is either impulsive or continuous, harmonic. The illumination is by a spark for impulse waves and stroboscopic for harmonic vibrations. The illumination is synchronized with the vibration source. Elastic waves of compression and decompression (longitudinal) became visible in transparent media and were photographed by the swill [E. Korolev] or shadow method.

The main shortcoming of methods in this group is the difficulty of investigating dynamic problems, viz. intensity and amplitude measurements and study of alteration of the form of vibrations. The intensity of the sonic field can be assessed for harmonic vibration by the swill method [P. Bazhulin, F. Korolev] light diffraction on ultrasound [S. Rytov] and by other methods. There is a great deal of literature on these problems (see, for instance, Malov, Mikhailov, Bergmann, Wood). However, for impulsive excitation, which is specially interesting for seismic research, qualitative information on the intensity of waves may be obtained by the listed methods. However, with present methods their wave form cannot be investigated.

Oscillographic methods. This includes methods that allow oscillograms and seismograms, records of the vibration of individual particles of a medium, where elastic waves propagate, can be directly obtained. Note that although this is possible in principle, with all the three previous methods (e.g., by appropriate photometry in the optical methods), they were intended to obtain wave patterns rather than seismograms, and therefore are more suitable for the former result. The opposite is true of the methods in the last group.

The fact that seismograms (oscillograms) are the main (perhaps the only) kind of primary material that can be obtained by observations with these methods, means that these laboratory methods are close to modern instrumental field observations in seismology.

These methods employ basically the same physical phenomena for simulating seismic waves as those that occur *in situ*, the same ways of observation of the phenomena, and the same type of primary material is obtained. The difference between the model and the natural conditions is one of scale alone. This means the model results can be compared with those *in situ* at every stage of processing of observational data, from a comparison of the general appearance of the primary material, i.e. seismograms. The practical advantages are evident here.

The general character of the device employed in these laboratory methods is the same as in field seismometry. However, due to the difference between model's linear scales and reality (sometimes a thousand times or more), the instruments and techniques of the experiment are very different. Laboratory equipment mainly consists of a medium in which the elastic waves propagate (this may be a gas, liquid or solid). The main instruments include a source or emitter of mechanical impulses or harmonic vibration ("earthquake focus", "explosion", "vibrator") and a receiver of these vibrations ("seismograph"). The receiver is connected to an oscillograph. In the case of harmonic vibration the oscillograph may be replaced by another recorder that allows the amplitude to be measured and, if necessary, the phase angle of the vibration. Devices for synchronizing the actions of the focus and the oscillograph are important for impulses, as in some types of seismic prospecting.

The basic principles of modern techniques in this field were first stated by S. Ya. Sokolov in the USSR, when developing methods for ultrasonic non-destructive testing of metal items [S. Sokolov, D. Shraiber].

Our search for rational methods of simulating seismic waves lead to these methods and, especially, to the design of devices similar to ultrasonic impulsive non-detective testers. A description follows.

11.1.2. Excitation of Vibration

Let us consider the basic theoretical models for elastic vibration sources in continuous media, and the general properties of the natural foci to be simulated.

Theoretical models of the foci. The following main types of point foci are used during theoretical investigation of seismic wave phenomena: (1) concentrated force (simple force); (2) dipole with moment (double force with moment); (3) dipole without moment (double force without moment); (4) centre of rotation, with an action equivalent to three sources of type (2) directed along the coordinate axes; (5) expansion center—it is a similar combination of type-three sources. These sources can act over infinitesimal times (i.e. produce impulses, shocks) or over long time. In the latter case the problem of the curve form arises that represents the source action in time. Further, we may consider distributed sources of the same types.

We may also consider point and distributed sources composed of a combination of sources of these elementary types, or make them more complex by finite transitions similar to the transition from type 1 (with a feature of the first order) to types 3-5 (with features of the second order) [Lyav, Shebalin].

Let us consider what types of theoretical foci correspond to basic kinds of real sources of seismic waves studied in seismology and seismic prospecting.

Wave sources in seismic prospecting. An explosion in seismic prospecting seems similar, at least at first sight, to a center of expansion with a short-term action. However, this is just a first approximation.

In fact, as far as the direction is concerned, an explosion is simulated well by the center of expansion (of the first kind). However this is only if the explosion is produced by a ball charge with central initiation and occurs in a homogeneous medium, far from the boundaries of the homogeneity discontinuity. It approximates to a small charge in a deep borehole. However, explosions in the ground, in shallow wells, shotholes, or explosions close to the surface, on the surface and in the air are characterized by a sharply directed action which separates them from the approximation of an expansion center and brings them nearer to the case of a concentrated force. The latter corresponds best to a shock from a dropped weight, a method sometimes used in shallow depth seismic prospecting.

A concentrated focus is an ideal distributed source. Three main zones may be singled out around the charge in the case of shooting: (1) zone of failure, (2) zone of residual strain, and (3) zone of elastic strain. The space inside the boundary between the second and the third zone may naturally be considered an area of distributed sources of elastic seismic waves. Due to the source distribution within the area, effects of the second kind arise (see [A. Tsvetaev]) and for reception of vibration [Gamburtsev].

Let us now consider the duration of the action of wave sources used in seismic prospecting and the forms of mechanical impulses or shocks in the source area. It would be natural to determine the duration of the action and the form of the initial shock generating seismic elastic waves by motion seismograms. In other words, to examine the displacement of points in the same boundary, between the zone of residual strain surrounding the physical source, e.g., earthquake source or explosion in seismic prospecting, and the zone of approximately elastic strain, where these waves propagate, rather than by processes in the source itself (detonation of the explosive). The duration of the source process may be much shorter than the dominant periods of the recorded seismic vibration. But seismometry does not consider this.

Experience shows that an explosion in seismic prospecting (as well as apparently, fracture, shear or other dislocation of the rock at the earthquake source) usually generates vibration close to the source. The duration of these is much longer than the duration of the initial process that caused the vibration. In some cases this is due to long-term processes in the source area. The phenomena of repeated shocks, well known in seismic prospecting [Epinatieva], are examples. However, even during short-term source processes (explosion, shock, etc.), a short, or perhaps one-directional, shock in the source changes form as it passes through surrounding zones of failure and residual strain, and its duration increases.

The shock form (or more accurately seismograms of the particle motion in the source zone) changes both due to non-linear phenomena in the surrounding zones and under the influence of inhomogeneity zones in the medium. These may have already existed or were generated by the source. The change in form caused by the inhomogeneity would also take place if the medium remained ideally elastic. Besides, even in the ideal simplest case of a point source of expansion in a homogeneous ideally elastic medium, the form of the shock changes near the source (though in this ideal case the shock does not become longer) simply because the wave is not flat.

In general, the event proceeds as if the zones surrounding the source act like a filter suppressing both low and high frequencies. The suppression of low frequencies may be explained qualitatively from the classic theory of elasticity. Thus, the static strain caused by a local distortion in a continuous elastic medium (in the case of concentrated force imposed to the flat boundary of an elastic semi-space, for instance) is known to reduce with increasing distance from the local cause much faster than the dynamic strain, viz. elastic waves. The suppression of high frequencies is first of all associated with non-ideal elasticity: high frequency vibration is absorbed in the medium, particularly in the failure zone, faster than low frequencies. Given all this, the initial shock generating seismic waves *in situ* is conceived to be a vibration process. The duration of this is equal to a considerable portion of the dominant period of the observed seismic vibration, rather than a one-directional shock of finite energy and vanishingly short duration which is the way it is sometimes treated in theoretical investigations.

Wave sources during earthquakes. The same considerations remain valid to a great degree for wave sources of natural earthquakes, though there are some distinguishing features.

The main feature is that there is a directionality of forces in the sources of the most frequent tectonic earthquakes at least of "normal" focal depth (about 30-50 km). The source, as a first approximation, may be theoretically simulated as a dipole with moment [Vvedenskaya, Keilis Borok, Kawasumi] rather than the center of expansion or a concentrated force, as in the case of explosion. This corresponds to a discontinuity in the medium at the surface and the relative displacement of dissociated parts of the. medium (fault, shear). In the case of volcanic earthquakes which make minor contributions, the mechanical effect of the source may, as a first approximation, be considered as a joint action of the concentrated force and the center of rotation.

As far as the distribution of sources in earthquake foci is concerned, there is some similarity with wave sources in seismic prospecting. In most cases where seismic waves can be observed at long distances, the source zone is often regarded to be a point. In some cases, particularly when studying vibration near a source, this is wrong, and its finite size should be considered.

The situation with duration of the source process is similar. The duration may provisionally be assumed to be of the same order as the dominant periods of first waves on the seismograms (longitudinal waves). The form of vibration in the focus, like that at the source of elastic seismic waves as a whole, may be characterized by the type of curved motion of the particles in the medium [in the $(t, \vec{u})$ coordinate system, where t is time and $\vec{u} = (u, v, w)$ is displacement] in the boundary area where the strain in the medium near the focus is approximately elastic.

Wave emitters in simulation. In order to simulate elastic waves generated by an explosion or shock in seismic prospecting or by focal motion during earthquakes, a source or emitter of elastic waves is generally needed that gives mechanical pulses. The emitter should be small in dimension and at the same time provide powerful vibration; the pulses should be short. The directivity of the forces at the source is important.

Besides the main requirements, which are mostly intended to ensure a similarity between the sources in the model and *in situ*, there are a number of other requirements related to representing and controlling the rapid processes in the model. The main requirements are for good reproducibility (with respect to force, duration, and form) of the individual shocks and that these shocks follow each other frequently and at regular intervals.

The following processes and devices that yield such shocks are considered, namely elastic shocks, electromagnetic shock units, piezoelectric and magnetostrictional emitters, and electric sparks.

Elastic shock. The simplest ways of exciting a short burst of concentrated force is to induce an elastic shock, e.g., the impact of a steel ball against the solid elastic surface of the medium in which the waves are to be generated. Such an emitter has no mass continuously connected with the medium, a property unique to this method and a possible advantage. Its drawback is that it is difficult to reproduce frequent and accurate shocks. That is why this method of excitation has been rejected.

Electromagnetic shock unit. Experiments with such a unit at relatively low (for simulation purposes) frequencies, of the order of a thousand Herz, have shown that the method is quite applicable at such frequencies. It corresponds to the action of a force distributed over a surface. A complication is the additional mass in the elastic medium in which the generated waves propagate.

Piezoelectric emitters. Most experiments have been done with piezoelectric emitters. To obtain short mechanical shocks short peak-shaped voltage oscillations are delivered to the piezoelectric emitter generating by special radio set. The duration and form of each mechanical shock, as well as the duration of the intervals between the shocks (from 0.1 to 0.01 s) can be varied by altering the generating circuit. Increasing the duration of the shock itself is not difficult. Concerning the minimum duration, we have tried to reduce it as much as possible, mainly by reducing the width of the voltage pulse sent to the emitter. The width can be changed over broad range, and in some circuits can be reduced to about 5×10^{-6} s. The duration of a shock in the model may be assumed approximately equal or less than the dominant period T of the oscillations on seismograms during the simulation. The shape of the shock produced by the piezoelectric emitter was such that the vibration recording of a point in the medium near the source, when the direct wave was passed (in water), has the same shape as a rapidly attenuating quazi-sinusoid with just two or three significant extrema, at best.

These considerations show that such piezoemitters produce shocks that satisfactorily simulate an explosion in seismic prospecting or an earthquake source, at least so far as vibration duration and shape are concerned. Model seismograms using these emitters have the same general appearance as *in situ* ones. We have also employed piezoelectric emitters to excite stationary sinusoidal elastic vibration. The reception and recording of such vibration does not offer any difficulties during simulation, but interpreting the interference wave patterns is rather difficult, as it is *in situ* seismic prospecting [I. Kosminskaya].

There is a problem of the dimensions of piezoelectric emitters to be considered. The source being "concentrated" or distributed seems to depend upon the area of the active surface. The dimension of the source in the model, as *in situ*, is assessed in relation to the dominant wave length and to the distance from the source to the observation station. Thus, for frequencies of 10 to 100 kHz, for instance, and a velocity of wave propagation of 2 km/s, the wave length was 20 to 2 cm, and the diameters of the bases of the emitters in operation were usually less than 1 cm though sometimes more. For the given relations between the lengths, the problem of the source being distributed or of the point type has to be solved individually given the experimental conditions.

The difficulty of making a concentrated and at the same time rapid (pulsed) piezoemitter depends on its capacity. Smaller emitter dimensions, on the one hand, result in a reduced capacity, which causes a number of difficulties, when sensing and recording the excited mechanical vibration.

Magnetostrictional emitters. These emitters possess characteristics close to those of piezoelectric ones. However, they are, as a rule, more sophisticated in construction, are larger and operate at lower frequencies. That is why we did not use them.

Spark. The shortest, most intense and concentrated mechanical shock is caused by an electric spark. A spark obviously corresponds to a centre of expansion, or to the explosion of a concentrated charge. A spark emitter as a model of an explosion in seismic prospecting under conditions similar to explosion near the ground surface may also yield directivity properties.

Now let us compare the emitters and come to same conclusions about their use for simulating seismic waves. As shown, obtaining a similarity between a model and an *in situ* object by trying to get one-directional pulses from emitters is not sensible since they can be obtained by shocks. Thus, this sort of radiation is unsatisfactory, while its generation entails much difficulty. The other emitters produce vibrational shocks. The difficulty of getting one-directional pulses arises because phenomena in the zone close to the vibration source in the model mainly repeat *in situ* vibration phenomena. Besides, emitters possessing mass continuously connected to the media in which the waves are generated initiate additional vibration in the source zone due to elastic contact between the emitter mass and the medium. All these vibrations are, however, permissible if the vibration is in the form of a rapidly attenuating curve, which can certainly be achieved.

As far as the possibilities of concentrating a big capacity in a small volume and obtaining a rapidly acting source are concerned, piezoemitters cannot compete with the spark. But piezoemitters are easier to operate. Their principal advantage over the spark, besides directivity of action (if necessary), is that the pulse shape can be controlled, even to reproducing, if required, standing harmonic vibration of arbitrary frequency.

Besides expanding and contracting piezoelectric emitters there are emitters with torsion, bending or shear action. They can be used, in principle, to reproduce different types of sources, or to simulate the force combinations acting in earthquake sources. The other types of emitters considered above (electromagnetic and magnetostrictional) add little in principle to facilities of piezoelectric and spark emitters.

Thus, experiments and the above considerations indicate that piezoelectric emitters are the best sources of shaking, when simulating seismic waves for seismology and seismic prospecting. Spark emitters are also reasonable when simulating seismic prospecting explosions.

11.1.3. Vibration Sensing

The main tasks of representing vibration. When simulating seismic waves, as in field seismometry, two different tasks may be posed for devices reproducing the vibration of particles in the medium in the form of seismograms.

The first task is to obtain undistorted recordings of the real motion of the particles in the medium. Developing instruments to solve this very problem was, for a long time, the predominant idea in the seismological study of earthquakes and fractures. The idea may also seem reasonable to theorists who try to solve direct dynamic seismic problems by computation (a source and a medium are given, and the motion in the medium must be determined). In fact, the easiest way to obtain an experimental proof to a theoretical result is to compare the latter with a corresponding undistorted observed seismogram.

The second task is to get the most effective resolution of seismic recording instruments. Very strong (sometimes) linear distortions in frequency are purposefully introduced in the real motion in order to make the waves on the seismogram (e.g., longitudinal) as sharp as possible and to suppress other waves (e.g., surface ones). We chose the second tendency as the main one—achieving a high resolution of the instruments from the very beginning.

Seismograms with linear distortion, that can nevertheless, represent the motion in a resolved form, when the operation of the instruments is stable and their frequency response and amplification are controlled can allow direct relative measurements of the amplitudes and periods of various waves to be made during the simulation, as in the field. Also investigations of the shape peculiarities of their recordings and the alterations in these factors over time and space can be made.

It is often possible during a simulation to keep to this first step. In some cases the next step is to go on to absolute measurements, to a determination of the real motion or to establishing a definite correspondence to the latter. This may be necessary, e.g., to compare the simulation data with the results of the theoretical solutions of dynamic problems in seismology.

If the character of the linear distortion is known—the frequency response is removed—and if the transfer functions are also known, then the real motion may be obtained from the distorted recording by computation (e.g., the inverse to those described in Antokol'sky) or by appropriate electrical correction circuits.

Requirements of the sensor. The same principal requirements are naturally made with respect to the sensors of seismic vibration during simulation, as for *in situ* seismographs. The requirement that the frequency response is certain is the main one.

It seems reasonable to employ sensors with a broad bandwidth including, somewhere in the middle (approximately at 10 to 100 kHz), an "operative" frequency band that is chosen given the conditions of the experiment. Frequency filtering, selection of the vibration with the needed frequencies from all vibrations received by the seismograph or the correction of the frequency distortions introduced by the apparatus may be accomplished in the electrical circuit to which the seismograph is connected.

The significance of the seismograph's dimensions is specific to simulation. During *in situ* observations the dimensions of the seismographs are always smaller than the lengths of the waves being recorded: the dimensions of both are comparable during simulation. The same considerations as for emitters remain valid: i.e., it should be remembered that seismographs during simulation are, generally speaking, distributed sensors (each seismograph presents a "continuous group"). The question as to whether the apparatus may be approximated as a point sensor or not has to be solved individually depending upon concrete conditions of the experiment.

Amplifiers. A piezoelectric seismograph in our experiments was connected to a broad band amplifier either directly or through a regulated bandpass frequency filter. Amplifiers used in 1947-1948 were of the three-stage resistance type. Frequency filters were mainly employed to suppress low-frequency noise. Most seismograms were recorded without filters.

Recording devices. Voltage from the amplifier output was delivered to a cathode-ray oscillograph either directly or through an interface in order to watch simultaneously two processes on a single-gun display: from two seismographs located at different distances from the emitter, or from the seismographs and the timer.

Observation technique. Only one seismograph was usually installed along the "seismic profile", unlike in modern seismic prospecting *in situ* when there are many (12-24 is usual) seismographs in simultaneous operation along the profile. As a result of a series of observations a number of seismograms were obtained corresponding to a number of points in the "profile" in the model. The seismograms were compiled yielding a composite quite similar to the usual "tapes" in seismic prospecting and to seismograms from multichannel seismic field stations.

11.1.4. Main Features of the Pulsed Unit for Simulating Seismic Waves

General requirements. The requirements when simulating earthquakes and seismic prospecting was first of all that the processes in the model and *in situ* were similar.

As far as simulation for seismic prospecting purposes is concerned, the relations between the sizes *in situ* and in the model were determined by the following guidelines. Let the linear dimensions of the field of the medium under investigation be presented *in situ* by a length of 1 km (the length of the seismic profile or the depth of the boundary under study, etc.), and that of the field in the model be one meter. Let us also assume that the velocity of propagation of longitudinal elastic waves is the same *in situ* and in the model. From this we have:

	in situ	in the model
length	1 km	1 m
time	1 s	10^{-3} s
frequency	1 Hz	10^3 Hz

Therefore, if the dominant frequencies of the registered vibrations are 10 to 100 Hz *in situ*, then in the model they should be 10 to 100 kHz; if the duration of the whole process *in situ* is 1 to 2 s, then in the model it is 0.001 to 0.002 s; if the accuracy of timing on a seismogram *in situ* is 0.001 s, then in the model it should be 1 μs; if the rate of tape movement in a seismogram recorder *in situ* is 1 m/s, then the rate of the oscillograph sweep should be 1 km/s (of the same order as the velocity of a bullet).

When simulating seismological problems (earthquakes), the characteristic length of the regional seismic network is 10 to 1000 km *in situ*, and the vibration frequency is 1 to 0.01 Hz (i.e., for a vibration period of 1 s to 2 min). If velocities in the model remain the same as *in situ*, then the requirements on the model are the same as for seismic prospecting.

In other words the same unit can meet basic requirements for simulating seismic prospecting with a linear scale of approximately 1:1000, and for regional seismology with a scale of 1:10 000 to 1:1 000 000.

It follows that simulations of this type demand short duration periods and require sonic and ultrasonic frequencies. Such a rapid sweep means that a single process cannot be monitored by an oscilloscope.

Instead, the multiple repetition of the process was chosen and a stationary pattern was obtained on the oscilloscope something that can easily be observed and photographed.

General information about the unit. The block diagram of our unit for simulating seismic waves is given in Fig. 99. The generator unit *G* issues a short electric pulse that goes to the emitter *E* (earthquake source or explosion) and is converted to mechanical shaking. The latter propagates as elastic waves in the model *M* (medium). After reaching the sensor *R* (seismograph) the mechanical vibration is again converted into electricity and is fed through the amplifier *A* to the oscilloscope *O*. At the other side, the sweep voltage is synchronized with the pulse reaching the emitter and is sent from the generator *G* to the oscillograph *O*. Time marks (which appear as a sinusoid on the display *D*) are also input from the generator to the oscillograph, as well as a short pulse indicating the explosion (it is fixed on the display at the beginning of the seismogram).

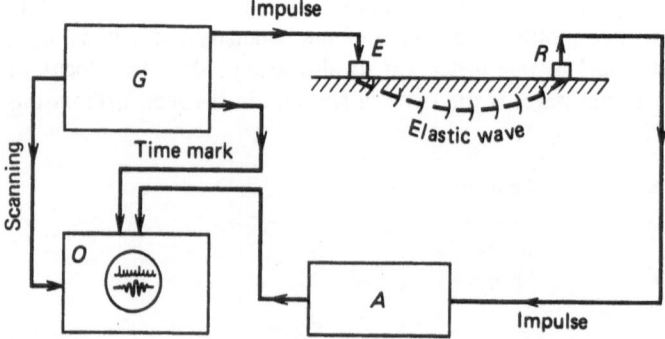

Fig. 99. Block diagram of a seismic curve simulation unit

11.2. Multiple Reflected and Transmitted Waves[*]

Multiple reflections in a multilayered medium during seismic exploration of structures has become a major topic in geophysical literature in recent years. However, most investigations take only one side of the problem, namely, when the number of interfaces with which multiple reflections are related is small. In the simplest case, this is a single strongly reflecting boundary at depth; the surface of the earth is the upper reflecting boundary (reflections from above can occur also from the base of the low velocity zone, etc.). In similar cases it is necessary to consider individual multiply reflected waves which in the main do not interfere with each other.

Multiple reflections in a medium containing many reflecting and refracting boundaries also constitute an important case. Then a complex interference pattern of many reflected or refracted waves from the many boundaries is recorded at the surface of the Earth. If the main reflecting boundary is overlain by a multi-layered medium, then the dispersed waves overlap the reflection from the main boundary and impede its identification. Such a multi-layered medium acts as a "turbid" medium. When the main boundary is deep in such a medium, the intensity of the reflected wave is weaker than the intensity of the background interference waves. As a result at a certain depth this boundary will be impossible to interpret in practice using the reflected wave method of prospecting. A similar topic is discussed below. Multiple reflections for dispersed interpreting waves were first considered theoretically in 1959 by Epinat'eva (also see Riznichenko).

The experimental results are compared with theoretical relations in Riznichenko.

11.2.1. Theoretical Considerations

Let us assume that a multi-layered medium consists of a homogeneous elastic half-space which contains thin layers with different elastic properties. The layers form a plane parallel periodic sequence. The distance between two neighboring layers is $h = const$. The source of short S-type seismic oscillations is applied to the free surface of a half-space OA. The seismic waves are recorded at point A on the same surface or on any of the thin layers (Fig. 100a).

The waves (created by the source) incident to any of the thin layers from above or below are reflected from it with an energy reflection coefficient r and are refracted when passed through the layer with an energy transmission coefficient p.

If a ray is normally incident to the boundaries and there is no absorption in the layers then we have a banalce of energy of the type $p + r = 1$. However, if the conditions are such that some of the energy q is retained in one way or other by a layer, then we have a balance $p + q + r + = 1$ and then $p + r < 1$. When waves are obliquely incident in a solid medium and only longitudinal or only transverse waves are analyzed, we deal with the second case in which the energy of the waves is transferred to waves of the opposite type.

The value q also includes absorption of the energy of the elastic waves within the layer. The absorbing waves in the surrounding medium can be included. Thus, q acts as an effective

---

[*] From [Riznichenko], G. Shamina, co-author.

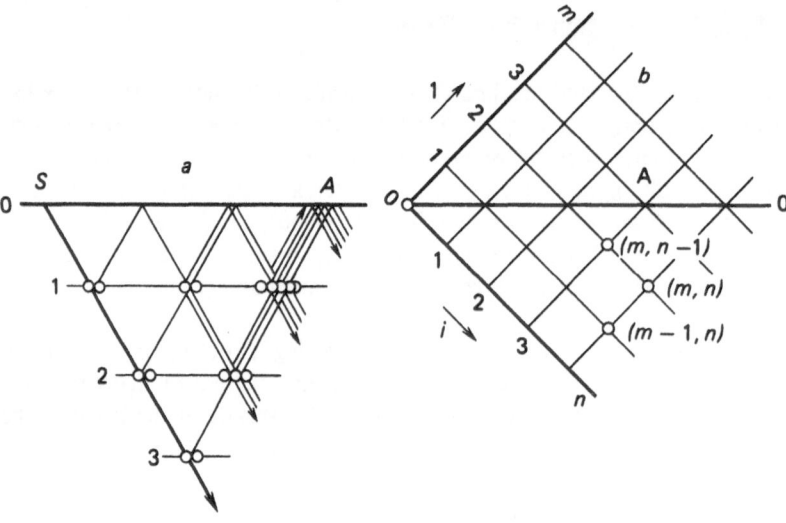

Fig. 100. Ray diagram in a multilayer medium

absortion coefficient. The coefficients p, q and r generally depend on the frequency, i.e. they can be expressed by the frequency characteristics of transmission, absorption, and reflection.

Total reflection from the free surface OA (surface of the Earth) changes compression into decomposition. However, if only the energy of the wave is considered, this fact is not important because the energy always remains positive. This analysis is concerned with energy, but not the displacements considering their signs. This approach is needed because in natural geological media the scattering layers in the overlying medium do not occur at equal intervals. Multiple reflections in such a medium, even if the deviations of layers from regular spacing are small, can change the conditions of interference considerably, e.g. in-phase summation is replaced by out of phase summation. Therefore, it seems more logical to apply averaging and smoothing in this energy analysis.

A reflected ray from the first layer l arrives at the recording point A on the earth's surface, it is followed by the ray reflected from the second layer and simultaneously by a double reflection from layer l (Fig. 100a) etc. With time the multiplicity and total number of the dispersed multiply reflected-refracted waves, arriving simultaneously with the reflection from the deepest horizon for the given reflection time layer, increases. The layer number can serve as a time measure.

The problem is to determine the number of homogeneous (for instance longitudinal waves) dispersed reflected-refracted waves arriving at recording point A simultaneously with the primary reflection from the deepest layer, and the intensity of the primary and the total intensity of all the dispersed waves. We assume that the dispersed waves are plane waves.

For calculational convenience, we will consider, in addition to the actual lower half-plane (Fig. 100) the upper mirror half-plane which may be represented as the path of the rays after reflection from the Earth's surface OA (Fig. 100b). To obtain the true distribution of the rays in the lower half-plane, the drawing is folded along OA (Fig. 100a). In such a representation total reflection at OA will appear diagramatically as a transition of a seismic ray (without changing its direction) through the OA line. The diagram in Fig. 100b will be used both for the kinematic, and energy analysis.

Two more assumptions will be introduced for simplicity. We firstly assume that OA in Fig. 100b represents the same thin layer as the other horizons. This is equivalent to the assumption that the incidence of the wave at OA develops two waves: one reflected (r) and one refracted (p). Secondly, for the symmetry of the diagram in Fig. 100b we assume that in addition to the wave directed downward from the source S, a symmetrical wave is directed upward into the mirrored half-plane. This will lead to a simple duplication of all the waves. For point A, this can be interpreted as the known seismic duplication of the amplitude of the wave incident to the free surface.

Let us show that under these assumptions and under the condition that $p + r = 1$, the phenomenon of total reflection from a free boundary OA is automatically considered. Actually, we assume that at the boundary OA a wave incident from below has an intensity E. According to the first assumption this wave generates two waves: one of these travels into the upper half-plane with an intensity E_p and the other wave is reflected into the lower half-plane with an intensity E_r. According to the second assumption, due to the symmetrical repetition of all waves, it must be assumed simultaneously that at the boundary OA the incident wave from above has the same intensity E. In the lower half-plane it produces wave E_p, and wave E_r in the upper half-plane. As the results in the lower and upper half-planes develop two equal summation reflected-transmitted waves with intensities $E(p + r) = E$, which corresponds, as stated, to total reflection.

If $p + r < 1$, i.e. if the effective absorption q is considered with the layer, then the same absorption q will also occur for our assumptions for the free surface OA. This version of the problem is not unreasonable from a geophysical point of view. Under true conditions, seismic waves incident from below at the Earth's surface (or at the lower surface of the low velocity zone, etc.) and reflected back from the Earth's surface certainly lose some of their energy due to non-ideal elastic near-surface deposits, and also due to the dispersion caused by the increased nonhomogeneity of these deposits. However since the absorption parameter q for each of the many thin dispersing layers can be considered to be small in comparison to p, the additional absorption at the free surface OA cannot substantially change the general result for the energy of the summed dispersed waves. Therefore for $q \ll p$, we again return, however, this time only approximately, to the case of total reflection from a free surface.

To determine the total intensity of dispersed waves for any point M of a real medium at any of the analyzed boundaries, including the surface OA, it is necessary to compute the total intensity of all such waves in a medium (Fig. 100b) consisting of two half-spaces, and then divide the result by two.

Computation. Let us use 1, 2, ..., m and 1, 2, ..., n to label the lines of the grid in Fig. 100b. The condition $m = n = 0$ corresponds to the earth's surface OA.

For any node $M(m, n)$ of the grid of Fig. 100b two waves arrive simultaneously: one from below, from $(m - 1, n)$ labelled in real units and one from above, from a point $(m, n - 1)$ and labelled in imaginary units. Then both waves arriving at (m, n) can be written symbolically as:

$$(m, n) = [m - 1, n] + i[m, n - 1], \tag{11.1}$$

where the parentheses symbolize the arriving waves, and the brackets the departing waves emitted by this point. For each segment of the ray between the nodes of the grid, a relationship exists between both waves; viz.

$$\text{Re}\,(m,\ n) = \text{Re}\,[m-1,\ n]$$
$$\text{Im}\,(m,\ n) = \text{Im}\,[m,\ n-1], \tag{11.2}$$

where the Re is the real and Im the imaginary part of the expressions as in (11.1).

Due to the mirror symmetry of the pattern, (Fig. 100b) this can be written:

$$\text{Im}\,[m,\ n] = \text{Re}\,[n,\ m], \qquad \text{Im}\,(m,\ n) = R(n,\ m)$$
$$\text{Re}\,[m,\ n] = \text{Im}\,[n,\ m], \qquad \text{Re}\,(m,\ n) = \text{Im}\,(n,\ m) \tag{11.3}$$

The transmission of a wave through a layer is characterized by a transmission coefficient p and reflection by a coefficient r. We can write the relation between the waves (m, n), arriving at the point M and the waves $[m, n]$ departing from the same point in the form:

$$[m,\ n] = p\,\text{Re}\,(m,\ n) + r\,\text{Im}\,(m,\ n) + i\{r\,\text{Re}\,(m,\ n) + p\,\text{Im}\,(m,\ n\}. \tag{11.4}$$

Finally, let us find the relationship between waves (m, n) arriving at the point M and the waves which arrive at points $(m, n-1)$ and $(m-1, n)$.

Following Fig. 100b we find:

$$(m,\ n) = p\,\text{Re}\,(m-1,\ n) + r\,\text{Im}\,(m-1,\ n)$$
$$+ i\{r\,\text{Re}\,(m,\ n-1) + p\,\text{Im}\,(m,\ n-1)\}. \tag{11.5}$$

This formula represents a recurrence relationship for waves at the nodes of our computational diagram with increasing ordinal numbers m and n. This applies for any node in Fig. 100b. The degree of medium transparency can be characterized by the ratio of the intensity of the useful primary reflected wave to the intensity of the interference waves, the dispersed radiation arriving at the recording point simultaneously. This relationship is not only determined by the parameters of the medium itself, but also by the incidence angle i and frequency f of the oscillations in the seismic pulses because $p = p(f)$ and $r = r(f)$ (we assume that v does not depend on f). The influence of these factors on the transmission and reflection of seismic energy from thin layers on the surrounding medium with other mechanical properties was studied by ultrasonic modeling.

11.2.2. Results

Theoretical computations and experimental investigations indicated that with sufficient complexity in the composition of the medium overlying the main reflecting boundary, the dispersed waves become more intensive than the main reflection, and a condition is created which prevents separation of this reflection. This is similar to the dispersion of light in a turbid medium: with sufficient dispersing ability and thickness of the turbid layer the contrast object becomes almost indistinguishable.

Under conditions of seismic prospecting with the reflected wave method an analysis of the relative intensity of the "useful" reflection and that of the masking reflections dispersed in the overlying medium is of greatest interest where this medium is characterized by low velocity and density differentiation, i.e., when it is not very "turbid". However, in our model, the overlying medium was characterized by considerable differentiation. The negative side of such a selection makes it impossible to make a direct comparison of the results, using the present modeling, with recordings in nature under similar conditions. Modeling under conditions in which the overlying medium has a sharp velocity and density differentiation has the advantage simplifying the simpler means of determining basic trends and characteristics of this phenomenon: to obtain a noticeable effect for this, it is sufficient to have a medium with fewer layers. In addition, it is natural to assume that the dispersing medium with sharp differentiation and fewer layers will behave in some degree similar to the medium with weak differentiation but with a large number of layers [Riznichenko].

11.3. Elastic Waves with Generalized Velocity in Two-dimensional Bimorphic Models[*]

Both in the USSR and abroad [Riznichenko, C. Horton, P. Narvarte] solid-liquid or liquid three-dimensional models of media have been used for simulations of seismic waves by supersonic impulses. Subsequently, a two-dimensional model [Riznichenko] using thin rigid plate (compared to the predominant wavelength) was tried out. In recent years, both in the USA [Oliver] and in the USSR. [Riznichenko], this type of model has been frequently applied.

To obtain a two-dimensional models of medium, plates made of metal, thermosetting plastic or other materials are joined side by side. This makes it fairly easy to obtain a medium with constant parameters (elasticity, density) in every layer.

At the same time, it is necessary for a number of important geophysical problems, to model a medium with parameters that change smoothly in space. Transition layers and waveguide channels with lowered propagation velocity of the elastic waves are particular examples, as is the interesting problem of a seismic waveguide in the upper part of the subcrustal layer of the Earth.

To model such problems, J. Oliver suggested "layered" two-dimensional models in which sheets of variable thickness are glued together in the same way that plywood is made. In contrast to the usual monomorphic two-dimensional models of layered media, in which every layer is represented by a homogeneous sheet, we consider models where a plate consists of two layers glued together as bimorphic and where the plates consist of several layers as polymorphic.

After conducting his first experimental work on a bimorphic model, Oliver observed body waves, mainly longitudinal ones, on models consisting of two sheets which get thicker in opposite directions. The experiments with bimorphic models of uniform sheets are described in [J. Oliver], where a formula for the generalized velocity of waves in a bimorphic model is deduced without derivation. However, in [Oliver], neither theoretical problems, nor experimental studies of the formation and oscillation intensity were discussed. F. Press observed flexural waves in bimorphic and monomorphic variable-thickness models attempting to model Rayleigh surface waves with a variable velocity. From the physical point of view, why he should do this is not entirely clear to us, if only because in the simplest case of a uniform elastic

---

[*] [Riznichenko, et al]. O. G. Shamina, P. V. Khanutina — co-authors.

half-space Rayleigh surface waves do not possess any dispersion, whereas flexural waves in a homogeneous plate—unbounded or bounded—strong negative dispersion is typical.

In order to control the elastic parameters of the materials being used in models particularly of media with smoothly changing properties, other methods were selected in the USSR. L. Rykunov realized a two-dimensional waveguide with soft boundaries, by utilizing the dependence of the propagation velocities of elastic waves in thermosetting plastics on temperature. Authors [Gilbershtein, et al., Ivakin] developed perforated and other structurally non-uniform models in which the size or spacing of the holes or structural elements change smoothly in the plane of the model.

In this section an attempt has been made to follow the approach in [F. Kokesh]. Elongated longitudinal, transverse and surface waves in bimorphic (or polymorphic in general) models are investigated theoretically and experimentally. This is to determine the practical possibilities and limitations of bimorphic or polymorphic models for seismology and seismic exploration. We restrict ourselves to oscillations whose displacement vector lies in the plane of a two-dimensional model. In contrast to flexural waves, these oscillations are analogous to longitudinal, transverse SV and the Rayleigh surface waves in a three-dimensional medium. The two-dimensional seismic models were designed to study these.

11.3.1. Theory

The elastic properties of quasi-anisotropic media consisting of isotropic layers with different properties were developed for a three-dimensional case in [Riznichenko]; in the same article other investigations on the same subject were listed. By applying the method described in [Riznichenko] let us do the analogous calculation for a two-dimensional case.

Assuming that the layers comprising a bimorphic plate are thin compared to the length of the elastic waves, we may assume that the stress distribution in an element of a plate is uniform within each uniform part. It is possible to calculate its elasticity K for longitudinal (K_P) and transverse (K_S) waves by solving the corresponding static problems.

We shall define the elasticity K to be the quantity which depends on the elastic parameters of the material of the layers and on their thickness. The latter quantity is a part of the following expression of the velocity v of the propagation of elastic waves:

$$v = \sqrt{\frac{K}{p}} \tag{11.6}$$

where p is the density. For longitudinal waves in a thin homogeneous plate, we have (see [Landau, et al., Riznichenko, Timoshenko]):

$$K_P = \frac{E}{1 - \sigma_d^2} \tag{11.7}$$

where E is Young's modulus and σ_d is Poisson's ratio for the plate material. This quantity differs from the elasticity of a solid $K = E(1 - \sigma)/(1 + \sigma)(1 - 2\sigma)$. For the transverse waves the elasticity of the plate is the same as for a solid [Landau] i.e.

$$K_S = \frac{E}{2(1 + \sigma)} = G, \tag{11.8}$$

where G is the shear modulus.

The problem is to calculate the corresponding elasticities K_P and K_S and velocities v_P and v_S for the bimorphic plate. This we do by examining deformations which originate in the plate during the propagation therein of longitudinal and transverse waves. We shall neglect flexural deformations of the plate.

The initial equations—Hooke's law—for isotropic bodies, form the basis of the procedure, viz.

$$e_{xx} = \frac{1}{E} X_x - \frac{\sigma}{E} (Y_y + Z_z)$$

$$e_{yy} = \frac{1}{E} Y_y = -\frac{\sigma}{E} (Z_z + X_x)$$

$$e_{zz} = \frac{1}{E} Z_z - \frac{\sigma}{E} (X_x + Y_y) \tag{11.9}$$

$$e_{yz} = \frac{1}{G} Y_{zi}$$

$$e_{xz} = \frac{1}{G} X_{zi}$$

$$e_{xy} = \frac{1}{G} X_y, \tag{11.10}$$

(here $Y_z = Z_y$, $Y_x - X_y$, $Z_x = X_z$), and the end conditions on the faces of the elementary parallelepiped cutout from the bimorphic layer when the longitudinal (Fig. 101a) or transverse (Fig. 101b) waves propagate along it. Equations (11.9) and (11.10) are true for each part of the elementary parallelepiped for each type of wave, but the boundary conditions at the ends, in each case, are different.

The mean density ϱ of the bimorphic plate remains, of course, the same in both cases and is determined from

$$(h_1 + h_2)\varrho = h_1 \varrho_1 + h_2 \varrho_2,$$

where h_1 and h_2 are the thicknesses of the first and second layers; and ϱ_1 and ϱ_2 are their densities.

$$\varrho = \frac{h_1 \varrho_1 + h_2 \varrho_2}{h_1 + h_2} \tag{11.11}$$

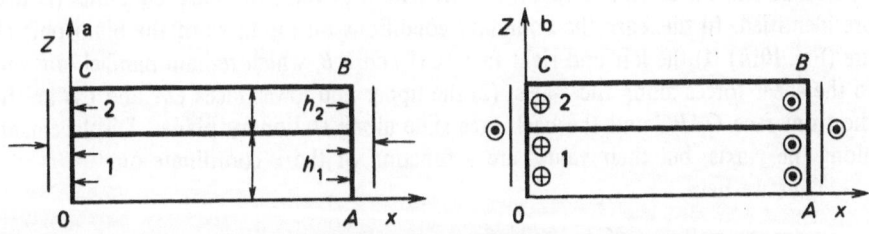

Fig. 101. Stress distribution in the elementary parallelepiped:

a—in the case of a longitudinal wave; b—in the case of a transverse wave

Longitudinal waves. In this case it is sufficient to examine (11.9) only, since there is no shear; (11.10) are identities. The boundary conditions in this case are (see Fig. 101a): (1) the elementary bimorphic parallelepiped is either compressed or extended between the parallel surfaces OC and AB at right and left; (2) the upper and lower faces OA and CB of the element are free; (3) the front face $OABC$ and the riverse face of the element slide along stationary surfaces.

Under these conditions, and keeping in mind condition (1), we obtain:

$$e_{xx1} = e_{xx2} = e_{xx}. \tag{11.12}$$

The mean stress X along the x-axis is determined from the equality:

$$(h_1 + h_2)X = h_1 X_{x1} + h_2 X_{x2}, \tag{11.13}$$

(here and below the values corresponding to the first and second parts of the bimorphic element are labelled by the subscripts 1 and 2). From conditions (11.7) we have:

$$K_P = \frac{X}{e_{xx}}, \tag{11.14}$$

which produces the mean elasticity of the bimorphic element for longitudinal waves.

From (11.9), taking (11.7), (11.8) into consideration, we obtain the elasticity of the bimorphic element for extension and compression, i.e. for longitudinal waves, i.e.,

$$K_P = \frac{X \, h_1 E_1/(1 - \sigma_1^2) + h_2 E_2/(1 - \sigma_2^2)}{e_{xx}}$$

$$= \frac{h_1 K_{P1} + h_2 K_{P2}}{h_1 + h_2}. \tag{11.15}$$

In this case K_{P1} and K_{P2} are expressed in terms of E_1, σ_1 and E_2, σ_2, which are analogous to (11.7). Lastly, in a manner analogous to (11.6), we find from (11.11) and (11.15) the propagation velocity v_P of long longitudinal waves in a thin bimorphic plate; i.e.

$$v_P = \sqrt{\frac{K_P}{\varrho}} = \sqrt{\frac{h_1 E_1 (1 - \sigma_1^2) + h_2 E_2 (1 - \sigma_2^2)}{h_1 \varrho_1 + h_2 \varrho_2}}. \tag{11.16}$$

This quantity depends solely on the ratio h_1/h_2 and not on the absolute values of the layer thicknesses.

Transverse waves. In this case it is sufficient to examine the shear equalities (11.10); (11.9) are identified. In this case the boundary conditions on the faces of the bimorphic element are (Fig. 101b) (1) the left and right faces OC and AB, which remain parallel, are subjected to the shear forces along the z-axis; (2) the upper and lower faces OA and CB are free; (3) the front face $OABC$ and the back face slide along stationary planes. Displacements exist along the y-axis, but their values are a function of the x coordinate only.

Thus we have

$$e_{xz} = e_{yz} = 0, \qquad X_z = Y_z = 0.$$

Only the last equation of (11.10) yields a non-zero quantity, thus, taking into consideration that $X_y = Y_x$, we have for the bimorphic element as a whole and for each of its parts;

$$e_{xy} = \frac{1}{G} Y_x, \qquad e_{xy1} = \frac{1}{G_1} Y_{x1}, \qquad e_{xy2} = \frac{1}{G_2} Y_{x2}, \tag{11.17}$$

while the analogue to (11.2) is:

$$e_{xy1} = e_{xy2} = e_{xy}. \tag{11.18}$$

The mean tangential stress Y_x (which is related to force directed in Fig. 101b along the y-axis) will be determined in a manner similar to (11.13) from the formula:

$$(h_1 + h_2)Y_x = h_1 Y_{x1} + h_2 Y_{x2}. \tag{11.19}$$

From (11.17)-(11.19) we find the elasticity of the bimorphic element for the shear, that is, for the transverse waves

$$K_S = G = \frac{h_1 G_1 + h_2 G_2}{h_1 + h_2} \tag{11.20}$$

and in manner analogous to (11.6) the velocity v_S of the transverse waves in a thin bimorphic plate is found using:

$$v_S = \sqrt{\frac{K_S}{\varrho}} = \sqrt{\frac{h_1 G_1 + h_2 G_2}{h_1 \varrho_1 + h_2 \varrho_2}}. \tag{11.21}$$

By comparing (11.16) and (11.21) for the propagation velocities of longitudinal and transverse waves in a bimorphic plate, we note that they have the same structure and may be written in a general form:

$$v_2 = \sqrt{\frac{h_1 K_1 + h_2 K_2}{h_1 \varrho_1 + h_2 \varrho_2}} = \sqrt{\frac{h_1 \varrho_1 v_1^2 + h_2 \varrho_2 v_2^2}{h_1 \varrho_1 + h_2 \varrho_2}}. \tag{11.22}$$

When applying (11.22) one should keep in mind that in these formulas all the velocities and elasticities should correspond to the same type of values, viz. either longitudinal or transverse.

It is also interesting to note that the formulae for the extension and compression elasticities (11.15), as well as for the shear elasticity (11.20) for a bimorphic plate are similar in character to those for the elasticity of a parallel arrangement of the two "concentrated elasticities" (see [Riznichenko]). For a three-dimensional layered, two-component medium of a "bimorphic solid", similar relationships are retained for the transverse waves as for the plate, whereas for the longitudinal waves more complex, and essentially different, relationships [Riznichenko] are obtained. The approximate formulae for longitudinal waves in a bimorphic solid have a structure which is analogous to (11.22).

Equations like (11.15) and (11.20) for the elasticities in a bimorphic plate, permit generalization to polymorphic plates consisting of n layers, and for a plate whose properties depend on the z-coordinate (Fig. 101): $K = K(z)$, $\varrho = \varrho(z)$.

Analogous generalizations can also be from (11.22) for the propagation velocities of longitudinal and transverse waves. For example, for an n-layered plate we have

$$v^2 = \sum_{i=1}^{n} h_i \varrho_i v_i^2 / \sum_{i=1}^{n} h_i \varrho_i,$$

for a heterogeneous layered plate, where $\varrho = \varrho(z)$ and $v = v(z)$ we will obtain:

$$v^2 = \int_0^h \varrho v^2 \, dz / \int_0^h \varrho \, dz. \tag{11.23}$$

Surface waves. In a thin plate with a free boundary—a "two-dimensional half-space"—Rayleigh surface waves are similar to the Rayleigh surface waves in a three-dimensional half-space.

For the three-dimensional case it has been demonstrated [468] that the propagation velocity $v = v_R$ of the Rayleigh waves is related to the velocities v_P and v_S by the equation

$$\left[\left(\frac{v_R}{v_S} \right)^2 - 2 \right]^4 = 16 \left[1 - \left(\frac{v_R}{v_{Pl}} \right)^2 \right] \left[1 - \left(\frac{v_R}{v_S} \right)^2 \right]. \tag{11.24}$$

The difference between the three-dimensional and two-dimensional cases is only the difference in the velocities v for a solid and a plate and Eq. (11.24) holds true. We have only

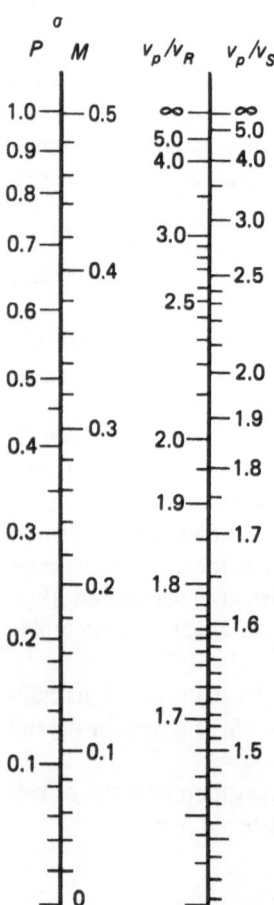

Fig. 102. Nomogram for a Poisson coefficient calculation in relation to v_P/v_R or v_P/v_S velocities

to assume that $v_P = v_{P_{Pl}}$ and $V_R = V_{R_{Pl}}$. If V_P and v_S are known, then from (11.24) we can determine $V_{R_{pl}}$ in the functions. It is probably preserved in a complex, i.e. biomorphic, polymorphic and in a general layered nonhomogeneous plate.

Equation (11.24) contains v_R^2 to the fourth power and its solution for this value is in general cumbersome. Therefore, following [Landau, et al., Timoshenko, Knopoff, Lyav], we restrict ourselves to the numerical solution of the equation and present it in the form of a nomogram which can be used in practice (Fig. 102).

It measures v_P/v_R or v_P/v_S velocities and finds the Poisson coefficient σ. This nomogram is useful for massif M and for plate $Pl\sigma$; the scales for these cases are different: scale M corresponds to a massif and Pl to plate.

A comparison of the σ scales for M and Pl allows us to assess the "effective" ratio σ for a three-dimensional seismic problem (solid) for its two-dimensional model (plate). For example, in order to model a solid with an effective $\sigma = 0.25$ (Fig. 102, scale M) it is necessary to select a plate made of a material with Poisson ratio of $\sigma = 0.33$ (Fig. 102, scale Pl).

The foregoing theory is true for waves longer than wavelength λ with the thickness of the plate h. This theory does not produce any value of h/λ, required if it is to be applied. Such an investigation was carried out experimentally using supersonic impulses [456-462].

11.3.2. Discussion of the Results

The experimental investigation of the formation of generalized waves in two-dimensional bimorphic models, together with physical consideration, enable us to state the following physical pattern. This process represents the development and complication of interference waves in partially bounded bodies, such as bars and plates [O. Silaeva, O. Shamina]. If we only consider sufficiently long-wave parts of all oscillations and neglect the oscillations of small intensity, the wave pattern is outwardly strongly simplified.

The combination of the subsequent twofold interference oscillations creates some outwardly simple long wavelength trains of longitudinal P, transverse S and, under some specific conditions, Rayleigh surface R waves of a new generalized character.

The formation process of the waves depends on frequency and, correspondingly, on the wavelength (as compared to the thickness of the model layers), on the path traveled by the waves, and on the contrast between the layers comprising the complex plate: the lower the frequency or greater the wavelength, the longer the wave path, and the less the contrast in the properties of the layers, the sooner generalized oscillations are established.

Sufficiently long generalized waves are well correlated by the phases. Their propagation velocities—generalized velocities—are determined by averaged elastic properties and the mean density of the complex plate as a whole. Their measured values agree adequately with the long-wave theory adduced above. Let us note that a comparison of the observed and calculated generalized velocities of the waves (but only longitudinal ones) was carried out in [Oliver, et al] and produced good results. In a homogeneous bimorphic plate, the velocities of generalized P, S and R waves, as in a homogeneous monomorphic two- or three-dimensional space (for R waves in a half space) did not noticeably disperse, if the elastic properties of the model's materials do not differ greatly from one another. Model results are compared with seismological data in [Riznichenko].

12. Seismoacoustic Investigations of the Stress State in Rocks

Penetration of geophysical methods into mining science is one manifestation of a general tendency in modern science and technology. This is the implementation of physical and mathematical methods of measurement, the processing of observational data and their interpretation. This implementation sometimes involves fundamental problems in the relevant field. Rock pressure is one such problem in mining.

Rock pressure, i.e., the stress state of the rocks in the mining region that most vividly shows itself as rock explosions in deep mines, is of much interest to miners and geophysicists.

Miners need to understand the nature of rock pressure in order to predict and to control its effects as much as possible. The practical aim of studying rock pressure is to ensure safe working conditions for people underground, make mining more efficient while maintaining safety and economic expendiency, extracting the maximum quantity of minerals underground without hindering operations in mines.

Geophysicists are interested in rock pressure because of the similarity between deformation and failure of rock in mines due to human interference and solid flow and rock fracturing in the Earth's crust due to natural tectonic forces, the most vivid of which are earthquakes. Rock bursts are similar to earthquakes, as, by nature, both are rock failure under stress. Geophysicists have a rare opportunity in mines of conducting observations very close to the source of fracture. The results of those observations may help in developing a theory of earthquake source and seismic regime, as well as finding ways of predicting earthquakes.

Miners have repeatedly turned to geophysicists for help in studying rock pressure. Mine survey methods, developed from ancient times, make assessments of the stress state in the rock in a massive (e.g., in a shaft pillar) on the basis of strain measurements near the walls of openings but they do not always ensure reliable results. The stress state near the walls, where it is measured, is severely disturbed. It would be ideal to be able to determine the stress in the rock mass without disturbing its continuity. Geophysical methods based on non-invasive investigations of the Earth's interior give new opportunities to asses stress.

Contemporary mine survey methods for studying rock pressure are based on strain measurements in mines. They yield a great deal of information about the mechanical processes near the openings but encounter considerable difficulty when investigating processes that take place some distance from the openings at the stope. Known mechanical and optical methods of simulating rock pressure yield important general behavior patterns for the pressure at the stope, but do not solve the problem as a whole. No laboratory experiments can substitute for observations *in situ*, that is in mines where the conditions are much more complicated. However, the processes in the rock mass at the stope are the main factors in catastrophic results of rock pressure.

For this reason it was decided to complement existing mine survey methods for investigating rock pressure in mines with other geophysical methods. These register rock pressure behavior in the massive and the changes in the physical parameters of coal and the enclosing rock massive or other information about their state, in such a way that, the measuring devices and processes do not distort the natural stress state of the rock. It was decided to develop and test two geophysical, or to be more exact, seismo-acoustic methods: (a) the mine pulse seismic method based on the artificial excitation of elastic waves, and (b) mine acoustic method based on the basis of natural elastic waves.

Seismic and especially acoustic observations in mines may also be aimed at assessing how hazardous a potential failure, rock burst or explosion, will be.

Both methods—seismic and acoustic in principle, yield opportunities to estimate the rock pressure using phenomena that occur both in the vicinity of the wall openings and deep in the rock mass. In this they differ considerably from common mine survey methods and may be a necessary complement to them.

12.1. Rock Pressure and Rock Burst[*]

In studying rock pressure and rock bursts both in mining and as part of tectonic motion and earthquakes in geophysics, we must consider rocks both on the macroscopic scale and as real imperfectly elastic, structurally inhomogeneous bodies with ealsticity, flow (plasticity), and collapse as their main mechanical properties[**].

Because they are continuously affected by external forces, rocks are always under stress. Gravity is the main continuously acting force. However, other more variable forces, either natural in origin or resulting from human interference, act sometimes.

Stress in a body is associated with a certain amount of elastic energy. Rocks forced from one stress state into another by a variable force accumulate or release some of this energy, i.e., they perform work. The work is related to motion of the particles of the medium. The motion may be both about the equilibrium state, as in elastic waves, or be irreversible. The latter may proceed slowly and without manifesting any evident discontinuity such as in a flow (tectonic flow, slow vibrations, plicative dislocations in geology and geophysics; slow rock displacement in mining) or proceed rapidly, with marked discontinuities. Under natural conditions a discontinuity exhibits itself as a land collapse, slides in mountain, failures in caves, failures and explosions in mines, or earthquakes.

We shall define rock pressure to be the stress state of the rocks in a mine. The rock pressure caused by human activities does not differ in physical nature from the stress state generated by natural factors, e.g., during tectonic motion of the Earth's crust and upper mantle. It seems natural, therefore, to employ the same approach in investigating stress-induced plastic flow and faulting in the upper strata of the Earth both in geophysics and geology.

A rock burst is vivid manifestation of rock pressure. Let us cite a description of the rock burst, as treated in mining science, from S. G. Avershin. Note that Avershin defined a rock burst in relation to coal mines, but in fact, this definition remains valid for other mines (ore mines, tunnels etc.).

---

[*] From [Riznichenko].

[**] The term 'flow' (or plasticity in a broad meaning) is used here for various types of continuous irreversible deformations of solid bodies: viscous flow and creep, ideal plasticity, etc. In mining and geophysical literature the terms "quasiplasticity" and "pseudoplasticity" are sometimes employed for the same idea. We shall further term a flow "quasiplastic" if it is accompanied by sufficient discontinuities and fracturing. Geophysics has sometimes to consider rock healing — ability of the rocks to "glue" fractures and restore their continuity and strength with time. — *Author's note.*

According to Avershin, "a rock burst is a practically instantaneous outburst of coal or rock, or both, resulting from the stress state exceeding limit. It is similar to the detonation of a large amount of explosive in coal or rock. A rock burst is accompanied by the sound of a blast and shaking of the surrounding rock, which is felt for several kilometers on the surface. Supports in the mine workings are damaged or destroyed, the working is filled with exploded coal and rocks of various sizes, coal dust, and in some cases large quantities of coal damp. The rocks, weakened by the explosion, lose their stability and fail, thus filling the working. The rock aggravates pressure over the coal pillar and the supports. After the burst the pressure growth becomes normal. The ground swells and sometimes coal pillars are squeezed out and fill up the whole open volume. An air wave—air burst—often occurs after a rock burst. Bursts can occur over distances from several to several hundreds of meters, both in stopes and preparatory shafts. The remaining coal pillar and the surrounding rock are fractured, cleavage cracks open, and the strength of the rock and coal in this region of the deposit is reduced.

The same phenomenon actually takes place if the equilibrium is disturbed due to tectonic forces. The material in the interior begins to move from areas of high pressure (at the same surface level with respect to gravity) to areas of low pressure.

Close to a mine working the main pressure difference responsible for movement equals the difference between the hydrostatic pressure in the rock at the depth for the undisturbed deposit and atmospheric pressure, which is established through the working. The presence of the working itself affects the approximate homogeneity of the stress in the volume, which had been established before the working was opened. The stresses concentrate around the shaft and are higher than when there was no working. As a result the restoration of isostatic equilibrium near the working must be more intense than it is elsewhere.

The main point here is that the stresses become very different from the hydrostatic stresses rather than that they increase near the working. Compression stresses vary in different directions due to nonzero tangential, and shear stresses and strains. Tensile stresses and strains may also arise. Compressive stresses do not threaten the rock's integrity, but shear and tensile stresses do.

Rock bursts are the most serious natural phenomena in mines and threaten miners' lives and disrupt the production process."

Let us now describe the mechanism of rock bursts in terms of geophysics.

We proceed from the assumption that the Earth as a whole is approximately isostatic, especially the material in the crust and upper mantle. Approximate isostasy is understood here in a broad sense, i.e., as a general tendency for the quasiplastic material of the Earth to move to equilibrium in the field of gravity. While the material in small volumes is macroscopically and structurally inhomogeneous, in large volumes it is assumed to have a volume density approximately depending upon depth alone. Ideal isostasy in this sense would mean that the density is entirely depended on depth, while a stable isostatic state would be ensured if the density did not increase with depth.

The rock is mass without faults and in the absence of strong modern tectonic movement, beginning from just under the Earth's surface the rock is close to hydrostatic equilibrium, i.e., uniform omnilateral compression. This is caused by the weight of the upper layers of rock. If a state is disturbed in the process of geological history, approximate equilibrium can be restored after a long time as a result of plastic or quasiplastic flow. By adjusting themselves to hydrostatic compression, the rocks obtain stable parameters, in particular, volume density, elasticity, and a store of elastic energy. Under these conditions there is no cause for fast movement by the material detectable within a human life.

Mining workings disturb the stable isostasy and natural factors challenge this action and start to restore the previous state. Motion of the material near the working, especially fresh ones, is usually of quasiplastic character: continuous plastic flow is accompanied by faulting and generation of cracks of various sizes. Rock bursts are the strongest fracture motions. They differ from failure by mostly occurring due to elastic energy, while failure is caused by gravitation energy, which is related to the weight of the fall mass and the height of the drop. Rock bursts and the whole family of associated motions near working—plastic and fracturing—take elastic energy from rocks all around the working but mainly from the most non-hydrostatically stressed field. After the rock burst its energy is exhausted, the faulted, fractured, weaker material near the working cannot withstand the stress of continuous plastic or fracturing movement in the surroundings. Accordingly, the zone of high non-hydrostatic stress moves deeper into the rock mass to a field of less faulted, stronger material. Near the mine working some of the stress is again taken up by the supports.

The stresses in this secondary zone of concentration (which has moved from the working into the virgin rock) are less than those near the working. Accordingly the most dangerous shear stress components are less. This state of the rock is clearly not so likely to fail dramatically than before. Thus an increase in pressure on the supports, which indicates stress release in the rock near the working, is a good sign. At the same time when mining is fast and the area of fractured unloaded material in front of the stope does not have enough time to develop, the pressure on the supports is not high and the rocks near the stope remain very stressed. These conditions lead to rock bursts. The stress in the rocks adjacent to the opening is not, of course, the only cause of rock bursts; they may also be provoked by the stress in rocks further away from the opening; for example, the rock in the main roof rather than rock in the immediate vicinity.

The behavior of the rock near a mine working which has intruded into untouched rock and disturbed the natural conditions there may be considered from another point of view. When conditions (pressure is the main one in our case) change, the rock which was once in balance, loses its stability and tends towards the new conditions prevalent near the Earth's surface. This is true of many of the properties of the rock, even its composition, which may start changing while interacting with the new external medium. In the simplest case, we may confine our considerations to the tendency to smaller volume densities and internal elastic energies in the rock when pressure is reduced. This change would be predicted by the classic linear theory of elasticity, although, in fact, the elastic modulus of the rock also changes with pressure even in the absence of fracturing. A rock carried from the depths to the surface will release elastic and other energy accumulated deep down. The material resembles an explosive in this respect. The explosive failure of coal during instantaneous outbursts of coal and gas in deep mines is a dramatic example. Moreover, for rock bursts, the fundamental cause of the destructive phenomena in mines is similar. Hence the analogy between a rock burst and an explosion proposed by S. G. Avershin has a deeper meaning than there seems at first.

Proceeding from the assumptions of mechanics, we may conclude that the total energy of the rock burst must not exceed the work done against gravity, needed to lift the material extracted from the shaft and the energy it contains (when extracted from the shaft, the equivalent amount of elastic energy is released into the environment). Unfortunately, this does not result in a direct way of assessing the energy balance of possible rock bursts, as only a small fraction of the total energy spent on changing the state of the rock mass during stoping is expended on the rock. Most of the energy and a fraction difficult to assess is either retained in elastic form for a long time, or is dispersed by plastic flow and non-catastrophic ruptures (fracturing). Besides, the energy balance of the motion in the mines may be disturbed due to an input of energy from tectonic processes from the outside. A relation between rock bursts in seismically active regions is quite possible: natural weak earthquakes are known to have foci at depths of about 3 km; e.g., of the same order as deep mines.

The problem of strength and failure of rock under mechanical stress requires further discussion. Let us present some general considerations.

From experience in ordinary short-term strength testing of artificial materials and rock, engineers have a firm conception of some limiting stress that depends upon the properties of the material at which the solid begins to deform irreversibly, flow quasiplastically (the elastic limit), and fail at a still higher pressure (strength limit). The quotation from S. G. Avershin mentions this state. Other works on mining (see, e.g., [Ruppeneit]) also use this as a basic concept.

Plastic failure, if it takes place after considerable plastic deformation, is distinguished from brittle failure, if deformation has had no time to develop before the failure. Solid rocks are in this sense categorized as brittle materials. A reader may get acquainted with these and the other concepts of rock failure in mining in many places, in particular, [Avershin, Ruppeneit], which contain many references to other works.

However, the concept of limiting stresses is not sufficient from the point of view of geophysics and mining. First of all, take the statement about the significance of plastic flow when loading continues for a long time. These flows begin at stresses much lower than the elastic limit for short-term brittle materials (glacier ice or rock salt). More stable rocks, even granite, may also be substantially plastically deformed under moderate but constant loading. There is no need even to consider the high temperatures and pressures deep in the Earth, or long geological periods of time. The behavior of the rock in granite formations during drilling of tunnels in the Alps last century is an example. Given the proper instruments and long observation periods, plastic flow can be found in practically every rock in mines, especially deep ones.

Another difficulty is that the rate of plastic deformation can change over time under constant stress. This is due to the ability of the material to change its properties during deformation. Thus, during quasiplastic deformation in rocks, which are structurally inhomogeneous materials, weak zones or microcracks between strong pieces may form "grains" of various sizes and these actually permit the pieces to move relatively during the low. At the same time the opposite process—gradual healing and disappearance of the cracks—must also take place. In the Earth's crust this is due to the molecular motion in the material, to the grinding action of the sediments in mineralized waters, to the influence of endogenous factors and to other geological causes. If the disturbance in the deposit that moved it away from the equilibrium were to cease, then the deposit would return to its consolidated state over geological time. If the disturbance continues the mode of fracture development in the deposit depends upon the intensity of the two processes. When the disturbance is active, fracturing predominates over flow, the resistance of the material to deformation is reduced, quasiplastic flow accelerates, and thus may finish with complete failure. The higher the stress, the sooner this will occur.

When investigating the strength of metals in the laboratory Zhurkov and Narzullaev found that under constant stress the time before failure decreases exponentially with stress increase. The temporal theory of strength is based on this principle, which states that failure occurs at any stress, if it acts for a sufficiently long time. The assumption of some threshold stress below which failure does not take place is a refinement of the theory. This may correspond to our case of quasiplastic flow of fracture mass when at low stress the process of fracturing is compensated by the process of healing. The modification and development of the temporal strength theory, in relation to problems of mining and geophysics is of great interest.

We have approached the concept of flow and failure of rock as an intermittent and as a continuous process.

On one hand, in mines prone to rock bursts, the bursts are simply the end result of a sequence of shocks of different magnitudes and are related to failure and fracturing. Rock bursts in their weaker developments progress continuously from being cracks in the massive, first audible to people as noises and then detectable only with sensitive instruments. On the other hand, the process of plastic (in a broad sense) flow in rocks may exhibit itself, at least partially, by fracturing and the relative motion of more solid pieces along the cracks. If the movement of rock masses in large volumes is considered, where zones of rock bursts are represented only by sections of moving volumes, then rock bursts are but individual elements of fracturing in a general process of "flow" or displacement of rock. This macroscopic aspect does not exhaust the problem. A study of the features of the process associated with the generation of each individual burst is itself of interest.

The obvious analogy between rock bursts and earthquakes may be found here as the latter may also be regarded as individual details of a macroscopic rock flow over a large region. It is not a superficial resemblance. Investigations of seismic regimes in individual regions [Riznichenko, Gutenberg], and in the Earth as a whole [Gutenberg, Richter] have shown that the small earthquakes are more frequent. The same is true for rock bursts and smaller magnitude shocks in mines and in still weaker events during failure of rock samples in the laboratory [Vinogradov].

From observations of earthquakes a definite correlation has been found between the number of events and the seismic elastic energy released at the focus. Observations in mines have shown that both for rock bursts and elastic impulses their energy distribution is similar, and even the quantitative parameters of the distributions are found to be close [Riznichenko].

Thus, both in seismology and in mining, seismic methods allow the fracturing seismic component to be singled out from the general fracture-and-continuous process of quasiplastic flow of rock mass, which bear a general statistical resemblance in both fields.

In seismology, detailed observations over seismic regimes have two main purposes. The first is to find seismic regionalization, i.e., a mapping of the hazard degree of a region with respect to the probability of the occurrence of an earthquake with a given intensity. The second task is to find a way to predict the time, location and energy of single earthquakes. The first problem has actually been solved and regionalization methods are being refined. The second problem is still at the initial stage of development.

In mining, seismic and seismo-acoustic methods promise solutions to similar problems: an indication of the general hazard for different areas in mines as concerns the possibility of rock bursts, i.e., regionalization of the rock burst hazard. The second problem is to find ways of predicting a single rock burst. The second problem is more complicated.

At present it is too early to consider prevention of such energy-intensive phenomena as catastrophic earthquakes. However, the prevention of less energy-intensive rock bursts is quite realistic. This, together with the development of regionalization methods and the prediction of rock bursts, is the ultimate aim of the investigation of rock bursts using geophysics.

Besides a difference in scale between earthquakes and rock bursts, there is also a difference in origins, viz. the boundary conditions of the corresponding dynamic problems. A rock burst arises because of hollows (mine workings) and the stress in the rock. There are no such holes near earthquake sources. This difference is clearly not enough to overcome the similarity between the physical characters of these events, which can be seen in the similar general statistical description of the fracturing seismic component of the processes. These considerations allow us to hope that the same geophysical approaches to rock bursts and earthquakes may give good results in both fields.

12.2. Methods of Investigation of Rock Pressure

The methods for investigating rock pressure may be divided into the categories of qualitative and quantitative methods, instrumental and non-instrumental methods *in situ* (field, mine) methods, laboratory methods, and theoretical methods.

The development of the theory of rock pressure started with non-instrumental *in situ* methods. These include visual observations—changes in the shape of walls, roof and soil of working mines; the effects of rock pressure on supports, their deformation and failure; the development of fractures in walls and the spalling of material from walls; observations of rock outbursts from walls—and also listening to the various cracklings and other noise in the massives resulted from rock pressure. These methods are not now the only ones in use, but they remain a useful source of information which complements the more reliable but as yet less extensive information obtained by quantitative methods. Nevertheless, it is quantitative methods that are now the basis of investigations of rock pressure. The quantitative instrumental methods include mine surveying methods which are now very widely used. They are older and thus more understandable for miners than new "geophysical" methods that have come from "outside".

Classical mine surveying locates individual points in mine workings and on Earth's surface using theodolites and levelling instruments. A comparison of such locations over time allows us to make a judgement about the displacements caused by rock pressure.

Over the past decade various methods and instruments have been developed around the world for measuring displacements and strains in mines, together with the forces and stresses to assess quantitatively the pressure effects in mine workings. A review of measuring instruments is given in [C. Vest] (as of 1957). These instruments use different principles for assessing displacements (strains) namely mechanical, optical, electrical, magnetic techniques or a combination of techniques. Mechanical (hydraulic, in particular), electrical and other meters were designed to determine the support strengths and stress in rocks. In addition to sensitive instruments designed to measure small changes, because of the low plastic strains in strong rocks under moderate loads, we also have to use instruments with lower precisions to detect the large changes that occur when there is a major quasiplastic flow, especially if it is accompanied by strong fracturing.

All these mine surveying instruments and methods were designed exclusively to study the mechanical phenomena in the rocks and all work statically.

The new geophysical methods for investigating rock pressure include inclination measuring, seismic, acoustic, ultrasonic, electrical, thermal and radiometric methods. Each of them is described below.

The first four methods are also mechanical in nature, and the first few are also static. The seismic, acoustic and ultrasonic methods are dynamic: all utilize the rapid propagation of elastic waves in rock. The remaining geophysical methods deal with very different physical phenomena, viz. the electrical, thermal and radioactive characteristics of rocks and the changes in the stress state of the rocks, together with their fracturing.

Geophysical methods were first applied in mining [Riznichenko and others] because mine surveying methods require workings into the rock massive penetrating directly the region where it is necessary to take measurements; but any new opening distorts the pre-existing state of the rock. Therefore, information about rock pressure obtained in this way is distorted, and the degree of the distortion cannot be estimated.

Laboratory experiments and theoretical considerations yield some indirect information about possible silication within the rock mass.

In the laboratory, we can conduct various tests on rock samples under artificial conditions of pressure and temperature, close to those observed in the rock mass, and we can simulate situations. Although both these laboratory investigations are practiced (see [Ermakov, Kuznetsov, Trumbachev and Fedotov]), these are considerable difficulties and restrictions. The main difficulty is that the material in rock samples extracted from the bulk may exhibit different properties than the same rock when it exists as part of the massive. This occurs because the properties of the massive may change as the working is approaching the place where the sample is taken.

In a simulation the greatest difficulty involves the rock properties and behavior *in situ* and the properties and behavior of the sample in the laboratory. What is important is that different processes—elasticity, plasticity and failure—have to be simulated simultaneously. The simulation of the force of gravity is also advisable but involves technical complications (centrifuging). Nevertheless, the last difficulty can be overcome in some situations. However, the main difficulty is to choose the material for the samples so that they have properties similar to the material in a rock mass with respect to all three characteristics. It is moreover difficult to study all these properties, at least for the rock mass. Finally, there is no correct formulation of the mathematics concerning the rock pressure as there is no recognized unified theory of elasticity, plasticity and failure in rock, either in mining or in geophysics. All things considered, we have at present to be content with an appoximate simulation of the individual aspects of the theoretical processes.

Difficulties in the theory are similar to those in simulation. The main problem is to account simultaneously for the elasticity, plasticity and failure of the medium. A number of interesting treatments of elastic [A. Dinnik, et al.], plastic, and elastoplastic deformations [A. Popov] are now available. Attempts were made to apply the theory of particulate media to the calculations [Protodyakonov, R. Fenner], but these cannot be associated with the conditions in consolidated rocks. A number of engineering calculations have been made [Dinnick, Protodyakonov, Rodin, Slesarev, Tsimbarevich, Chevykov, Fenner], but sometimes these solutions are more theoretical than attempts to get a better knowledge of the true situation *in situ*. However, all these attempts are still very far from a comprehensive description of the processes, especially as concerns changes in the material properties near the mine workings during deformation (quasiplasticity and larger fractures).

At present laboratory and theoretical methods for investigating this topic continue to be developed, as well as combinations of these techniques or these techniques with *in situ* observations. However, it is *in situ* measurements that are mainly being improved, because these methods alone give reliable primary information about all aspects of rock pressure. In this respect, miners in different countries try to apply geophysical methods to the investigation of rock pressure in mines. These methods are refined to obtain information about the structure, properties and even movements of matter at depths without any direct penetration into the Earth's interior, i.e., without disturbing the continuity of the rock mass. They may help to overcome the difficulties that have confronted mining science. Geophysicists have only taken the first steps in meeting the requirements of miners.

12.3. Physical Basis for Seismo-acoustic Methods of Investigating Stress State[*]

When assessing the potential of different geophysical methods with respect to the study of rock pressure, or more exactly, the mechanical stress state of rocks (both deposits of coal and rock walls), it would be natural to consider, first of all, methods based on the mechanical properties of rocks and mechanical processes in them. This is the seismic or seismo-acoustic method. Employing other geophysical methods, e.g., gravitational, magnetic, electrical, seems problematic.

Seismic methods are widely used in the USSR. Two main categories may be distinguished: those using artifical and the other fractural seismic fields.

Methods involving artificial nonstationary impulse fields are now being developed in seismic prospecting (studies of the geological structure of regions, and the search for mineral deposits). Elastic impulses (seismic waves) are generated artificially, usually by means of explosions. Impulsive ultrasonic methods of simulating seismic waves and investigating the elastic properties of rock samples are similar to the methods of seismic prospecting, as well as non-destructuve testing methods in technology. Stationary fields (sinusoidal elastic vibration) have also been used, but possible applications are more limited.

Methods involving natural fields are being developed in seismology. Natural elastic impulses, i.e., the seismic waves generated during earthquakes, are observed and studied. Faults occur as a result of stress accumulation when there is intense orogenesis.

Both types of seismic methods (those using artificial and those using natural fields) may, in principle, be employed to study rock pressure. An artificial seismic field may be generated in a rock body by an appropriate means of exciting elastic pulses. As to natural seismic fields they also exist in mines due to the fracture of rocks under pressure, which generates elastic waves in the massive, and the corresponding acoustic phenomena, viz. noise, crackle, etc. A rise in the intensity of these phenomena often precedes rock failure, instantaneous coal or gas outbursts, and rock bursts. They are very similar to earthquakes in their physical nature, while the most intense of such phenomena and rock bursts themselves are minor earthquakes.

---

[*] From [Riznichenko]. O. I. Sulaeva, O. G. Shamina et al are co-authors.

Thus, our investigations of rock pressure are closely related to a number of other fields of investigation, namely, seismic prospecting, seismology, the study of mechanical properties of materials, etc. This means that experience gained in these fields can be used in other fields: equally results of an investigation, as described below, may be of interest for adjacent fields.

The term "rock pressure" is widely used in mining and is understood there in a very broad sense. We shall treat "rock pressure" or just "pressure" as the stress state of the rock. We know that the stress state of a volume element of a solid (e.g. rock) is determined by a tensor, that in the simplest case of isotropy it is a value with six components. Therefore, such expressions as "increase" or "decrease" in pressure cannot have unique meanings as these terms characterize changes in scalar values that is those determined by only one number. Nevertheless, in descriptions it is not ambiguous, so we shall speak about pressure and changes in it as if it were a scalar. If required the topic will be made clearer.

12.3.1. Basic Principles of Impulse Seismic Methods in Mines

This method uses the dependence between the mechanical and elastic properties of rocks and pressure.

Many of the data discovered in seismology, seismic prospecting, physics and other materials depend upon pressure. The elastic properties we are most interested in are the velocity or propagation of elastic waves (mostly longitudinal) and the absorption of elastic energy during wave propagation.

Dependence of propagation velocities and absorption on pressure. It is well known in seismic prospecting and seismology that the velocities v for various rocks depend on their lithological compositions, geological age, degree of metamorphism, and the associated parameters, viz. structure, porosity, fracture, and humidity. They also depend on the depth beneath the surface of the rocks [E. Denton, J. Evans, Riznichenko, Sologub, L. Faust, D. White]. This dependence is due mainly to the increase in pressure over the rock area by the weight of the overlying rock masses. Corresponding data are obtained in seismic prospecting from field observations of reflected and refracted waves, and by seismic logging, i.e., seismic observations in bore holes. Similar data, though less detailed and accurate, are obtained in seismology by observing the earthquake waves and interpreting travel times of seismic waves.

The dependence of velocities v of pressure also follows from laboratory experiments on samples at different pressure to determine the elastic properties of rocks and other materials. Such observations have been made in different ways using static and dynamic methods and stationary and impulse ones, and under different pressure conditions, e.g., uniaxial compression hydrostatic pressure, etc. [M. Volarovich, F. Birch, Hughes, M. Ide et al., W. Zisman]. Special investigations of how the velocity depends on pressure were conducted by an impulse ultrasonic method at the Geophysical Institute of the Academy of Sciences [Riznichenko].

All these observations lead to the conclusion that the velocity v, as a rule, increases with increasing pressure. As far as the absorption of elastic energy is concerned, it follows from experience in seismic prospecting and some laboratory observations (see, for example, Riznichenko) that absorption is reduced with increasing pressure.

Causes for the dependence of elastic properties on pressure. Rocks like other test materials, e.g., concretes, are heterogeneous, structurally inhomogeneous substances that consist of components with varying degree of mechanical pliancy. The less pliable components include parts of the material's solid skeleton and the most pliable are pores and cracks filled with gas or liquid. The porosities of sediments and some metamorphic rocks are usually in tens of percent (see, for example, E. Denton), while magmatic rocks, even "dense" ones such as granites, have porosities of about 1% [B. Zalessky, et al., Riznichenko]. Under the influence of increasing pressure on structurally inhomogeneous substances the varyingly pliable components are regrouped. This primarily results from displacements and elastic deformations of the solid's skeleton, and then from plastic deformations, and then from partial local failures, viz. crushing of individual solid particles, destruction of the interpore material, etc. In macroscopic terms this is also exhibited as some "plastic" deformation (quasiplasticity).

This process of skeletal reformation affects the configuration and sizes of pores and cracks: voids are partially bridged and there is a consequent increase in the number, dimensions and efficiency of solid links and elastic contacts between the less pliable components of the skeleton. As a result, the whole skeleton becomes more rigid. The volumetric elasticity of the liquids and gases that fill the pores increases due to this reduction in the pore volume. Some of the air may dissolve in water, which strongly affects the elastic properties (as well as density) of the pore contents. All this results in a rise of the effective moduli of elasticity of the rock as a whole, namely its Young's modulus E and shear modulus G. Some of these processes are reversible while others are irreversible.

The increase in volumetric weight or density ϱ of the rock increases simultaneously with the rise in the moduli of elasticity under compression. But not that E and G rise much faster than the density ϱ does. This follows from experiments and some theoretical considerations. Thus, calculations based on Hertz's theory of elastic contiguity were done for the simple case of a granular medium with perfectly elastic spherical grains (for references see [Riznichenko, p. 126, and 474]. The effective E and G appear to vary proportionally with the cubic root of the pressure in an ideal case, while the density ϱ varies negligibly.

Propagation velocities of longitudinal v_P and transverse v_S elastic waves are given by formulae of the type $v_P \approx \sqrt{E/\varrho}$, $v_S \approx \sqrt{G/\varrho}$. Accordingly the faster rise in the elastic moduli E and G as compared to the density ϱ, as the pressure P increases leads to an increase in the velocities v. In the above, ideal, case the dependence between v and p is $v \sim P^{1/6}$. But in a real situation the way the velocities depend on pressure as obtained from experiments are more complicated [Riznichenko].

A decrease in the absorption of impulsive elastic waves with increasing pressure can also be formally explained within the conception of perfect elasticity. In structurally inhomogeneous media a passing elastic pulse is attenuated due to a partial reflection and dispersion of its energy at the boundaries of the heterogeneous components. Meanwhile, the degree of inhomogeneity of such media falls with increasing pressure. This is caused both by the partial closure of cracks and pores and by the rise in density and elasticity of the filler materials such that they approach the skeletal properties. As a result, the energy of the reflected wave diminishes while the energy of the passing wave grows, i.e., the effective absorption of the elastic pulse falls. Of course, under real conditions the processes are also much more complicated; they cease to be perfectly elastic.

Apparatus and observation methods. The idea that geophysical methods can be applied to the study of rock pressure is not new. In the USSR attempts were made in 1934-1937 [Tapponier, Molnar], but those experiments were conducted using primitive apparatus and observation techniques, and thus were not developed further. Similar investigations were later performed abroad [Buchheim, Obert]; however, no firm information obtained in mines has been published, as far as we know. Mine investigations using the pulse seismic method as conducted at the Geophysical Institute of the Academy of Sciences (GIAS) were first reported in [Riznichenko].

These seismic observations in mines were obtained by generating pulsed seismic waves and recording them in the coal mass or enclosing rock. Their propagation from the emitter to the receiver, located relatively close to each other in the massive is studied. The time taken for the waves to travel between the emitter and the receiver (or between two receivers) is observed and registered, and the amplitude and form of the pulse are recorded. A variation in these observed factors in time (or space) is associated with variations in the stress state and mechanical properties of the rock (fracturing etc.).

A mine seismoscope developed at the GIAS, the main instrument involved in this method, is based on the same principles of pulsed ultrasonic instruments [Sokolov] as a laboratory seismoscope developed at the same institute and which is used to simulate seismic waves and determine the elastic properties of rocks on samples [Riznichenko, et al.]. The mine seismoscope consists of a radio circuit that generates short electric pulses, which are fed at a certain frequency to a piezocrystal emitter, in which they are converted into ultrasonic pulses and fed into the rock mass. After the pulses have passed through the rock as elastic waves (ultrasound), they are picked up by a piezocrystal sensor, where they once again are converted into electric oscillations and passed to seismoscope where they are displayed on a cathode ray tube with a true scale. They may be observed directly or photographed to yield seismograms for analysis. Together with the received pulse, a short pulse corresponding to the emittence moment of the ultrasonic pulse is part of the input. This allows the time t of propagation of the ultrasonic pulse in the rock to be measured on the seismogram. Given distance d between the emitter and the receiver—the measurement base—we may find the velocity $v = d/t$ of propagation of the elastic pulse in the mass. The seismogram can thus be used to find the amplitude A of the pulse, its dominant period T (and accordingly the frequency $f = 1/T$), and its shape.

The basic principle in mines is to make repeated measurements over certain time intervals to assess the seismic characteristics in the whole rock volume, namely the time t of the propagation of the elastic pulse (and correspondingly its velocity v), and the amplitudes A of the pulse (and accordingly its absorption). A comparison of these data yields a qualitative judgement about variation in pressure and the state of the rock mass. The seismic observations are compared over time with other mine parameters, i.e., stoping, roof sagging, etc.

Additional observations to elucidate the dependence of the seismic characteristics on pressure are significant in this technique and give us the opportunity for a quantitative interpretation of the data from seismic observations of variations in rock pressure. Such observations may be conducted both in the laboratory and in mines. In the laboratory they are performed on rock samples, coals and enclosing rocks, which are pressure tested in a press. In mines the observations are made by the hydraulic cushion method, which is practical in mine surveying, as it permits a variable controlled pressure to be artificially produced in the wall rock of a shaft. Seismoscopes—mine or laboratory—are the main measuring instruments in all these seismic observations.

Most of the investigations to develop the impulse seismic method for studying rock pressure, particularly in mines were combined with acoustic observations and strain measurements as used in mine surveying.

12.3.2. Basic Principles of the Acoustic Method in Mines

This method of investigating rock pressure is based on the fact that many bodies affected by external forces crackle before failure. In mines acoustic phenomena such as crackles and shocks have always been considered characteristic effects of rock pressure, and an increase in intensity of these phenomena is usually associated with rise in pressure. So, the possibility of using acoustic phenomena in mines as a method of investigating rock is a natural idea.

The acoustic phenomena in mines, and at higher scales (e.g., for earthquakes), arise because bodies subjected to increasing stress usually undergo minor local fractures which generate the cracks before a major general fracture occurs. These fractures are accompanied by fast movements of material and for a short time the fractures become a natural source of elastic vibration in their surroundings. An observer perceives the elastic vibration as a sound, which in turn is evidence of a fracture.

Basic stress fractures. Experience of failure in different materials under mechanical stress from various fields of technology and in physics indicates that the process may proceed in different ways depending upon two oppositely-acting factors. These are: (a) the rate of increase in the external forces that cause destructive internal stress, and (b) the rate the stress is dissipated due to non-ideal elasticity in the substance (flow etc.).

If the rate of increase in the external force is higher than the dissipation of the internal stress, the material experiences brittle failure without any plastic deformation. If the rate of increase in the force is slower there is enough time for plastic deformation and the material undergoes plastic failure. Finally, if the force acts still more slowly, the body may undergo plastic deformation without discontinuity until all the energy of the external forces is exhausted and the deformation ceases.

Tensile and shear failure [Ya. Fridman] are the basic types of material fracture and both are, generally speaking, possible in all materials. Brittle tensile and shear breaks are both equally possible. A shear break is more likely in the event of plastic yielding. The resistance to tensile and shear break is a strength parameter of a material which depends upon temperature, rate of deformation etc. The type of failure is determined by the relationship between the strengths and values defining the stress state [P. Uzhik].

A shear break may occur in some cases of constant stress instead of a tensile one (or vice versa) by changes in the temperature or the rate of deformation, which alter the relationship between the strength characteristics. In other cases, the type of failure may be determined for stable strength parameters by changes in the stress state.

The state of tension resulting in tensile break and various states of triaxial compression, unequal along the axes or of compression and tension with shear components that cause shear break are the most hazardous stress states for most materials including rocks. The state of "hydrostatic" compression (compression with equal stress along all three axes) is usually not dangerous even at very high stress.

Basic types of stress and fractures in a rock mass with mine openings. Under natural conditions without the non-hydrostatic stresses caused by tectonic processes, the stress state of volume elements in the deposit and the enclosing rock far from the mining stopes is mainly determined by the weight of the overburden and is close to uniform triaxial compression. That is why coal and its enclosing rocks do not experience failure when removed from the stope.

The hydrostatic compression close to the stope is disturbed. It is known that there is a zone of high pressure with asymmetric compressive stress some distance from the stope. Such stresses are associated with shear. Shear break should be dominant in this zone. Closer to the stope these may also be dangerous tensile stresses due to macro- and micro-inhomogeneity in the medium (layers with different mechanical properties, porosity, fracturing, etc.); shear stresses also continue to occur. Thus both forms of failure—tensile and shear break—can be observed in this given zone. Fracturing and loosening of the coal in this zone must be due both to opening of existing cracks and the generation of new shear cracks, some of which open immediately, and tension cracks, which open from the moment they originate (if they are not immediately filled with intruding material).

The surrounding rocks are usually much stronger than coal, which is why they start to fail at higher stresses and may occur in other places than where the coal seam fails.

The general failure, both of small separate pieces of a material (samples), and large parts of the massive starts with a growth in the crack dimensions and an increase in their number (for the morphology of this process see [V. Beloussov, Gzovskii]). Thus, in most cases both the number of cracks and also the number of new cracks that are generated per time unit should naturally increase as the moment of general failure, i.e., the major fracture that splits the sample into separate parts, splits the seam right through and forces single pieces out from the mass.

However, intense fracturing does not always cause general failure; it may be temporary. The quasiplastic deformation of the mass associated with the fracturing may exhaust most of the accumulated energy of the elastic stress that causes fractures, releasing the stresses, smoothing their distribution in space and preventing general failure.

Elastic vibration in fractures. A fracture (shear break or tensile) is accompanied by a release of energy into the medium in various forms, especially in the form of elastic vibration. If the frequency of this vibration is within the range of sonic frequencies, they are perceived by the ear as crackle, rustle, etc. (if they are loud enough) or by appropriate instruments. The energy of the vibration is clearly larger the greater the magnitude of the elastic stress accumulated in the zone, and the larger the size of the crack. Therefore, the amplitude and the number of crackles (per unit of time) are measures of the general character of the stress state and the time remaining before a major fracture. Observations of the features of the elastic vibration caused by fractures must also help distinguish when the stress is dissipated as a result of quasiplastic deformation, and the hazard of general fracture diminishes.

The generation of each crack is limited in time, so there is a corresponding continuous frequency spectrum of excited elastic vibration. If parts of the medium nearest to the crack come into a new state of equilibrium without a large number of vibrations after a fracture, then the frequency spectrum must be broad. If the transition to a new equilibrium state is accompanied by vibrations of parts of the medium, the spectrum may have a maximum in the frequency range of these vibrations, the peak being sharper the larger the number of vibrations accompanying the fracture.

The location of the energy maximum in the vibration frequency spectrum may primarily depend upon the type of fracture (shear or tensile), the elastic properties of the medium of fracture (shear or tensile), the elastic properties of the medium adjacent to the fracture, and upon the size and shape of the fracture. The frequency of proper vibrations falls with increasing fault size [Keilis-Borok]. It may be expected, therefore, that during the failure and development of a large fracture, at which point the action of all these factors changes, the location of the energy maximum in the frequency spectrum must also change. When approaching a general fracture, the sizes of the preliminary cracks, as a rule, become larger so there is even reason to expect that the energy maximum will shift towards lower frequencies, and the main fracture must have the lowest frequency.

The spectrum of the vibrations generated by the fractures are recorded some distance from the crack and so may differ from the spectrum obtained in the vicinity of the crack as the pulse's slope changes as it propagates in the inhomogeneous and non-ideal elastic medium between the source (crack or fracture) and the vibration sensor. But, if the preparatory processes and the final fracture are in the same zone, and the elastic vibrations from every source travel approximately equal distances to the sensors, then it would be natural to suppose that they are distorted in the same way. In this case, by analyzing the received vibrations the relative temporal changes in the shape and spectra of the vibrations which accompany the origin and occurrence of a major fracture, can be followed.

Associating fractures and acoustic phenomena with rock pressure and outbursts. As a first approximation single fractures in the massive may be considered to occur where the non-hydrostatic pressure is strongest and the weakest parts of the massive are affected. Cracking and fracturing together with the associated acoustic activity increase, in general, with a growth in pressure. The process may end with a general failure—a rock burst etc.

However, a more thorough investigation shows the problem is more complicated. Let us consider the failure process in rocks as that in a structurally inhomogeneous material for a continuous growth in pressure. First, the weakest bonds are destroyed in such a material. Stronger bonds remain for some time, which initially leads to a general strengthening of the material. As the pressure grows further, the stronger bonds will start to break too. If the material has a property such that the bonds are distributed strengthwise in a stepped manner, this will lead to a similar acoustic activity: phases of intense activity will be followed by quiescence phases. The final stage in this process, viz. general failure (rock burst etc.), may also occur after some quiescence.

It is natural to suppose that the more homogeneous the material, the more concentrated in time must be the process of its failure (meaning the degree of homogeneity of the material immediately before the failure). At the limit of an absolutely homogeneous material and homogeneous stress in the volume, a general failure must be instantaneous and consist of one "stage" without any precursors. The failure should be characterized by a significant seismic effect, as it involves the instantaneous release of all the elastic energy accumulated in the volume during the period leading up to the failure.

In order to assess correctly the relations between rock pressure, fracturing, and the associated acoustic activity, the "acoustic characteristics", the acoustic activity as a function of pressure for different loading regimes, should first be determined by experiment for each rock.

In spite of the strengthening of the material due to initial single fractures, these same fractures increase the hazard of further failure in the same zone because the ends of cracks are sites of stress concentration so that the weaker external forces are able to cause further failure than in an unfaulted zone. A fractured zone thus may become the source of larger, even catastrophic, faulting in the material, i.e., rock bursts, instantaneous explosions, etc. However, fracture (cracking, exfoliation, loosening of the material) may, as we know proceed calmly, e.g., a coal bulge.

Fractures are both the result and the cause of changes in rock pressure. A fracture first leads to a stress reduction in the local zone where it occurred. The pressure is partially transmitted to neighboring zones where new fractures occur, leading to new changes in pressure, etc. Therefore, a local fracture may result in a developing process of rupture and the redistribution of rock pressure. Migration, displacement of sources of acoustic activity in space and time may thus be expected.

Shaking related to the generation of a fracture in some area may play a significant role. Sufficient shaking combined with zones of weakness in other places may cause short-term additional stresses and result in fractures. This process may assume an avalanche character, and then a major fracture—a rock burst—occurs. The shaking may also cause an instantaneous outburst of coal and gas which brings about its specific mechanism, which in turn is known to depend on both rock pressure and gas desorption.

The relation between fracturing and instantaneous outbursts is still more complex than that which relates fracturing with rock bursts. Thus, weak fractured zones in coal, where the bulk strength is lowered and the gas leaking increased, are known to be the most dangerous as far as outbursts are concerned. New fracturing increases the danger and may sometimes even initiate an outburst. However crack generation and opening also create favorable conditions for slow gas release, which leads to a degasification of the seam and may help to prevent outbursts.

Therefore, there is no unambiguous or universal connection between fracturing, associated acoustic activity and rock pressure or between rock bursts and instantaneous outbursts of coal and gas. The relation between all these phenomena may take sufficiently different forms under different conditions. This impedes the use of acoustic observations in mines which were once interpreted in a most primitive way, viz., intensifying crackles were considered a direct indicator of an increase in pressure and, accordingly, an indication of a growing hazard of rock bursts and outbursts. Numerous examples of how such a simple approach is unfounded testify to the necessity of a more thorough study of the acoustic phenomena.

12.4. On Impulsive Ultrasonic Seismic Logging[*]

Seismic observations in wells have been performed so far to solve two kinds of problems: (1) parametric determination of propagation velocities of seismic waves to supplement methods of seismic prospecting (reflection and refraction surveys); (2) direct solution of in-

---

[*] From [T. Hanks, M. Wyss]. V. A. Glubov as the co-author.

dividual prospecting problems concerning underground structure in well location zones. In both cases observations were carried out with instruments of the same type, as in ordinary seismic prospecting, where elastic impulses are used with dominant recording frequencies of about 100 Hz. The resolution of these methods is comparatively low: they allow inhomogeneities (e.g., individual layers) of usually no less than some tens of meters in size (thickness) to be detected. However, the methods of electric logging and some other kinds of logging, practiced in well geophysics with favorable conditions provided allow layers of about one hundred times thinner to be singled out. Thus, as far as resolution is concerned, seismic methods in wells yielded to other methods of geophysics.

A new impulsive ultrasound method of seismic logging based on the use of apparatus of the type of impulsive ultrasound defectoscopes by S. Ya. Sokolov employ oscillation frequencies of the order of 10^4-10^5 Hz, i.e., 100-1000 times as high as in ordinary seismic prospecting. The resolution of the method increases by the same factor approximately. It is quite compatible with resolutions of electric and other types of logging.

This permits problems of the third type to be examined in wells by seismics in their new ultrasonic variant: a detailed separation of the well cross-section by mechanical properties of the rocks. These characteristics include elastic properties, particularly, velocities of propagation of longitudinal and transverse waves, as well as absorption of the elastic waves. High resolution of this seismic log method may be used in a number of areas: in field seismic prospecting (to supplement reflection and refraction methods), in well geophysics, in engineering geophysics, etc.

This section summarizes the methods of seismic observations in wells including the new method of impulsive ultrasound seismic log which is a natural completion of the research in increasing resolution of the well seismometry.

12.4.1. Seismic Methods in Wells

Ordinary seismic log. Seismic log parametric observations for the assessment of propagation velocities v of elastic waves in rock layers used to be performed long ago and have been performed now in field seismic prospecting. Using these observations blasts are the sources of elastic waves, which are most often fired at the surface near the well head, and a special well seismic receiver, usually of electromagnetic type, is installed in the well at different depths successively (see, for instance, [Gamburtsev, Gurvich]). The seismic oscillations are transformed by the receiver into electric ones, then they are amplified and registered with an oscillograph containing mirror galvanometers with natural frequency of the order (1-5) 10^2 Hz, if the photofilm rate is about 1 m/s. Only the times of the first onsets of direct logitudinal waves from the explosion to the receiver in the well are mainly used to make geophysical conclusions. These times are determined with an accuracy upto 10^{-3} s.

If we consider that propagation velocities of longitudinal waves in rocks are usually several kilometers per second, then it is easy to determine the assessment of the wave propagation velocity v by such observation techniques sufficient for practical purposes (errors of no more than 5-10%) at the minimum bases ΔH of about 50-100 m. Thus, the resolution of this method (it may be evaluated by $1/\Delta H$) is not high. It often becomes still less due to errors of time determination t as a result of poor maintenance of conditions after repeated firing. The repeated blasting is necessary to obtain a successive counting of times t at different depths

H with using a single seismic receiver. The errors become still larger due to decrease of oscillation arrivals at seismograms at great depths H. It is connected with absorption of (mostly) high-frequency oscillation components in the rocks which are responsible for sharpness of the arrivals.

Therefore, this ordinary seismic logging method allows us to determine sufficiently reliably only average velocities $\bar{v}$ from the soil surface to various depths H as the bases ΔH appear rather large. These velocities are used to obtain sections by means of travel times of reflections and refractions in field seismic prospecting. Layer velocities v_l, that characterize thick layers only, are determined with a lower accuracy which does not always meet practical requirements. Singling out the comparatively thin layers, in particular, the layers with high boundary velocities v_b which play a primary role in the correlation refraction method (CRM), by means of velocity values, is practically impossible within this method. This greatly hinders the comparison of data from such ordinary seismic logging with the results of investigating velocity sections by the reflection method [Riznichenko]. As a rule, a direct comparison of ordinary seismic logging data with the reflection survey is barely possible. The reflecting horizons obtained by this method often correspond to thin layers that cannot be detected by means of ordinary seismic logging.

The problems of methodology and interpretation in ordinary seismic logging, and particularly the problems of geological and geophysical generalization of results of the layer velocity assessment method are described in a number of papers [C. Dix, W. Hafner, N. Haskel, J. Legge, J. Rupnik, P. Naevarte, W. Olson, M. Slotkin, et al., E. Stulkin, R. Wells].

Other similar methods. Methods of determination of the velocities v by means of ordinary seismic logging and other methods where the explosion and the receiver replace each other are related. The explosions are performed at different depths H in the well, and the receivers (one or several) are installed at the Earth's surface [A. Claudet, N. Smith, J. Henderson, R. Brewer, G. Kokesh].

There were attempts to employ somewhat higher frequency apparatus than in ordinary "mid-frequency" seismic prospecting, both in ordinary seismic logging and in similar methods. However, the potentialities of high frequencies remained limited in all these methods due to a fast attenuation of high-frequency oscillations over long bases H which reach the full depth of the well under investigation.

Seismic observations in wells with the use of similar apparatus, and firings on the Earth's surface or in wells, were also conducted to solve other special problems. These included: well bending, low velocity zones near the Earth's surface, conditions of exciting seismic waves at blastings in different rocks and at different depths, as well as research in some physical problems of elastic wave propagation [Sokolov, Horton, F. Ittke, R. Jolly].

Determination of the underground structure in the well area. Seismic methods in wells, that are aimed at investigating special location of refracting boundaries crossed by the well or situated close to it, are of special importance. These methods were the most successful in studying the boundaries of complicated forms with high angle dip, as in salt domes, but they have been also applied to study boundaries with low angle dip. The techniques of either the source of oscillation—a charge (seismic blasting), or seismic receivers, are used in all these investigations. In both cases it was again the times t of wave propagation from definite points in the well to points of the Earth's surface, that were directly measured. Ordinary apparatus for seismic prospecting were employed in these investigations. The publications describe only the interpretation of such observations [M. Habuty, E. McCollum, La Rue, Gamburtsev, Riznichenko, F. Agnich, L. Gardner].

Common in all the above-mentioned seismic methods in wells is their relatively low resolution due to the use of apparatus and techniques intended for recording oscillations in a rather low-frequency range, developed for ordinary seismic prospecting (30-100 Hg). That is why time intervals Δt adn velocities $v = \Delta H/\Delta t$ could be only determined with large bases ΔH (of the order of some tens of meters).

To increase the resolution and the accuracy of seismic logging, attempts have been made in the following basin directions. [Vinogradov, V. Voyutsky, A. Mozzhenko, Ostrovsky, Popov, L. Athy, Riznichenko].

Use of multichannel probes in wells. Two or more seismic receivers are simultaneously sunk into the well, which constitute a multichannel probe, and the charges are blasted at the soil surface (see for example [R. Beers, Mozzhenko]). This method is aimed at elimination of the errors in determining Δt of oscillation arrival times to the points at different depths H_i, H_{i+1} ($H_{i+1} - H_t = \Delta H$) due to the use of a single receiver with repeated blasting and poor maintenance of exciting vibration conditions. The method of the multichannel probe allows some increase in the accuracy and efficiency of seismic logging observations, which is its important technical merit. But in principle this method does not offer much advantage over the "uni-channel" one, as the frequencies are the same in both methods and thus the resolution of seismic logging does not change considerably. Location of the vibration excitation point (explosion) at the soil surface in the "multichannel" method is an insurmountable obstacle for the employment of higher frequencies, as the latter are absorbed by the medium at long distances H, the waves have to pass from the surface source to the receivers in wells.

Measurements in wells with short bases. Ordinary, relatively low-frequency, seismic prospecting apparatus with low rates of oscillation time scanning become useless in determining short travel times Δt of elastic waves with short bases ΔH (for seismic logging, in particular). For such measurements V. S. Voyutsky and A. E. Ostrovskii proposed and tested a number of non-inertial devices, including circuits with a ballistic galvanometer and a cathode oscillograph [Voyutsky]. However, no measuring circuits gave or could give positive results, while the common path of elastic waves from the source to the receiver remained long—this caused unclear wave arrivals.

A. E. Ostrovsky was the first to use a probe for both the seismic receivers and an impulse source of elastic waves in seismic logging with short bases. The source consisted of a cartridge with several detonators that could be fired one after another. The length of the probe with the source and two receivers totalled 4-6 m. Oscillations were recorded on photographic paper on a revolving oscillograph drum which provided the rate of time scanning upto 10 m/s. In spite of primitiveness (from contemporary point of view) of this method of velocity measurement in rocks over short bases A. E. Ostrovsky managed to obtain good results: by [Ostrovsky], the accuracy of velocity v determination at the bases ΔH = 1-2 m was 3-5%. The measurements met with difficulties caused by reciprocal influence of the electric circuits with long wires in deep wells. The necessity of frequently extract the probe from the well for to replace the detonators limited the efficiency and was a particular disadvantage of the method.

Use of stationary oscillations. Sources of stationary sinusoidal oscillations were proposed and tested for seismic logging with short bases besides impulsive ones (see, for example, [Voyutsky, Athy]). Either the degree of oscillation absorption from the source to the receivers, or the difference between oscillation phases at two receivers was measured. However, the experiments in this direction have not given any considerable results.

Note that methods of stationary oscillations are characterized in seismic surveying by lower resolution than impulsive methods, as well as in a number of other fields (ultrasonic flow detection, hydro-acoustics, etc.).

Impulse ultrasonic seismic logging. As was mentioned, this developing method is based on the principle of the impulse ultrasonic defectoscope proposed by S. Ya. Sokolov.

The first attempts to employ apparatus of this type in seismics were made in 1937-1938, when V. S. Voyutsky proposed the use of circuits with scanning cathode oscillographs in velocity measurements of elastic wave propagation in rocks over short bases, in particular, for seismic logging [Voyutsky]. However, there were no effective ways of exciting seismic waves with sharp fronts at that time, which delayed realization of the proposal.

Special seismic impulse ultrasonic apparatus with effective piezoelectric emitters used in investigations of elastic waves over short bases was developed in the GeoPhIAN, USSR in 1947-1948 (published in 1951 in [Riznichenko, see also Riznichenko]. This instrumentation was employed to simulate seismic waves and to study elastic properties of rock samples. Similar seismic investigations were later conducted in the USA [J. Evans, et al., E. Howes et al., Nortwood, P. O'Brien] as well as in other institutions in the USSR [Yu. Timoshin, C. Chubarova]. The same apparatus was also employed both in the USSR and abroad in seismic investigations in mines [Riznichenko, W. Buchheim, I. Malecki, et al.]. The same types of apparatus were also widely used in impulse ultrasonic seismic logging.

Implementation of similar instrumentation in well seismics was first fulfilled in the USA [S. Summers, Vogel] in 1951-1952, and then in Canada [489]. The American investigations in the impulse ultrasonic seismic logging made use of spark or magnetostrictive emitters and piezoelectric receivers. The probes were 2-4 m long. Registration or visual examination of the oscillation form was by means of an electron-ray tube which allowed wave onsets to be detected and their form to be observed, as well as to control the oscillation amplitude. Major attention was however paid to fixing times Δt of first arrivals of longitudinal waves, so that their propagation velocities v could be found. Measurements were performed during the probe movement in wells which ensured high efficiency. The data obtained were compared with those from electric logging, with velocity estimates by ordinary mid-frequency seismic logging and observations based on the reflection survey. Good agreement of various data was recognized.

In the USSR, impulse ultrasonic seismic logging was first tested in the Geophysical Institute of the USSR Academy of Sciences in 1954-1955 [Riznichenko].

V

Structural Seismology

An important part of geophysics is the study of the Earth's internal structure. This does not cover everything under the heads of geophysics. Another, no less significant task concerns profound questions about the Earth, namely, the establishment of its origin, present life, and future fate. However, the first task is general and typical of geophysics as a whole. It is also the one that underlies all other methods geophysics has for reaching a comprehensive understanding of our planet. Human beings live a very short time, and even the whole of mankind has existed for but a moment in comparison to the duration of the Earth. We have to assess the entire history of the Earth to great extent on the basis of what its present condition is*.

_____

* From [Riznichenko].

13. Why Seismology?

The seismic method seems to be a "privileged" physical method for studying the Earth's interior compared to others such as gravitational, magnetic, electrical, seismic, thermal, and various electromagnetic and nuclear methods. Seismic studies attract more people, more money is spent, they are written and spoken about more, and are referred to more often in studies of the Earth by geophysicists, geologists, geochemists, etc. Even the International Geophysical Association founded to study the Earth is called "The Association of Seismology and Physics of the Earth's Interior".

Why? Why such inequality? The other geophysical methods may be "offended"! Injustice may be suspected. Perhaps, if we were more attentive, we would prefer a different method. But perhaps the question "what is more essential, what is more important" is inappropriate and incorrect: no method will replace any other. The seismic method cannot be used, for instance, to study the Earth's heat balance, or to determine the electrical or magnetic properties of the material composing the Earth. Which is the more important method or which predominates is an absurd question.

And yet there is something in it. We cannot simply discard it, we have to discuss it.

Let us look deeper: what is the relation between these methods in our geophysical society? What has received more rights and what must be made more responsible? Who would do more if the most favorable conditions were created for every one to an equal extent?

But enough of questions. Let us look at some replies. Although I shall refer to some simple and clear matters, they are apparently not clear to everyone because from time to time they generate heated debate, especially recently, and I do not expect the debate to stop after my explanations. Perhaps even the contrary may occur. Well, let us argue. Truth is born in argument.

The distribution of work between the different methods of geophysics is chiefly affected by the objective possibilities of the different methods in solving geophysical problems posed by science and practice.

True, the possibilities we know today do not coincide with the fundamentally inexhaustible possibilities, although there are limitations imposed by nature itself on the processes and phenomena being studied. Understanding continues, the horizons are opening up more and more, but we know that miracles do not occur. To avoid being on a false path tomorrow, we must proceed from what we already know today. Naturally, in the future, when new values appear, reassessment is possible. But this is another matter.

Studying the internal structure and properties of the Earth is a major task of seismology and has great scientific and practical significance. What attracts us to the structure debate is that, the relative possibilities of various geophysical methods in performing a common task, can be compared more easily against such a background. We know that any comparison is only possible using common indices. If the latter are absent, a comparison will give way to a joint consideration of different things. We shall compare different geophysical methods by a property belonging to all of them to some extent, namely other resolving power, i.e.

the ability to cover and describe a common object with increasing detail and accuracy.

We shall mainly deal here with the methods of general geophysics as related to studies of the Earth as a whole or at least to a great depth. But we shall also pay attention to the experience gained by the methods of exploration geophysics used to study the comparatively thin subsurface layer of our planet. Almost all the exploration methods have their general analogues, and a very close relation is retained between them. The real possibilities of exploration methods, in particular their resolving power, have been shown in practice by drilling. Hence, this knowledge can, to some extent, be transferred to corresponding methods in general geophysics. We know that experience is a criterion of truth, while the development of the national economy is almost the most weighty and convincing criterion.

To make further narration clear we should, first of all, know what a resolving power means, how it is determined and on what discriminating factors it is based.

Let us consider examples from physics and engineering. The best example is the microscope. The magnification of a microscope can be increased to enable one to distinguish smaller detail, but cannot be increased infinitely. The limiting achievable resolution or the detail that can still be distinguished is established by the wave length of light. Detail smaller than the wavelength is missed by the waves; after passing an object or being reflected from it, a wave returns to its initial configuration and fails to inform the observer about the detail. This is why the resolving power of a microscope is higher in violet light at the end of the spectrum, i.e., high-frequencies, short-wave lengths, than at low-frequency longwave light. It is even higher in an electron microscope because electrons use shorter waves than visible light.

The light we use in a microscope can be represented as an infinite sequence of sinusoidal oscillations of various frequencies. If the light is monochromatic, we can distinguish the details in the field of vision of the microscope by the intensity of the light. If the light is white, we can see the details in their natural colours and distinguish them both by intensity, and by colour, i.e., the frequency of the oscillations. Instead we could examine a series of monochromatic images of an object obtained at various wavelengths. At any rate, in order to distinguish detail, we use two distinguishing factors, or discriminators, namely, the intensity of the effect (light, sound, etc.) and the frequency of oscillation. The details are distinguished by the contrast of the intensity and the frequency of oscillations (color), while the resolving power is determined by the frequency of oscillations or the wavelength.

This holds true both for light and sound, and in general for elastic oscillations and waves (viz. those in seismology and to a lesser extent in gravimetry). We must mention, by the way, and this is true for physics and engineering, but not geophysics, and a sonic, more exactly, an ultrasonic microscope, is known whose resolving power depends on the same circumstances as that of a light microscope.

Pulse radio and ultrasonic locators, i.e., radar and sonar, are another example from physics and engineering. Instead of a continuous sequence of sinusoidal oscillations, these devices emit and receive, after reflection from an object, oscillating pulses of a definite frequency and separated by quite large intervals. If the detail being sought, e.g., aircraft or submarine, is some distance from a locator, the signals reflected from them will occur at different times without superposition, which allows them to be distinguished. Hence, here a new important discriminator, viz. time, appears.

Moreover, another discriminator is available, namely, the direction of the light beam. This is similar to what light engineers use in an ordinary searchlight and astronomers in a radio telescope. But in the last two cases, time is absent as a discriminator.

In principle, the same discriminators can be used in laser devices, which are now being employed for surface and space communications, e.g., to study the Moon's surface. The resolving power of the laser in comparison with other electronic equipment is associated with the fact that lasers employ shorter (light) waves. Ultrasonic flaw detectors and seismoscopes (and, as we shall see, the seismic method in general) rely on the pulse location principle. The intensity of the effect, oscillation frequency, direction, and time—a complete set of discriminators—may all be used.

The same discriminators are used in various combinations in geophysical methods of studying the structure of our planet. Let us see the relative possibilities of the various methods.

Gravimetry. This method is integral. The point of observation is simultaneously acted upon by all masses surrounding it, although this diminishes with distance.

In the simple static case considered in gravitational exploration, there is no discrimination of effects over time. There is also no discrimination in the frequency at which the field changes with time—in statics it is zero. As a result, only one scalar characteristic of the field has to be studied, viz. its potential, and its derivatives—the intensity and amplitude. These quantities are distributed smoothly in space. There are no clear fine details that could reveal relevant detail in the structure of the medium. True, gravimetrists sometimes try, by calculation, to uncover the "sharp corners" of the gravitational field in a medium and ascribe the physical meaning of details of its structure to them. But such a problem, apart from ambiguity (which can be limited by choosing an appropriate model of structure of the medium) is also extremely unstable.

Information about the Earth's structure that can be obtained by this method also follows from the small details and the smoothness of the change in the quantities being measured in space. They are to a great degree also only integral, generalized, blurred, and deprived of fine details. Most explorative geophysicists have now stopped being deluded by a few formal possibilities of detailed quantitative interpretation and clearly recognize the low practical accuracy of the interpretation. Now there are only rare attempts to interpretate all the details of a gravitational field quantitatively. Investigations limit themselves to a qualitative treatment and compare the situation with the general regional structural features. Lately, statistical correlations of gravitational field features with other geophysical data and with geological structure are popular. Now quantitative conclusions again become possible, but they are attended by a sensible appraisal of the degree of confidence in their "quantitative aspect".

Gravitational exploration is often preferred not because it is accurate and reliable, but because it is much cheaper than, say, seismic exploration, which is only conducted where it is indeed required. If it is desired that gravitational exploration provide quantitative conclusions on the depths of boundary surfaces, and on the size and shape of subterranean bodies, its data are most often supported by seismic data obtained at a limited number of spots over the entire territory being studied.

In general geophysics, gravimetrists have to a certain extent overcome the narrow restriction of the method. They have included the time factor, or, more exactly, the oscillation frequency. Hence the elasticity and plasticity of a material, i.e. properties dealt with seismology, could also be considered in addition to the density of the material. Here I mean the employment of gravimetry to study the solid tides associated with the Sun and Moon affecting the rotating Earth. These phenomena periodically repeat in space and time, and this is what enables us to use a new discriminator—the oscillation frequency. The order of magnitude of these frequencies is one or two oscillations a day. This is very low in comparison with seismology; the resolving power of low-frequency, long-wave oscillations is low. The method can be somewhat enhanced using the frequency discriminator by increasing the number of periods observed, i.e., prolonging the observation time. This improves the accuracy, but this relates to averages and it is difficult to improve the detail in this way.

The achievements of gravimetry in studying tides and other motions of the Earth associated with its rotation are well known. M. S. Molodenskii was awarded the Lenin Prize in this field for his work. Gravimetrists have narrowed the ambiguity of our knowledge on the average distribution of the density of the material inside the Earth and provided important general estimates of the distribution of the Earth's rheological parameters. But we must stress the general nature, viz. poor localization and detail in comparison with what is provided by other methods, in particular by seismology.

For example, by analyzing surface surveys and also from observations of Earth satellites, gravimetrists can deduce inhomogeneities in asymmetry of the Earth's mantle. But these inhomogeneities can be studied by seismology with much greater accuracy and localization. An analysis of the tidal motions of the Earth revealed that it should have a liquid core. This fact was torn by gravimetrists from nature at the price of tremendous effort and only very recently. At the same time, seismologists learned this long time ago and under more difficult conditions, i.e., when nothing was known beforehand. Gravimetrists, on the other hand, knew what they were seeking. Even now the boundary of the core can be determined by seismology much more accurately than by gravimetry. Gravimetrists meanwhile do not even claim to study the properties of the core material or say anything about its own properties, though now seismologists have discovered it and are now attempting to study it further.

This naturally does not imply that gravimetry may simply be replaced by seismology, that the latter is "in general" more important than gravimetry. Each field, like the other fields of geophysics, is important and needed. It is obvious, for instance, that in some cases gravimetry has greater possibilities than seismology, in particular for deducing the density of the Earth's interior, while seismology can determine its elasticity more effectively, though both these fields can yield information on these parameters and on the non-ideal elasticity of the Earth's materials. The nature of the information obtained and used by the various methods of geophysics differs, and only a set of methods enables us to characterize the various properties of the Earth. In particular, we cannot assert *a priori* that the rheological properties of the Earth's material should be identical for the frequencies of tidal motion (gravimetry) and elastic waves (seismology). It is possible, in principle, that for tectonic tidal and seismic motions of the Earth, a single common model of a rheological body sufficiently close to actual conditions will be found, though we do not yet feel a need for this. But to select such a model, joint efforts are needed.

There should be no competition between the different banches of geophysics. On the contrary, close collaboration of the various methods in studying our planet must prevail. The poor resolving power of gravimetry when studying the internal structure of the Earth should be compensated by seismology's greater possibilities. This is how matters stand.

Magnetometry. The magnetic effect, unlike the gravitational one, has polarity, i.e., a certain direction and this provides magnetometrists with an additional discriminator. However, this adds little to the resolving power of the method. Generally, gravitometry and magnetometry explorations are close in their potential. Chiefly they yield qualitative information on regional structure. In magnetometry, statistical correlations for forming quantitative conclusions with appraisals of their variances are possible.

There are also fundamental possibilities for quantitatively interpreting all observations, even calculating the distribution of "magnetic masses" in the Earth given a general model of the Earth's structure. But the degree of reliability of these solutions is as low as in gravimetry. Consequently, quantitative conclusions about subterranean structure from magnetic data only becomes possible in practice when basic data from elsewhere are available. They may be data from electrical or seismic exploration. The magnetic method is naturally effective for detecting accumulations of ferromagnetic material, e.g., magnetite iron ore. On the other hand, the depth to which this method can be employed for exploration is restricted by the existence of a temperature at which ferromagnetic properties vanish (the Curie point). This temperature is 365 °C for nickel, 768 °C for iron, and 1150 °C for cobalt. A temperature of 1000 °C is typical at depths of about 100-150 km. Hence, the magnetic method is limited to the Earth's crust or, perhaps, the upper mantle.

Although the magnetic method has a comparatively low resolving power when studying the deep structure of the Earth, we must not shade its importance or indispensability. The direction of the magnetic field and the existence of the Curie point enable us to study the thermal history of the formation and behavior of blocks of rock. The blocks may vary in size from very small ones to entire continents. The global magnetic field and its changes during the centuries are related to the rotation of the Earth and to its having a liquid core. This is clearly why the liquid core contains a "dynamo" that generates the global magnetic field. On the other hand, the rock solidifying in the Earth's crust, and passing through the Curie point or precipitating from water becomes magnetized in the direction of the global magnetic field at the moment of its formation or metamorphization. If the direction of rock magnetization does not correspond to today's magnetic field either the Earth's field has changed, or the rock has moved, or both. Both events can occur, so it is not simple to discover just what happened at different stages of the Earth's history. Moreover traces of some stages may have been erased and lost. Some possibilities for analysis nevertheless remain, and geomagnetists are busy explaining all this.

These important and complicated problems must not eclipse the main question about the resolving power of various methods; that of the magnetic method is comparatively low.

Electromagnetic and seismic methods. (When speaking of seismology, I do not refer to the study and prediction of earthquakes here.) A feature of these methods in either natural or artificial cases is the possibility of using fields varying in time, either stationary (or quasi-stationary) or pulse. Constant fields (e.g., the electrical direct current method) return us to static gravimetry and magnetometry. Variable fields, on the other hand, yield new discriminating factors such as the frequency of oscillation, direction, and time. The coefficient of intensity naturally remains.

In essence, the magnetic method and direct current electrical method are separate aspects or particular cases of the electromagnetic methods corresponding to a zero oscillation frequency.

The appearance of a frequency factor when using an oscillating stationary field does not eliminate the integral phenomenon that hinders structural detail being distinguished in a medium. With any stationary field, be it electromagnetic, seismic, or thermal, signals from all directions and from all (both near and remote) elements of the medium arrive simultaneously at the receiver. The introduction of a beam direction factor eliminates integration by the angle of vision without eliminating integration along the beam direction.

The strongest discriminator—time—only comes into play when there is a pulsed field. It eliminates the integral nature in the function of the distance.

The broadest possibilities when studying the structure of a medium appear when using all the above i.e., intensity, frequency, direction, and time. The effectiveness of the factors often increases in this order. Intensity and frequency are comparatively weak discriminators, while direction and time are the strongest. Depending on the circumstances, different factors dominate.

When discussing the various discriminating factors, we used the analogy of a microscope and radar. To see the role of these factors, we turn to an analogy with the finest human sense, the eye.

Using the intensity factor only, as in gravimetry, magnetometry, and the constant electric field method, is like someone with cataracts who cannot distinguish color. He only sees light—weak or strong, i.e., he can distinguish its intensity but nothing more. The addition of a frequency factor would allow the person to recognize color as well. But nothing more. Clarity of vision is still absent.

Such are the very limited possibilities possessed by geophysical methods employing low-frequency stationary or quasi-stationary oscillations whose frequency composition is easy to analyze, but whose directions are difficult to establish, e.g., dynamic gravimetry study of the Earth's tides. Electromagnetic methods of this type include frequency electromagnetic profiling and sounding with an interpretation procedure based on the propagation of plane sine waves of various frequencies in the Earth. In seismology, this category includes sounding the Earth's crust and mantle and observing the spread of the velocities of seismic waves in quasi-stationary oscillations associated with microseisms or produced by explosions and earthquakes. These possibilities also feature seismogravitational methods for studying the various modes of natural oscillations in the Earth, as a whole, produced by very strong earthquakes.

Should geophysical methods, which must be satisfied with a below "normal vision", be pitied? Yes, if "pity" is a suitable word. These methods have become adapted to inadequate "vision" and do not notice it, as it were, unless someone explains reality. Someone with inadequate vision, if observant, active, and clever, can compensate the lack by force of experience, work, and deduction. The richness of feeling and perception of the world and our ideas about it, though related, do not always develop in proportion. To see does not mean to understand. An eagle sees better and further than man, but the latter sees more, compensating weakness of vision by force of intellect. Something similar holds for geophysical methods.

Finally, the different geophysical methods "see" better or worse but live in a society where collaboration and aid reign, where physical shortcomings are compensated for by the merits of others.

Continuing our biogeophysical analogy, we discuss a method that corresponds to a person with normal vision. In addition to the discriminating factors of intensity and frequency—light intensity and color—there is also direction. It is this factor that allows us to distinguish detail in the colorful world around us, and gives us sharpness and clarity of vision. This is the result of the group reception of oscillations (the mosaic of the retina) at a low wavelength. If the wavelength is much longer than the length of the group no direction is obtained. Mosaic will function as a single point, and we return to previous blurred perception.

Geophysicists employing electromagnetic and seismological methods can use the group emission and reception of oscillations (although they do not always do so), in particular, if they deal with large wavelengths.

It is interesting that "normal vision" methods are in general used little in geophysics. On the other hand, vision in biology, has stopped at the stage of "normal vision" and has not progressed further. There is superhearing, e.g., in bats and dolphins. The short ultrasonic pulses emitted and received by special organs in these animals enable them to orient themselves better in complete darkness, than is possible in the light, with normal vision. Recently doctors and physicists have begun to develop similar hearing devices to replace vision in blind people.

We can imagine how the richness of our perception of the world would grow if the human eye became a "supereye" and we could directly sense the distance to every visible object! Binocular vision, which we have become so accustomed to, only partially compensates for this natural shortcoming and for small distances, in the same ways as a stereoscopic telescope in navigation is a poor substitute for radar. Academician S. I. Vavilov in *The Eye and the Sun* (in Russian) noted that this shortcoming of human and animal eyes is not due to poor natural design in constructing the visual organs of the Earth's inhabitants, but rather due to a more general factor, namely, the stationary nature of the Sun's light field. If our eye has a time discriminator as in radar, we could only use this discriminator for illumination by laser flashes. If the stars were suddenly to become lasers, then perhaps in a million of years a bodily organ of radar type would appear.

Let us return to geophysics. Both the electromagnetic and seismic methods can in principle be used with the best technique of locating detail, viz. radar involving all the discriminators, i.e., intensity, frequency, direction, and time. In this, it differs from all the other geophysical methods. Methods poor in discriminators also include ones which we have not discussed, thermal and radiometric methods. Their resolving power, at least for depths, is very low.

One point, however, is that one method is still theoretical while the other is the real possibility depending on the conditions. We shall only compare the general features and relative real possibilities of electromagnetic methods and seismology for a detailed study of the Earth.

The techniques are limited by the transparency of the medium to electromagnetic and elastic oscillations. If the Earth were made of an optically transparent material, we would be able to see its subterranean wealth with our own eyes, but this is not the case. Hence visible frequency electromagnetic waves are not employed in geophysics.

The Earth, especially hard rock, is more transparent to ultrasound than to light, but waves can, nevertheless, only travel several scores of meters. This is why ultrasound in geophysics is restricted to shaft, borehole, and laboratory distances. We shall deal below with electromagnetic and sound oscillations of lower frequencies than light and ultrasound. The problem of transparency still remains.

This problem poses a dilemma. On one hand, the more transparent a medium the more it can be seen. On the other hand, detail has to be contrasted as much as possible against its surroundings to study it thus, its transparency must be low in general. It should differ from its environment, at least by having a different refraction index.

A transparent object detected when isolated, can at the same time, screen a more remote one. If a medium contains a multitude of details, it becomes turbid. An example is a multilayer medium where there are multiple reflections and refractions of the waves used. A signal reflected from an object being studied, e.g., boundaries may vanish in a general background of oscillations associated with multiple reflections and refractions from other boundaries. The problem of stable interference in geophysics is more complicated than the problem of unstable, time-varying, chance interference, for example, like that which has to be overcome in long distance radio communication. In radio communication, the problem can be overcome by multiple repetition and addition of the stable signal to the unstable interference, when the signal is in a more advantageous position. In the stationary, multilayer, turbid medium of the Earth where interference is stable, nothing can be achieved in this way.

Another factor besides turbidity, that hinders electromagnetic or elastic waves penetrating deeper, involves the irreversible loss of wave energy into other energy, mostly heat.

Therefore the more transparent a medium is as a whole and the less contrasting the constituent layers and details, i.e., the less it absorbs the waves, the deeper waves can penetrate, and the greater the total detail detected. To improve our ability to investigate a medium like the Earth means to penetrate it more directly and to the maximum depth.

The energy loss of direct pulsed signal due to *effective absorption*, both absorption proper and scattering caused by turbidity, leads to waves being rapidly and exponentially attenuated with distance. The wavefront also *diverges* with distance and this is superimposed on absorption. Divergence also leads to a damping of intensity, but it leads to weaker polynomial attenuation and so absorption always dominates eventually.

For simplicity, we shall confine our considerations to flat waves with no divergence of the wave front. Then signal attenuation with distance is determined by the effective absorption of the medium. The coefficient of absorption per unit of time usually depends upon the wavelength and, accordingly, upon the frequency: the medium becomes less transparent for high frequencies. The highest frequencies are rapidly attenuated in the Earth: light and ultrasound are simply particular cases. Let us see how electromagnetic and elastic waves with lower frequencies are attenuated in the Earth.

Electromagnetic oscillations are absorbed in rock much more intensely than elastic oscillations of the same wavelength. The same is true of sea water, which is why high frequency radar is ineffective underwater or underground, while elastic waves—acoustic ones underwater and seismic waves underground—dominate.

Electromagnetic oscillations are absorbed in the Earth and in water because of their good electric conductivity, with low electric resistance ϱ. This may seem paradoxical as electromagnetic waves propagate freely in a vacuum, where the resistance is very large ($\varrho = \infty$), while in metals, where resistance ϱ is low, these waves attenuate practically instantly; metals "screen" them. The Earth is somewhere in-between.

Absorption also strongly affects the velocity of an electromagnetic wave, i.e., the velocity of light in a vacuum is faster than that in air. In the Earth it decreases so much that it approaches the velocity of elastic seismic waves. Thus, for sedimentary rock the velocity of electromagnetic waves at the frequencies used in geophysics is only several kilometers per second. Since attenuation rises with frequency, only electromagnetic waves with moderate frequency have any chance of being used in the geophysics. What does "moderate" mean?

Waves of the frequencies which are used for pulsed radar attenuate in the Earth completely within meters or tens of meters. Radiowaves at lower frequencies, such as those employed in broadcasting, penetrate the Earth no deeper than tens or hundreds of meters.

Electromagnetic oscillations with frequencies from about one oscillation per second are useful for geophysical purposes. In spite of the low propagation, their wavelengths in the Earth are so long that pulsing or direct application of the time factor to discriminate detail at different distances from the observer is not sensible. Directionality is also practically impossible. Intensity and frequency remain the only discriminators, such as in blurred vision.

These two discriminators have to be employed in frequency electromagnetic sounding or profiling, which seems to be the most promising electromagnetic method for investigations deep in the Earth. A developing method is based on measuring the variations in the natural electromagnetic field of the Earth, viz. magnetotelluric sounding (MTS). These variations can be seen in changes in the intensity of ultraviolet and particle radiation from the Sun as it enters the Earth's ionosphere. Variations with periods from several seconds to several minutes are used in MTS to study depths of the order of a kilometer, and with periods of several hours or even days to study deeper variations in the Earth's interior.

Magnetotelluric sounding is, generally, similar to the seismic method of surface waves. Surface waves also attenuate rapidly, exponentially with depth, and the higher the oscillation frequency the faster the attenuation, although the attenuation mechanism of elastic surface waves is different from that of electromagnetic ones. In elastic waves attenuation is associated with surface effects; as in waves on the surface of a liquid, waves damp quickly with depth, rather than with energy absorption. But the cause of the attenuation makes no difference in our comparisons of penetration depths of various waves. What matters is the result. From the same point of view, difference in the propagation direction for electromagnetic and elastic surface waves: downwards from the top for electromagnetic waves and horizontally for seismic surface waves, is not important either.

The two methods similar in that harmonic oscillations at different frequencies are combined. There is exponential damping with depth (the quicker, the higher the frequency), and the oscillations are treated as long waves with intensity and frequency being the only discriminators. This is also a reason for a common approach to the quantitative interpretation of observations. In particular, the properties and depths of the strata in the Earth are calculated on the assumption that they are horizontal.

Now, we can compare directly depth of penetration and resolution of the electromagnetic MTS method and the seismic method of surface waves. Later we shall compare MTS with the seismic method for body waves. The idea of the effective depth (z_e) of investigation or wave penetration, namely the depth at which the wave amplitude reduces by $e = 2.7$ times, is convenient in these comparisons.

Length (in km) of an electromagnetic wave in rock is $\lambda = \sqrt{10\varrho T}$, where ϱ is electrical resistivity (in Ohm·m) and T is the period (in s). The amplitude of an electromagnetic wave decreases with depth as $A = A_0 \exp(-2\pi z/\lambda)$, so that the effective depth (z_e) is $z = \lambda/2\pi = 0.16\lambda$. Numerical estimates for λ and z_e using these formulae are given in Table 21.

These effective depths are just for orientation, because the sedimentary layer in the Earth is usually no thicker than several kilometers; only in some areas it is 10 km and in places 20 km. Crystalline rock occurs at geater depths. Even so, Table 21 shows that although one minute waves cannot penetrate, electromagnetic waves with periods of an hour penetrate both the sedimentary cover and the entire crust (10-50 km thick) and even enter the mantle to a considerable depth.

The analogous estimates for seismic Rayleigh surface waves for a normal value of Poisson's ratio $\sigma = 0.25$ concerns the amplitude of the vertical displacement, which is

Table 21. Penetration of Electromagnetic Waves of Different Periods in Sedimentary and Crystalline Rocks

Oscillation period, T	Sedimentary rocks, $\varrho = 10$ Ohm $\cdot$ m		Crystalline rocks, $\varrho = 1000$ Ohm $\cdot$ m	
	λ, km	Z_l, km	λ, km	Z_l, km
1	10	2	100	20
1 min	80	10	800	100
1 h	600	100	6000	1000

$A = (A_0/0.7320) \times [1.7320 \exp(-0.3933\, 2\pi/\lambda) - \exp(-0.8475\, 2\pi z/\lambda)]$. Whence we find that the amplitude attenuates by $e = 2.7$ times $(A/A_0 = 1/e)$ by a depth of $z_e = 0.1\lambda$. On the other hand, the wavelength λ is related to the propagation velocity V_R and the period T as $\lambda = V_R T$. As a result we obtain Table 22.

It can be seen from these guiding values of the depth z_e that seismic surface waves penetrate the Earth's crust and enter the mantle to a considerable depth if they have periods of about a minute or more.

Comparing Tables 21 and 22 indicates that electromagnetic waves with periods of the order of hours or days and seismic waves with a period of about 10 s penetrate to the same depth. This difference in periods cannot, however, be directly used to compare the resolutions of the methods; the wavelengths have also to be compared for the same depth of investigation.

The number of waves with length λ fitting into the effective penetration depth z_e may serve as a characteristic of the resolution. The larger z_e/λ, the higher the resolution of the method. Calculations have shown that this ratio is 0.16 for electromagnetic waves, and 0.7 for surface seismic waves. Therefore, seismic waves have a resolution five times that of electromagnetic waves. This is a considerable but not overpowering superiority. It sometimes can be modified by other circumstances considered in our very schematic approach to the comparison of the resolutions.

Table 22. Penetration of Seismic Surface Waves of Different Periods in Sedimentary and Crystalline Rocks

Oscillation period, T	Sedimentary rocks, $V_R = 1.5$ km/h		Crystalline rocks, $V_R = 3$ km/h	
	λ, km	z_e, km	λ, km	z_e, km
10 s	15	10	30	20
1 min	90	60	180	100
10 min	900	600	1800	1000

Using the same approach body seismic waves, viz. longitudinal and transverse ones, are even better compared to electromagnetic waves. Let us do the calculations for such waves. Experience of deep seismic sounding in the USSR shows that the effective loss in absorption for the wavelength λ in sedimentary rock is of the order of 10^{-1}, for crystalline rock it is 10^{-2}, and for the mantle material is 10^{-3}. This decrement ϑ is included in the formula for the amplitude $A = A_0 \exp(-\vartheta z/\lambda)$, from this $z_e = \lambda/\vartheta$. Let us take approximate values for propagation velocities V_P of longitudinal P seismic waves in sedimentary rock $V_P = 3$ km/s, in crystalline rock $V_P = 6$ km/s, and in mantle material $V_P = 8$ km/s. Then we obtain the following table.

It can be seen from Table 23 that seismic longitudinal waves with a period of 1 s effectively penetrate ($z_e = 30$ km) any thickness of sediments. In crystalline rock, waves with period of 0.1 s and even better 1 s penetrate the whole of the Earth's crust. As concerns the mantle, waves with a period of 1 s and better again 10 s can penetrate the entire Earth. Note that in the case of seismic surface waves, oscillations with periods of some tens of minutes correspond to the Earth as a whole, while for electromagnetic waves such oscillations must have periods of a day.

The resolution z_e/λ for seismic longitudinal waves, using the assumed absorption loss ϑ, is $z_e/\lambda = 10$, 100 for crystalline rock, and 1000 for the mantle.

As compared with seismic surface waves in sediments, for seismic longitudinal waves the resolution is $10/0.7 = 15$ times higher, in crystalline rock 150 times higher and in the mantle material 1500 times higher. As compared with MTS electromagnetic waves, seismic longitudinal waves possess resolutions $10/0.16 = 60$ times higher for sediments, 600 times higher for crystalline rock, and 6000 higher for the mantle. This is an overwhelming superiority.

Seismic transverse waves in hard rock have about the same loss of absorption as longitudinal ones. This is why in this approach they have resolutions as good as longitudinal waves for seismic surface waves and MTS electromagnetic ones, and also for seismic longitudinal waves, namely, according to the circumstances of ten, hundred and thousand times.

Longitudinal seismic waves are somewhat better than transverse ones, as they have higher propagation velocity. Longitudinal waves are the first to arrive from a pulsed source to an area, not littered with interfering oscillations, which are usually characterized by lower velocities of propagation.

Table 23. Penetration of Seismic Longitudinal Waves of Different Periods in Sedimentary Rocks, Crystalline Rocks and in a Mantle Material

Oscillation periods, T	Sedimentary rocks, $V_P = 3$ km/s		Crystalline rocks, $V_P = 6$ km/s		Mantle material, $V_P = 8$ km/s	
	λ, km	z_e, km	λ, km	z_e, km	λ, km	z_e, km
0.1 s	0.3	3	0.6	60	0.8	800
1 s	3	30	6	600	8	8 000
10 s	30	300	60	6 000	80	80 000

Thus, seismic methods based on body waves leave all other geophysical methods far behind as far as resolution is concerned. Body seismic waves can actually have shorter wavelengths and yet penetrate to the same depth as other waves. This, in turn, also allows us to use (in body wave seismics) the best pulsed radar technique, viz. "elastic supervision", which makes use of all the known discriminators: intensity, frequency, directivity, and time. This yields the most subtle details to be distinguished in the Earth's structure, and allows them to be studied with great accuracy and reliability.

To sum up: seismic methods, or more precisely, methods based on observations over body seismic waves—longitudinal and transverse—appear to be the most powerful in studies of the Earth's structure due to the physics of waves. This makes seismology superior to other geophysical methods and makes it first among equals.

This does not mean, of course, that seismology can replace the other main physical methods of the Earth sciences: gravitational, magnetic, electrical, electromagnetic, thermal, and nuclear methods. Each is necessary, and each will find its place in the common cause: but because it has a higher resolution than other methods seismology has a special position among the basic geophysical methods. The quantitative conclusions of the other methods about the Earth's structure have to be compared with those obtained in seismology. Comparison is a reciprocal process, but because seismic data are accurate and reliable, they should be given more weight.

Since the community of geophysical methods should follow the principle of reciprocal assistance rather than competition, all geophysical methods will benefit from seismological advantages, in the long run.

Geophysical methods together with seismic surveying is an enormous force in overall knowledge of the structure of our planet's interior and its importance in the mind and necessities of man.

14. General Properties of Systems of Seismic Wave Travel Times*

G. A. Gamburtsev and I. S. Berson introduced the interesting theories "complete" and "incomplete" systems of seismic wave travel times. They concentrated on *incomplete* systems, defining them as systems that are insufficient for obtaining the unique solution of an inverse problem** with given initial conditions. The general properties of *complete* systems were of less interest to them. It was assumed in [Riznichenko], when considering main waves that no travel time systems with undetermined initial points can be used to estimate the velocities in the layer covering the refracting boundary. However, this was not proved.

I. S. Berson considered in great detail inverse problem for a head wave and a flat boundary between two homogeneous layers. She assumed that the layer velocity V_1 in the covering medium was known. But she did not consider whether V_1 could be determined by solving a similar problem.

Here we try to answer several associated questions which to my mind were not posed clearly enough. The investigation will be in two main directions.

1. To define more clearly the idea of a complete system of seismic wave travel times for waves of any type, head waves, in particular, and to study the relations between the complete system and individual travel times in the system.

2. To study the possibilities for complete solutions of the inverse problem—two-dimensional and spatial—for head wave travel times. The basic problem is to ascertain whether the layer velocity in the medium covering the refracting boundary can be determined.

Note that estimating layer velocities is the most involved of the inverse problems in seismics whatever the method: reflection survey, refraction survey and, partially, seismometry in wells. When the velocities are known, the inverse problem can, as a rule, be solved without much difficulty.

14.1. Complete System of Travel Times

Let us assume the medium to be isotropic.

Definitions. A set of travel time curves for waves of the same class together with a solution of the inverse problem is called a *system* of travel time curves. Waves of the same type (e.g., longitudinal head waves, transverse reflected, etc.) associated with the same seismic boundary are said to belong to the same class.

Travel time curves are categorized as I—*linear*, and II—*surface*.

Linear travel time curves are subdivided into (1) *longitudinal* when the shotpoint is on the profile; in this case the problem is two-dimensional, i.e., it is assumed that all rays are

---

* Manuscript prepared in 1944. In the 1970s this problem was considered in detail by N. N. Puzyrev.
** The inverse problem in geophysics is to interpret seismic observations or to determine the structure of a medium from seismic wave travel times or time fields—(*Author's note*).

on the same plane passing through the profile, and (2) *non-longitudinal*, when the shotpoint is not on the profile, and the problem is three-dimensional. *Transverse* travel time curves are conventionally categorized as (2) (see [Berzon]). Surface travel time curves—G. A. Gamburtsev calls them "isochron charts"—have not yet been subdivided. Systems of travel time curves may be classified in a similar way.

We shall not do this, as we are going to concentrate on just two classes of travel time systems, namely, (a) *longitudinal systems*, corresponding to the general two-dimensional case, and (b) *surface systems*, which correspond to the general three-dimensional case.

System function. In geometrical seismics an observation is an estimate of the time t, a wave takes to propagate between two set points: shotpoint-seismograph. Let their position in space be determined in the fixed coordinate system by vectors ϱ and $\mathbf{r}$, respectively.

In respect of $t\,(\mathbf{r}, \varrho)$ the following equation expressing the reciprocity principle is valid:

$$t(\mathbf{r}, \varrho) = t(\varrho, \mathbf{r}). \tag{14.1}$$

Assume, that the vector ϱ = const, and the end of the vector $\mathbf{r}$ runs across all points of a set section of the profile or a set area of the surface. Then (14.1) is a linear function or a surface travel time at the fixed shotpoint (ϱ). We may mention similarly a travel time function for a fixed seismograph ($\mathbf{r}$).

Let now both vectors $\mathbf{r}$ and ϱ change, and their ends run through all points of a given line or surface; in this case the function in (14.1) is called the *function of the complete travel time system*, or shorter, *system function*. It should be defined for a fixed area by changing the arguments, i.e.,within a fixed section of the profile or within a fixed area of the surface. Setting the function of the complete system in a three-dimensional region does not seem to be physically sensible*.

Complete system. Let us imagine, in the abstract, a set of observations consisting of an infinitely large number of travel time curves, so that shotpoints are at every point of a fixed area of the profile or the surface, and for each shotpoint seismographs are also located at every point of the same area. This set of observations fully determines the function of the complete system (14.1). A "dense" detailed system of travel time curves corresponds to this set in practice.

A *complete system* is one that for general premises concerning the wave propagation and structure of a medium (the premises themselves depend on the actual formulation of the problem) fully determines the function of the complete system in the same area. A system that does not possess this property will be called an *incomplete system.*

It is important that this definition of a complete system does not require the system to consist of an infinitely large number of travel time curves or to be dense. We shall show below that, in certain cases, a complete system may be represented by just a small number of properly chosen travel time curves. This definition does not require a complete system to be sufficient for a unique solution of the inverse problem. We shall show further, that in some important cases, such a requirement would have no meaning. Let us look at the corollaries of this definition; we are going to base out further considerations upon them.

---

* When this paper was written, observation techniques for wave fields inside a medium had not been developed.—(*Editor's remark*).

1. If a system is complete, it is sufficient for determining any travel time curve or any other incomplete or complete system of travel time curves for waves of the same class, irrespective of the concrete structure of the medium.

2. If a system is complete, then the addition to it of any number of travel time curves or extension of any other system of travel time curves for waves of the same class cannot give any more information about the structure of the medium.

Clearly an important prerequisite is that the general premises concerning wave propagation and the structure of the medium are in agreement with reality.

Space of the system. Let the problem be two-dimensional, so the travel time curves and their systems will be linear, i.e. longitudinal. The profile may be taken as a coordinate axis by firing origin and the positive direction. Then the scalar arguments χ and ξ, the coordinates of the seismograph and shotpoint, respectively, may be included in the function of the complete system (14.1) instead of the vector arguments $\mathbf{r}$ and ϱ, thus, the function may be written as

$$t = t(\chi, \xi) = t(\xi, \chi). \tag{14.2}$$

It may be thought of as a field of the scalar t in an imaginary "plane of the travel time system", with dimensions χ and ξ (Fig. 103). The values χ and ξ may be considered as the coordinates of individual observations. This field may be presented as a graph with level lines $t = t_i = $ const of isochrons. This presentation was proposed for the first time by G. A. Gamburtsev.

If we consider a three-dimensional problem, when shotpoints and seismographs are located on a surface, i.e., the two-dimensional analogue of a point, then for scalar arguments the function of the complete system will depend upon four arguments:

$$t = t (x, y; \xi, \eta) = t(\xi, \eta; x, y), \tag{14.3}$$

where (x, y) are the coordinates of the seismograph and (ξ, η) are coordinates of the shotpoint. The function may be conceived as a field of the scalar t in "the space of the travel time system", four-dimensional in our case. A simple graphic representation of the field defined by (14.3), unlike the field defined by (14.2) in the previous case, is not possible.

Generalizing both cases we refer to a "space of the complete travel time system", or shorter, the "space of the system" in the sense of a point analogue of x, ξ for a flat problem and the analogue of x, y, ξ, η for a spatial problem.

The field of the time t in the space of the system possesses the following properties:

(a) It has "mirror symmetry" with respect to $x = \xi$, $y = \eta$; so to define the field completely it is sufficient to define it for half the system space rather than for the whole space (for two-dimensional problems in one of two triangles, e.g., in the lower one, $0 \leqslant \xi \leqslant x \leqslant x_m$, rather than in the square $0 \leqslant x$, $\xi \leqslant x_m$ [see Fig. 103]).

(b) Reciprocal points coincide when one of the areas is reflected onto the other. So in the space of the system any linear travel time system with a continuous correlation (along the profiles and among reciprocal points) is represented by a sequence of points making up a continuous outline*.

---

* Continuous movement of a point of reflection or outcome of seismic radiation along seismic boundary also corresponds to a continuous correlation and the continuous outline in the space of the system.—(*Author's remark*).

Fig. 103. The ray of the BC curve

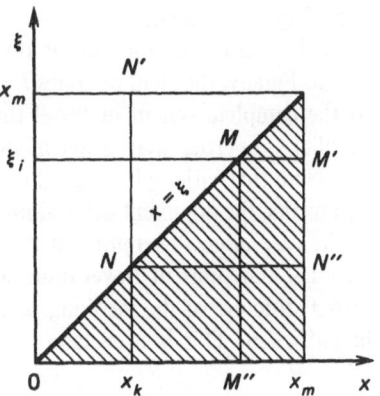

Property (b) is good in practice when choosing and analyzing complex correlation systems of longitudinal travel time curves, since the regions of existence and optimal traceability of waves of different types can easily be considered.

Time fields. The time field in geometrical seismics may be interpreted in various ways.

Let the number of dimensions of the space in the medium in which a seismic wave is propagating be n; $n = 2$ for two-dimensional problems; $n = 3$ for three-dimensional ones. Then the numerous points that include some travel time curves—the travel time space—are determined by the dimension $h \leqslant n - 1$. Thus, linear travel time curves alone $h = n - 1 = 1$ are considered in two-dimensional problems; linear travel time curves $h = 1 < n - 1$ and surface ones (isochron charts), $h = 2 = n - 1$—in three-dimensional problems. The manifold point that includes complete systems of travel time curves, the space of the system, has dimension $s = 2(n - 1)$. Thus, $s = 2$ in two-dimensional problems, $s = 4$ in three-dimensional ones. The behavior of the same value—the travel time t of a perturbation between some two points of the space (1) is studied in all three spaces, viz. those of (1) the medium, (2) the travel time curves, and (3) the system. This value in every case is considered to be a function of a point in a corresponding space (1), (2) or (3). It is reasonable to speak about time t fields in all three cases.

1. The time t^* field in the space of the medium characterizes movement of the wave front inside the medium. It directly depends on the properties of the medium. The main property of the field is expressed by the equation $(\text{grad } t)^2 = 1/V^2$, where V is the waves' true velocity at any point of the medium.

2. The field of times $t = g^{(h)}$ in the space of the travel time curve, on a line or on a surface, expresses the function of the travel time curve. It is related to the properties of the medium to a lesser degree. Its differential characteristic for the case $n = 2$ (linear travel times) is given by the apparent velocity $V \geqslant V^{(h)}$, and for $n = 3$ (surface travel times, isochron

---

* Functions of t are continuous, but their derivatives, generally speaking, may have discontinuities at a finite number of points (for $n = 2$) or at a finite number of curves (for $n = 3$). Functions of t may be both one-valued and many-valued.—(*Author's remark*).

charts) by the gradient of the travel time function (grad $t^{(h)})^2 \leqslant 1/V^{(h)}$, where $V^{(h)}$ is velocity V in the vicinity of the line or the surface where the travel times are defined.

3. Finally, the field of times $t = t^{(s)}$ in the space of the system determines the function of the complete system of travel time curves. Its general properties are related by (2); the peculiar properties were given in the previous section.

Note that although, in all three cases, the functions of the time t fields have common arguments, the argument determining the penetration of the waves into the medium is present in the case alone. Therefore, at a certain stage of the solution of problems in geometrical seismics, functions of travel times and their systems may be considered irrespective of the actual structure of the medium, so a general knowledge of its character and the waves may be sufficient. This is used later.

14.2. Longitudinal Systems of Travel Time Curves of Head Waves

Let us assume the problem is flat, i.e., the space of the medium is a plane containing the profile and all the seismic rays. The profile, a curve actually, will be taken for the coordinate axis x.

We now define the general premises mentioned in Sec. 14.1. Let the structure of the medium be such that head waves, which do not turn into refracted (piercing) waves or diffracted ones under an angle other than the limit angle, arrive in the section of the profile x of the refracting boundary R; a ray is represented in Fig. 104 by the curve BC. This is formed in the medium covering R when there is a travelling wave in the layer under R. The ray A_1, A_2, B, which moves with boundary velocity V_2, has higher velocity than the layer velocity V_1 in the upper layer. The travelling wave in the layer under R has first arrivals in points of R which correspond to rays coinciding with R. These premises exclude from our consideration penetration and diffraction at an upward salient angle.

Differential equation of the system function. Let two shotpoints O_1 and O_2 have coordinates ξ_1 and ξ_2, with $\xi_2 - \xi_1 = \Delta\xi$ (see Fig. 104). The corresponding travel time curves for

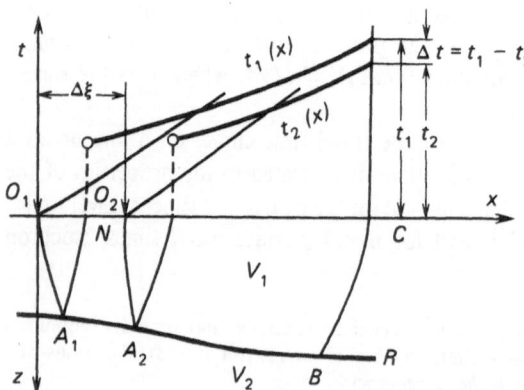

Fig. 104. ξ_1 and ξ_2 coordinates of two shotpoints O_1 and O_2

head waves $t_1(x)$ and $t_2(x)$ are expressed by means of the system function $t(x, \xi)$ as:

$$t_1(x) = t(x, \xi_1); \quad t_2(x) = t(x, \xi_2), \tag{14.4}$$

where ξ_1 and ξ_2 are constants.

For the same value of the argument x, the travel time curves $t_1(x)$ and $t_2(x)$ of head waves are separated in time by the interval $\Delta t = t_1 - t_2$, which depends on ξ_1 and ξ_2 only, but does not depend on x. This statement can easily be seen in Fig. 105.

Let us now assume that ξ_1 and ξ_2 are variables. We compose a relation $\Delta t/\Delta \xi$ and make $\Delta \xi$ tend to zero. As a result we have:

$$\partial t(x, \xi)/\partial \xi = \psi(\xi), \tag{14.5}$$

similarly,

$$\partial t(x, \xi)/\partial x = \varphi(x). \tag{14.6}$$

Differentiating with respect to the second argument, both (14.5) and (14.6) may be substituted for by a smaller equation, which is equivalent to both but more convenient being symmetrical:

$$\partial^2 t(x, \xi)/\partial x \partial \xi = 0. \tag{14.7}$$

Equation (14.7) expresses the features of the function of a longitudinal system of head wave travel time curves. However, it may be treated more broadly, namely as a differential equation of the function $t(x, \xi)$ for the given case. We shall show that (14.7) allows the value of the funciton $t(x, \xi)$ to be restored at each point of the area of the system x, ξ space, two-dimensional if the function is fixed for some one-dimensional point manifold $\Xi(x, \xi) = 0$, belonging to the same domain (a Cauchy boundary value problem). The boundary of the head wave's existence area containing the initial points of the travel time curves is not determined. This boundary is not of interest here; it is presumed that the travel time curves of head waves may be extended to corresponding shotpoints by means of overlapped systems.

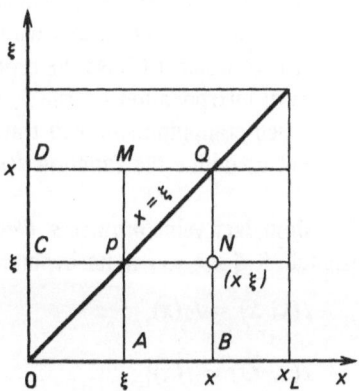

Fig. 105. The dependence of $\Delta t = t_1 \simeq t_2$ interval on ξ_1 and ξ_2

General properties. Before we formulate the boundary value problems, let us discuss the general properties of the function t that directly results from (14.7).

We shall integrate (14.7) first with respect to ξ, $\partial t(x, \xi)/\partial x = f_1(x)$, and then with respect to x, $t(x, \xi) = \int f_1(x) dx + \theta_2(\xi)$. Here $\theta_2(\xi)$ is an arbitrary function of ξ.

Let us introduce the variable $\theta_1(x) = \int f_1(x) dx$, from which the system function $t(x, \xi)$ will be finally written as:

$$t(x, \xi) = \theta_1(x) + \theta_2(\xi). \tag{14.8}$$

So, in this case the system function may be expressed as the sum of two functions, one of which depends on x alone, and the other on ξ alone.

Assuming that $x = x_i$, then $x = x_k$ in (14.8) and taking the difference of the resultant expressions, we get:

$$t(x_i, \xi) - t(x_k, \xi) = \theta_1(x_i) - \theta(x_k) = \Phi_1(x_i, x_k). \tag{14.9}$$

We do a similar operation with respect to $\xi = \xi_l$ and $\xi = \xi_m$:

$$t(x, \xi_l) - t(x, \xi_m) = \theta_1(\xi_l) - \theta_2(\xi_m) = \Phi_2(\xi_l, \xi_m). \tag{14.10}$$

Expressions (14.9) and (14.10) present, in a generalized form, the property of the "parallelism" of the branches of overlapped travel time curves of head waves (the same property in a narrower form was used to derive the differential equation (14.7)). These expressions allow us to formulate the following rule for operations with the functions $t(x, \xi)$ for the head wave.

If two functions of a longitudinal system of travel time curves for head waves have a difference and have one equal argument, an arbitrary value may be assigned to the argument, and the difference will not be changed:

$$t(a, b) - t(c, b) = t(a, N) - t(c, N). \tag{14.11}$$

Here N is an arbitrary number from the range of the arguments $0 \leqslant N \leqslant x_L$. Note that the order of the arguments is not important, i.e., $t(a, b) = t(b, a)$ etc.

The following interpretation may be given to (14.11) for the space (plane) of the system x, ξ (see Fig. 105). We draw two "vertical" lines, i.e., parallel to the ξ axis. For each pair of such lines the difference of times t for any pair of points lying on these lines at equal "height" ξ is a constant. A similar statement is valid for the "horizontal" lines, viz. the difference of times between points with the same abscissa x remains equal.

It is convenient to use the property (14.11) expressed for the function $t(x, \xi)$ and its geometrical interpretation on the x, ξ plane in operations with head-wave travel time curves, e.g., when compiling summed travel time curves.

Let us define the function $t(x, \xi)$ for given boundary conditions.

Boundary value problems. *Case 1.* Find the function $t(x, \xi)$ at any point $N(x, \xi)$ of the triangle $0 \leqslant x \leqslant x_L$, if its hypotenuse is $x = \xi$ and horizontal is:

$$t(x, x) = t_0(x) \tag{14.12}$$

$$t(x, O) = \overrightarrow{t}(x). \tag{14.13}$$

In other words, the problem is given the times $t_0(x)$ at shotpoints ("densely" located) along the entire length of the profile $0 \leqslant x$; $\xi \leqslant x_L$ and the times $t(x)$ along the straight travel time graph with shotpoints at the origin of the profile (O), find the time $t(x, \xi)$ at which the head wave arrives at any point x of the profile at any position of the shotpoint ξ.

Let us look at Fig. 105. We label values of the function $t(x, \xi)$ at the points A and B by the same letters A and B. $A = C$, $B = D$, P and Q, M and N are given according to the problem. $N(t, \xi)$ is to be determined. Using (14.11) we take the differences $N - P = B - A$, $N - Q = C - D = A - B$. Hence we get $N = B - A + P = A - B + Q$ or in coordinates:

$$t(x, \xi) = t(x, O) - t(\xi, O) + t(\xi, \xi) = t(\xi, O) - t(x, O) + t(x, x). \qquad (14.14)$$

Eventually, using the notations of (14.12) and (14.13) the last expressions take their final form:

$$t(x, \xi) = \vec{t}(x) - \vec{t}(\xi) + t_0(\xi) = t(\xi) - \vec{t}(x) + t_0(x). \qquad (14.15)$$

Note that this result is in good agreement with (14.8): the function of system (14.15) is equal to the sum of two functions one of which depends on x alone, and the other on ξ alone. *Case 2.* Find the function $t(x, \xi)$ for the entire triangle, if its two sides are O, $O \leqslant x \leqslant x_L$, x_L, $0 \leqslant \xi \leqslant x_L$

$$\vec{t}(x, O) = \vec{t}(x) \qquad (14.16)$$

$$t(x_L, \xi) = \vec{t}(\xi). \qquad (14.17)$$

In order words, given the times along the profile $O \leqslant x$, $\xi \leqslant x_L$ by the direct $\vec{t}(x)$ and the reverse $l(x)$ travel time curves in shotpoints at the ends O and x_L, respectively, of the profile (keep in mind that $t(x_2, \xi) = t(x, x_L)$). What is the time t at any point x of the profile at any position of the shotpoint ξ?

We shall apply the same solution as above. According to Fig. 105, using (14.11) we write $N - B = E - L$, $N = B + E - L$, which in coordinates gives:

$$t(x, \xi) = t(x, O) + t(x_L, \xi) - t(x_L, O). \qquad (14.18)$$

We use the notation $T = t(x_L, O) = t(O, x_L)$. This value is the time at reciprocal points for travel time curves with shotpoints at O and x_L, i.e. at the ends of the profile.

In the notation of (14.16), (14.17) and (14.18) we get:

$$t(x, \xi) = t(x) + \overleftarrow{t}(\xi) - T. \qquad (14.19)$$

This solution has a general form, as mentioned above, viz. (14.8). For the special case when $\xi = x$, we get from (14.19), taking into account (14.12), a formula, well known in seismic prospecting, i.e.,

$$t_0 = \vec{t}(x) - \overleftarrow{t}(x) - T.$$

Reverse travel time curves as a complete system. The result in (14.19) is of principal importance. It shows that a function of a complete longitudinal system of travel time curves for head waves is fully determined irrespective of the actual structure of the medium just by two reverse travel time curves (14.16). Hence by definition, *two head wave reverse travel time curves with shotpoints at the ends of the profile make up a complete system.*

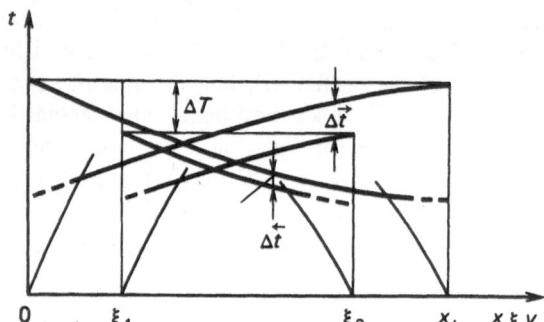

Fig. 106. The equality of the sum of carriers $\Delta \vec{t}$, $\Delta \overleftarrow{t}$ and the carry $\Delta \overleftarrow{t}$ of their reciprocal points

Carry-over rule. Let us also deduce from first principles a rule for reciprocal points to be carried over when passing from one pair of reverse travel time curves to another. This rule is as follows: the sum of carries $\Delta \vec{t}$ and $\Delta \overleftarrow{t}$ of reverse travel time curves is equal to the carry ΔT of their reciprocal points (Fig. 106).

So we have a relation between Δt, $\Delta \overleftarrow{t}$ and ΔT (see Fig. 106): $\Delta \vec{t} = \vec{t}(x, O) - t(x, \xi_1)$; $\Delta \overleftarrow{t} = t(y, x_L) - t(y, \xi_2)$; $\Delta T = t(x_L, O) - t(\xi_1, \xi_2)$.

Let us make up the sum of the first two differences and according to (14.11) substitute the value ξ for the equal arguments in the first difference and zeroes in the second one:

$$\Delta \vec{t} + \Delta \overleftarrow{t} = t(\xi_2, O) - t(\xi_2, \xi_1) + t(O, x_L) - t(O, \xi_2).$$

Exchanging the arguments in $t(O, \xi_2)$ and $t(\xi_2, \xi_1)$ we get finally $\Delta \vec{t} + \Delta \overleftarrow{t} = \Delta T$, QED.

14.3. Complete Inverse Two-dimensional Problem

General considerations. Inverse problems of two kinds are considered in geometrical seismics of layered media [Riznichenko]. The first kind is when the layer velocity in the medium covering the boundary is given; it is required to construct the boundary on the basis of travel times for waves of the given class. Another requirement may be added in a reflection survey, namely to determine the boundary velocity. The second kind of problem is when the seismic boundary is required and the velocity in the covering medium is determined on the basis of travel times for waves of the given class without any outside data. We shall call the problems of the first kind *bounded*, and those of the second kind *complete*.

Bounded problems for complete systems of travel time curves of different wave types may be solved with the general time-field method and offer no difficulty, in principle. However, complete problems have no general method of solution; each concrete problem has to be solved by a specific technique*.

The simplest problem of the second kind is solved by the reflection survey. In almost all cases the velocity in the covering medium is constant. For a refraction survey, problems

---

* There is a solution for the problem of velocity determination in the upper medium by conjugate points of refracted (head) waves [S. Goldin] etc.—(*Editor's remark*).

of this kind were considered by Wiechert where the velocity depends on depth alone and increases with depth; the travel times of refracted waves were exclusively used, rather than those of head waves. To formulate a complete problem for head waves we need to find a way to determine at least some effective ("intermediate") velocities V_1 = const in the covering medium, when the true velocity, for the same area, changes with depth in an arbitrary way.*

The impossibility of estimating layer velocities by longitudinal travel time systems without initial points. Tracing the beginning of head wave travel time curves is difficult in practice today in seismic prospecting; it can rarely be performed systematically with sufficient accuracy for quantitative interpretation. So we shall presume that the initial points of the travel time curves are not given.

We use the same premises concerning the general character of wave propagation and structure of the medium as in Sec. 14.2. The law of layer velocities and the form of the refracting boundary remain arbitrary with some insignificant constraints.

Let us prove that *no longitudinal system of head wave travel time curves allows the layer velocity in the covering medium to be estimated if the initial points of the travel time curves are not given.* In other words, the complete problem has no certain solution under these conditions.

We shall use the conclusions of Sec. 14.2 about general systems of head wave travel time curves. Instead of any preassigned longitudinal system, we shall consider a complete system incorporating the given one. Thus, the system does not contain more data for the solution of the inverse problem than the complete system. If our statement is valid for the complete system then it will surely be valid for any incomplete system.

Let us realize the complete system according to Sec. 14.2 as two reverse travel time curves extended to reciprocal points and shotpoints at the ends of the profile. We assume an arbitrary law of layer velocities V_1 in the medium, but it should be such that, in the vicinity of the line where the travel time curve is given $V_1 \leqslant |V^*_{max}|$, where V^*_{max} is the maximum by absolute value apparent velocity by reverse travel-time curves. Then we pose a boundary problem: construct the refracting boundary and determine the boundary velocity at each point on the basis of two reverse travel time curves and the law of layer velocities. It follows from solving the problem by the time field method that the solution does exist (and is unique) for any given law of layer velocities under the condition $V_1 \leqslant V_{max}$. None of the given laws of layer velocities is contradictory to the given reverse travel time curves under this condition. From this the validity of our statement follows.

A restriction on the domain of existence of the solutions. Solution of the complete problem is not arbitrary, at least due to the condition $V_1 \leqslant |V^*_{max}|$. It would be interesting to find boundaries for the region of possible solutions for the velocity V_1 and the position of the refracting boundary in space.

The condition $V_1 \leqslant |V_{max}|$ delineates a small area of the medium in the immediate vicinity of the curve, where the travel times are given. This is why it does not characterize the

---

* There is a note (given here) in a paper by G. A. Gamburtsev that concerns this question: "An interesting idea of M. Vasiliev should be mentioned here. He showed that a layer with lower velocity may be detected by travel times of refracted waves (a two-dimensional problem seems to be meant here— *Yu.R*) when the surface of the underlying layer with high velocity has some sharply expressed unevennesses." We shall see to what degree M. Vasiliev's idea may be considered valid.—*(Author's remark)*.

behavior of V_1 throughout the covering medium. To make the problem more sophisticated let us introduce an additional condition. Let $V = $ const, i.e., we introduce into consideration the constant effective (or "average") velocity.

Under the condition $V_1 = $ const $\leqslant |V^*_{max}|$ and with respect to the evident condition $V_1 > 0$, the region where the solutions can exist limited. From a strictly formal point of view, further restrictions should be recognized as groundless, but physical considerations allow us to go further.

The requirements of a general character may first of all be made to the result, as follows: the boundary should not have any return points or cross itself, the boundary velocity should not extend to the limits 0, ∞. In some cases, these conditions may considerably narrow the region of existence. But in practice much stronger restrictions may be imposed, considering the actual conditions of seismic prospecting. Sometimes they appear so large that the solution will be "practically unique". However these problems outstep the limits of this paper.

Travel time curves with initial points. If initial points of head wave travel time curves are given, then under certain conditions, the complete problem obtains a unique solution. For the sake of completeness let us consider this case too.

Let a complete longitudinal system of head wave travel time curves be given, together with the position N of the initial point of the corresponding travel time curve $t = t$, (x) the shotpoint being located on O_1 (see Fig. 104). Assume additionally that the velocity V_1 in the covering medium is constant. These conditions are enough for the unique solution of the complete problem.

In fact, we define the travel time $\overleftarrow{t}$ with the shotpoint in N, opposite to the one given (not shown in Fig. 104). The point O_1 is initial for it. Elements of head wave travel time curves t and $\overleftarrow{t}$ in the initial points coincide with elements of travel time curves of reflected waves for the same boundary. The points O_1 and N are reciprocal. Thus we have defined elements of travel time curves of reflected waves in reciprocal points. But as is proved in [Riznichenko], these data are sufficient to estimate uniquely the velocity $V = $ const in the covering medium. The statement is proved.

Note that there is no formal need in the complete system to solve the problem in question. The availability of elements of two head wave reverse travel time curves t and $\overleftarrow{t}$ at their initial points $\overrightarrow{x}$ and $\overleftarrow{x}$, with shotpoints of the travel time curves t and $\overleftarrow{t}$ located in points $\overrightarrow{x}$ and $\overleftarrow{x}$, would be enough.

14.4. Linear and Surface Systems of Head Wave Travel Time Curves

Let us consider the three-dimensional problems (medium x, y, z).

Let surface travel time curves for waves of any type be traced on the plane $z = 0$ (ground surface) in the area $O \leqslant x \leqslant x_m$, $y \leqslant y \leqslant y_m$. Then the function of the complete system of surface travel time curves depends on four components:

$$t = t(x, y; \xi, \eta), \qquad\qquad (14.20)$$

where x, y are the coordinates of the seismograph, and ξ, η are those of the shotpoint, $O \leqslant x$, $\xi \leqslant x_m$; $0 \leqslant y$, $\eta \leqslant y_m$.

Linear system. Let us single out from the set of all travel time curves (14.20) a set related to a line $y = f(x)$ on the ground

$$t = t[x, f(x); \xi, f(\xi)] = \tau(x, \xi). \tag{14.21}$$

The function t, (the function of the linear system) has the form of a complete longitudinal system, which is considered in flat problems (medium x, z). One might think that all the conclusions valid for longitudinal systems should also be valid for the linear system of travel time curves determined by this function in the case of the three-dimensional problem (medium x, y, z). However, this is not so. Indeed, in the case of longitudinal systems, the main rule is that all rays lie on one plane x, z that passes through the profile line, for any profile. But for linear profiles $y = f(x)$ on the surface $z = 0$ of the three-dimensional medium, this statement is not true. When the profile $y = f(x)$ is straight, this is also no exception.

Therefore, the function of a linear system such as (14.21) cannot be considered a function of a complete longitudinal system. All the previous conclusions concerning complete systems and, in particular, the specific features of the function of longitudinal systems of head wave travel time curves for three-dimensional problems are not applicable.

General properties of the surface system function. Let us consider some properties of (14.20). The length element in the system space x, y, ξ, η will be defined by

$$dl = \sqrt{dx^2 + dy^2 + d\xi^2 + d\eta^2}. \tag{14.22}$$

If the shotpoint is stationary, ξ, $\eta = $ const, then dl becomes the element ds of the surface $z = 0$; $dl = ds = \sqrt{dx^2 + dy^2}$, and if the seismograph is stationary, x, $y = $ const— into the element of the same surface $dl = d\sigma = \sqrt{d\xi^2 + d\eta^2}$.

Let us introduce the vector:

$$\text{grad } t(x, y, \xi, \eta) = \{\partial t/\partial x, \partial t/\partial y, \partial t/\partial \xi, \partial t/\partial \eta\}, \tag{4.23}$$

where grad t is the gradient of the surface system function; its components along the corresponding axes are written on the right-hand side.

We fix two points in the system's space: $A(x_A, y_A, \xi_A, \eta_A)$ and $B(x_B, y_B, \xi_B, \eta_B)$. Let values of the function t equal to t_A and t_B correspond to these points. We draw a line L between the points A and B. Then

$$\int_A^B \text{grad } t \, d\mathbf{l} = t_B - t_A, \tag{14.24}$$

where $d\mathbf{l}$ is an element of the integration outline L.

Let us now assume that $\xi = $ const and $\eta = $ const, i.e., the shotpoint is stationary. The projection grad t_{xy} of the vector grad t on the plane x, y (ξ, $\eta = $ const) is the gradient of the function of the surface travel time curve with a stationary shotpoint:

$$\text{grad } t_{xy} = \{\partial t/\partial x, \partial t/\partial y\}. \tag{14.25}$$

It is well known that $|\text{grad } t_{xy}| = 1/V_{max}^*$, where V_{max}^* is the maximum of the apparent velocity for the surface travel time curve. The projection of τ_S vector of grad t to S direction in xy plane is the value opposite to the apparent velocity in this direction:

$$\tau_S = |\text{grad } t_{xy}|_S = 1/V_S^*. \tag{14.26}$$

Similar relations may also be found for x, y = const: the seismograph is stationary, and the shotpoint moves.

Surface travel-time curves of head waves. Equations (14.20)-(14.21) are applicable to travel time curves of any type. Let us consider the case of head waves. We restrict our considerations to the specific case, when the velocities V_1, V_2 in the layers (V_1), (V_2) are constant, and the boundary between the layers is flat and oriented in space in an arbitrary way. A feature of this case is that each radial straight profile (crossing the shotpoint or the stationary seismograph) is longitudinal, and the apparent velocity V^*- remains constant along the profile.

Individual travel time curves. Let us fix the position of the shotpoint at the point O on the ground surface xy (Fig. 107). We define the directions of the normal $\mathbf{n}$ to the refracting boundary k and the profile λ by unit vectors:

$$\mathbf{n} = \sin \varphi \cos \alpha_0 - \sin \varphi \sin \alpha_0; \quad \cos \varphi; \quad \lambda = \cos \alpha; \quad \sin \alpha_0. \qquad (14.27)$$

Here φ is the dip angle of the boundary R; α_0 is the dip azimuth; α is the profile azimuth.

The gradient τ = grad $t(x, y, z)$ of the field of times t of the head wave in the layer (V_1) equals in modulus the inverse value of the velocity V_1; $|\tau| = 1/V_1$; its direction at each point coincides with ray directions. The latter are orthogonal to the isochrones $t = t_M$ = const of the head wave—the surface of a circular cone with axis n and crossing the shotpoint O. The vertex M of the cone evenly slides along the straight line $\mathbf{n}$, with time t_M increasing. The isochrone t_M on the ground surface is a cone section, i.e., an ellipse, parabola or hyperbola, with azimuth α_0. The vertex angle of the cone is $\pi - i$; $\sin i = V_1/V_2$.

Every vector τ crosses the straight line $\mathbf{n}$. The angle of crossing is i, so $(\mathbf{n}, \tau/\tau) = \cos i$.

Let us define the apparent velocity V^* of the head wave along the profile λ: $1/V^* = \tau_\lambda = \tau\lambda$, by expressing it through V_1, V_2, n, λ. We keep in mind that every vector τ crossing the λ is complanar with $\mathbf{n}$ and λ (Fig. 108). Hence:

$$(\tau/\tau, \lambda) = (1/\tau)(\tau, \lambda) = V_1/V^*. \qquad (14.28)$$

On the other hand, $(\tau/\tau, \lambda) = \cos \delta = \cos (\pi - \gamma - i) \cos (\psi - i)$, where $\Psi = \pi - \gamma$. Let us express ψ in terms of the function of the given vectors, or in components according

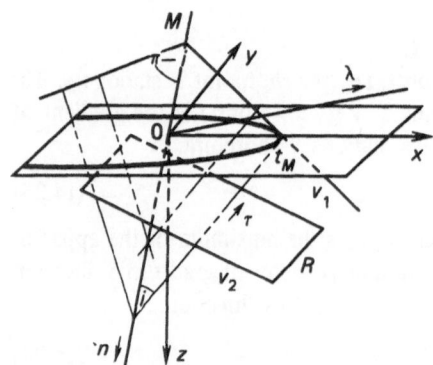

Fig. 107. The fixation of shotpoint at the point O on the ground surface xy

Fig. 108. The crossing of τ vectors with λ,
complanar with $\mathbf{n}$ and λ

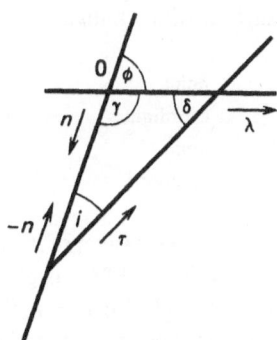

to (14.27):

$$\cos \Psi = \sin \varphi \cos (\alpha - \alpha_0). \tag{14.29}$$

From (14.28) we find the necessary final formula for the apparent velocity:

$$V^* = V_1/\cos (\psi - i) \tag{14.30}$$

Here ψ is determined by the equation (14.29). Equation (14.30) expresses the dependence of the apparent velocity V^* on the azimuth α of the radical profile $V^* = V^*(\alpha)$. In particular, for $\alpha = $ const we have $V^* = $ const, the apparent velocity is constant for each radial profile.

Let a point on the ground surface xy have coordinates (x_i, y_i), and another point have coordinates (x_k, y_k). The distance l between these points is $l_{ik} = \sqrt{(x_k - x_i)^2 + (y_k - y_i)^2} = l_{ki}$. The azimuth α of the direction from the point (x_i, y_i) to the point (x_k, y_k) is $\alpha_{ik} = \arctan (y_k - y_i)/(x_k - x_i) = \alpha_{ki} \pm \pi$. Let the time t of the head wave's arrival from (x_i, y_i) at (x_k, y_k) be $t_{ik} = t(x_k, y_k; x_i, y_i) = t_{ki}$. Then there is a relation between the times t and the apparent velocities V^* along the radial profiles

$$t_{ik} = t_{ii} + l_{ik}/V^*(\alpha_{ik}), \tag{14.31}$$

or:

$$t_{ki} = t_{kk} + l_{ki}/V^*(\alpha_{ki}). \tag{14.32}$$

The equations (14.31) and (14.32) in our notation are in this case equations of surface travel time curves ("izochrone charts") with shotpoints at the points (x_i, y_i) and (x_k, y_k) respectively. The values t_{ii} and t_{kk} are values of the times at corresponding shotpoints.

Surface system function. We show that in the case of a flat boundary R between two homogeneous layers (V_1) and (V_2), the function of the complete surface system of head wave travel time curves is fully determined by one surface travel time curve belonging to this system.

Assuming $V^* = V^*(\alpha)$, which is the dependence of the apparent velocity on the azimuth, also that the time t_0 of the arrival of the head wave to the seismograph (x_0, y_0) from the shotpoint (ξ_0, η_0), $t_0 = t(x_0, y_0; \xi_0, \delta_0)$ is known. We shall try to find the function of the complete system $t = t(x, y; \xi, \delta)$. In a shorter notation we shall designate points on the ground surface with xy digits:

point coordinates	x_0, y_0	ξ_0, η_0	x, y	ξ, η
point notations	1	2	3	4

Then we can easily use the notation adopted earlier in (14.30)-(14.31), supposing that $i, k = 1$, 2, 3, 4. Thus, $V^*(\alpha)$ is given in the new notation, it is known that $V^* = V^*(\alpha_{ik})$, $i, k = 1$, 2, 3, 4, and also $t_0 = t_{ik}$; $t = t_{34}$ is to be found. Let us take the general formula (14.24). The coordinates of the points A and B in the space x, y; ξ, δ are written in the new notations as (1, 2) and (3, 4), and the formula (14.24) takes the form

$$t_{34} = t_{12} + \int_{(1,2)}^{(3,4)} \operatorname{grad} t \, dl.$$

In order to be able to use the property V^* which remains constant at each of the radial profiles $(ii)(k, k)$, we shall complete the integration with respect to one of the polygons with apexes at the points (12), (11), (13), (33), (34), or (12), (22), (23), (33), (34), or (12), (11), (14), (44), (43), or (12), (22), (24), (44), (43), We choose, for example, the first outline.

Due to (14.24) and (14.31) we have:

$$\int_{(12)}^{(11)} \operatorname{grad} t dl = t_{11} - t_{12} = l_{12}/V^*(\alpha_{12});$$

$$\int_{(11)}^{(23)} \operatorname{grad} t dl = t_{13} - t_{11} = l_{13}/V^*(\alpha_{13});$$

$$\int_{(13)}^{(33)} \operatorname{grad} t dl = t_{33} - t_{31} = l_{31}/V^*(\alpha_{31});$$

$$\int_{(33)}^{(34)} \operatorname{grad} t dl = t_{34} - t_{33} = l_{34}/V^*(\alpha_{34}).$$

Making the following:

$$\int_{(12)}^{(34)} = \int_{(12)}^{(11)} + \int_{(11)}^{(13)} + \int_{(13)}^{(33)} + \int_{(33)}^{(34)} = t_{34} - t_{12} =$$

$$- l_{12}/V^*(\alpha_{12}) + l_{13}/V^*(\alpha_{13}) - l_{31}/V^*(\alpha_{34}) + l_{34}/V^*(\alpha_{34})$$

thus, we have:

$$t_{34} = t_{12} - l_{12}/V^*(\alpha_{12}) + l_{13}/V^*(\alpha_{13}) - l_{31}/V^*(\alpha_{31}) + l_{34}/V^*(\alpha_{34}).$$

Finally, considering that $l_{13} = l_{31}$, and returning to the initial notation, we find the required function of the complete system:

$$t(x, y, \xi, \eta) = t(x_0, y_0, \xi_0, \eta_0)$$

$$= \sqrt{(\xi_0 - x)^2 + (\eta_0 - y_0)^2}/V^*/(\alpha_{12}) + (1/V^*(\alpha_{13})$$

$$- 1/V^*(\alpha_{31})\sqrt{(x - x_0)^2 + (y - y_0)^2}$$

$$+ \sqrt{(\xi - x)^2 + (\eta - y)^2}/V^*(\alpha_{34}), \tag{14.33}$$

where

$$\alpha_{12} = \arctan\left[(\eta_0 - y_0)/(\xi_0 - x_0)\right]$$

$$\alpha_{13} = \arctan\left[(y - y_0)/(x - x_0)\right], \quad \alpha_{31} = \alpha_{13} \pm \pi$$

$$\alpha_{34} = \arctan\left[(\eta - y)/(\xi - x)\right]. \tag{14.34}$$

Integration with respect to other outlines may be carried out in the same manner. Note that (14.33) and similar equations may be derived from considerations about correlation of times by individual profiles and by reciprocal points; this is common in seismic prospecting.

So, in accordance with Section 14.1, for a flat boundary between two homogeneous layers, one surface travel time curve of head waves makes up a complete system.

Corollary (1) in Sec. 14.1 is realized here as follows: the function of the complete system $t(x, y; \xi, \eta)$ may be restored provided there is one surface travel time curve; this is fulfilled by means of (14.33). But any travel time curve or any system may be derived from the function $t(x, y; \xi, \eta)$. Indeed, if we fix one pair of factors, e.g., $\xi = \xi_i$, $\eta = \eta_i$ in the expression (14.33), it produces the equation of a single surface travel time curve, an isochrone chart, with the shotpoint (ξ_i, η_i). If we further assume that x is linearly dependent on y, for instance $y = ax + b$, where a, $b = $ const, then (14.33) gives the equation of the travel time curve at a linear profile. The latter may be longitudinal or non-longitudinal, transverse, in particular, depending upon the chosen values of a and b.

We shall show in the next section that with the general premises we have adopted an individual surface travel time curve may, in turn, be fully determined provided a finite number of linear travel time curves is given, so long as they do not make up a longitudinal system.

14.5. Complete Inverse Spatial Problem

General considerations. 1. It was shown in Sec. 14.3 that for a two-dimensional problem with any structure of the medium, it is impossible, in principle, to find the layer velocity in the covering medium for all longitudinal systems of head wave travel time curves without initial points.

2. It was shown in]Riznichenko] that things are different in the three-dimensional problem: there exist travel time systems that allow the layer velocity to be found.

3. Finally, it was shown in Sec. 14.4, if we restrict the three-dimensional problem to a flat boundary between two homogeneous layers (later), then just one surface travel time curve

of the head wave is a complete system and that, consequently, any specific travel time system—linear or surface—cannot in, principle, provide more independent data than one surface travel time curve.

It follows from (1), (2) and (3) that unlike all longitudinal systems, one surface travel time curve of head waves for this medium structure contains sufficient data to solve the problem of determining the layer velocity in the covering medium. Thus, the existence of a solution is not questioned. The problem remains to find this solution.

Problem formulation. Main premises: the boundary is flat, the velocities are constant (V_1, i, φ, α_0 = const). The inverse spatial problem consists, in this case, in determining five independent parameters which fully characterize the structure of the medium: V_1, i, ψ, α and h_0. The meaning of the first four values is known. The fifth parameter h_0 determines the position of the refracting boundary by depth below the given point on the ground surface.

It was shown in Sec. 14.4 that the complete system of head wave travel time curves is determined in this case, first, by the function:

$$V^* = V^*_{(\alpha)} = V^*(V_1, i, \varphi, \alpha_0, \alpha),\tag{14.35}$$

and, second, by the value t_0 at a certain point (x_0, y_0, ξ_0, η_0).

If V_1, i, ψ and α_0 are known, the parameter h_0 can easily be expressed as a function of t_0; but we should not do that. The main difficulty of the reverse problem consists in determining just the parameters V_1, i, φ, α_0, i.e., factors of the function V^* (14.35). That is what we are going to do. The parameter V^* does not depend on t_0 and is not related to h_0. Thus, the complete set of data to be a basis for our solution is represented by the function $V^*_{(i)}$.

It is natural to pose the problem in one of the following formulations:

Problem 1. Given a system of four (the number of unknowns) equations like (14.35) $V^*_k = V^*(V_1, i, \varphi, \alpha_0, \alpha_k)$, where V^*_k and α_k are observed values, i.e., numbers represented by points, the system must be solved for the unknowns V_1, i, φ, α_0 and it is necessary to determine their exact values.

Problem 2. Given a system $n > 4$ of an equation such as (14.35) (the measured values V^*_i, α_i, $i = 1, 2, \ldots, n$) may have errors. Find the most probable values of four parameters: V_1, i, ψ and α_0.

The second formulation fits better the practical requirements. The first one is more convenient for developing clear questions of a principle character. We shall consider both problems. We shall try to obtain an analytical solution for the first problem, and a graphical one for the second problem. Let us consider the first.

Problem using four radial profiles. Given a system of four equations like (14.35), that is (see 14.4):

$$V^*_k = V_1(\cos \Psi_{k-1}), \quad k = 1, 2, 3, 4,\tag{14.36}$$

where

$$\cos \Psi = \sin \varphi \cos (\alpha_k - \alpha_0).$$

V_1, i, φ and α_0 are to be expressed in the function V^*_k, k; 1, 2, 3, 4.

We shall solve the system by the sequential exclusion of the unknowns. The value of V_1 is excluded in a term-by-term division of the equations from the system (14.36).

We use $V_{kl} = V_k^*/V_l^*$, $k, l = 1, 2, 3, 4$. The v_{kl} are clearly to be considered preassigned. Using (14.36) we may write the following equations with three unknowns: i, φ, and α_0:

$$v_{kl} = \frac{\cot i \cos (\alpha_l - \alpha_0) + \sqrt{1/\sin \varphi - \cos^2 (\alpha_l - \alpha_0)}}{\cot i \cos (\alpha_k - \alpha_0) + \sqrt{1/\sin \varphi - \cos^2 (\alpha_k - \alpha_0)}}. \tag{14.37}$$

From (14.37) find cot i. Then we assume, at first, $l = 1$, $k = 2$, after that, $l = 1$, $k = 3$, and equate the resultant expressions for cot i. Finally, we perform a similar operation for $l = 1$, $k = 2$ and $l = 1$, $k = 4$. As a result, we get two equations with two unknowns φ and α_0:

$$\frac{\sqrt{\Phi - \omega_1^2} - v_2\sqrt{\Phi - \omega_2^2}}{v_2\omega_2 - \omega_1} = \frac{\sqrt{\Phi - \omega_1^2} - v_3\sqrt{\Phi - \omega_3^2}}{v_3\omega_3 - \omega_2} \tag{14.38}$$

$$\frac{\sqrt{\Phi - \omega_2^2} - v_2\sqrt{\Phi - \omega_2^2}}{v_2\omega_2 - \omega_1} = \frac{\sqrt{\Phi - \omega_1^2} - v_4\sqrt{\Phi - \omega_4^2}}{v_4\omega_4 - \omega_1}$$

where the following notation was adopted:

$$\Phi = 1/\sin^2 \varphi; \quad v_k = v_{kl\ l=1}; \quad \cos (\alpha_k - \alpha_0) = \omega_k. \tag{14.39}$$

In order to exclude Φ (unknown φ) from (14.38), we determine it from each of the equations in (14.38) separately. We find:

$$\Phi = (-b_{kl} - \sqrt{b_{kl}^2 - 4k_lc_{kl}})/2a_{kl}, \tag{14.40}$$

where k and l are assigned as follows: the first time $k = 3$, $l = 2$, and the second time $k = 4$, $l = 2$, in addition:

$$a_{kl} = \beta_{kl}^2 - 1; \quad b_{kl} = \omega_k^2 + \omega_l^2 + 2\beta_{kl}\gamma_{kl}; \quad c_{kl} = \gamma_{kl}^2 - \omega_k^2\omega_l^2,$$

$$\beta_{kl} = \frac{(v_{kl} - 1)^2 - \gamma_{kl}v_2^2 - v_k^2}{-2v_{kl}v_lv_k},$$

$$\gamma_{kl} = \frac{-(v_{kl} - 1)^2\omega_1^2 + v_{kl}^2v_l^2v_k^2 - v^2\omega_k^2}{-2v_{kl}v_lv_k},$$

$$v_{kl} = (v_k\omega_k - \omega_1)/(v_l\omega_l - \omega_1). \tag{14.41}$$

Now we can formulate an equation with one unknown. In the designations of (14.36)-(14.41) this equation will be:

$$\frac{-b_{32} - \sqrt{b_{32}^2 - 4a_{32}c_{32}}}{a_{32}} = \frac{-b_{42} - \sqrt{b_{42}^2 - 4a_{42}c_{42}}}{a_{42}}. \tag{14.42}$$

It determines the required α_0 as an implicit function of the given values V_1^*, V_2^*, V_3^*, V_4^*, α_1, α_2, α_3, α_4. If its solution with respect to α_0 is obtained in some way, the rest of the unknowns from the basic system (14.36) can easily be found.

Determining α_0 from (14.42), in general, leads to bulky calculations and is practically difficult. The solution of the basic system becomes easier if we assign certain fix angles between the profiles along which apparent velocities are to be determined.

Two pairs of mutually perpendicular lines and reverse profiles. Let us assume, for example, $\alpha_1 = 0$; $\alpha_2 = \pi/2$; $\alpha_3 = \pi$, $\alpha_4 = 3\pi/2$. Then according to (14.39), $\omega_1 = \cos \alpha$; $\omega_2 = \sin \alpha_0$; $\omega_3 = -\cos \alpha_0$; $\omega_4 = -\sin \alpha_0$. Further, due to (14.36):

$$\cos \Psi_1 = \sin \varphi \cos \alpha_0; \quad \cos \Psi_2 = \sin \varphi \sin \alpha_0;$$

$$\cos \Psi_3 = -\sin \varphi \cos \alpha_0; \quad \cos \Psi_4 = -\sin \varphi \sin \alpha_0.$$

If we designate $\sin \varphi \cos \alpha_0 = p$, $\sin \varphi \sin \alpha_0 = q$, $\cot i = r$, the basic system (14.36) may be written in this case as:

$$V_1/V_1^* \sin i = pr + \sqrt{1 - p^2}; \quad V_1/V_2^* \sin i = qr + \sqrt{1 - q^2}$$

$$V_1/V_3^* \sin i = -pr + \sqrt{1 - p^2}; \quad V_1/V_4^* \sin i = -qr + \sqrt{1 - q^2}. \qquad (14.43)$$

Using (14.36), where $v_{kl} = V_k^*/V_l^*$, move to the system:

$$(pr + \sqrt{1 - p^2})/(-pr + \sqrt{1 - p^2}) = v_{31}$$

$$(qr + \sqrt{1 - q^2})/(-qr + \sqrt{1 - q^2}) = v_{42}$$

$$(pr + \sqrt{1 - p^2})/(qr + \sqrt{1 - q^2}) = v_{21}.$$

It may be solved without difficulty, provided p is found from the first equation and q from the second, e.g.,

$$p^2 = \left[\left(\frac{v_{31} + 1}{v_{31} - 1}\right)^2 r^2 + 1\right]^{-1}$$

$$q^2 = \left[\left(\frac{v_{42} + 1}{v_{42} - 1}\right)^2 r^2 + 1\right]^{-1}, \qquad (14.44)$$

and their values are substituted into the third equation, which gives an equation with one unknown r. A result of the solution is shown below. It is a formula for the first unknown: angle $i = \arcsin V_1/V_2$:

$$r^2 = \frac{[v_{41}(v_{31} - 1)]^2 - [v_{31}(v_{42} - 1)]^2}{[v_{31}(v_{42} + 1)]^2 - [v_{41}(v_{31} + 1)]^2} = \cot i. \qquad (14.45)$$

The rest of the unknowns can be found as easily. The formula for the dip angle φ of the refracting boundary has the form $\sin^2 \varphi = p^2 + q^2$; similarly, for the azimuth α_0 of the direction of the boundary dip, if the azimuth $\alpha_1 = 0$ of the profile 1 is chosen as zero azimuth: $\cos \alpha_0 = p/\sin \varphi$, $\sin \alpha_0 = q/\sin \varphi$. Here p and q are expressed through r (14.45) via (14.44).

Finally, the last unknown, i.e., the velocity V_1 in the covering medium, may be found from any equation of the system in (14.43), or from their combinations. Thus, from the first and the third, and from the second and the fourth equations (14.43) we get:

$$2V_1^* V_3^* pr/[(V_3^* - V_1^*)\sqrt{r^2 + 1}] = 2V_2^* V_4^* qr/[(V_1^* - V_2^*)\sqrt{r^2 + 1}] \qquad (14.46)$$

So the problem is solved. The solution may be interpreted in two ways: (1) as a solution of the inverse problem, i.e., determination of the structure of the medium on the basis of seismic data, and (2) as a solution of the following problem: from observations of head wave travel time curves at four radial profiles determine, in conformity with Sec. 14.4, the function of the complete system, or, which is the same, show that with the premises $(V_1, i, \varphi, \alpha = \text{const})$ four radial linear travel time graphs make up a complete surface system.

Let us briefly investigate the result. Assume $\varphi = 0$. Then in (14.46), at least, either $V_3^* \neq V_1^*, p \neq 0$, or $V_4^* \neq V_2^*, q \neq 0$ ($m = \cot i \neq 0$ always) and, consequently, V_1 is always finite. The solution exists and is unique.

Now assume $\varphi = 0$. Then $V_3^* = V_1^*, p = 0$; $V_4^* = V_2^*, q = 0$. Consequently, the solution is indeterminate. This conclusion we already reached proceeding from somewhat different considerations.

So, we can restate: it is impossible, in principle, to find the layer velocity $V_1 = \text{const}$ in the covering medium above a horizontal refracting boundary ($\varphi = 0$) by head wave travel time graphs without initial points. The statement is true for all cases (at $V_1 = \text{const}$), i.e., both for two-dimensional and three-dimensional problems. If $\varphi \neq 0$, V_1 may be determined in a three-dimensional case.

14.6. Theoretical Graphs

We continue by considering the complete inverse spatial problem for head waves given V_1, i, φ, $\alpha_0 = \text{const}$.

Formulation of the problem for graphic solution. Let be given n values α_k, V_k (14.35), (14.36), $k = 1, 2, \ldots$, which are measured values of azimuths α of radial profiles and the apparent velocities V^* of head waves. The number of observations is $n \geq k$ and disposition of the radial profiles is now arbitrary. Solve, as previously, the system of equations of the type (14.36) with respect to the constant parameters V_1, i, φ, and α_0, that determine the structure of the medium. Since the number of equations n is in general more than the number of unknowns, the question is to find the most probable result (see Sec. 14.5, Problem 2).

We shall solve the inverse problem by comparing graphs plotted by observed values α_k, V_k^* with a series of "theoretical graphs", calculated beforehand on the basis of the direct problem's solution.

Choosing a graph system. The first question is: how many graphs do we need? Or more accurately, what is the necessary number of parameters ("dimensions") in the graph system?

It follows from the problem that each experimental graph should contain α_k as a function of V_k^*. Consequently, it is represented by one curve. In order to allow comparison with the experimental graphs, theoretical ones should also be presented in the form of corresponding curves.

Each time we fill in the plane of the area, a one-parameter set of curves for a one-valued function may be placed at every theoretical graph (abacus). Consequently, in our case each graph-abacus may reflect changes in just one parameter of the four; taken for $(V_1, i, \varphi, \alpha_0)$.

Changes in three more parameters are needed. This can be performed by moving from one graph to another. In this case a three-dimensional graphical system is necessary. However, such a system would lack any practical sense: it is too bulky. Other ways have to be sought. Fortunately for relations like (14.36), ways do exist and consist of using motion on the plane.

Theoretical graph system. Let us use the following technique to plot the theoretical graphs: the coordinate system is rectangular and the abscissa A is a linear scale of azimuths $(\alpha - \alpha_0)$ and the ordinate V axis is a logarithmic scale of apparent velocities $\log V^*/V_1$ marked with numbers. The chart plane is the field of the function V_2/V_1, which is represented by its isolines (level lines); the parameter φ remains constant for every graph.

Using this approach each graph embraces all cases of changing three parameters: $i = \arcsin V_1/V_2$, α_0 and V_1 already, rather than those of first one. The first of the parameters is changed by passing from one isoline of the function V_2/V_1 to another and is determined by the form of the curve. The two others are changed by means of translation: they are determined by the position of the curve. Translating along the azimuth A axis corresponds to changes of α_0. Shifting along the velocity V axis corresponds to changes of absolute values of the velocities V_1, V_2, the fourth parameter φ alone has to be changed by transmission from one graph to another. Note that use of the third degree of freedom of the motion on the plane with the same purpose is impossible here in principle.

The complete graph system, embracing all four parameters $(V_1, i, \varphi, \alpha_0)$, is reduced to a "one-dimensional" system which is the simplest for problems of this type. This system can clearly be realized in practice.

Solution of the inverse problem now reduces to the following series of operations. An experimental graph (α_k, V_k^*), $k = 1, 2, 3, \ldots$, is constructed by points in the coordinate system $\alpha = A$, $\log (V^* = V)$. Then, by superimposing the experimental graph onto various theoretical graphs and translating it along the axes A and V the theoretical curve is found that coincides best of all with the observed one within error limits. Finally, the values sought for are counted.

The azimuth α_0 of the refracting boundary dip is counted in the crossing point of the scale α in the experimental graph with the straight line $A = 0$ of the theoretical one. Value V_1 of the layer velocity in the covering medium is counted in the crossing point of the scale V^* of the experimental graph and the straight line $V^*/V_1 = 1$ of the theoretical one. Ratio of the velocities V_2/V_1 is directly read on the chosen curve of the theoretical graph, and the dip angle φ of the refracting boundary—by a special mark on this graph.

The solution is obtained. The result is the most probable in the sense, that the theoretical curve to represent it is chosen under the condition of the best coincidence with points of the experimental graph within error limits.

14.7. On Stability of Solution of the Complete Inverse Problem for Head Waves

General information. The accuracy of a solution to a complete inverse problem, particularly of estimation of the velocity V_1 in the covering medium, mainly depends upon the following circumstances: values of parameters, that characterize the medium, the adopted system of seismic observations (in this case upon directions of the radial profiles) and accuracy of observations. The depth of the refracting boundary does not matter. Besides it is important to find out the degree of correspondence between the assumed premises and reality.

Accuracy of calculations is of no importance for us, as it can always be increased sufficiently by well-known means.

We do not intend to investigate the accuracy and stability of the solution as a whole, but rather try to consider individual questions of particular interest. It was shown above, that the solution becomes indeterminate, if the dip angle φ of the refracting boundary approaches zero. That is why the following is of interest: what is the lower limit of the angle φ, when determining V_1 is still practically possible?

Dependence of the accuracy on the boundary dip angle. To make the calculations simpler, we restrict our considerations to the case $\alpha_1 = \alpha_0$, $\alpha_2 = \alpha_0 + \pi/2$, $\alpha_3 = \alpha_0 + \pi$, $\alpha_4 = \alpha_0 3/2\pi$, then

$$V_1^* + V_1/\sin(i + \varphi) = \vec{V}, \quad V_2^* = V_4^* = V_1/\sin i = V_2,$$

$$V_3^* = V_1/\sin i - \varphi = \overleftarrow{V}*. \tag{14.47}$$

The accuracy in measuring azimuths is much higher than that in measuring seismic velocities, so the entire error in calculating V_1 may be put down to errors in $\vec{V}$, V_2, V. Note that assuming α_0 to be given, we make the error in the result somewhat smaller.

For our specific case, we use the formula deduced in [Riznichenko]:

$$V_1 = V_2 \frac{\sqrt{1 - [1/2V_2/\vec{V}* + V_2/\vec{V}*]^2}}{\sqrt{1 - (V_2/\vec{V}*)(V_2/\overleftarrow{V}*)}}. \tag{14.48}$$

Equation (14.48) has the general form $V_1 = F(V_2, \vec{V}, \overleftarrow{V})$, so the error ΔV_1 of the function V_1 is:

$$\Delta V_1 = \pm\sqrt{(\partial F/\partial V_2)^2(\Delta V_2)^2 + (\partial F/\partial \vec{V})^2(\Delta \vec{V})^2 + (\partial F/\partial \overleftarrow{V})^2(\Delta \overleftarrow{V})^2}, \tag{14.49}$$

where ΔV_2, $\Delta \vec{V}$, $\Delta \overleftarrow{V}$, are errors in the factors, the measured values.

Find:

$$\partial F/\partial V_2 = V_1/V_2 + \frac{1}{(1/V_2^2 - 1/\vec{V}\overleftarrow{V})}[V_1/V_2^3 - 1/V_1 V_2]$$

$$\partial F/\partial \vec{V} = \frac{1}{2(1/V_2^2 - 1/\vec{V}\overleftarrow{V})}[1/4V_2^2/V_1(1/\vec{V} + 1/\overleftarrow{V}) - V_1/\vec{V}]$$

$$\partial F/\partial \overleftarrow{V} = \frac{1}{2(1/V_2^2 - 1/\vec{V}\overleftarrow{V})}[1/4V_2^2/V_1(1/\vec{V} + 1/\overleftarrow{V}) - V_1/\overleftarrow{V}] \tag{14.50}$$

Equations (14.49) and (14.50) solve the problem in a general form. Let us give a numerical example. Assume that all the velocities $V = L/T$, $\{\vec{V}, V_2, \overleftarrow{V}\}$ are measured at equal bases $L = 1$ km; the entire error V is put down to inaccurate measuring of the differences between the times T. The errors ΔT for T are the same for all V and equal $\Delta T = 0.01$ s: $(\Delta V)^2 = L^2(\Delta T/T)^2$.

Suppose, finally, that the velocities in the layers above and below the refracting boundary are $V_1 = 2$ km/s, $V_2 = 4$ km/s, so that $i = \arcsin V_1/V_2 = 30°$.

Calculations according to the formulae (14.47)-(14.50) lead to the following:

$\varphi°$	0	10	15	20	25	30
$\pm\Delta V_1$, km/s	∞	1.95	0.85	0.45	0.27	0.08
$\Delta V_1/V_1$, %	∞	98	42	23	14	4

Hence it is clear that accuracy in determining the layer velocity V_1 by head wave travel time curves becomes practically acceptable at large dip angles φ of the refracting boundary. It is rather high only with angles φ approaching the limit angle of refraction for the given ratio of velocities V_1 and V_2.

However it should be borne in mind that the existence of large dip angles (φ) of layers in nature is very often associated with the changeability of the elements of the bedding (φ, $\alpha_0 = \text{const}$) or with discontinuities in the layers (plicative or disjunctive dislocations), as well as with changes in state of layer materials (V_1, $V_2 \neq \text{const}$). Therefore, the solution at large angles may become unreliable for geological reasons. However, exclusions are certainly possible in the last case.

On the use of theoretical graphs. A more complete investigation of solution stability might be done by means of theoretical graphs. This may help us to investigate the influence of all factors, provided at least the premise V_1, i, φ, $\alpha = \text{const}$ is strictly fulfilled. To do this, efforts should be made to achieve coincidence within admissible error limits of the experimental curve under study with various theoretical curves. The range of the variables, where such coincidence is possible, is the answer to the question.

One may find that under certain conditions, in some area of the phase space φ, areas of solution existence take an extended form approaching some curves. In these cases it seems possible to refer to an approximate "principle of equivalence" similar to that known in the theory of curve interpretation in vertical electrical sounding for electric prospecting by the direct current method.

Note, as an example of equivalence, that when (a) $V_2/V_1 = 1.5$, $\varphi = 20°$ and (b) $V_2/V_1 = 2.5$, $\varphi = 10°$ the functions $V^*/V_1 = f(\alpha - \alpha_0)$ are almost the same. The ratio of absolute values of the velocities V_1 in cases (a) and (b), is 1.6.

Stability conclusions. We finish by considering whether the facts we have are enough to conclude that the solution stability of our problem is not high. So, the use of reflection survey remains the main, though not the only, as thought previously, method to determine the velocities V_1 in the covering medium, if they do not increase with depth, in areas where no wells are available.

Let us sum up the basic considerations. A complete system of seismic travel time curves for waves of any type is a set of travel time curves that completely determines the function of the complete system under the assumed premises about character of wave propagation and structure of the medium. The function t of t $(\bar{\eta}, \vec{\varrho})$ of the complete system expresses the travel time t of a seismic wave of the class between any two set points $(\bar{r})$, $(\bar{\varrho})$ in the area of the profile or the surface.

For longitudinal systems of head wave travel time curves (two-dimensional problem), two reverse travel time curves comprise a complete system given the arbitrariness of changing both layer velocities V_1 in the covering medium and boundary velocities V_2 in the underlying layer. It should be added that under the same conditions, overlapped travel time curves make for a complete system only if location of shotpoints is infinitely dense; if V_1, $V_2 = \text{const}$ a complete system can comprise two overlapped travel time curves. For surface travel time curves of head waves (three-dimensional problem) with a flat boundary and constant velocities a complete system is made up of one surface travel time curve or a set of four radial ones.

A consideration of possibilities of solving the complete inverse problem for head waves, when the only data about the medium are provided by travel time curves of waves from this very class, has led to the following conclusions.

The problem is insoluble, in principle, for the two-dimensional case unless positions of the initial points of these curves are known. In the three-dimensional case, unlike the two-dimensional one, the complete problem has a solution in a concrete case of a flat boundary between two homogeneous layers. This proves, in principle, the possibility of determining the velocities V_1 in the covering medium by head wave travel time curves: it does not seem insoluble in more complicated cases of the three-dimensional problem either.

An analysis of solution stability for a complete spatial problem for head wave travel time curves shows that it is not high. This is why the bounded (rather than complete) inverse problem, almost alone remains a practical interest for seismic prospecting by reflection survey, i.e., the interpretation of travel time curves when the velocity in the covering medium is already determined by another method, e.g., by reflection survey or observations in wells. The stability of the bounded problem in reflection survey makes this certain.

15. Study of the Earth's Crust and Upper Mantle

15.1. Basic Sections of the Earth*

Using boreholes Man has only been able to penetrate directly into the Earth to a depth less than 15 km. Geological speculations about the Earth's interior are just hypotheses. Positive information about the Earth's deep structure from the Earth's crust to the core and subcore comes from the physical and mathematical sciences, viz. geophysics, geodesy, and astronomy (observations of the orbits of satellites).

Seismic methods, i.e., observations of elastic seismic waves from earthquakes and explosions are the most important of the geophysical methods (seismic, gravitational, magnetic, electrical). Seismic waves penetrate the entire Earth and, in principle, give us the potential to obtain detailed and exact information about its interior.

For the last 20 years the Jeffreys-Bullen travel time tables have been the main foundation for seismic conclusions, and consequently the starting point for any other data about the Earth's deeper structure. The tables give travel times of seismic waves of various types from the source epicenter (earthquake source, explosion) to the observation point (seismic station) as a function of epicentral distance. They are in fact averages of travel time curves and were constructed by combining many observed travel times, only high quality material being selected. The tables include data about many sources and receivers arranged in a variery of ways.

Information about the Earth's interior structure obtained from the tables is actually information about some "average" Earth. This Earth is presumed to possess central symmetry, or the symmetry of a rotation ellipsoid and has no local peculiarities.

The parameters of the average Earth's structure is often believed to be determined with a high accuracy. Thus, the radius of the Earth's core, R_c, was estimated by Jeffreys with an error of just $0.0004\,R$, where $R = 6338$ km (i.e. the Earth's radius), viz. $R_c = 0.5480\,R + 0.0004$. In absolute terms this is $R_c = 3473 \pm 2.5$ km. If the precision were in fact 2.5 km for determination of the core boundary some 3000 km below the surface, i.e., we could affirm that the core in fact does not differ from a ball of the indicated radius by any more than ± 2.5 km, this result would be the envy of seismologists and seismic prospectors. We have only just managed to determine the same absolute error in estimating the position and relief of the lower boundary of the Earth's crust, namely the Mohorovičić discontinuity, which lies just tens of kilometers below the surface. Note also that the method of deep seismic sounding in the USSR is commonly accepted to be the most accurate technique available for doing this; it is much more accurate than the classical methods of large scale seismology which were also used by Jeffreys.

Another example of the danger of using averages and optimistic estimates of their accuracy comes from the traditional determinations of the thickness of the Earth's crust using the dispersion of group velocities of surface Rayleigh and Love waves. What is the real mean thickness of the crust at, say, the Aleutian Islands, Tbilisi, the Caucasus, or Africa, where

---

* See [V. Slesarev].

the crust thickness varies by twenty kilometers, if the variations are several times the error given by the investigators for the mean value (±5 km, by Rayleigh)?

Geophysics is now changing emphasis away from general averages for the Earth's structure to studies of local peculiarities. Problems of primary interest today are the geophysical differences in the structure of the crust and the covering layer of continents and oceans, platforms and geosynclines, the differences in the deep structure of regions, mountain roots, transition zones from the continent to the ocean, deep water hollows, volcanism sources, seismic and aseismic zones etc. These topics lay a scientific foundation for better understanding of the origin of the Earth's present state including its interior.

Old methods of general statistical averaging are no use. When investigating such peculiarities, attempts to confine their range to local areas do not give better results. Difficulties have been encountered in the construction and interpretation of regional travel time curves. The problem is that travel times in fact depend on both the region, i.e., the place and the direction of the measuring bases. Besides, the structure of the same region may only arbitrarily be uniform everywhere.

At present, when vast amounts of observation data are becoming available, we cannot be satisfied with averages and their apparent accuracy. The old assumptions of homogeneity and the symmetry of deep structure are being challenged, and this should be continued. Methods should be sought that can selectively process and interpret data so as to reveal inhomogeneities both in the crust and deeper. It is important to avoid restricting the research to obtaining separate local results, but rather to get results valid for the entire planet. Besides, due to the large amount of material and its qualitative unevenness, there is a problem of selecting data to ensure a consistent result, at least in the main direction, while retaining accuracy and solution rapidity.

Thus, the present objectives of general geophysics resemble those of prospecting geophysics for small areas, remote from one another. At the beginning of the century, prospecting geophysics evolved from general geophysics by modifying its traditional methods. Later the methods of prospecting geophysics developed much faster, and now its methods, in seismology particularly, are in many aspects better than those of general geophysics. It is now time to re-exchange experience between these two fields.

Prospecting geophysics, as a result of its experience of practical applications for national needs, has developed rational systems of observations to obtain certain and precise information about the deep structures of objects. These also consider changes in its properties in horizontal directions.

In seismic prospecting these systems are based upon profile observations which are accompanied by less accurate but cheaper aerial surveys. Linear longitudinal profiles are fundamental to seismic prospecting, as both the sources and recorders of seismic waves are located using them. Coordinated reverse and overlapped systems of travel time curves, as well as systems of observations of seismic wave dynamics are obtained from these profiles. The interpretation of these systems reduces the solution to flat two-dimensional problems that are simpler than three-dimensional ones and allow more definite solutions. Vertical sections across a region are obtained by interpreting the observations. These sections are basic for the construction of spatial picture using other observations.

We believe it reasonable to transfer this general system of acquisition and processing of observational material from prospecting geophysics to general geophysics in view of new problems to be solved. We have some experience of using longitudinal profiling to process observational data near earthquakes. Our results show that this method can be expanded to more remote earthquakes and, accordingly, to deeper zones within the Earth.

For the globe as a whole, arcs of the great circles on the Earth's surface are the analogs of longitudinal profiles in prospecting, and circular sections across its center are the analogs of flat vertical sections. Further detail in terms of its ellipsoid and geoid form may be introduced as required.

Something similar has been attempted in the Earth sciences earlier. The famous "degree measurements" using triangulations along significant arcs of meridians marked a historic stage in general study of the Earth. They laid the basis for determining the dimensions and form of the Earth, and for determining the main unit of the metric system. Long trips along arcs of a great circle have been made many times by oceanologists. Satellites were the first to follow those trajectories around the Earth; at least their trajectories would be very close to great circles, if the Earth did not rotate underneath them. But all these investigations yield little information in understanding the Earth's interior, which is why the solid Earth is best studied with the most powerful methods, viz. those of seismology.

If we assume it is reasonable to study certain sections of the Earth then we have to determine their concrete location on the Earth from the very beginning. Such a big program will demand the use of large scientific forces on an international basis; at first, a small number of such sections will be studied.

These sections and the corresponding lines on the Earth's surface—the basic profiles— must be of interest geologically and geophysically. Further, they should be favorable for the study. From the seismological point of view this means the availability of a sufficient number of good seismic stations on the sections, as well as of severe earthquake foci. Practically speaking, we do not mean here that the stations and the epicenters should be located exactly on the circle, but within a narrow band along them. The band width should be narrower at short epicentral distances, while at long ones it may be widened, say, to 10° of the great circle arc, i.e., 1000 km. This also determines the approximate width of the areas to be studied in greater detail by geophysical, geological, volcanological and other methods, which require aerial observations. Such methods can make big contributions to solutions of the general problem. This is the structure of the basic sections of the Earth and especially of understanding the interpretation of complex results.

We suggest the following great circles as the basic sections of the Earth to begin with (Fig. 109).

Circle *1* and its band covers Europe and Asia along the seismic zones of the Tethys belt: across the Pyrenees, Alps, Carpatheans, Crimea, Caucasus, Pamir, Himalayas, and through Indochina. Then it crosses the seismically active area of the Indonesian islands, passes across seismically inactive Australia, then across seismically active New Zealand, spans the Pacific Ocean and passes through the seismic zone of Peru. It continues across Brazil and Guiana to the Atlantic Ocean, with its mid-ocean seismic ridge, and returns to Europe. There are many good seismic stations within this band in Europe, Asia, Indonesian Islands, Australia and New Zealand. The profile passess across various zones of the Earth that are of interest geologically and geophysically; namely the geosynclinal regions of Europe and Asia, extremely dislocated and tectonically active areas in New Zealand, and through the ancient platforms of Australia. It crosses continent-ocean transition zones several times.

Circle *2* and its band crosses Arabia, then through the rather seismically active zones of Iran and Afghanistan to the Soviet Union and follows the longest seismic section in the

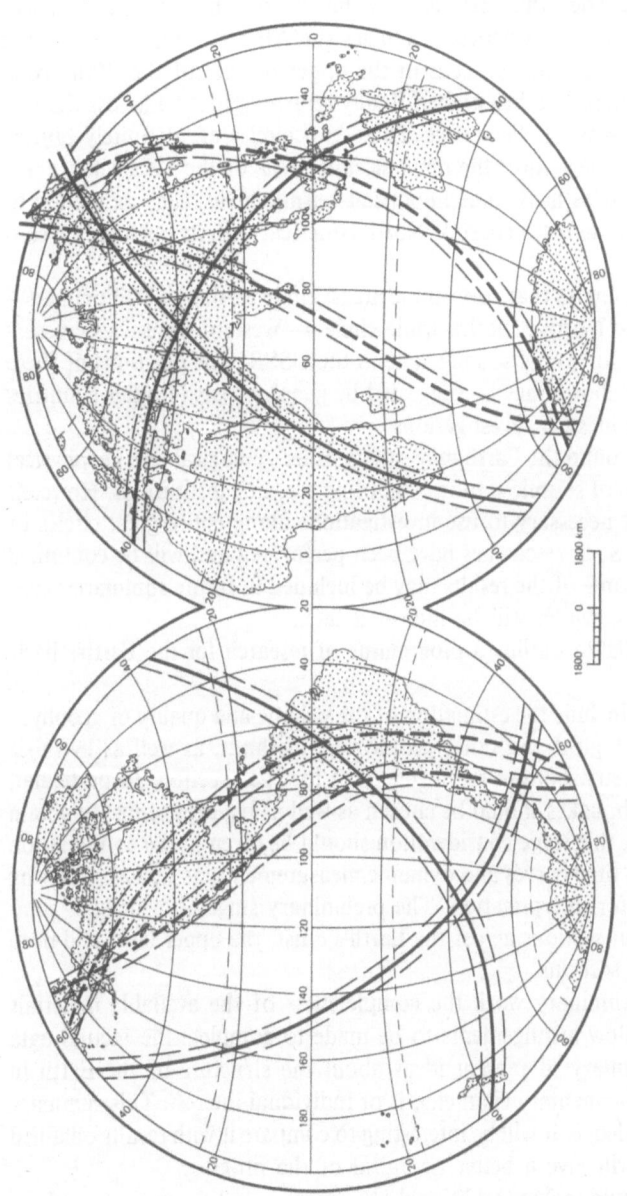

Fig. 109. Projection of the Earth's basic sections

USSR across the zones of the Tethys band, and the mountain structures of the Tien Shan, Altai, and Baikal regions. It continues to Kamchatka, where it crosses the Pacific seismic zone and here, across volcanic areas and northern Kuril-Kamchatka deep water trenches over to the platform of the Pacific. Then the circle crosses the Pacific, the Hawaiian Islands, then to the Tierra del Fuego in South America. It crosses the African Continent, over the South African platform, an area of large grabens in the upper reaches of the White Nile, through Abyssinia and the southern Red Sea graben finally returning to Asia across the volcanic area in the south of the Arabian Peninsula. These two circles are definitely objects for the proposed investigations. They cover the main seismic areas of the USSR and many other countries. However, the circles miss some important seismic areas, such as the arc of the Japanese Islands, and also western sea coast of North America, with its outstanding seismicity.

The figure shows two other circles that cover these areas: circle *3*—Japan-Philippinian—crosses Indonesia and South and North America while circle *4*—Western America—crossess India, China and Verkhoyansk—an Arctic seismic zone in the USSR. The positions of these circles are not so clear. A better choice may be suggested by geophysicists in other countries if they would like to take part in such investigations.

Many other basic profiles around the Earth may satisfy other geological and geophysical interests better or contain groups of seismic stations and other research institutions. However, to solve local problems it is not necessary to use investigations along entire great circles or even along arcs of great circles. Such researches have been performed and will be continued irrespective of our suggestion. Some of the results may be included in future summaries concerning other planetary sections, which will be indicated later.

In conclusion, we should like to outline a programme of research for the Earth's basic sections.

A preliminary stage should include the estimation of the volume and quality of geophysical material obtained by seismological and other geophysical methods, as well as by those in geodesy, oceanology (bathymetry, bottom sediments, etc.) for each section. In particular, seismic stations and known earthquakes should be chosen as well as big explosions, to obtain seismograms for comparison. At this stage consideration should be given to the coordination of observational systems—travel time curves and dynamic measurements, as well as gathering other materials to be the basis for interpretation. The preliminary stage also includes summarizing the available data about sections across the Earth's crust, the upper layer and deep zones of the Earth along each sections.

As an outcome of this preliminary work the completeness of the available materials should be assessed. This will allow arrangements to be made to complete the factual data with extra observations. A summary of present ideas about the structure of the Earth in certain sections, in spite of its fragmentary character, is of individual interest. This summary would describe our present knowledge. It will be interesting to compare it with results obtained after the investigations which will give a better appraisal of the progress.

The main investigations should include additional observations and theoretical research, as well as the interpretation of the material and construction of refined sections.

Pushing forward the front of our knowledge and accelerating new investigative and theoretical research on the one hand, and constructing sections on the other hand, may for the most part be done simultaneously, and only in some cases will it be necessary to carry out the two parts of the study in sequence.

The final stage would be to describe the results obtained by different methods, and compare them and then summarize the work with a geophysical and geological interpreation. This stage should also lead to recommendations for the development of the Earth's basic sections and employ them for studies of the Earth as a whole. The materials should be made available to the broad community of Earth scientists. This is why the publication of the results will require careful attention and concern of the scientists who will participate in this significant project.

We think it will take about 3-4 years to accomplish the program.

It would be desirable if other investigations of the Earth could be carried out along these circular sections at the same time. It is very important to refine data about geological structure obtained by geological and geophysical methods, especially by deep seismic sounding; to refine data about gravitational and magnetic fields; to conduct experiments on electromagnetic sounding of the Earth; to study weak seismicity; to perform volcanological and geothermal investigations; observations of slow motions of the Earth by geophysical, geological and geomorphological methods; and to study ocean floor relief, bottom sediments, underwater tectonics in the oceans.

A study of the Earth's basic sections will be useful in solving a number of adjacent problems of geophysics. For instance, in seismology—for improving data about the propagation of seismic waves in the Earth, and in this connection—for refining methods of the seismic energy in earthquake sources. Finding deviations of the deep boundaries from spherical (ellipsoidal) must be of importance for geodesy and gravimetry, as well as for magnetism and geothermal studies. The same data, together with localized information on the absorption of elastic waves may be useful for studying the state and properties of matter inside the Earth.

These investigations will certainly be significant for geology. They would provide refined and localized data on the deep structure of the Earth which may be of general geological interest. They would foster a stronger basis for such branches of geology, as theoretical geotectonics or the general theory of magnetism, the results of the investigation are of particular importance for discovering the deep geological structure of the regions crossed by the basic profiles and bands. These data may be used to develop general concepts about the distribution of mineral deposits. Investigations into the framework of this programme have just begun in the USSR*.

We are hoping scientists in other countries will participate in this work not only "on the whole" but also to find local peculiarities as we suggest. This program, if successful, will yield a reliable foundation of sound, accurately determined and localized data about the structure of the Earth's interior that will be a basis for further physical and geological studies of our planet.

---

* In 1970-1980 the idea of basic sections was partially realized in long (some thousands kilometers) profiles of deep seismic sounding using powerful industrial explosions, as well as in international DSS profiles. These investigations have not yet been combined with observations of severe earthquakes from seismic stations along the global sections.—(*Editor*).

15.2. Stratification of the Earth's Crust
and Upper Mantle*

Many sections of the Earth's crust have been obtained recently in various regions of the USSR
[Gambutsev, S. Zverev] and abroad. They reveal two man features.

The first one is the block character of the crust, and often of the upper mantle, with
blocks of different order of sizes. Under continents and oceans, plains and mountains, different blocks occur and there is a block structure of consolidated basement for individual areas.

The second feature is the horizontal stratification of the crust and upper mantle, at least
within large blocks and uniting smaller blocks. This stratification is also of different orders:
from a basic rough division into the crust and mantle to their subdivision into smaller individual layers.

Once there was a tendency to contrast the idea of a structurally inhomogeneous, "granular" and macroscopically homogeneous crust with the idea a layered crust. The stratification
of the consolidated continental crust was considered doubtful. However, direct evidence for
this and many concrete examples of "seismic boundaries" and, accordingly, layers in the consolidated crust and the upper mantle have appeared in recent years (beginning from 1957,
when continuous profiling using deep seismic sounding was completed for the first time);
the Mohorovičić discontinuity is just one such layer.

If both of these seemingly exclusive structures, viz. block character ("granularity") and
layered structure of the crust and upper mantle actually coexist, how does this phenomenon
come about? Are the deep seismic boundaries associated with changes in rock composition
or are they caused by changes in state due to the action of temperature and pressure? Finally,
are the boundaries similar to stratigraphic boundaries in sediments and result from the successive laying down of individual layers in the crust?

All these possibilities are being discussed now, but the lack of factual material hinders
definite conclusions. Still some explanations may be rejected as unfounded. First of all, any
general explanation of deep seismic boundaries in the crust (even more so in the mantle)
as due to successive deposition of the layers [I. Rezanov] should be rejected. It is hardly worth
considering the possibility of a layer by layer fallout of heterogeneous material to the ground
from space. It would be enough to consider just internal factors. There are examples of the
local accumulation of considerable portions of the Earth's crust as a result of volcanic emission and layer by layer intrusions, such structures are still of relatively local distributions.
However, the main point is that approximately horizontal deep strata with different physical
properties exist with rock complexes and complicated dislocations, with complicated folding
and faulting, with contacting heterogeneous blocks over vast areas of ancient shields and
in consolidated platform basements of different ages. If there was a layer by layer formation
of the crust somewhere, then these "layers" were later disturbed many times due to tectonic
motion. So modern smooth details—seismic boundaries—were imposed later upon the already formed block or "granular" structure of the medium. Note that, contrary to the above,
less intensive stratification of the crust and mantle is observed in modern volcanic areas.

We cannot agree, either with the idea that stratification in the crust and upper mantle,
especially upon formation of the Moho, is a result of modern pressure and temperature distri-

---

* From [Riznichenko]. I. P. Kosminskaya is the co-author.

butions in a macroscopically homogeneous medium of the crust and upper mantle. For the crust this was shown by Birch in the USA, whose arguments are confirmed by Soviet observations. It is also evidenced by large differences in the thickness of the crust as a whole and of individual layers in various regions for the same mean density and velocity compositions, at least in the upper consolidated crust.

The possible origin of the Moho [S. Stishov, B. Sologub, et al] has been discussed in the literature from geophysical, geochemical and petrological points of view. It may be possible to explain the origin of the discontinuity mainly by solid phase transition of rock-forming minerals under conditions of thick continental crust (gabbro-eclogite transition). This explanation seems unfounded for the thin oceanic crust, where changes in composition have clearly to be assumed at the Moho. However typically continental and typically oceanic crusts are just end members of a continuous sequence of different types of crust structure. Besides, the point is in explaining the origin of both the Moho and other boundaries inside the crust, where modern values and gradients of pressure and temperature may appear insufficient for considerable phase transitions.

Finally we consider it impossible to accept that all known deep seismic boundaries are interfaces between layers of different composition, formed, say, by the successive melting out of acid rocks and then of rocks with increasing alkalinity from the mantle material, which settles in appropriate "floors" of the Earth's crust. This does not agree with complex block constitution of the basement, where rocks of different composition may lie at the same level, but the horizontal stratification still continues to exist in this structural inhomogeneity.

Therefore, we think that deep seismic boundaries and, accordingly layers, result from joint influence of changes in both composition and state under varying pressure and temperatures, and the motion of matter inside the Earth. They are, in our opinion, *metamorphism fronts*, characteristic of certain, rather narrow, depth ranges, at least under stable platform conditions. These boundaries have very different origins from the stratigraphic boundaries common in geology, or, on the other hand, from disjunctive boundaries between different types of complexes, i.e., metamorphic and other rocks, or between magmatic massives with different compositions or origin times. If these geological features are regarded as primary elements, then seismic boundaries should be considered secondary elements imposed upon the structural details of the medium.

It also seems clear that the present positions of deep boundaries contain traces of past formation conditions, whose action gradually disappears. True, the relaxation time of this past influence is not long in the geological scale—it is shorter than one era. Evidence for this is the same depth approximate of the present stratification of platform basements irrespective of age, and this is why the position and form of deep boundaries (in contrast to structures in sediment layers) reflect crustal motion from the recent geological past to the present.

Thus, in spite of the "non-geological" character of these "geophysical" boundaries, their study should lead to solutions to important geological problems associated with origin and development of the earth's crust and upper mantle.

15.3. Study of the Shadow Zone in Simulating the Earth's Upper Mantle*

At the 12th General Assembly of the International Union of Geodesy and Geophysics in 1960, the study of the Earth's upper mantle and its connection with the earth's crust was called (V. V. Beloussov) the central problem of the physics of the Earth for the next few years. Various seismic methods are the most important for investigating this problem. They involve body (longitudinal and transverse) waves, surface waves, and long-period natural oscillations of the Earth as a whole. However, the most certain and detailed results should come from a study of body seismic waves.

Field observations should be accompanied by theoretical research and laboratory simulation study. The general objective of a simulation is, in our opinion, first to test possible models of mantle structure.

The first results on the seismology of the upper mantle were produced by Golitsyn, Gutenberg, Jeffreys, and Lehmann. Later Caloi and other scientists made considerable contributions to the problem of seismic velocity distribution in the mantle. In particular, the discovery of the existence of an "athenosphere low velocity channel" in it, is the most important problem that still waits for complete solution.

Gutenberg was the first to suggest the existence of this channel, which resulted from a consideration of how body wave amplitudes depend upon epicentral distances. Gutenberg believes this dependence is not monotonous and that there is a distinct minimum in the P and S wave amplitudes—a seismic shadow zone between 7° and 14°. Its existence may be explained by some reduction in velocity at the depth of about 100 km followed by an increase, and weak oscillations that still can be observed in the shadow zone—by arrival of diffracted waves. Note that this is not the only explanation of the observed reduction in the amplitudes of body waves at these distances.

The seismic shadow, Gutenberg believes, observed for every earthquake with a normal focal depth. The shadow zone reduces with increasing focal depth and shifts somewhat towards the epicenter, and disappears completely at depths of 300 km. This is explained qualitatively by the idea of a low velocity channel.

Some peculiarities in the recording form in case of waveguide mantle are described in [Riznichenko, O. Shamina]. Here we restrict our considerations to amplitude dependences. The following cases were simulated to reveal properties of the wavepath: (a) increase the velocity with depth, (b) decrease the velocity with depth, (c) the existence of a distinct wavepath.

15.3.1. Scheme of the Wave Phenomena

A kinematic description of the shadow zone is given by B. Gutenberg and others, while M. Bath calculates the body wave amplitudes for an appropriate velocity section. Using the ray path method, he calculated the amplitudes in the neighborhood of a geometrical shadow zone, but not inside it, where diffracted oscillations alone can exist. At present, given the

---

* From [Riznichenko]. Co-author—O. G. Shamina.

lack of a developed theory of elastic wave diffraction, a comprehensive investigation of the shadow zone proper is only possible on the basis of model observations.

The expected wave pattern near and inside the zone is shown schematically in Fig. 110. The presumed velocity section $V(z)$ is presented as the curve at the lower left-hand corner. The seismic rays in the medium are given to the right; the case of a surface source is taken as an example. The corresponding travel time curves are shown above and amplitude graphs for "simple" (propagating along geometrical seismics) and diffraction waves are given at the top.

Here a, b_1 and b_2 are simple refracted waves: a is a wave in the layer above the low-velocity zone, b_1 and b_2 are waves refracted at the layer under the zone. Both branches b_1 and b_2 correspond to the same wave whose front has an inversion rib and whose travel time curve has an inflection point at the boundary furthest from the source of geometrical shadow zone.

Waves a', b', and b'' are diffractions. All of them are continuations of the corresponding simple waves in the space area "forbidden" for geometrical rays. Still we show the approximate position of the corresponding "rays" at the lower part of Fig. 110. The wave a' continues the wave a, which propagates in the upper layer, to the shadow zone from the source. The b' wave continues the $b_1 + b_2$ wave in the shadow zone towards the focus from the inversion rib of this wave. The b'' wave continues the b wave from the source outside the shadow zone. It is a Caloi quasi-head wavepath wave [Bath] which propagates for long distances with constant apparent velocity corresponding to a "mean" velocity in the wavepath zone.

It may be interesting to draw an analogy between the waves that originate in a wavepath medium with a smooth change in velocity with depth, and waves that would exist in a medium with a similar velocity structure, although layered, the velocities in each layer being constant. A corresponding diagram of velocities is shown with a broken line in the lower left-hand

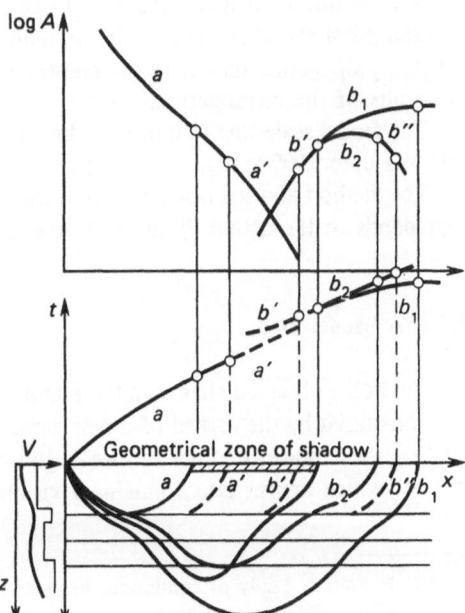

Fig. 110. Scheme of body waves in the presence of a low velocity channel in the mantle:

seismic rays (*below*); travel time curves (*middle*); amplitude graphics (*above*); $V(Z)$ on the left corner in Fig. 110. In a layered medium the a wave and its diffraction continuation a' would become an ordinary head wave with a high velocity layer p; the $b_1 + b_2$ wave would become an ordinary head wave with a high velocity layer r, which lies under the low-velocity layer q; the b' wave together with the b_1 and b_2 waves would be changed by reflections from the both interfaces of the low-velocity layer (for the continuation of b_1, viz. the reflection from the lower interface, for b_2 from the upper interface); the b'' wave would turn into a Zaitsev[*] "degenerate" head wave associated with the low-velocity layer q which lies under the high-velocity layer p. This degenerate wave would be an analog of the Caloi wavepath wave.

So far we have discussed the possible behavior of all waves in a zone near the front (or at a quasi-front zone for diffracted waves that do not have a sharp front). The situation would be like this, if the oscillation pulses were of high frequency and short duration. If the oscillations had long periods and duration, then some of the waves would interfere with each other. This would happen to the waves a' and b' within the shadow zone, and to the waves b_1, b_2 and b'' outside the zone, but close to the zone. This condition was encountered in our model experiments. We had to restrict our investigations to an area in front of and inside the shadow zone where the waves a, a' and b' can be observed. The model length was not sufficient to reveal the wave's behavior for long epicentral distances, where the waves b_1, b_2 and b'' would be well formed.

Note also that the diffraction wave's amplitude depends greatly upon the frequency (wavelength) of the recorded oscillations. With a reduction in the wave track and an increase in the frequency resolution of the wave pattern should increase. But the same cause leads to lower amplitudes of the diffractions, hindering their singling out and tracing. Therefore, strong interference seems to be inherent in this given diffraction problem.

The crust was not simulated in our model, but to avoid open surface effects, the model had a low-velocity layer of a thickness corresponding to the crust thickness. Velocities in the layer are higher than those in the crust, so that there is no velocity jump at the boundary with the mantle. If then the source is located in the crust the model does not yield waves P_n refracted at the Mohorovičič discontinuity, but, since in this problem we are interested in waves propagating mostly in the mantle, the absence of P_n does not significantly affect the results of the investigation.

The model scale was 10 mm $\approx$ 20 km, 1 μs ~ 1 s. Depths upto 300 -500 km, horizontal distances upto 2000 km, and oscillation periods of about 5 s were simulated.

The method allowed observations at the "profile" with a receiver step of 0.5 cm, which corresponds to the actual distance of 10 km.

15.3.2. A Model Study

We see that the observed change in the P and P_r waves in the shadow zone is very convenient. It is not caused by the arrival of a new wave P_r to the area of true first onsets, but by the fact that the P wave becomes too weak and cannot actually be traced, while the stronger wave P_r, which arrives later, remains a visible first result.

---

[*] L. P. Zaitsev. Study of continuous head seismic waves: Thesis for Master of Phys. and Math. IFE AS USSR, 1963. — (Author).

The place where this happens depends upon many factors including an apparently insignificant parameter as the effective sensibility of the apparatus. That is why if the amplitude graphs are constructed for the first distinct phase on the record, then both the epicentral distance, at which the minimum amplitude is observed, and the depth of the amplitude would cause the curve "minimum" to lose its meaning. Thus, at low apparatus sensitivity the first phase of the P wave cannot be traced from the point A_1 (Fig. 111); at higher sensitivities at the point A_2; respectively, we have the amplitude curve aA_1B_1b in the first case, and aA_2B_2b in the second (see Fig. 111). The amplitude jump at the transition of the amplitude curve a to the curve b changes by half an order. It seems more objective to consider the point A_0 where the graph of a intersects with the continuation of graph b at shorter epicentral distances to be the point of the first onset of minimum amplitudes.

Model observations allow some features of the wave pattern to be interpreted in terms of the oscillation phase correlation. For field seismological observations, these conditions are most often not maintained. Construction of amplitude graphs for visible first onsets is too prone to failure. If the maximum amplitudes in wave groups (trains) change rather than the first ones, the situation becomes better, viz. the necessity for a phase correlation vanishes, and the impossibility of duplicating by seismograms from separate stations should not affect

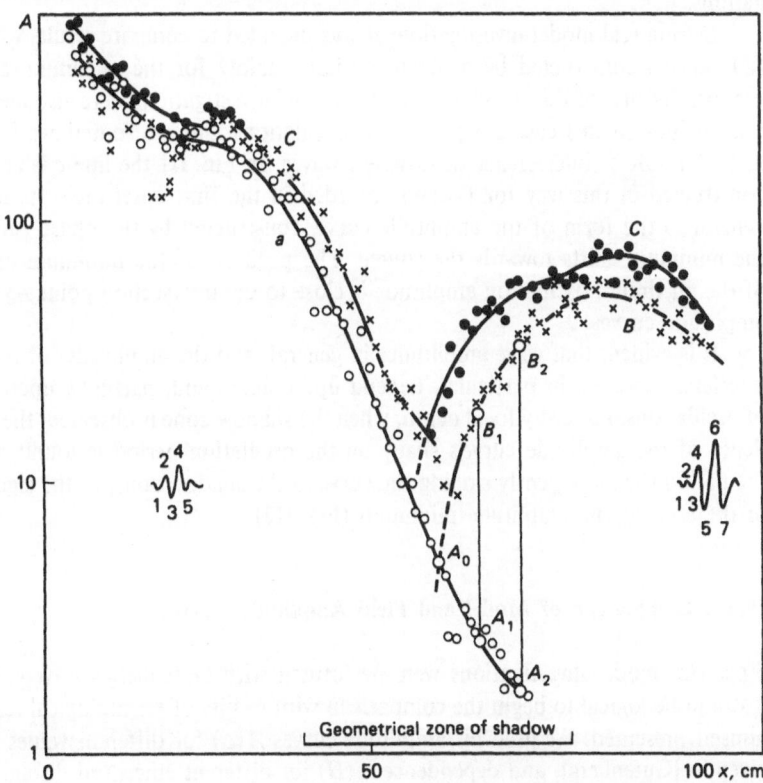

Fig. 111. Amplitude curves of longitudinal waves

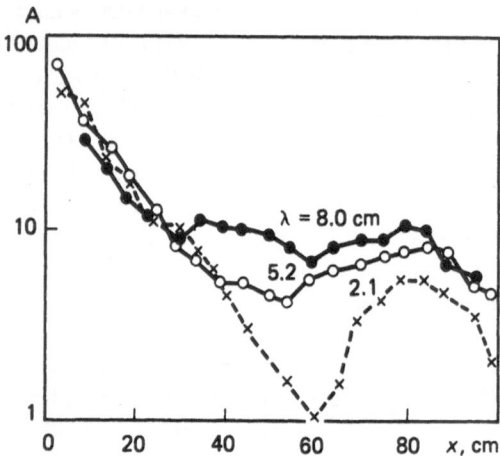

Fig. 112. Amplitude curves for different wave-lengths

the results. The wave correlation is certainly possible. Now interference remains the main difficulty when discussing observation materials in field seismology, as in model investigations.

During real model investigations it was intended to compare results with natural amplitude curves constructed by different authors mainly for the maximum amplitudes in the groups. Because of this, similar values for model investigations were also measured in groups, that include in this case 2-3 periods of oscillations and are sometimes known beforehand to be a result of interference of different waves. In Fig. 111 the line c is an amplitude curve constructed in this way for P-waves recorded in the first onset area. Its form is similar in general to the form of the amplitude curve, constructed by the phase curves a and b, but the minimum shifts towards the source. The position of the minimum on the length axis of the maximum oscillation amplitude is close to the intersection point A_0 of the first onset amplitude curves.

It is evident that wave amplitudes in general, and the amplitudes of the diffraction and interference waves, in particular, depend upon shape and, partially, upon the composition of oscillations. For every focal depth, when the shadow zone is observed, the expressed dependence of the amplitude curve's shape on the oscillation period is found, i.e., the longer is the period the more gently sloping the curve in the shadow zone, as the signal level increases in the area of the amplitude minimum (Fig. 112).

15.3.3. Comparison of Model and Field Amplitude Curves

Since our model investigations were performed with Gutenberg's wavepath mantle model, it would be logical to begin the comparison with results of seismological research Gutenberg himself presented. We used his amplitude curves $A(\Delta)$ for different values of the source H, given in [Gutenberg], and dependence $A(H)$ for different epicentral distances, which we derived with splines for magnitude corrections. From Gutenberg's method to obtain these amplitude dependences it follows, that they are averaged for various earthquakes and stations over

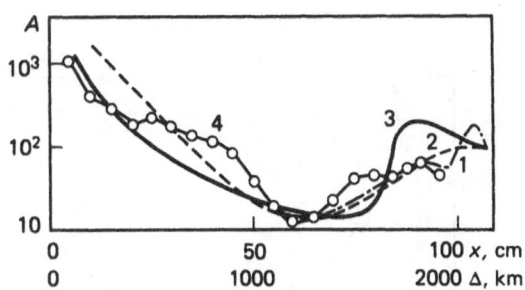

Fig. 113. Amplitude curves for minor earthquakes:

1—Vanck and Stelzner; *2*—Ruprechtova; *3*—Gutenberg; *4*—model curve for near-surface source

large paths of the Earth. They do not belong to a certain, narrow enough, period range, as records were used with different seismographs with periods from 1 s to 10 s. Still a comparison with Gutenberg's data makes some sense because in the "classical" period of seismology he managed to collect the richest statistical material, which remains unique.

There is satisfactory agreement between the model and actual data, sufficient at least to reveal the basic features. The amplitude minima in the model and in nature fall in the same epicentral distances. As the focal depth increases, the shadow zone shifts towards the epicenter and reduces simultaneously, disappearing at a focal depth of about 200 km.

Figure 113 shows amplitude curves for minor earthquakes by J. Vanek and I. Stelzner and L. Kuprechtova, and a model curve for a source located near the "Earth's" surface. Both the natural curves are constructed using data from European stations. As can be seen in the figure, the positions of the minimum in the model and in the natural curves coincide even better than with the corresponding amplitude curve by Gutenberg. Unfortunately we cannot make any judgement about the coincidence of the curves outside the shadow zone due to the model's constraints.

Figure 114 compares Gutenberg's data with simulation data for the oscillation amplitude, measured at a point on the Earth's surface at a fixed epicentral distance depending on the focal depth. The model and natural curves are the same in their main features: there is a distinct amplitude minimum at a certain depth range, the minimum diminishing with increasing epicentral distance. The qualitative similarity in form between the velocity-depth curve and the amplitude curve in the shadow zone about 1000 km from the epicenter is striking. Note that this similarity was also observed when investigating our first wave model, as described in [Uzhik] and it seems characteristic of a wavepath structure of the medium. This feature may possibly be used in seismology as a criterion for the presence of wavepaths in the mantle.

To summarize we may say that it has been shown that in this model of longitudinal wave amplitudes in a wavepath mantle, a characteristic feature is the existence of zones of low amplitudes A as a function of both the epicentral distance Δ (the observed "shadow zone") and the focal depth h at fixed Δ. The relations obtained using the model are in general agreement with averaged field observations.

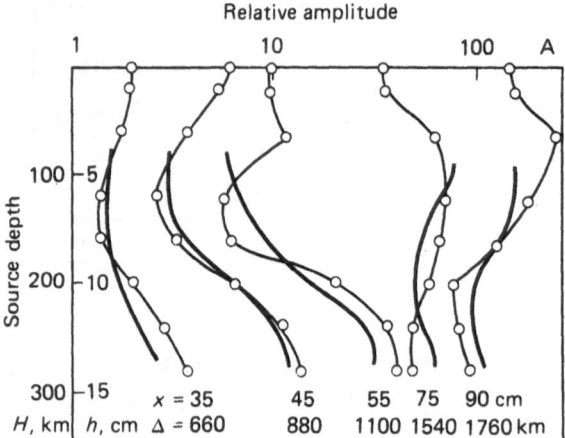

Fig. 114. Dependence of P-wave amplitude on source depth at a fixed epicentral distance. According to Gutenberg (*thick lines*) and in the model (*thin lines*)

   The model investigation showed that the amplitude drop, viz. the depth of the minimum in the curves $A(\Delta)$ and $A(h)$, greatly depends on the frequency for which the observations are conducted. The minimum depth, naturally, reduces with increasing recorded period (and wavelength, respectively). It is hoped that the character of the curves $A(\Delta)$ and $A(H)$ for different oscillation frequencies will yield information about the fine structure of the upper mantle in the area of a possible low-velocity channel: kinematic data are too scarce to determine this.
   The distinct dependence of oscillation amplitude on predominant frequency found in these experiments proves a statement made in considerations of earthquake magnitudes: the amplitude minimum in the shadow zone for a high-frequency source is more distinct than for a low-frequency one.

16 . Instrument Theory

16.1. The First Onset of the Sinusoidal Wave*

The onset of a seismic wave is simulated by that of a sinusoidal wave. The displacement x of particles of the medium in the wave is determined as folllows: $t \leqslant 0$; $x = 0$; $t \geqslant 0$; $x = x_m \sin \omega t$, where x_m is amplitude, ω is circular oscillation frequency of particles in the medium.

The equation of motion for a seismograph with direct recording has the form

$$y'' + 2\varepsilon y' + n^2 y = -Wx'', \tag{16.1}$$

where $y(t)$ is the seismogram ordinate; $x(t)$ is the ordinate of ground displacement in the seismic wave; ε, n and W are seismograph constants (damping coefficient, circular frequency and static, i.e., indicator magnification).

If a sinusoidal wave starts, the equation of motion will be:

$$y'' + 2\varepsilon y' + n^2 y = Wx_m \omega^2 \sin \omega t, \tag{16.2}$$

with initial conditions $t = 0$; $x_0 = 0$; $x' = \omega x_m$; $y_0 = 0$; $y_0' = -Wx_m \omega$.

1. If $0 < \varepsilon < n$, $y_1 = A \exp(-\varepsilon t) \sin(pt + \varphi)$—damped oscillations. Here $p = (n^2 - \varepsilon^2)^{1/2}$. At $\varepsilon = 0$ we get $y_1 = A \sin(nt + \varphi)$—undamped oscillations.
2. If $\varepsilon = b$, then $y_1 = (c_1 + c_2 t) \exp(-nt)$—aperiodicity boundary.
3. If $\varepsilon > n$, then

$$y_1 = C_1 \exp[-(q - \varepsilon)t] + C_2 \exp[-(q + \varepsilon)t]. \tag{16.3}$$

This is for aperiodic motion. Here $q = (\varepsilon - n^2)^{1/2}$.

The second integral (y_2) has the form $y_2 = N \sin(\omega t + \delta)$ for any ε and n. The values A, φ, C_1 and C_2 in the term y_1 are arbitrary constants of integration. They are determined from the initial conditions.

The constants N and δ, included in the term y_2, depend upon ε/n alone. They can be determined from each of the three above cases by the method of indeterminate coefficients. If for simplicity we assume $Wx_m = 1$ then the initial conditions take the form:

$$t_0 = 0; \quad \begin{cases} x_0 = 0 \\ x_0' = \omega \end{cases}; \quad \begin{matrix} y_0 = 0 \\ y_0' = -\omega. \end{matrix}$$

We introduce new constants: $g = \omega/n$; $\gamma = 2\varepsilon/n$, and now $y(t)$ may be found for any ratios ε/n under the assumed initial conditions.

We shall not give all the workings but just present the final result. Considering that damping in modern seismographs does not usually exceed the aperiodicity boundary, we re-

---

* From a diploma theses by Yu. V. Riznichenko "Determining True Ground Motion from Records of a Mechanical Seismograph" (1935). The thesis is in three parts: (1) a method of determining ground motion; (2) a method of term-by-term integration; (3) a determination of the periods and amplitudes of the first waves. This is the first publication of the original solution from part (3), which differs from the classical solutions by H. Berlage and S. Nakamura (1927) — (*Editor*).

strict our considerations to three cases:

(1) $\varepsilon = 0$, (2) $0 < \varepsilon < n$ and (3) $\varepsilon = n$

1. $\varepsilon/n = 0$; $y = y_1 + y_2 = A \sin(\omega t + \varphi) + N \sin(\omega t + \delta)$

$$\left. \begin{array}{ll} N = -q^2(q^2 - 1) & A = q/(q^2 - 1) \\ \delta = 0 & \varphi = 0 \end{array} \right\} \qquad (16.4)$$

2. $0 < \varepsilon < n$; $y = y_1 + y_2 = A e^{-\varepsilon t} \sin(pt + \varphi) + N \sin(\omega t + \delta)$

$$N = q^2/\sqrt{(1 - q^2) + \gamma^2 q} \qquad (16.5)$$

$$\mathrm{tg}\, \varphi = -N\sqrt{1 - (\gamma/2)^2} \sin \delta / - [q + N(q \cos \delta + (\gamma/2) \sin \delta]$$

$$\mathrm{tg}\, \varphi = -\gamma q/(1 - q^2); \quad A = -N \sin \delta/\sin \varphi$$

3. $\varepsilon/n = 1$; $y = y_1 + y_2 = e^{-\varepsilon t}(C_1 + C_2 t) + N \sin(\omega t + \delta)$

$$N = q^2/(1 + q^2); \quad C_1 = -N \sin \delta, \qquad (16.6)$$

$$\mathrm{tg}\, \delta = -2q/(1 + q^2); \quad C_2 = -n[N(q \cos \delta - \sin \delta) - q].$$

Properties of the solutions: 1. We supposed in formulae (16.4)-(16.6) that $Wx_m = 1$.
In order to consider x_m and W, it is necessary to multiply the solutions of y by the product Wx_m.

2. To prove that magnification (or diminution) of all the parameters ε, n, ω of the initial equation (16.2) by certain k does not change the graph of the function $y(t)$, but just reduces or increases its size along the t axis by the same k times.

The equation (16.2) may be rewritten as

$$d^2 y/dt^2 + 2\varepsilon(dy/dt) + n^2 y = Wx_m \omega^2 \sin \omega t.$$

We introduce a new variable θ for measuring time, such that $t = k\theta$. Substituting it into the previous equation:

$$d^2 y/k^2 d\theta^2 + 2\varepsilon(dy/kd\theta) + n^2 y = Wx_m \omega^2 \sin \omega k\theta,$$

and multiplying the equation by k^2 and labelling differentiation with respect to θ with points above the derivatives, we have:

$$\ddot{y} + 2(\varepsilon k)\dot{y} + (nk^2)y = Wx_m(\omega k)^2 \sin[(\omega k)\theta].$$

The last equation is identical to (16.2), which means that their solutions are also identical. Labelling the solution of the equation with parameters εk, nk, ωk by y_k, we may write

$$y(t) = y_k(\theta) \quad \text{or} \quad y(t) = y_k(t/k) \qquad (16.7)$$

Therefore, the statement is valid.

By choosing the values measured in the seismogram appropriately, formulae (16.4)-(16.6) as well as rules 1 and 2 allow the function $y(t)$ to be found if the function $x(t)$ is given

Fig. 115. The choice of T and L values

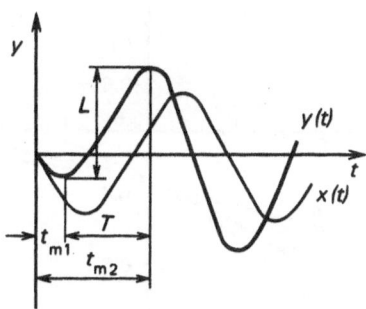

in the form $t_0 \geqslant 0$; $x = x_m \sin \omega t$. Now we solve the inverse problem. Using the curve $y(t)$ on the seismogram, find the period τ and amplitude x_m of the ground motion $x(t)$.

In order to determine two unknowns (τ, x_m) it is sufficient to measure two values on the seismogram. We choose T and L (Fig. 115). In most cases they may be obtained from the seismogram with certainty and precision. The time interval T is the difference between the times t_m of two successive extrema $y(t)$. The length L is the sum of corresponding extremum deviations $y(t_m)$ from the equilibrium.

Determining the period τ of ground motion. We differentiate (16.7) and make the substitution $t = t_m$: $y'(t_m) = (1/k)y'k(t_m/k) = 0$. Hence, the time:

$$t_m/k = t_{mk}, \tag{16.8}$$

is an extremum of the function y_k.

We find a similar function for $x(t)$. The period τ of the function $x = \sin \omega t$ is $\tau = 2\pi/\omega$, and the period τ_k of the function $x_k = \sin(\omega k)t$ is $\tau_k = 2\pi/\omega k$, from this:

$$\tau/k = \tau_k. \tag{16.9}$$

Dividing (16.8) by (16.9), we get $t_m/\tau = t_{mk}/\tau_k$ (it does not depend on k).

This rule is also valid for the relation T/τ. Therefore, T/τ depends upon the parameter ratio $\epsilon : n : \omega$ only, and not on their absolute values:

$$T/\tau = T_k/\tau_k = \Phi_1(\epsilon : n : \omega). \tag{16.10}$$

The function Φ_1 with arguments ω/n and ϵ/n is given in Fig. 116. The relation ω/n is plotted along the abscissa. The isolines correspond to $\epsilon/n = 0$, 0.5, and 1.0. The diagram is plotted with a log-log scale. The notations along the axes are in numbers.

The direct determination of given T in the diagram in Fig. 116 is impossible as Φ_1 is given with respect to the factor ω/n, which depends upon τ, which in turn is what we are determining, viz.

$$\tau = 2\pi/\omega = 2\pi/n \cdot 1/(\omega/n). \tag{16.11}$$

We find the following from (16.10) after substituting (16.11):

$$T = 2\pi/n\Phi_1/(\omega/n). \tag{16.12}$$

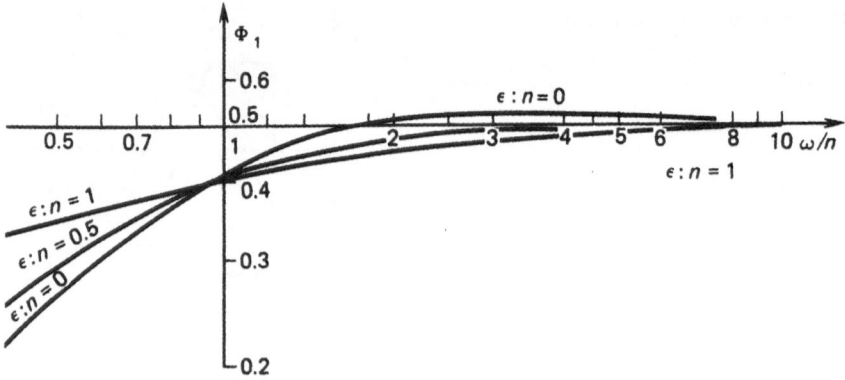

Fig. 116. Φ_1 function with arguments ϵn and ω/n

In order to find τ as an explicit function of T, we let $n = 1$. Then by (16.11) τ may be found for any ω/n, and by (16.12) and Fig. 116 T may be found with respect to the same argument.

Now it is sufficient to plot ω/n along the abscissa in T and to obtain a nomogram $\tau = f_1(T)$ with double scales. These nomograms are shown in Fig. 117 for three values of

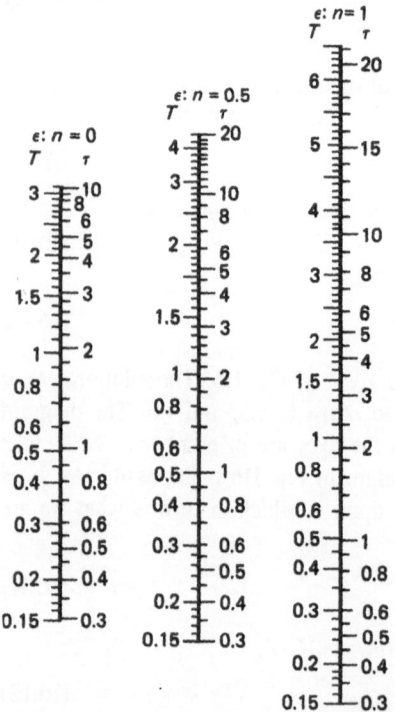

Fig. 117. Nomograms for three values of damping

damping $\epsilon/n = 0$, 0.5, and 1.0. There is no need to explain how to use them, the only thing to be considered is that T and τ are given under the assumption of $n = 1$. In order to have another value of n it is necessary before using the nomogram to multiply by n the T measured on the seismogram in seconds, and after determining from the nomogram, to divide it by n.

A comparison of the nomograms for the three ratios ϵ/n shows the advantage of the aperiodic seismograph: $\epsilon/n = 1$. For slow ground motion with period $\tau > \tau_0$, where $\tau_0 = 2\pi/n$ is the period of eigen-oscillation of the seismograph without damping the accuracy of determining τ falls. For $\epsilon/n = 0$ determining τ with sufficient accuracy is possible upto about $\tau \leqslant (1.5\text{-}2)\tau_0$. For $\epsilon/n = 0.5$ this limit shifts to $(3\text{-}3.5)\tau_0$, and for $\epsilon/n = 1$ still further. The nomogram is calculated upto $\tau = 3.5\tau_0$.

Determining the amplitude x_m of ground motion. We substitute in (16.7) the extremum t_m of the function y and take into account (16.8): $y(t_m) = y_k(t_{m_k})$ and at any k, i.e., the corresponding extremum value of y are a functin of the ratio $\varepsilon : n : \omega$ rather than their absolute values. The same property is possessed by the sums of two successive deviations:

$$L = L_k = \Phi_2(\epsilon : n : \omega) \tag{16.13}$$

Figure 118 shows the function Φ_2 of the arguments ω/n and ϵ/n at $Wx_m = 1$. The value of ω/n is plotted along the abscissa and the value ϵ/n designates the isolines. The diagram is constructed on a log-log scale. The notation at the axes is numerical.

If $Wx_m = 1$, then the length L_i measured on the seismogram (corresponding to L) is, according to 1 and (16.3), $L_i = Wx_m L = Wx_m \Phi_2$. Whence we find

$$x_m = L_i / W\Phi_2. \tag{16.14}$$

After determining τ in this way, we may find ω/n and Φ_2. After substituting Φ_2 in (16.14), we determine x_m, but for the sake of simplicity we shall take another route.

We find

$$\lambda = 1/\Phi_2 = Wx_m/L_i, \tag{16.15}$$

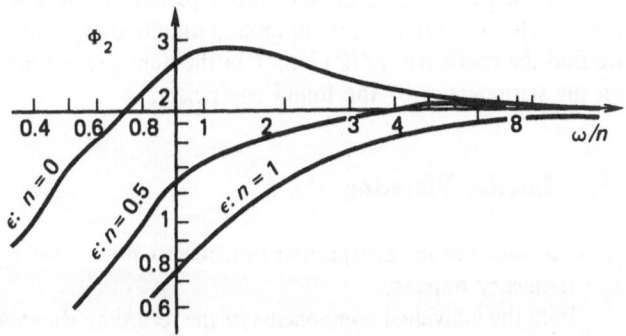

Fig. 118. The function Φ of the ω/n and ϵ/n arguments at $Wx_m = 1$

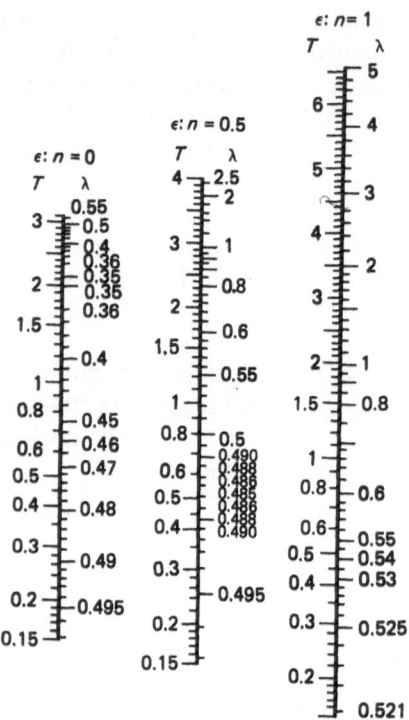

Fig. 119. Nomograms $\lambda = f_2(T)$

as a function of T. To do this we compare the nomograms $\lambda = f_2(T)$ (Fig. 119). Here, as above (see $\tau = f_1(T)$) we assume $n = 1$. From (16.15) we find

$$x_m = \lambda L_i / W \tag{16.16}$$

The process of determining τ and x_m may be simplified by adapting the nomograms to processing data from one seismograph with given constants ϵ, n, W.

In the nomogram $\tau = f_1(T)$, we divide both scales by n. In the nomogram $\lambda = f_2(T)$ we divide the scale T by n and divide the λ scale by W.

Then determining the ground motion period τ reduces to simple counting with the nomogram $f_1(t)$. To determine ground motion amplitude x_m two operations must be fulfilled: (1) we find the coefficient λ/W given T in the nomogram f_1 and (2) we multiply L_i, measured on the seismogram, by the found coefficient.

16.2. Inertial Vibrating Table

Introduction. Seismic prospecting instruments must meet rigorous standards of sensitivity and frequency response.

Both the individual components of the recording channel, viz. the seismograph, the amplifier, and the galvanometer must be tested and the channel as a whole tested too. The generation of controlled small precise amplitude oscillations is very important when testing sensitive instruments. Sources of electrical oscillations—mainly photogenerators*—are used to test the amplifiers and galvanometers.

Controlling small amplitude electric pulses is not difficult, but directly testing seismographs when small amplitude mechanical pulses are involved is much more difficult. The problem is aggravated when testing the seismic recording channel as a whole for amplitudes (sinusoidal or non-sinusoidal) of the order of the instrument's normal operation.

1. The first devices for testing seismographs were made for large seismology instruments. The first vibrating table for severe motion was designed in Japan at the end of the century. A little later (in 1902) B. Golitsyn made his vibrating table in Russia at the Pulkovo observatory. Seven years later the first vibrating table appeared in Germany.

In general, the vibration was generated by an engine. The drive rotor is connected eccentrically to a lever (or lever system) that is connected to the table. The design was often repeated.

2. In 1932 R. Köhler suggested a new method of seismograph testing. A wheel with an eccentrically located mass is connected to the seismograph mass; the seismograph mass gets shaken when the wheel rotates.

3. In 1937 M. L. Antokolsky at the seismic laboratory of the ALl-Union Office of Geophysical Prospecting in Moscow constructed a different sort of table. The wheel with the eccentric mass was not fixed to the seismograph mass, but to a heavy table suspended to a spring. The seismograph was installed on the table. A similar unit was designed by D. Kelley in America in 1938. Both units were designed to test prospecting seismographs.

A disadvantage of these systems is the difficulty of generating and controlling small amplitude vibration. Besides, only sinusoidal vibrations may be generated.

At present attempts are being made to generate electromechanical rather than mechanical vibration. This is better as firstly, the amplitude of the table vibration may be made as small as required, and, secondly, the motion depends on electric pulses fed to the transformer.

However, a very important problem is how to get accurate measurements and yet control the vibrating table's motion. Indeed, the direct measurement of table motion is very difficult due to their small amplitude.

The size of the table displacement could be assessed from the current or voltage applied to the electrical transformer. But there is an inconvenience here: transformer operation depends upon the frequency of the input oscillation. The problem is aggravated in seismograph testing for more complicated oscillations instead of non-stationary harmonic oscillations. If the absolute values of the displacement, and not just relative ones, must be measured, the problem becomes extremely difficult.

The electromechanical vibrator is created by applying the mechanical force to the table mass, and, on the other hand, to some immovable point. This way of excitation may be called the "force" method. It corresponds to mechanical vibrating tables described above.

There is another way of obtaining an electromechanical vibrator. The mechanical force may be applied between the main table mass and some additional mass, while the main mass is elastically suspended to an immovable point. A similar design was used in tables described in item 3. The technique may be called inertial, which can be seen from the following.

The main design ideas and calculations are considered below for an inertial vibrating table, which can generate mechanical pulses of any form, as well as check and measure them when the displacement values are small.

---

* Generators of seismic frequencies are used at present. — (*Editor.*)

Inertial vibrating table (IVT). The main design of the table (for vertical vibration) is shown in Fig. 120. The main mass M of the table is fixed by elastic suspension K. A system of magnets is rigidly fastened to the mass M_1. The vibrator M_2 is a coil in the magnetic field. The vibrator coil is connected to the table mass by a system that only allows one direction of motion. The center of the platform mass M_1 and that of the vibrator M_2 are situated along the same vertical line, which also passes through the suspension point A.

In order to obtain horizontal vibration the design has to be somewhat changed: the centers of the masses M_1 and M_2 should be placed along one horizontal line while the table suspension must be made rigid.

If an alternating current is passed through the vibrator coil, an alternating mechanical force is induced between the vibrator mass and the table mass M_2. It sets both masses in motion along the line connecting their centers. The vibrator's mass displacement with respect to the table mass is measured mechanically or electrically.

Let us first assume that the system of masses M_1 and M_2 is isolated, i.e. free from outer forces. We should assume that the masses M_1 and M_2 are not affected by gravity, there is no connection K with an immovable point not included in the system. Assume also that the masses M_1 and M_2 are only connected together by the force F.

Let the line passing through the mass centers (O_1 and O_2) of the vibrator table be the coordinate axis z. The coordinate origin will be placed at the system's center of inertia. Consider the motion of both masses along the z axis. The problem then reduces to motion in one coordinate of an isolated system with two points at O_1 and O_2 with masses M_1 and M_2 and a force F acting between them.

We may write in this case $M_1 z_1(t) = M_2 z_2(t) = 0$, where t is time, from this:

$$z_1(t)/z_2(t) = -M_2/M_1 = C_1 \qquad\qquad (16.17)$$

$$z_1(t)z_1(t) - z_2(t) = -M_2/(M_1 + M_2) = C_2. \qquad\qquad (16.18)$$

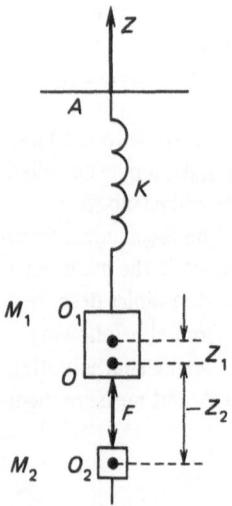

Fig. 120. IVT principle scheme:

M_1—table; M_2—vibrator; K—suspension; O_1—table M_1 mass center; O_2—vibrator M_2 mass center; O—inertia center of the system

Here $z_1(t)$ and $z_2(t)$ are absolute displacements of the masses M_1 and M_2, and $(z_1 - z_2)$ is relative displacement of these masses; C_1 and C_2 are constants.

If $\dot{M}_1 \gg M_2$, we get $|z_1 - z_2| \approx \tau_2$ and $C_2 = C_1$.

Assume that the center of inertia is immovable. Then the absolute displacements (and, consequently, velocities, accelerations, etc.) of both masses are inversely proportional to the masses. If M_1 is more than M_2, then displacements of the first repeat the displacements of the second but on a reduced scale. The reduction is determined by the ratio of the masses. Thus, our ideal isolated system has a reducing lever that assigns linearly and reduces the displacements of both masses.

If we succeeded in constructing a real mechanical system possessing the indicated properties of linear "inertial reduction" with a good approximation, then the problem of obtaining small amplitude vibration with a simple and accurate control and measurement would be solved.

Indeed, we could easily achieve any reduction in practice using inertial reduction; it is sufficient to choose an appropriate mass ratio. If the displacement $z_2(t)$ of the minor mass (of the vibrator) M_2 is recorded, dividing it by the ratio M_1/M_2 yields the displacement $z_1(t)$ of the main mass (of the table) M_1.

The displacement $z_2(t)$ may be accurately measured as it can be increased. The ratio M_1/M_2 can also be determined accurately by ordinary weighing. Therefore small displacements of the table mass M_1 can easily be measured with accuracy. To obtain the curve $z_1(t)$ of the table's displacement, it is sufficient to label the z axis of the vibrator's motion, $z_2(t)$, in another scale.

We give below a calculation of a vibrating table whose design principle was described above. The motion of M_1 and M_2 is shown to follow (16.17) and (16.18), which are valid for the isolated system, within certain frequency ranges. The table parameters may be chosen so that these frequency ranges and the approximation meet practical requirements.

Differential equation of IVT motion. Figure 121 shows the basic mechanical components of an IVT. We use the method of mechanical circuits developed by G. A. Gamburtsev. Figure 122 shows an IVT mechanical circuit, composed on the zero resistance plane. The one-pole masses are exchanged for two-pole ones. Figure 123 gives the corresponding mechanical circuit on the infinite resistance plane. The latter may be calculated in the same way as a corresponding electrical circuit.

Let x_1 and x_2 be relative pole displacements of the elements K and F. The notation x_1 is for the absolute displacement of the table M_1, x_2 is the displacement of the vibrator M_2 with respect to the table M_1.

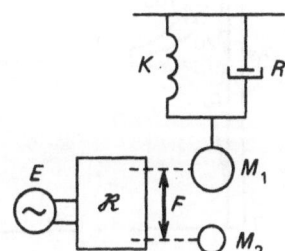

Fig. 121. IVT principle scheme:

a—mechanic part; M_1—platform mass; M_2—vibrator mass; K—platform suspension rigidity; R—damping constant; F—mechanic force; b—electric part; E—electric oscillation source; R—vibrator electromechanic scheme: amplifier and electromechanic transformer to generate mechanic force F at the outcome

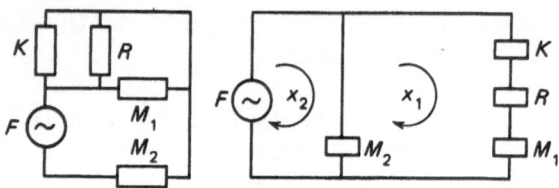

Fig. 122. IVT circuit:

a—on zero resistance plane; *b*—on infinite resistance plane (for notations, see Fig. 124)

To obtain an equation relating x_2 and x_1 it is sufficient to write one circuit velocity equation for the circuit with parameters

$$[(M_1 + M_2)D + R + K/D\dot{x}_1 - M_2 D\dot{x}_2 = 0. \qquad (16.19)$$

Here D is the differential operator such that

$$D^n U = d^n U/dt^n; \quad 1/D^n U = \underset{n}{\int \dots \int} U(dt)^n, \quad \text{where } U = U(t).$$

We rewrite (16.19) in another form:

$$(M_1 + M_2)\ddot{x}_1 + R\dot{x}_1 + Kx_1 = M_2\ddot{x}_2. \qquad (16.20)$$

Using the notation

$$h = R/2(M_1 + M_2), \quad n_0^2 = K/(M_1 + M_2), \quad V = M_2/(M_1 + M_2). \qquad (16.21)$$

we reduce (16.20) to form similar to the seismograph differential equation:

$$\ddot{x}_1 + 2h\dot{x}_1 + n_0^2 x_1 = V\ddot{x}_2. \qquad (16.22)$$

The expression (16.22) in (16.20) may be called the differential equation of motion of an inertial vibrating table. It gives the relation between the absolute displacement x_1 of the table and the relative displacement x_2 of the vibrator mass with respect to the table.

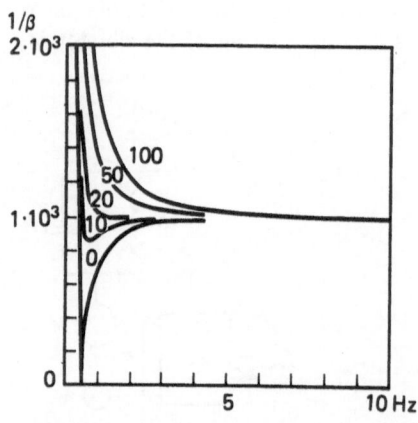

Fig. 123. Frequency response $(1/\beta)$ of the IVT inertial diminution at $M_1 = 2 \cdot 10^4$ g, $M_2 = 20$ g, $K = 2 \cdot 10^5$ CGS:

Curve family parameter $(R/M_2)^2$. Notations in 10^6 CGS units

Note that remarkable fact that the function $x_1(t)$ is fully determined by $x_2(t)$ and the system parameters M_1, M_2, R and K. In order to calculate $x_1(t)$ it is not necessary to know the true value of the force F, whose determination would involve a complicated electro-mechanical calculation. This property of the system shows the advantage of inertial excitation over force excitation, for which $F = F(t)$ must be known.

Let us again look at the function $x_2(t)$. Its graph should be constructed experimentally in the form of an oscillogram.

The function $x_1(t)$ may be determined by integration of (16.20) or (16.22). For practical purposes this operation needs to be as simple and quick as possible. In this respect it is of especial interest for the last two terms on the left-hand side of (16.22) to be insignificant as compared to the first term. If we assign values to the parameters h and n_0, then the frequency spectrum of the function $x_1(t)$ should contain sufficient high-frequency components.

Under these conditions we have the approximation $\ddot{x}_1 \approx V\ddot{x}_2$, whence from (16.21) we get:

$$\ddot{x}_1/\ddot{x}_2 \approx x_1/x_2 \approx M_1/M_1 + M_2. \tag{16.23}$$

This expression corresponds to (16.18), which was assuming an isolated system.

In order to find deviations in the motion of the system from the motion of an ideal isolated system, let us consider stationary harmonic vibration.

The frequency response of an IVT. The solution of (16.22) for a stationary sine vibration is known. We find the ratio $x_1/x_2 = \mathbf{B}(\omega)$, where ω is the circular vibration frequency. We shall call this ratio in our case the coefficient of dynamic inertial reduction

$$\mathbf{B}(\omega) = |V\omega^2/\sqrt{(n_0^2 - \omega^2)^2 + 4\omega^2 h^2}|. \tag{16.24}$$

$\mathbf{B}(\omega)$ corresponds to the coefficient of dynamic amplification of the seismograph, whose equation of motion is (16.22).

The angle φ of the vibration phase shift x_1 and x_2 is determined by

$$\cos \varphi = (n_0^2 - \omega^2)/\sqrt{(n_0^2 - \omega^2)^2 + 4\omega^2 h^2} \tag{16.25}$$

$$\sin \varphi = 2\omega h/\sqrt{(n_0^2 - \omega^2)^2 + 4\omega^2 h^2}.$$

Equations (16.24) and (16.25) have been already analyzed.

Let us make (16.24) more definite to be suitable for a vibrating table.

When choosing parameters for an IVT system we shall try to make the coefficient of inertial reduction $\mathbf{B}(\omega)$ as independent of the frequency within our chosen range as possible. This interval for seismic prospecting instruments is usually from $f_{\min} = 10$ to $f_{\max} = 200$ Hz where $f = \omega/2\pi$.

Let the reduction coefficient $V = \mathbf{B}_\infty = \lim\limits_{\omega \to \infty} \mathbf{B}(\omega)$ be $1 : 1000$, and the mass be $M_1 = 20$ kg. Then we find the vibrator mass $M_2 \approx 20$ g.

We choose a rigidity B of the suspension such that the frequency of free vibration $f_0 = n_0/2\pi$ of the system (at which unwanted resonance effects appear) is less than the frequency $f_{\min}$ of interest. On the other hand, too low a rigidity K is no use for design considerations as the static extension Δl of the suspension under the table's weight may be too large. If we assume $K = 2 \times 10^5$ CGS, we get $f_0 \approx 0.5$ Hz, i.e., $f_0 < f_{\min}$. On the other hand, $\Delta l = g/(2\pi f_0)^2 \approx 1$ t, which is acceptable.

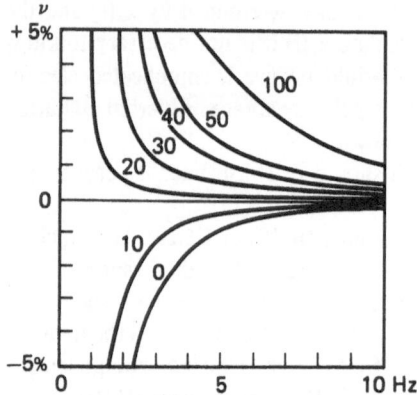

Curve family parameter as in Fig. 123

Figure 123 shows the frequency responses of an IVT for different values of R, calculated from (16.24). The curves on the diagram correspond to the function $1/\mathbf{B}(\omega)$. $(R/m_2)^2$ is the parameter of the curve family (the parameter values are given in 10^6 cos units).

Figure 124 shows relative deviations ν of the coefficient of inertial reduction $\mathbf{B}(\omega)$ from the coefficient $V = \mathbf{B}_\infty$ for infinitely quick vibrations.

Now we have to determine the value of the last parameter R. The damping R should be chosen from the requirement that the coefficient $B(\omega)$ should differ from V as little as possible over the largest possible frequency range. Such a damping is called "optimal". It corresponds to the coefficient $h = n_0/\sqrt{2}$. Hence from (16.21) we get $R = \sqrt{2}n_0(M_1 + M_2)$. In our case $R \approx 8 \cdot 10^4$ [g/s].

In Fig. 123 curves labelled 20 correspond to optimal damping. Figure 123 shows that the optimal curve 20 at frequency $f = 5$ Hz deviates from the straight line $\nu = 0$ by only about 0.1%; at $f = 2$ Hz the deviation is 0.4%; at lower frequencies it increases quickly.

So, the frequency response $\mathbf{B}(\omega)$ of the system at the chosen damping satisfies the condition. Distortions may be neglected for any frequencies higher than $f_{\min} = 2$ Hz.

These calculations show that a proper choice of parameter for an inertial vibrating table may ensure a practically constant coefficient of inertial reduction $\mathbf{B}(\omega)$ for every frequency higher than a threshold (2 Hz in our case). At frequencies of about 2 Hz or less modern seismic prospecting instruments have very low coefficients of amplification. Therefore we may regard the table as having a constant coefficient $\mathbf{B}(\omega)$, viz. $\mathbf{B}_\infty$, for practically all common frequencies. This allows small amplitude table vibrations to be controlled, measured and recorded simply and accurately.

The inertial vibrating table can generate both sine and more complicated vibrations. Practice will show how suitable this unit will appear for testing and studying modern sensitive seismic prospecting instruments.

References

Agnich, F. T. (1949) *Geophysics*, **14**, 4, 486.

Aki, K. (1956) *Bull. Earthquake Res. Inst. Tokyo Univ.*, **8**, 4, 205-228.

Aki, K. (1966) *Bull. Earthquake Res. Inst. Tokyo Univ.*, **44**, 1, 23-72.

Aki, K. (1967) *J. Geophys. Res.*, **72**, 4, 1212-1233.

Aki, K. (1972) In *The upper mantle developments in geotectonics*, pp. 423-446. Amsterdam: Elsevier.

Ambrayses, N. N. (1969) In *4th Int. Conf. Seismic Resistant Structures*. Santiago, Chile.

Ambrayses, N. N. and Zatopek, A. F. (1968) *Bull. Seismol. Soc. Amer.*, **58**, 1, 47-102.

Ananyin, I. V. (1973) In *Seismogennye struktury i seismodislokatsii* (Seismogenic structures and seismodislocations), pp. 91-94. Moscow: VNII Geofizika.

Antokolskii, M. L. (1947) *Prikl. Geofizika*, 3, 34.

Archambeau, C. B. (1973) *Geophys. J. Roy. Astron. Soc.*, **31**, 4, 361-363.

Artamonov A. M. (1975) In *Theor. Forschungsmeth. Geophys., Geol. und Astrophys., Kurzfassungen der Fortrage, Int. Symp. KAPG, Eisenach, 13-18. I.1975*, pp. 15-17. Berlin.

Artamonov, A. M. (1976) In *Issledovaniya po fizike zemletryasenii* (Studies in the physics of earthquakes), pp. 127-131. Moscow: Nauka.

Artemyev, M. E., Buné, V. I. , and Kambarov, N. S. (1972) *Izv. AN SSSR. Fizika Zemli*, 11, 8-27.

Artuyshkov, E. V. (1979) *Geodinamika* (Geodynamics). Moscow: Nauka.

Athy, L. F., Prescott, H. R., and Hughes, R. F. (1941) *U.S. Patent No. 2 265 768*.

Atomic Energy Commission release on Hardtack bomb tests (1959), No. 2, March.

Avershin, S. G. (1955) *Gornuye udary* (Mountain impacts). Moscow: Ugletekhizdat.

Averyanova, V. N. (1976) *Glubinnaya seismotektonika ostrovnykh dug* (Depth seismotectonics of island arcs). Moscow: Nauka.

Azizbekov, S. A., Riznichenko, Yu. V., Radtabov, M. M., *et. al.* (1973) *Izv. AN AzSSR. Ser. Nauk o Zemle*, **4**, 50-55.

Bailey, L. F. and Romney, C. F. (1958) *Bull. Geol. Soc. Amer.*, **69**, 1962.

Balakina, L. M., Vvedenskaya, A. V., Golubeva, N. V., *et al.* (1972) In *Seismologiya* (Seismology), No. 8. Moscow: Nauka.

Baldin, M. A. (1937) *Materialy k soveshchaniyu po probleme upravleniya krovlei* (Materials for conference on the problem of roof control). Moscow: Gostekhizdat.

Baratov, R. B. and Gaiskii, V. N. (eds.) (1962) *Geologiya i seismichnost' raiona Nurekskoi GES* (The geology and seismicity of the region of the Nurek hydroelectric plant). Dushanbe: Izd. AN TadzhSSR.

Bath, M. (1953) *Gerlands Beitr. Geophys.*, **63**, 3, 173-208.

Bath, M. (1956) *Bull. Seismol. Soc. Amer.*, **46**, 3, 217-218.

Bath, M. (1957) *Trans. Amer. Geophys. Union*, **38**, 4, 529-538.

Bath, M. (1960) *IUGG Monogr.*, 1, 1-24.

Bath, M. and Benioff, H. (1958) *Bull. Seismol. Soc. Amer.*, **48**, 1, 1-15.

Bazhulin, P. A. (1941) *DAN SSSR*, **31**, 2.

Beers, R. G. (1940) *Geophysics*, **5**, 1, 15.

Belikov, B. P. (1952) *Tr. IGD AN SSSR. Ser. Petrogr.*, vyp. 146, 42, 32-38.

Belikov, B. P., Aleksandrov, K. S., and Ryzhova, T. V. (1970) *Uprugie svoistva porodoobrazuyushchikh mineralov i gornykh porod* (Elastic properties of rock-forming minerals and rocks). Moscow: Nauka.

Beloussov, V. V. (1952) *Tr. Geofiz. In-ta AN SSSR*, 17 (144), 145.

Beloussov, V. V. (1975) *Osnovy geotektoniki* (Fundamentals of geotectonics). Moscow: Nedra.

Beloussov, V. V. (ed.) (1978) *Tektonosfera Zemli* (The Earth's tectosphere), Moscow: Nauka.

Benioff, H. (1949) *Bull. Geol. Soc. Amer.*, **60**, 6, 1837-1856.
Benioff, H. (1951a) *Bull. Geol. Soc. Amer.*, **62**, 331-338.
Benioff, H. (1951b) *Bull. Seismol. Soc. Amer.*, **41**, 31-62.
Benioff, H. (1951c) *Trans. Amer. Geophys. Union*, **32**, 4, 508-514.
Benioff, H. (1952) *Bull. Cal. Div. Mines and Geol.*, 171, 281-289.
Benioff, H. (1964) *Science*, **143**, 3613, 1399-1406.
Ben-Menachem, A. (1961) *Bull. Seismol. Soc. Amer.*, **51**, 3, 401-435.
Ben-Menachem, A. (1962) *J. Geophys. Res.*, **67**, 1, 345-350.
Ben-Menachem, A. and Töksoz, M. N. (1962) *J. Geophys. Res.*, **67**, 5, 1943-1956.
Ben-Menachem, A. and Töksoz, M. N. (1963) *Bull. Seismol. Soc. Amer.*, **53**, 5, 905-919.
Ben-Menachem, A., Smith, S. W., and Teng T. (1965) *Bull. Seismol. Soc. Amer.*, **55**, 2, 203-235.
Berckhemer, H. and Jakob, K. H. (1967) *Tectonophysics*, **4**, 3, 279.
Berckhemer, H. and Jakob, K. H. (1968) In *Final Sci. Rep. AF 61(052)—801/Air Force Cambridge Res. Lab.*, pp. 1-85. Bedford, Mass.
Berg, J. W., Gaskell. R., and Rinehart, V. (1964) *Bull. Seismol. Soc. Amer.*, **54**, 2, 777-784.
Bergman, L. (1954) *Der Ultraschall and seine Anwendung in Wissenschaft and Technik.* Zurich.
Berlage, H. P. (1930) *Handb. Geophys.*, L 2(5), 368-372.
Berzon, I. S., (1947a) *Tr. ITG AN SSSR*, **11**, 2, 87-107.
Berzon, I. S. (1947b) *Tr. ITG AN SSSR*, **11**, 2, 22-85.
Berzon, I. S. (1955) *Izv. AN SSSR. Ser. Geogr.*, **4**, 299-302.
Berzon, I. S., Epinatyeva, A. M., Pariiskaya, G. N., and Starodubrovskaya, S. P. (1962) *Dinamicheskie kharakteristiki seismicheskikh voln v real'nykh sredakh* (Dynamic characteristics of seismic waves in real media). Moscow: Izd. AN SSSR.
Bilik, S. M., Korablev, A. A., Panov, A. D., and Slobodov, M. A. (1958) *Pribory i apparatura dlya iss-ledovaniya proyavleniya gornogo davleniya* (Instruments and apparatus for studying the signs of rock pressure). Moscow: Ugletekhizdat.
Birch, F. (1943) *Bull. Geol. Soc. Amer.*, **54**, 263.
Birch, F. and Bancroft, P. (1938) *J. Geol.*, **46**, 59, 113.
Birch, F., Shairer, J., *et. al.* (eds.) (1942) *Handbook of Physical Constants.* New York.
Blake, A. (1941) *Bull. Seismol. Soc. Amer.*, **31**, 3, 787-791.
Borisov, A. A. and Shenkareva, G. A. (1972) *Byul, MOIP. Otd. Geol.*, **47**, 6, 5-16.
Borovik, N. S. (1970) *Izv. An SSSR. Fizika Zemli*, 12, 3-9.
Borovik, N. S. (1972) In *Voprosy seismichnosti Sibiri* (Questions of seismicity of Siberia) pp. 59-64. Novosibirsk: Nauka.
Brune J. N. (1968) *J. Geophys. Res.*, **73**, 2, 777-784.
Brune, J. N. (1970) *J. Geophys. Res.*, **75**, 26, 4997-5009.
Brune, J. N. and Allen, C. R. (1967) *Bull. Seismol. Soc. Amer.*, **57**, 3, 501-514.
Buchheim, W. (1953) Freiberg. Forschungsh., H. 7, 41.
Bulanzhe, Yu. D., Guseva, T. V., Pevnev, A. K., *et al.* (1982) In *Sovremennye dvizheniya zemnoi kory* (Modern movements of the Earth's crust), pp. 17-28. Kishinev: Shtinitsa.
Bullen, K. (1953) *Trans. Amer. Geophys. Union*, **34**, 1, 107-109.
Buné, V. I. (1956) *Tr. TISSS*, 1, 3-27.
Buné, V. I. (1959) *Dokl. AN Tadzh SSR*, **2**, 3, 31-38.
Buné, V. I. (1962) In *Geologiya i seismichnost' raiona Nurekskoi GES* (The geology and seismicity of the region of the Nurek hydroelectric plant), pp. 57-87. Dushanbe; Izd. AN TadzhSSR.
Buné. V. I. (1964) *Tr. IFZ AN SSSR*, 33 (200), 100-117.
Buné, V. I. and Butovskaya, E. M. (1955) *Tr. Geofiz. In-ta AN SSSR*, 30 (157), 142-153.
Buné, V. I. and Polyakova, T. P. (1975) In *Voprosy kolichestvennoi otsenki seismicheskoi opasnosti* (Questions of a quantitative appraisal of seismic danger), pp. 9-31. Moscow: Nauka.
Buné, V. I. and Reiman, V. M. (1960) *Tr. TISSS*, 7, 3-26.
Buné, V. I., Kirillova, I. V., Ananyin, I. V., *et al.* (1971) In *Voprosy inzhenernoi seismologii* (Questions of engineering seismology), vyp. 14, pp. 3-29. Moscow: Nauka.
Burridge, B. (1969) *Philos. Trans. Roy. Soc. London* A, **265**.
Burridge, B. and Knopoff, L. (1967) *Bull. Seismol. Soc. Amer.*, **57**, 3, 341-372.
Butovskaya, E. M. and Sokolova, I. A. (1972) *Uzb. Geol. Zhurn*, 3, 3-7.
Butovskaya, E. M., Zakharova, A. I., and Iodko, V. K., *et al.* (1964) Seismichnost' Uzbekistana, 2.

Butovskaya, E. M., Atabaev, Kh. A., and Flenova, M. G. (1971) In *Glubinnoe stroenie zemnoi kory territorii Uzbekistana* (The depth structure of the Earth's crust on Uzbekistan territory), pp. 9-27. Tashkent: Fan.

Butovskaya, E. M., Flenova, M. G., and Atabaev, Kh. A. (1974) In *Zemnaya kora Uzbekistana* (The Earth's crust in Uzbekistan), pp. 7-17. Tashkent: Fan.

Chetvertichnaya tektonika Pamira i Tyan-Shanya (1979) (The Quaternary tectonics of the Pamir and Tian Shan). Moscow: Nauka.

Chinnery, M. A. (1961) *Bull. Seismol. Soc. Amer.*, **51**, 3, 355-372.

Chinnery, M. A. (1969) *Bull. Seismol. Soc. Amer.*, **59**, 5, 1969-1982.

Chubarova, S. I. (1954) Izuchenie rasprostraneniya seismicheskikh voln metodom modelirovaniya (Studying the propagation of seismic waves by modeling). *Synopsis of Candidate's thesis*. Moscow State University.

Claudet, A. P. and Smith, N. J. (1950) *Oil and Gas J.*, Sept. 14.

Constantinescu, L., Cornea, I., and Lazarescu, V. (1973) *Neotectonic map and pliocene tectonic map* (Synthesis using published data and professional reports UNDP UNESCO Balkan Region Seismicity Project).

Cramer, H. (1946) *Mathematical Methods of Statistics*. Stockholm.

Davies, G. F. and Brune, J. N. (1971) *Nature. Phys. Sci.* **229**, 4.

Denton, E. R. (1955) *Oil Canada*, **7**, 17, 26.

Dinnik, A. N. (1925) *Inzh. Rabotnik*, No. 7.

Dinnik, A. N., Morgaevskii, A. B., and Savin, G. N. (1938) In *Tr. soveshch. po upravleniyu gornym davleniem* (Transactions of conference on rock pressure control). Moscow: Izd. AN SSSR.

Dix, C. H. (1946) *Geophysics*, **11**, 4, 45.

Dobrev, T. B. and Shchukin, Yu. K. (1974) Geofizicheskie polya i seismichnost' vostochnoi chasti Karpato-Balkanskogo regiona (Geophysical fields and the seismicity of the eastern part of the Carpathian-Balkan region). Moscow: Nauka.

Drozdovskii, B. A. and Fridman, Ya. B. (1960) *Vliyanie treshchin na mekhanicheskie svoistva konstruktsionykh stalei* (Influence of cracks on the mechanical properties of structural steels). Moscow: Metallurgizdat.

Drumya, A. V. and Stepanenko, N. Ya. (1972a) *Izv. An SSSR. Fizika Zemli*, 10, 77-78.

Drumya, A. V. and Stepanenko, N. Ya. (1972b) *Izv. AN SSSR. Fizika Zemli*, 5, 60-64.

Drumya, A. V., Popov, V. M., and Reshetilov, A. I. (1969) *Geofiz. sb. AN SSSR*, 36, 46-54.

Drumya, A. V., Popov, V. M.,. Reshetilov, A. I., and Stepanenko, N. Ya. (1971a) In *Izuchenie seismicheskoi opasnosti* (Studying seismic danger), pp. 55-62. Tashkent: Fan.

Drumya, A. V., Popov, V. M., and Stepanenko, N. Ya. (1971b) In *Izuchenie seismicheskoi opasnosti* (Studying seismic danger), pp. 62-65. Tashkent, Fan.

Dzewonsky, A., Bloch, C., and Landisman, H. (1969) *Bull. Seismol. Soc. Amer.*, **59**, 2, 427-441.

Dzhibladze, E. A. (1971a) *Izv. AN SSSR. Fizika Zemli*, 9, 72-75.

Dzhibladze, E. A. (1971b) In *Izuchenie seismicheskoi opasnosti* (Studying seismic danger), pp. 50-54. Tashkent: Fan.

Dzhibladze, E. A. (1975a) In *Voprosy kolichestvennoi otsenki seismicheskoi opasnosti* (Questions of the quatitative appraisal of seismic danger), pp. 59-62. Moscow: Nauka.

Dzhibladze, E. A. (1975b) *Izv. AN SSSR. Fizika Zemli*, 1, 76-80.

Dzhibladze, E. A. (1980) *Energiya zemletryasenii, seismicheskii rezhim i seismotektonicheskie dvizheniya Kavkaza* (Earthquake energy, seismic conditions and seismotectonic movements of the Caucasus), Tbilisi: Metsniereba.

Dzhibladze, E. A. and Riznichenko, Yu. V. (1973) *Izv. AN SSSR. Fizika Zemli*, 1, 9-10.

Epinatyeva, A. M. (1951) *Izv. AN SSSR. Ser. Geofiz.*, 4, 43-60.

Epinatyeva, A. M. (1959) *Izv. AN SSSR. Ser. Geofiz.*, 8, 1089-1102.

Ermakov, I. I. (1960) *Tr. VNIMI*, sb. 38.

Eshelby, J. D. (1957) *Proc. Roy. Soc. London A*, **241**, 376-396.

Evans, J. F., Hadley, C. F., Eisler, J. D., and Silverman, D. (1954) *Geophysics*, **19**, 2, 220.

Faust, L. Y. (1951) *Geophysics*, **16**, 2, 192.

Fedorov, F. I. (1965) *Teoriya uprugikh voln v kristallakh* (The theory of elastic waves in crystals). Moscow: Nauka.

Fedotov, A. P. (1960) *Tr. VNIMI*, sb. 38.

Fedotov, S. A. (1965) *Tr. IFZ AN SSSR*, 36 (203), 66-98.

Fedotov, S. A. (1968) In *Seismicheskoe raionirovanie SSSR* (Seismic division of the USSR), pp. 121-150. Moscow: Nauka.

Fedotov, S. A. and Shumilina, L. S. (1971) *Izv. AN SSSR. Fizika Zemli*, 9, 3-15.

Fenner, R. (1938) *Glückauf*, No. 32/33.

Flenova, M. T. (1969) In *Tr. III Vsesoyuz. simpoz. po seismicheskomu rezhimu 3-7 uyulya 1968 g. Novosibirsk* (Transactions of 3rd All-Union symposium on seismic conditions, July 3-7, 1968, Novosibirsk), part II, pp. 54-64. Nauka.

Florensov, N. A. and Solonenko, V. P. (1963) *Gobi-Altaiskoe zemletryasenie* (The Gobi-Altai earthquake). Moscow: Izd. AN SSSR.

Fridman, Ya. B. (1952) *Mekhanicheskie svoistva metallov* (Mechanical properties of metals). Moscow: Oborongiz.

Fukuo, Y. (1970) *Bull. Earthquake Res. Inst. Tokyo Univ.*, 48, 707-727.

Gaiskii, V. N. (1950) *Tr. Geofiz. In-ta AN SSSR*, 12 (139), 57-65.

Gaiskii, V. N. (1970) *Statisticheskie issledovaniya seismicheskogo rezhima* (Statistic studies of seismic conditions). Moscow: Nauka.

Gaiskii, V. N. and Katok, A. P. (1965) In *Dinamika zemnoi kory* (Dynamics of the Earth's crust), pp. 9-13. Moscow: Nauka.

Gaiskii, V. N., Katok, A. P., and Bilman B. M. (1960) *Tr. TISSS*, 7, 89-108.

Gaiskii, V. N., Ladynin, A. V., and Presnyakova, L. P. (1974) In *Voprosy kolichestvennoi otsenki seismicheskoi opasnosti* (Questions of the quantitative appraisal of seismic danger), pp. 63-72. Moscow: Nauka.

Gamburtsev, G. A. (1953a) *DAN SSSR. N.S.*, **88**, 5, 787-789.

Gamburtsev, G. A. (1953b) *DAN SSSR*, **92**, 4, 747-749.

Gamburtsev, G. A, (1955) *Byul. Soveta po Seismologii AN SSSR*, 1, 7-14.

Gamburtsev, G. A. (1959) *Osnovy seismorazvedki* (Fundamentals of seismic prospecting). Moscow: Gostoptekhizdat.

Gamburtsev, G. A. (1960) *Izbrannye trudy* (Selected works). Moscow: Izd. AN SSSR.

Gamburtsev, G. A. and Galperin, E. I. (1954a) *Izv. AN SSSR. Ser. Geofiz.*, 2, 184-189.

Gamburtsev, G. A. and Galperin, E. I. (1954b) *Izv. AN SSSR. Ser. Geofiz.*, 1, 3-10.

Gardner, L. W. (1949) *Geophysics*, **14**, 1, 29.

Gilbershtein, P. G. and Gurvich, I. I. (1960) *Izv. Vuzov. Geologiya i Razvedka*, 1, 139-156.

Goikhman, G. I. (1937) In *Sb. trudov po upravleniyu krovlei* (Collected works on roof control), pp. 174-182. Kharkov: DNTVU.

Goldin, S. V. (1982) *Izv. AN SSSR*, 10, 11-16.

Golitsyn, B. B. (1960) *Izbrannye trudy* (Selected works), Vol. 2. Moscow: Izd. AN SSSR.

Gorbunova, I. V. (1962) *Tr. IFZ AN SSSR*, 25 (192), 312-324.

Gorbunova, I. V. (1964) *Tr. IFZ AN SSSR*, 32, 138-147.

Gorbunova, I. V. (1969) *Izv. AN SSSR. Fizika Zemli*, 11, 3-14.

Gorbunova, I. V. and Riznichenko, Yu. V. (1965) *Izv. AN SSSR. Fizika Zemli*, 7, 22-29.

Gotsadze, O. D., Keilis-Borok, V. I., Kirillova, I. V., *et al.* (1957) *Tr. Geofiz. In-ta AN SSSR*, 40 (166), 146.

Grudeva, N. P., Levshin, A. L., and Pisarenko, V. F. (1972) In *Teoreticheskaya i vychislitel'naya geofizika* (Theoretical and computational geophysics), pp. 5-17 (No. 1). Moscow: Nauka.

Gubin, I. E. (1960) *Zakonomernosti seismicheskikh proyavlenii na territorii Tadzhikistana* (Laws of seismic appearances on the territory of Tadzhikistan). Moscow: Izd. AN SSSR.

Gurevich, G. I., Nersesov, I. L., and Kuznetsov, K. K. (1960) *Tr. TISSS*, 6, 41-88.

Gurvich, I. I. (1954) *Seismorazvedka* (Seismic prospecting). Moscow: Gosgeoltekhizdat.

Gutenberg, B. (1945a) *Bull. Seismol. Soc. Amer.*, **35**, 1, 3-12.

Gutenberg, B. (1945b) *Bull. Seismol. Soc. Amer.*, **35**, 3, 612-628.

Gutenberg, B. (1946) *Bull. Seismol. Soc. Amer.*, **36**, 327.

Gutenberg, B. (1953) *Proc. Nat. Acad. Sci. US*, 39, 849.

Gutenberg, B. (1954) *Review geofic. pura e appl.*, **29**, 1-10.

Gutenberg, B. (1960) *Science*, **131**, 3405, 959-965.

Gutenberg, B. and Richter, C. (1941) *Seismicity of the Earth*. Princeton.

Gutenberg, B. and Richter, C. F. (1946) *Trans. Amer. Geophys. Union*, 27, 776.

Gutenberg, B. and Richter, C. F. (1949) *Seismicity of the Earth and Associated Phenomena*, 2nd ed. Princeton: Univ. Press.

Gutenberg, B. and Richter, C. F. (1956a) *Bull. Seismol. Soc. Amer.*, **46**, 2, 105-145.
Gutenberg, B. and Richter, C. F. (1956b) *Ann. Geofis.*, **9**, 1, 1-15.
Gvozdev, A. A., Kuznetsov, V. V., and Parkhomenko, I. S. (1967) *Izv. AN SSSR. Fizika Zemli*, 12, 30-39.
Gzovskii, M. V. (1953) *Izv. AN SSSR. Ser. Geofiz.*, 2, 101.
Gzovskii, M. V. (1954) *Izv. AN SSSR. Ser. Geofiz.*, 6, 527.
Gzovskii, M. V. (1957) *Izv. AN SSSR. Ser. Geofiz.*, 2, 141-160; 3, 273-283.
Gzovskii, M. V. (1963) In *Sovremennye dvizheniya zemnoi kory* (Modern movements of Earth's crust), vyp. 1, pp. 149-158. Moscow: Izd. AN SSSR.
Gzovskii, M. V. and Chertkova, E. I. (1953) *Izv. AN SSSR. Ser. Geofiz.*, 6, 481.
Gzovskii, M. V. and Nikonov, A. A. (1969) In *Problemy sovremennykh dvizhenii zemnoi kory* (Problems of modern movements of the Earth's crust), pp. 405-410. Moscow: Izd. AN SSSR.
Gzovskii, M. V., Krestnikov, V. N., and Reisner, G. I. (1959) *Izv. AN SSSR. Ser. Geofiz.*, 8, 1146-1156.
Habuty, M. T. (1933) *Bull. Amer. Assoc. Petrol. Geol.*, **20**, No. 5.
Hafner, W. (1940) *Bull. Seismol. Soc. Amer.*, **30**, 4, 309.
Hanks, T. S. and Thatcher, W. R. (1972) *J. Geophys. Res.*, **77**, 23, 4393-4405.
Hanks, T. C. and Wyss, M. (1972) *Bull. Seismol. Soc. Amer.*, **62**, 2, 561-589.
Hanson, M. E., Sanford, A. R., and Sheffer, R. J. (1974) *J. Geophys. Res.*, **39**, 2, 365-376.
Hartzell, S. (1979) *Bull. Seismol. Soc. Amer.*, **69**, No. 2.
Haskell, N. A. (1941) *Geophysics*, **6**, 4, 318.
Haskell, N. A. (1964) *Bull. Seismol. Soc. Amer.*, **54**, 6, 1811-1842.
Haskell, N. A. (1966) *Bull. Seismol. Soc. Amer.*, **56**, 1, 125-140.
Hausrath, H. (1913) *Phys. Ztschr.*, **14**, No. 21.
Henderson, J. B. and Brewer, R. (1953) *Geophysics*, **18**, 2, 324-337.
Honda, H. (1962) *J. Phys. Earth*, **10**, 2, 1-97.
Honda, H. and Ito, H. (1939) *Geophys. Mag.*, **23**, 2, 155-161.
Horton, C. W. (1943) *Geophysics*, **8**, 3, 290.
Housner, G. W. (1955) *Bull. Seismol. Soc. Amer.*, **45**, 3, 197-218.
Howes, W. and Randolph, D. (1953) *J. Acoust. Soc. Amer.*, **25**, No. 5.
Howes, E., Tejada-Flores, and Lee, R. (1953) *J. Accoust. Soc. Amer.*, **25**, No. 5.
Hughes, D. S. and Cross, S. H. (1951) *Geophysics*, **16**, 577.
Hughes, D. S. and Jones, H. J. (1950) *Bull. Geol. Soc. Amer.*, **61**, 842.
Ida, G. and Aki, K. (1972) *J. Geophys. Res.*, **77**, 11, 2034-2044.
Ide, M. J. (1936) *Proc. Nat. Acad. Sci. U.S.*, **22**, 2, 81.
Iida, K. (1959) *J. Earth Sci. Nagoya Univ.*, **7**, 2, 98-107.
Iosif, S., Prochaskova, D., and Iosif, T (1979) *Tectonophysics*, 53, 3/4, 195-201.
Ittner, F. (1939) *Techn. Publ. Amer. Inst. Min. Met. Eng.*, **1059**, No. 15.
Ivakin, B. N. (1950) *Tr. Geofiz. In-ta AN SSSR*, 9 (136), 84-121.
Ivakin, B. N. (1960) *Izv. AN SSSR. Ser. Geofiz.*, 8, 1149-1167.
Jolly, R. N. (1953) *Geophysics*, **18**, 3, 662.
Kallaur, T. N. (1971) In *Izuchenie seismicheskoi opasnosti* (Studying seismic danger), pp. 42-46. Tashkent: Fan.
Kamenobrodskii, A. G. (1974) *Izv. AN SSSR. Fizika Zemli*, 6, 29-38.
Karataev, G. I. (1969) *Geologiya i Geofizika*, 10, 33-50.
Karnik, V. (1965a) *Stud. geophys. et geod.*, **9**, 3, 236-249.
Karnik. V. (1965b) *Stud. geophys. et geod.*, **9**, 4, 341-349.
Karnik, V. (1968) In *Seismicity of the European Area*, Pt. 1, pp. 139-144. Acad. Publ. House, Czechosl. Acad. Sci.
Kasahara, K. (1957 *Bull. Earthquake Res. Inst. Tokyo Univ.*, **35**, 473-572.
Kasahara, K. (1959) *Bull. Earthquake Res. Inst. Tokyo Univ.*, **37**, pt. 1, 39-52.
Kasahara, K. (1967) *Proc. Jap. Acad.*, **43**, 6, 483-488.
Kasahara, K. (1969) *Bull. Dominion Obs.*, **37**, 7, 187-189.
Katok, A. P. (1965) In *Dinamika zemnoi kory* (Dynamics of the Earth's crust), pp. 15-26. Moscow: Nauka.
Kawasumi, H. (1937) *Publ. Bur. Centr. Seismol. Intern. Ser. A.*, **15**, Pt. 2.
Keilis-Borok, V. I. (1950a) *Tr. Geofiz. In-ta AN SSSR*, 9 (136), 3-19.
Keilis-Borok, V. I. (1950b) *Tr. Geofiz. In-ta AN SSSR*, 9 (136), 20.
Kelley, D. (1939) *Geophysics*, **4**, 1, 69-75.

Keylis-Borok, V. I. (1959) *Ann. geofis.*, **12**, 205-214.
Kharin, D. A. and Masarskii, S. I. (1954) *Tr. Geofiz. In-ta AN SSSR*, 25 (152), 97-112.
Kharin, D. A., Keilis-Borok, V. I., and Kogan, S. D. (1953) *Tr. Geofiz. In-ta AN SSSR*, 21 (148), 27-42.
Khattri, K. N. (1969) *Bull. Seismol. Soc. Amer.*, **59**, 615-630.
Khattri, K. N. (1972) *J. Geophys. Res.*, **77**, 11, 2062-2071.
Knopoff, L. (1952) *Bull. Seismol. Soc. Amer.*, **42**, 4, 307-308.
Knopoff, L. (1958) *Geophys. J.*, **1**, 44-52.
Knopoff, L. (1971) *Revs. Geophys. and Space Phys.*, **9**, 1, 175-188.
Knopoff, L. and Burridge, R. (1966) *Geol. Soc. Amer. Spec. Pap.*, 87, 265.
Knopoff, L. and Gilbert, F. (1959) *Bull. Seismol. Soc. Amer.*, **49**, 2, 163-178.
Kochetkov, V. M. (1977) In *Rol' riftogeneza v geologicheskoi istorii Zemli* (The role of rifting in the Earth's geological history), pp. 125-129. Novosibirsk: Nauka.
Kochetkov, V. M., Borovik, N. S., Leontyeva, L. R., *et al.* (1977) In *Seismichnost' i seismologiya Vostochoi Sibiri* (Seismicity and seismology of Eastern Siberia), pp. 62-73. Novosibirsk: Nauka.
Kogan, S. Ya. and Parkhomenko, I. S. (1971) In *Interpretatsiya i obnaruzhenie seismicheskikh voln v neodnorodnykh sredakh* (Interpretation and detection of seismic waves in heterogeneous media), pp. 107-125. Moscow: Nauka.
Kogan, L. A. and Shakirova, G. D. (1974) In *Voprosy kolichestvenoi otsenki seismicheskoi opasnosti* (Questions of a quantitative appraisal of seismic danger), pp. 132-139. Moscow: Nauka.
Köhler, R. (1939) *Ztschr. Geophys.*, 1, 69-75.
Kokesh, F. P. (1952) *Geophysics*, **17**, 3, 560.
Kondorskaya, N. V., Solovieva, O. N., Zakharova, A. I., *et al.* (1979) *Tectonophysics*, **53**, 203-215.
Koridalin, E. A. (1939) *Izuchenie stroeniya zemnoi kory seismicheskimi metodami* (Studying the structure of the Earth's crust by seismic methods). Moscow: Izd. AN SSSR.
Korolev, F. A. (1937) *DAN SSSR*, **15**, No. 1.
Korolev, F. A. (1938) *DAN SSSR*, **20**, No. 7/8.
Korolev, F. A. (1941) *Zh. Eksp. i Teor. Fiz.*, **11**, No. 1.
Kosminskaya, I. P. (1952) *Izv. AN SSSR. Ser. Geofiz.*, 4, 33-54.
Kosminskaya, I. P. (1968) *Metod glubinnogo seismicheskogo zondirovaniya zemnoi kory i verkhov mantii* (A method of deep seismic sounding of the Earth's crust and mantle top). Moscow: Nauka.
Kosminskaya, I. P. (ed.) (1982) *Razvitie idei Gamburtseva v geofizike* (Development of Gamburtsev's ideas in geophysics). Moscow: Nauka.
Kostrov, B. V. (1974) *Izv. AN SSSR. Fizika Zemli*, 1, 23-40.
Kostrov, B. V. (1975) *Mekhanika ochaga tektonicheskogo semletryaseniya* (Mechanics of the source of a tectonic earthquake). Moscow: Nauka.
Kövesligethy, N. (1907) *Gerlands Beitr. Geophys.*, **8**, 22-29.
Kuchai, O. A. (1978) In *Rezul'taty kompleksnykh geofizicheskikh issledovanii v seismoaktivnykh zonakh* (Results of complex geophysical studies in seismically active zones), pp. 159-180. Moscow: Nauka.
Kuchai, O. A., Kuchai, V. K., and Soboleva, O. V. (1975) *Dokl. AN TadzhSSR*, **18**, 9, 24-27.
Kuchai, V. K. (1980) In *Tr. Mezhdunar. geol. kongr., XXVI ses. Dokl. sov. geologov* (Transactions of Int. geol. congr., 26th ses. Reports of Soviet geologists), pp. 78-86, Moscow: Nauka.
Kuchai, V. K. and Trifonov, V. G. (1977) *Geotektonika*, 3, 91-105.
Kuchai, V. K., Pevnev, A. K., and Guseva, T. V. (1979) *Izv. AN SSSR. Fizika Zemli*, 8, 36-44.
Kuliev, F. T. and Kasparov, V. A. (1974) In *Regional'nye issledovaniya seismicheskogo rezhima* (Regional studies of seismic conditions), pp. 232-238. Kishinev: Shtiintsa.
Kuznetsov, G. N., Butko, M. N., Filippova, A. A., and Shklyarskii, M. F. (1959) *Izuchenie proyavleniya gornogo davleniya na modelyakh* (Studying the signs of rock pressure using models). Moscow: Ugletekhizdat.
Kuznetsova, K. I. (1969) *Zakonomernosti razrusheniya uprugo-vyazkikh tel i nekotorye vozmozhnosti prilozheniya ikh k seismologii* (The laws of failure of visco-elastic bodies and possibilities of applying them to seismology). Moscow: Nauka.
Kuznetsova, K. I. (1976) In *Issledovaniya po fizike zemletryasenii* (Studies in earthquake physics), pp. 114-126. Moscow: Nauka.
Kuznetsova, K. I., Aptekman, Zh. Ya., Shebalin, N. V., and Shteinberg, V. V. (1976) In *Issledovaniya po fizike zemletryasenii* (Studies in earthquake physics), pp. 94-113. Moscow: Nauka.
Landau, L. D. and Lifshits, V. M. (1953) *Mekhanika sploshnykh sred* (Continuum mechanics), 2nd ed. Moscow: Gostekhteoretizdat.

Landau, L. D. and Lifshits, V. M. (1965) *Teoreticheskaya fizika* (Theoretical physics), Vol. 7. Theory of elasticity. 3rd ed. Moscow: Nauka.

Latter, A. L., Martinelli, E. A., and Taller, E. (1959) *Phys. Fluids*, **2**, 3, 28-82.

Legge, J. A., and Rupnik, J. J. (1943) *Geophysics*, **8**, No. 4.

Linde, A. T. and Sacks, I. S. (1972) *J. Geophys. Res.*, **77**, 8, 1439-1451.

Lukk, A. A. and Yunga, S. L. (1980) *Izv. AN SSSR. Fizika Zemli*, 4, 39-50.

Lukk, A. A., Nersesov, I. L., Pevnev, A. K., *et al.* (1980) *Izv. AN SSSR. Fizika Zemli*, 5, 32-41.

Lyav, A. (1953) *Matematicheskaya teoriya uprugosti* (The mathematical theory of elasticity). Moscow: ONTI.

Malecki, I. and Koltonski, W. (1955) *Arch. gornitstwa i hutnictwa*, **3**, 2, 157.

Malov, N. N. (1935) *Usp. Fiz. Nauk*, **15**, No. 1.

Maruyama, T. (1963) *Bull. Earthquake Res. Inst. Tokyo Univ.*, **41**, pp. 467-486.

Maruyama, T. (1964) *Bull. Earthquake Res. Inst. Tokyo Univ.*, **42**, pp. 289-368.

Masarskii, S. I. and Gorbunova, I. V. (1964) In *Eksperimental'naya seismika* (Experimental seismicity), pp. 94-137. Moscow: Nauka.

McCollum, E. V. (1933) *U.S. Patent 1 923 107.*

McCollum, E. V. and La Rue (1931) *Bull. Amer. Assoc. Petrol. Geol.*, **15**, 12, 1409.

Medvedev, S. V. (1947) *Tr. Seismol. In-ta AN SSSR*, 119, 83-85.

Medvedev, S. V. (1962) *Inzhenernaya seismologiya* (Engineering seismology), Moscow: Gosstroizdat.

Medvedev, S. V. (1968a) In *Seismicheskoe raionirovanie SSSR* (Seismic division of the USSR), pp. 151-163. Moscow: Nauka.

Medvedev, S. V. (ed.) (1968b) *Seismicheskoe raionirovanie SSSR* (Seismic division of the USSR). Moscow: Nauka.

Mikhailov, I. G. (1949) In *Sovremennye problemy fiziki* (Modern problems of physics). Moscow.

Mikhailova, R. S. (1979) In *Seismicheskaya sotryasaemost' territorii SSSR* (Seismic tremors in the USSR by territory), pp. 113-124. Moscow: Nauka.

Mikhailova, R. S. and Bibarsova, D. G. (1977) *Izv. AN TadzhSSR. Otd. Fiz.-Mat., Khim. i Geol. Nauk*, 1, 39-48.

Mikumo, T. (1969) *J. Phys. Earth*, **17**, 169-192.

Mikumo, T. (1971) *J. Phys. Earth*, **19**, 303-320.

Misharina, L. A., Solonenko, N. V., and Leontyeva, L. R. (1975) In *Baikal'skii rift* (The Baikal rift), pp. 9-21. Novosibirsk: Nauka.

Mitropolskii, A. K. (1961) *Tektonika statisticheskikh vychislenii* (The tectonics of statistical calculations). Moscow: Fizmatgiz.

Mogi, K. (1962) *Bull. Earthquake Res. Inst. Tokyo Univ.*, **40**, 125-173.

Mohorovičic, A. (1910) *Jahr Meteorol. Observatorium Zagreb für das Jahr 1909*, Vol. 4. Zagreb.

Molnar, P. (1972) *Geophys. and Planet Phys.*

Moskvina, A. G. (1969) *Izv. AN SSSR. Fizika Zemli*, 6, 3-10.

Mozzhenko, A. N. (1955) *Razvedochnaya i promyslovaya Geofizika*, vyp. 13, 40-42.

Narvarte, P. E. (1946) *Geophysics*, **11**, 1, 66.

Nersesov, I. L. (1967) In *Problemy geofiziki Srednei Azii i Kazakhstana* (Problems of geophysics of Central Asia and Kazakhstan). Moscow: Nauka.

Nersesov, I. L., and Riznichenko, Yu. V. (1959) In *Seismicheskie i glyatsiologicheskie issledovaniya v period Mezhdunarodnogo geofizicheskogo goda* (Seismic and glaciologic studies during the International Geophysical Year), pp. 31-38. Moscow: Izd. AN SSSR.

Nersesov, I. L. and Simbirtseva, I. G. (1969) In *Tr. III Vsesoyuz. simpoz. po seismicheskomu rezhimu* (Transactions of 3rd All-Union symposium on seismic conditions), pp. 102-127. Novosibirsk: Nauka.

Nersesov, I. L., Grin, V. P., and Dzhanuzakov, K. O. (1960) *O seismicheskom raionirovanii basseina reki Naryn* (On the seismic division of the Naryn River basin). Frunze: Izd. AN KirgSSR.

Nikolaev, N. I. (ed.) (1973) *Karta sovremennykh vertikal'nykh dvizhenii zemnoi kory Vostochnoi Evropy. M-b 1:250000* (Map of modern vertical movements of Earth's crust in Eastern Europe. Scale 1:250000). Moscow: GUGK USSR.

Nortwood, T. D. and Anderson, D. V. (1953) *Bull. Seismol. Soc. Amer.*, **43**, 3, 239.

Novozhilov, V. V. (1958) *Teoriya uprugosti* (The theory of elasticity). Moscow: Sudpromgiz.

Obert, L. (1939) *Report of Investigations. U.S. Bureau of Mines*, April, p. 3444.

O'Brien, P. N. (1955) *Geophysica*, **20**, 2, 227.

Oliver, J. (1956) *Earthquake Notes Seismol. Soc. Amer.*, **27**, 4, 427.

Oliver, J., Press, F., and Ewing, M. (1954) *Geophysics*, **19**, 2, 328.

Olson, W. S. (1941) *Bull. Amer. Assoc. Petrol. Geol.*, **25**, 7, 1343.

Ostrovskii, A. E. (1937) *DAN SSSR*, **17**, 7, 353-356.

Ostrovskii, A. E. (1944) *DAN SSSR*, **45**, 5, 196-199.

Pasechnik, I. P. (1956) *Izv. AN SSSR. Ser. Geofiz.*, 3, 285-289.

Petrescu, G., and Radu, C. (1961) *Stud. si cercet. astron. si seismol.*, 1, 225-247.

Pevnev, A. K., Guseva, T. V., Odinev, P. P., and Saprykin, G. V. (1978) In *Sovremennye dvizheniya zemnoi kory* (Modern Movements of Earth's crust), pp. 86-92. Novosibirsk: Nauka.

Plotnikova, L. M. (1968) *Izv. AN UzSSR. Ser. Tekhn. Nauk*, No. 4.

Polyakov, S. V. (ed.) (1973) *Sovremennoe sostoyanie teorii seismostoikosti i seismostoikie sooruzheniya* (Modern state of siesmic resistance theory and seismic resistant structures). Moscow: Stroiizdat.

Popov, A. D., Ruppeneit, K. V., and Liberman, Yu. M. (1959) *Gornoe davlenie v ochistnykh i podgotovitel'nykh vyrabotkakh* (Rock pressure in finishing and preparatory drifts). Moscow: Gosgortekhizdat.

Press, F. (1957) *Geophysics*, **22**, 2, 284.

Press, F. (1967) In *Proc. VESIAC Conf. on Shallow Seismic Events in the 3 to 5 Magnitude Range*. Vela Center. Univ. Michigan.

Protodyakonov, M. M. (1931) *Delenie gornykh porod i rudnichnoe kreplenie* (Separation of rock and mine props). Moscow: GONTI.

Pustovitenko, B. G., Kamenobrodskii, A. G., and Kulchitskii, V. E. (1974) In *Voprosy kolichestvennoi otsenki seismicheskoi opasnosti* (Questions of a quantitative appraisal of seismic danger), pp. 38-43. Moscow: Nauka.

Puzyrev, N. N. (1979) *Vremennye polya i effektivnye parametry otrazhennykh vvoln* (Time fields and effective parameters of reflected waves). Novosibirsk: Nauka.

Puzyrev, N. N. (1981) *Geofiz. Zhurn.*, 1, 41-49.

Puzyrev, N. N. (1982) *Izv. AN SSSR. Fizika Zemli*, 10, 5-10.

Radu, C. and Apopei, I. (1977) *Publ. Inst. Geophys. Pol. Acad. Sci.*, Na-5 (116).

Randal, M. J. (1971) *Bull. Seismol. Soc. Amer.*, **61**, 5, 1321-1326.

Ratnikova, L. I. (1973) *Metody rascheta seismicheskikh voln v tonko-sloistykh sredakh* (Methods of calculating seismic waves in thin stratified media). Moscow: Nauka.

Ratnikova, L. N. and Levshin, A. L. (1967) *Izv. AN SSSR. Fizika Zemli*, 2, 41-53.

Rautian, T. G. (1960a) *Tr. TISSS*, 7, 41-96.

Rautian, T. G. (1960b) In *Metody detal'nogo izucheniya seismichnosti* (Methods of the detailed studying of seismicity), pp. 75-114 (No. 176). Moscow: Izd. AN SSSR.

Rayal, A., Duglas, B. M., Melon, S. D., and Savidge, W. I. (1972) *Izv. AN SSSR. Fizika Zemli*, 12, 12-24.

Reid, H. F. (1910) In *Report of the State Investigation Commission 2 (192)*. Washington: Carnegie Inst.

Reid, H. F. (1933) *Bull. Nat. res. Counc.*, No. 90.

Reinhardt, H. C. (1954) *Freiberg Forschungsh. C*, **15**.

Report of the Special Committee on the Geophysical Study of Continents, 1952-1954 (1955) *Trans. Amer. Geophys. Union*, **36**, No. 4.

Rezanov, I. A. (1962) *Byul. MOIP. Otd. Geol.*, No. 37, 1, 25-42.

Richter, C. F. (1935) *Bull. Seismol. Soc. Amer.*, **25**, 32-42.

Richter, C. F. (1948) *Publ. Bur. Central Seismol. Int. Ser. A*, **17**, 217-224.

Richter, C. F. (1958) *Elementary Seismology*. San Francisco: W.H. Freeman & Co.

Ritsema, A. R. (1954) *Indones. J. Natural Sci.*, No. 1/3.

Ritsema, A. R. (1967) *Tectonophysics*, **4**, 247-259.

Ritsema, A. R. (1969) *Geol. Rdsch.*, **59**, 1, 36-56.

Ritsema, A. R. (1974) In *UNESCO survey of the seismicity of the Balkan region, UNDP Project REM/70/172*. De Bilt.

Riznichenko, Yu. V. (1939a) *Izv. AN SSSR. Ser. Geogr. i Geofiz.*, 3, 247-266.

Rizhichenko, Yu. V. (1939b) *Izv. AN SSSR. Ser. Geogr. i Geofiz.*, 3, 267-274.

Riznichenko, Yu. V. (1940) *Izv. AN SSSR. Ser. Geogr. i Geofiz.*, 4, 751-758.

Riznichenko, Yu. V. (1944a) In *Prikladnaya geofizika* (Applied geophysics), pp. 62-76. Moscow: Gostoptekhizdat.

Riznichenko, Yu. V. (1944b) *Izv. AN SSSR. Ser. Geogr. i Geofiz.*, 2/3, 87-92.

Riznichenko, Yu. V. (1945) *Izv. AN SSSR. Ser. Geogr. i Geofiz.*, 1, 11-20.

Riznichenko, Yu. V. (1946) *Tr. ITG AN SSSR*, **2**, 1, 114.

Riznichenko, Yu. V. (1947a) *Izv. AN SSSR. Ser. Geogr. i Geofiz.*, 4, 311-335.

Riznichenko, Yu. V. (1947b) *Izv. AN SSSR. Ser. Geogr. i Geofiz.*, 2, 153-172.

Riznichenko, Yu. V. (1949a) *Sposob teoreticheskikh godografov dlya opredeleniya skorosti po nablyudeniyam otrazhennykh voln* (A method of theoretical hodographs for velocity determination by observing reflected waves). Moscow: Gosgeolizdat.

Riznichenko, Yu. V. (1949b) *Izv. AN SSSR. Ser. Geogr. i Geofiz.*, 2, 115-118.

Riznichenko, Yu. V. (1949c) *Izv. AN SSSR. Ser. Geogr. i Geofiz.*, 6, 518-544.

Riznichenko, Yu. V. (1951) *Izv. AN SSSR. Ser. Geofiz.*, 3, 9-15.

Riznichenko, Yu. V. (1952a) *Izv. AN SSSR. Ser. Geofiz.*, 1, 12-20.

Riznichenko, Yu. V. (1952b) *Vestn. AN SSSR*, 5, 16-20.

Riznichenko, Yu. V. (1955) *Izv. AN SSSR. Ser. Geofiz.*, 6, 538.

Riznichenko, Yu. V. (1956) *Tr. Geofiz. In-ta AN SSSR*, 35, 9-41.

Riznichenko, Yu. V. (1958a) *Izv. AN SSSR. Ser. Geofiz.*, 4, 425-437.

Riznichenko, Yu. V. (1958b) *Izv. AN SSSR. Ser. Geofiz.*, 9, 1057-1074.

Riznichenko, Yu. V. (1959a) *Ann. geofis.*, **12**, 2, 227-237.

Riznichenko, Yu. V. (1959b) *DAN SSSR*, **126**, 4, 759-762.

Riznichenko, Yu. V. (1960a) *Tr. IFZ AN SSSR*, 15, 53-87.

Riznichenko, Yu. V. (ed.) (1960b) *Tr. IFZ AN SSSR*, No. 9.

Riznichenko, Yu. V. (1961) *Vestn. AN SSSR*, 8, 37-44.

Riznichenko, Yu. V. (1962) *Tr. IFZ AN SSSR*, 2, 5-15.

Riznichenko, Yu. V. (1964a) *Izv. AN SSSR, Ser. Geofiz.*, 7, 969-977.

Riznichenko, Yu. V. (1964b) *DAN SSSR*, **159**, 2, 322-323.

Riznichenko, Yu. V. (1964c) *DAN SSSR*, **157**, 6, 1352-1354.

Riznichenko, Yu. V. (1965a) *Izv. AN SSSR. Fizika Zemli*, 10, 7-16.

Riznichenko, Yu. V. (1965b) In *Dinamika zemnoi kory* (Dynamics of the Earth's crust), pp. 56-63. Moscow: Nauka.

Riznichenko, Yu. V. (1965c) *Izv. AN SSSR. Fizika Zemli*, 11, 1-12.

Riznichenko, Yu. V. (1965d) *DAN SSSR*, **161**, 1, 97-99.

Riznichenko, Yu. V. (1966a) *Izv. AN SSSR. Fizika Zemli*, 2, 3-24.

Riznichenko, Yu. V. (1966b) *Izv. AN SSSR. Fizika Zemli*, 5, 16-32.

Riznichenko, Yu. V. (1966c) In *Geoakustika* (Geoacoustics), pp. 3-8. Moscow: Nauka.

Riznichenko, Yu. V. (1967a) In *Problema geofiziki Srednei Azii i Kazakhstana* (Problems of geophysics of Central Asia and Kazakhstan), pp. 36-51. Moscow: Nauka.

Riznichenko, Yu. V. (1967b) In *Papers 9th Assembly ESC, 1-7 Aug. 1966 in Copenhagen*, pp.307-312. Cobenhavn Acad. Forl.

Riznichenko, Yu. V. (1967c) In *Papers 9th Assembly ESC*, pp. 403-411. Cobenhavn Acad. Forl.

Riznichenko, Yu. V. (1968a) *Izv. AN SSSR. Fizika Zemli*, 5, 3-19.

Riznichenko, Yu. V. (1968b) In *Seismicheskoe raionirovanie SSSR* (Seismic division of the USSR), pp. 112-120. Moscow: Nauka.

Riznichenko, Yu. V. (1969a) In *Tr. III Vsesoyuz. simpoz. po seismicheskomu rezhimu* (Transactions of 3rd All-Union symposium on seismic conditions), part I, p. 28. Novosibirsk: Nauka.

Riznichenko, Yu. V. (1969b) *Izv. AN SSSR. Fizika Zemli*, 7, 3-20.

Riznichenko, Yu. V. (1969c) *Zemlya i Vselennaya*, 4, 3-11; 5, 16-24.

Riznichenko, Yu. V. (1970a) *Izv. AN SSSR. Fizika Zemli*, 4, 33-48.

Riznichenko, Yu. V. (1970b) *Bull. geofis. teor ed appl.*, **12**, No. 48.

Riznichenko, Yu. V. (1971a) *Zemlya i Vselennaya*, 5, 2-10.

Riznichenko, Yu. V. (1971b) *Ot maksimal'noi ball'nosti zemletryasenii k spektral'no-vremennoi sotryasaemosti* (From maximum earthquake scale numbers to spectral-time tremoring). Tashkent: Fan.

Riznichenko, Yu. V. (1973a) *DAN SSSR*, **210**, 1, 82-84.

Riznichenko, Yu. V. (1973b) *Umschau*, **73**, 7, 217-218.

Riznichenko, Yu. V. (1975a) *Izv. AN SSSR. Fizika Zemli*, 2, 4-8.

Riznichenko, Yu. V. (1975b) *Izv. AN SSSR. Fizika Zemli*, 10, 21-30.

Riznichenko, Yu. V. (1975c) In *Voprosy kolichestvennoi otsenki seismicheskoi opasnosti* (Questions of a quantitative appraisal of seismic danger), pp. 5-8. Moscow: Nauka.

Riznichenko, Yu. V. (1976a) In *Issledovaniya po fizike zemletryasenii* (Studies in earthquake physics), pp. 236-262. Moscow: Nauka.

Riznichenko, Yu. V. (1976b) In *Issledovaniya po fizike zemletryasenii* (Studies in earthquake physics), pp. 9-27. Moscow: Nauka.

Riznichenko, Yu. V. (1976c) *DAN SSSR*, **226**, No. 2.

Riznichenko, Yu. V. (1977) *Izv. AN SSSR. Fizika Zemli*, 10, 34-47.

Riznichenko, Yu. V. (ed.) (1979) *Seismicheskaya sotryasaemost' territorii SSSR* (Seismic tremors of the USSR territory). Moscow: Nauka.

Riznichenko, Yu. V. (1981) *Izv. AN SSSR. Fizika Zemli*, 11, 3-9.

Riznichenko, Yu. V. and Artamonova, A. M. (1975) *Izv. AN SSSR. Fizika Zemli*, 12, 35-42.

Riznichenko, Yu. V. and Bagdasarova, A. M. (1974) In *Regional'nye issledovaniya seismicheskogo rezhima* (Regional studies of seismic conditions), pp. 50-65. Kishinev: Shtiintsa.

Riznichenko, Yu. V. and Bagdasarova, A. M. (1975) *Izv. AN SSSR. Fizika Zemli*, 11, 14-32.

Riznichenko, Yu. V. and Dzhibladze, E. A. (1972) *Izv. AN SSSR. Fizika Zemli*, 1, 35-49.

Riznichenko, Yu. V. and Dzhibladze, E. A. (1974) *Izv. AN SSSR. Fizika Zemli*, 5, 64-85.

Riznichenko, Yu. V. and Dzhibladze, E. A. (1975) *Izv. AN SSSR. Fizika Zemli*, No. 12.

Riznichenko, Yu. V. and Dzhibladze, E. A. (1976) *Izv. AN SSSR. Fizika Zemli*, 1, 23-31.

Riznichenko, Yu. V. and Glukhov, V. A. (1956) *Izv. AN SSSR. Ser. Geofiz.*, 11, 1258-1268.

Riznichenko, Yu. V. and Gorbunova, I. V. (1967) In *Seismicheskoe raionirovanie SSSR* (Seismic division of the USSR), pp. 71-83, Moscow: Nauka.

Riznichenko, Yu. V. and Kosminskaya, I. P. (1963) *DAN SSSR*, **153**, 323-325.

Riznichenko, Yu. V. and Myachkin, V. I. (1955) *DAN SSSR. N.S.*, **102**, 3, 507.

Riznichenko, Yu. V. and Nersesov, I. L. (1960) *Byul. Soveta po Seismologii AN SSSR*, 8, 36-59.

Riznichenko, Yu. V. and Nersesov, I. L. (1961) *Ann. geofis.*, **15**, 2, 173-186.

Riznichenko, Yu. V. and Seiduzova, S. S. (1972) *Izv. AN SSSR. Fizika Zemli*, 11, 3-7.

Riznichenko, Yu. V. and Seiduzova, S. S. (1974) In *Regional'nye issledovaniya seismicheskogo rezhima* (Regional studies of seismic conditions), pp. 174-188. Kishinev: Shtiintsa.

Riznichenko, Yu. V. and Seiduzova, S. S. (1975a) *Izv. AN SSSR. Fizika Zemli*, 9, 10-16.

Riznichenko, Yu. V. and Seiduzova, S. S. (1975b) In *Voprosy kolichestvennoi otsenki seismicheskoi opasnosti* (Questions of a quantitative appraisal of seismic danger), pp. 100-104. Moscow: Nauka.

Riznichenko, Yu. V. and Seiduzova, S. S. (1976a) In *Geofizicheskie issledovaniya* (Geophysical studies), pp. 36-58. Tbilisi: Metsniereba.

Riznichenko, Yu. V. and Seiduzova, S. S. (1976b) *Izv. AN SSSR. Fizika Zemli*, 3, 28-43.

Riznichenko, Yu. V. and Seiduzova, S. S. (1984) *Spektral'no-vremennaya kharakteristika seismicheskoi opasnosti* (Spectral-time characteristic of seismic danger). Moscow: Nauka.

Riznichenko, Yu. V. and Shamina, O. G. (1957) *Izv. AN SSSR. Ser. Geofiz.*, 7, 855-873.

Riznichenko, Yu. V. and Shamina, O. G. (1960). *Izv. AN SSSR. Serv. Geofiz.*, 12, 1689-1706.

Riznichenko, Yu. V. and Shamina, O. G. (1963a) *Izv. AN SSSR. Ser. Geofiz.*, 2, 223-247.

Riznichenko, Yu. V. and Shamina, O. G. (1963b) *Byul. Soveta po Seismologii*, 15, 11-24.

Riznichenko, Yu. V. and Shamina, O. G. (1964) *Izv. AN SSSR. Ser. Geofiz.*, 8, 1129-1141.

Riznichenko, Yu. V. and Silaeva, O. I. (1955) *Izv. AN SSSR. Ser. Geofiz.*, 3, 193-197.

Riznichenko, Yu. V. and Vinogradov, S. D. (1967) *Issledovanie gornogo davleniya geofizicheskimi metodami* (Studying of rock pressure by geophysical methods). Moscow: Nauka.

Riznichenko, Yu. V. and Zakharova, A. I. (1971) *Izv. AN SSSR. Fizika Zemli*, 3, 29-38.

Riznichenko, Yu. V., Ivakin, B. N., and Bugrov, V. R. (1951) *Izv. AN SSSR. Ser. Geofiz.*, 5, 1-30.

Riznichenko, Yu. V., Ivakin, B. N., and Bugrov, V. R. (1952) *Izv. AN SSSR. Ser. Geofiz.*, 3, 58-69.

Riznichenko, Yu. V., Ivakin, B. N., and Bugrov, V. R. (1953) *Izv. AN SSSR. Ser. Geofiz.*, 1, 26-32.

Riznichenko, Yu. V., Silaeva, O. I., Shamina, O. G., et al. (1956) *Tr. Geofiz. In-ta AN SSSR*, 34 (161), 74-163.

Riznichenko, Yu. V., Shamina, O. G., and Khanutina, R. V. (1961) *Izv. AN SSSR. Ser. Geofiz.*, 4, 497-519.

Riznichenko, Yu. V., Zakharova, A. I., and Seiduzova, S. S. (1967) *DAN SSSR*, **174**,, 4, 830-832.

Riznichenko, Yu. V., Zakharova, A. I., and Seiduzova, S. S. (1969a) *Izv. AN SSSR. Fizika Zemli*, 6, 11-20.

Riznichenko, Yu. V., Pshennikov, K. V., and Zorin, Yu. A. (1969b) *Izv. AN SSSR. Fizika Zemli*, 10, 10-27.

Riznichenko, Yu. V., Bune, V. I., Zakharova, A. I., and Seiduzova, S. S. (1969c) *Izv. AN SSSR. Fizika Zemli*, 8, 3-15.

Riznichenko, Yu. V., Zakharova, A. I., and Seiduzova, S. S. (1969d) *Bull. geofis. teor. ed appl.*, **11**, 43/44, 227-237.

Riznichenko, Yu. V., Zakharova, A. I., and Seiduzova, S. S. (1970) *Izv. AN SSSR. Fizika Zemli*, 7, 3-19.
Riznichenko, Yu. V., Zakharova, A. I., and Seiduzova, S. S. (1971) In *Tashkentskoe zemletryasenie* (The Tashkent earthquake), pp. 370-384. Tashkent: Fan.
Riznichenko, Yu. V., Drumya, A. V., Stepanenko, N. Ya., *et. al.* (1973) *Izv. AN SSSR. Fizika Zemli*, 10, 23-41.
Riznichenko, Yu. V., Drumya A. V., and Stepanenko, N. Ya. (1974) In *Regional'nye issledovaniya seismicheskogo rezhima* (Regional studies of seismic conditions), pp. 10-19. Kishinev: Shtiintsa.
Riznichenko, Yu. V., Drumya, A. V., and Stepanenko, N. Ya. (1975a) In *Voprosy kolichestvennoi otsenki seismicheskoi opasnosti* (Questions of a quantitative appraisal of seismic danger), pp. 32-37. Moscow: Nuaka.
Riznichenko, Yu. V., Butovskaya, E. M., Zakharova, A. I., *et al.* (1975b) In *Voprosy kolichestvennoi otsenki seismicheskoi opasnosti* (Questions of a quantitative appraisal of seismic danger), pp. 73-81. Moscow: Nauka.
Riznichenko, Yu. V., Drumya, A. V., Stepanenko, N. Ya, and Onofrash, N. I. (1975c) *Izv. AN SSSR. Fizika Zemli*, 5, 3-15.
Riznichenko, Yu. V., Dzhibladze, E. A., and Bolkvadze, I. N. (1976) In *Issledovaniya po fizike zemletryasenii* (Studies in earthquake physics), pp. 74-86. Moscow: Nauka.
Riznichenko, Yu. V., Seiduzova, S. S., and Matasova, L. M. (1977a) *Izv. AN SSSR. Fizika Zemli*, 3, 8-20.
Riznichenko, Yu. V., Kochetkov, V. M., Misharina, L. A., and Gileva, N. A. (1977b) *Izv. AN SSSR. Fizika Zemli*, 11, 41-53.
Riznichenko, Yu. V., Drumya, A. V., Stepanenko, N. Ya., *et al.* (1980) *Izv. AN SSSR. Fizika Zemli*, 11, 10-21.
Riznichenko, Yu. V., Kuliev, F. T.., Ismail-Zade, T. A., and Rutman, A. N. (1982a) *Izv. AN SSSR. Ser. Nauk o Zemle*, 1, 25-31.
Riznichenko, Yu. V.. Soboleva, O. V., Kuchai, O. A., *et al.* (1982b) *Izv. AN SSSR. Fizika Zemli*, 10, 90-104.
Rodin, I. V. (1957) *Tr. Dal'nevostochnogo politekhn. In-ta* (Transactions of the Far-Eastern polytechnical institute), Vol. 1, vyp. 1.
Romney C. F. (1959) *J. Geophys.*, **64**, 10, 1489-1498.
Rozova, E. A. (1939) *Tr. Seismol. In-ta AN SSSR*, 94, 75.
Rozova, E. A. (1947) *Tr. Seismol. In-ta AN SSSR*, 123, 119.
Ruppeneit, K. V. (1957) *Davlenie i smeshchenie gornykh porod v lavakh pologopadayushchikh plastov* (Pressure and displacement of rock in downwalls of gently dipping beds). Moscow: Ugletekhizdat.
Ruprechtova, L. (1960) *Geofyz., sb.*, 99, 125.
Rykunov, L. N., Khorosheva, V. V., and Sedov, V. V. (1960) *Izv. AN SSSR. Ser. Geofiz.*, 11, 1601.
Rytov, S. M. (1935) *Zh. Eksp. Teor. Fiz.*, 5, No. 9.
Rytov, S. M. (1936) *DAN SSSR*, 3, (12), No. 4 (99).
Rytov, S. M. (1937) *Izv. AN SSSR. Ser. Geofiz.*, No. 2.
Sadovskii, N. A. (ed.) (1960) *Tr. IFZ AN SSSR*, 15, 108.
Sadovskii, M. A. (ed.) (1980) *Seismicheskoe raionirovanie territorii SSSR* (Seismic division of the USSR territory). Moscow: Nauka.
Savage, J. C. (1972) *J. Geophys. Res.*, **77**, 3788-3795.
Savage, J. C. and Hastie, L. M. J. (1966) *Geophys. Res.*, **71**, 20, 4897-4904.
Savarenskii, E. F. (1956) *Izv. AN SSSR. Ser. Geofiz.*, 7, 745-754.
Savarenskii, E. F. (ed.) (1961) *Zemletryaseniya v SSSR* (Earthquakes in the USSR). Moscow: Izd. AN SSSR.
Savarenskii, E. F. (ed.) (1964-1969) *Zemletryaseniya v SSSR* (Earthquakes in the USSR). Annual editions for 1962-1966. Moscow: Nauka.
Savarenskii, E. F. and Kirnos, D. P. (1955). *Elementy seismologii i seismometrii* (Elements of seismology and seismometry). Moscow: Gostekhteoretizdat.
Schumberger, C. (1940) *U.S. Patent No. 2 191 119*.
Seiduzova, S. S. (1976) *Izv. AN SSSR. Fizika Zemli*, 9, 21-32.
Seiduzova, S. S. and Zakharov, A. I. (1971) In *Izuchenie seismicheskoi opasnosti* (Studying of seismic danger), pp. 21-28. Tashkent: Fan.
Shamina, O. G. and Silaeva, O. I. (1958) *Izv. AN SSSR. Ser. Geofiz.*, 302-316.
Shebalin, N. V. (1955) *Izv. AN SSSR. Ser. Geofiz.*, 4, 377-380.
Shebalin, N. V. (1957) *Byul. Soveta po Seismologii AN SSSR*, 6, 122-126.

Shebalin, N. V. (1959a) *Tr. IFZ AN SSSR*, 5, 100-113.

Shebalin, N. V. (1959b) *Union geodes. et geophys. Intern., Ser. A, Trawaux Sci.*, 20, 31-37.

Shebalin, N. V. (1968) In *Seismicheskoe raionirovanie SSSR* (Seismic division of the USSR), pp. 95-111.

Shebalin, N. V. (1971a) In *Seismicheskie issledovaniya dlya stroitel'stva* (Seismic studies for construction), pp. 50-78. Moscow: Nauka.

Shebalin, N. V. (1971b) *Izv. AN SSSR. Fizika Zemli*, 6, 12-20.

Shebalin, N. V. (1971c) In *Tashkentskoe zemletryasenie 26 aprelya 1966 g.* (The Tashkent earthquake of Aptil 26, 1966), pp. 149-163. Tashkent: Fan.

Shebalin, N. V. (1972) In *Seismichnost', seismicheskaya opasnost' Kryma i seismostoikost' stroitel'stva* (Seismicity, seismic danger in Crimea and seismic resistance in construction), pp. 14-20. Kiev: Nauk. Dumka.

Shengeliya, I. S. (1975) *Tr. In-ta Geofiziki AN SSSR*, 37,

Sherman, S. I. (1974) *Neotektonika i seismichnost' Pribaikal'ya* (Neotectonics and seismicity of Baikal region). Irkutsk.

Shevyakov, L. D. (1953) *Ugol'*, No. 12.

Shirokova, E. I. (1974) *Izv. AN SSSR. Fizika Zemli*, 11, 22-36.

Shirokova, E. I. (1979) *Izv. AN SSSR. Fizika Zemli*, 10, 44-57.

Shraiber, D. S. (1948) *Zavod. Lab.*, 14,, No. 3.

Shteinberg, V. V. (1971) In *Seismicheskoe raionirovanie Ulan-Batora* (Seismic regioning of Ulan Bator), pp. 132-193. Moscow: Nauka.

Shteinberg, V. V., Levshin, A. L., Aptekman, Zh. Ya., and Grudeva, N. P. (1974) *Izv. AN SSSR. Fizika Zemli*, 2, 3-14.

Silaeva, O. I. and Shamina, O. G. (1958) *Izv. AN SSSR. Ser. Geofiz.*, 1, 32-45.

Simonova, N. A. (1975) In *Novye dannye po seismichnosti i tektonike territorii Moldavskoi SSR* (New data on the seismicity and tectonics of the Moldavian SSR territory), pp. 43-51. Kishinev: Shtiintsa.

Slesarev, V. D. (1948) *Mekhanika gornykh porod* (Rock mechanics). Moscow: Ugletekhizdat.

Slotnick, M., Brooks, J. A., and Redding, V. L. (1950) *Geophysics*, 15, 4, 663.

Smirnov, N. V. and Dunin-Barkovskii, I. V. (1959) *Kratkii kurs matematicheskoi statistiki dlya tekhnicheskikh prilozhenii* (A brief course in mathematical statistics for technical applications). Moscow: Fizmatgiz.

Soboleva, O. V., Shklyar, G. P., and Blagoveshchenskaya, E. E. (1974) In *Poiski predvestnikov zemletryasenii na prognosticheskikh poligonakh* (Searches for foreshocks on prognostic grounds), pp. 65-70. Moscow: Nauka.

Soboleva, O. V., Kuchai, O. A., Shklyar, G. I., and Blagoveshchenskaya, E. E. (1980) *Katalog fokal'nykh mekhanizmov zemletryasenii Tadzhikistana i Severnogo Afganistana* (Catalogue of focal mechanisms of earthquakes in Tadzhikistan and Northern Afghanistan), manuscript dep. in VINITI, No. 3567-80 Dep.

Soboleva, O. V., Bibarsova, D. G., and Vakhidova, Z. M. (1981) *Raschet parametrov seismotektonicheskoi deformatsii* (Calculation of parameters of seismotectonic deformation), manuscriprt dep. in VINITI, No. 5402-81 Dep.

Sokolov, P. T. (1931) *Tr. Gl. Geol.-Razved. Upr. VSNKh SSSR*, vyp. 17.

Sokolov, S. Ya. (1948) *Zavod. Lab.*, 14, No. 11.

Sokolov, S. Ya. (1954) Priroda, 3, 21-34.

Sollogub, V. B. (1954) *Geol. Zhurn. AN UkrSSR*, 14, 1, 71.

Sologub, V. B., Guterkh, A., Prosen, D., *et al.* (1978) *Stroenie zemnoi kory i verkhnei mantii Tsentral'noi i Vostochnoi Evropy* (Structure of Earth's crust and upper mantle in Central and Eastern Europe). Kiev: Nauk. Dumka.

Solonenko, V. P., Khromovskikh, V. S., Pavlov, O. V., *et al.* (1968) *Seismotektonika i seismichnost' riftovoi zony Pribaikal'ya* (Seismotectonics and seismicity of the Baikal region rift zone). Moscow: Nauka.

Solonenko, V. P. (ed.) (1981) *Seismogeologiya i detal'noe seismicheskoe raionirovanie Pribaikal'ya* (Seismogeology and detailed seismic division of Baikal region). Novosibirsk: Nauka.

Solovyev, S. L. (1955) *Tr. Geofiz. In-ta AN SSSR*, 30, 3-21.

Somov, V. I. (1974) *Geotektonika*, 6, 97-104.

Spivak, G. V. (1945) *Spetsial'nyi fizicheskii praktikum* (Special practical work in physics), Vol. 1. Moscow: Gostekhizdat.

Sponheuer, W. (1953) *Freiberg. Forschungsh. Reihe C*, Heft 7, 15-25.

Sponheuer, W. (1960) *Freiberg. Forschungsh. C*, **88**, 117.

Sponheuer, W. and Maaz, R. (1971) *Communa Observ. roy Belg. Sér. géophys.*, 101, 85-87.

Stishov, S. M. (1963) *Izv. AN SSSR. Ser. Geofiz.*, 1, 42-48.

Stulken, E. J. (1941) *Geophysics*, **6**, 4, 327.

Summers, G. C. and Broding, R. A. (1952) *Geophysics*, **17**, 3, 598.

Tan Gotsyuan (1961) *Tr. IFZ AN SSSR*, 17, 74-86.

Tapponier, P. and Molnar, P. (1976) *Nature*, **264**, 5584, 319-324.

Thatcher, W. R. (1972) *J. Geophys. Res.*, **77**, 3, 1549-1569.

Thatcher, W. and Hanks, T. C. (1973) *J. Geophys. Res.*, **78**, 35, 8547-8576.

Thomson, K. C. and Haskell, N. A. (1972) *Bull. Seismol. Soc. Amer.*, **62**, 3, 675-697.

Timoshenko, S. P. (1934) *Teoriya uprugosti* (The theory of elasticity). Moscow: Gostekhteorizdat.

Timoshin, Yu. V. (1954) Krivolineinye otrazhayushchie granitsy v seismorazvedke (Curved reflecting boundaries in seismic survey). *Synopsis of Candidate's thesis*. Lvov Polytechnical Inst.

Tocher Don (1958) *Bull. Seismol. Soc. Amer.*, **48**, 2, 147-153.

Treskov, A. A. (1948) *DAN SSSR*, **61**, No. 2.

Trifunac, M. D. (1971) *Bull. Seismol. Soc. Amer.*, **61**, 2, 343-356.

Trifunac, M. D. (1972a) *Bull. Seismol. Soc. Amer.*, **62**, 721-750.

Trifunac, M. D. (1972b) In *Analysis of strong motion earthquake accelograms*, vol. 3, Pt. A., p. 272. Pasadena: Cal. Inst. Technol. Earthquake Eng. Res. Lab.

Trumbachev, V. F. (1956) *Raspredelenie napryazhenii vokrug gornykh vyrabotok* (Stress distribution around underground workings). Moscow: Ugletekhizdat.

Tsareva, N. V. (1956) *Izv. AN SSSR. Ser. Geofiz.*, 9, 1044-1053.

Tsimbarevich, P. M. (1948) *Mekhanika gornykh porod* (Rock mechanics). Moscow: Ugletekhizdat.

Tsuboi, C. (1940) *Proc. Imp. Acad. Jap.*, **16**.

Tsuboi, C. (1956) *J. Phys. Earth*, **4**, 2, 63-66.

Tsuboi, C (1959) *J. Phys. Earth*, **6**, 2, 51-55.

Tsuboi, C. (1964) *J. Phys. Earth*, **12**, 2, 25-36.

Tsvetaev, A. A. (1948) *Prikl. Geofizika*, vyp. 5.

Tvaltvadze, G. K. (1965) *Izv. AN SSSR. Ser. Geogr. i Geofiz.*, No. 1.

Usami, T. (1966) *Bull. Earthquake Res. Inst. Tokyo Univ.*, **44**, Pt. 2, 1571-1622.

Uspenskaya, T. A. (1971) In *Izuchenie seismicheskoi opasnosti* (Studying seismic danger), pp. 79-83. Tashkent: Fan.

Utsu, T. and Seki, A. (1955) *Zisin*, **11**, No. 7.

Uzhik, P. V. (1950) *Soprotivlenie otryvu i prochnost' metallov* (Cleavage strength and durability of metals). Moscow.

Vanek, J. and Stelnzer, I. (1962) *Gerlands Beitr. Geophys.*, **71**, 2, 105-119.

Vasilyev, Yu. I. and Shcherbo, M. N. (1965) *Izv. AN SSSR. Fizika Zemli*, 10, 63-71.

Vincent, J. H. (1898) *Philos. Mag.*, **46**, No. 4.

Vinogradov, S. D. (1959) *Izv. AN SSSR. Ser. Geofiz.*, 12, 1850-1852.

Vinogradov, S. D. (1962) *Izv. AN SSSR. Ser. Geofiz.*, 2, 171-180.

Vinogradov, S. D. (1963) *Izv. AN SSSR. Ser. Geofiz.*, 4, 501-512.

Vinogradov, S. D., Mirzoev, K. M., and Solomov, N. G. (1973) *Izv. AN SSSR. Fizika Zemli*, 3, 32-39.

Vogel, C. B. (1952) *Geophysics*, **17**,, 3, 586.

Volarovich, M. P. and Balashov, D. B. (1956) *Tr. Geogiz. In-ta AN SSSR*, 34 (161), 164-178.

Voyutskii, V. S. (1937) *Byul. Neft. Geofiziki*, 4, 18-24.

Voyutskii, V. S. (1938) *Tr. VKGR*, 12 (19), 132-138.

Vvedenskaya, A. V. (1950) Ob osobennostyakh ochagov i godografov chetyrekh krupnykh zemletryasenii Srednei Azii (On the features of the sources and hodographs of four major Central Asian earthquakes). *Synopsis of Candidate's thesis*. Moscow.

Vvedenskaya, A. V. (1969) *Issledovanie napryazhenii i razryvov v ochagakh zemletryasenii pri pomoshchi teorii dislokatsii* (Studying stresses and ruptures in earhquake sources by dislocation theory). Moscow: Nauka.

Vvedenskaya, N. A. (1954) *Izv. AN SSSR. Ser. Geofiz.*, 6, 497-514.

Vvedenskaya, N. A. (1961a) In *Zemletryaseniya v SSSR* (Earthquakes in the USSR), pp. 278-313. Moscow: Izd. AN SSSR.

Vvedenskaya, N. A. (1961b) *Tr. IFZ AN SSSR*, 17 (184), 119-127.
Vvedenskaya, N. A. (1962) *Tr. IFZ AN SSSR*, 22 (189), 25-45.
Wells, R. J. (1949) *Geophysics*, **14**, 3, 346.
West, G. D. (1934) *Proc. Phys. Soc.*, **46** (2), 253.
West, S. (1953) In *Voprosy seismicheskoi razvedki* (Questions of seismic survey), pp. 84-91. Moscow: Izd. In. Lit.
White, D. E. and Sengbusch, R. L. (1963) *Geophysics*, **18**, 4, 1001-1019.
Wood, A. B. (1946) *A textbook of Sound*. London.
Wu, F. T. and Ben-Menachem, A. (1965) *J. Geophys. Res.*, **70**, 3943-3949.
Wyss, M. A. (1970) *J. Geophys. Res.*, **75**, 8, 1529-1544.
Wyss, M. and Brune, J. N. (1968) *J. Geophys. Res.*, **73**, 14, 4681-4694.
Wyss, M. and Hanks, T. S. (1972) *The Source Parameters of the Borrago Mountain, California, Earthquake*. U.S. Geol. Surv. Prof. Pao.
Yakovleva, I. B. (1974) In *Issledovanie seismicheskogo rezhima* (Studying seismic conditions), pp. 27-36. Kishinev: Shtiintsa.
Zakharova, A. I. (1964) *Dokl. AN UzSSR*, 3, 43-47.
Zakharova, A. I. (1972) *Raschet parametrov seismicheskogo rezhima na EVM* (Calculations of seismic condition parameters in computer). Tashkent: Fan.
Zakharova, A. I. (1974) In *Issledovanie seismicheskogo rezhima* (Studying seismic conditions). Kishinev: Shtiintsa.
Zakharova, A. I. and Seiduzova, S. S. (1969a) In *Seismicheskii rezhim* (Seismic conditions). Dushanbe: Donish.
Zakharova, A. I. and Seiduzova, S. S. (1969b) *Izv. AN SSSR. Fizika Zemli*, 7, 66-69.
Zakharova, A. I. and Seiduzova, S. S. (1971) In *Izuchenie seismicheskoi opasnosti* (Studying seismic danger), pp. 28-31. Tashkent: Fan.
Zakharova, A. I., Ibragimov, R. N., Iodko, V. K., *et al*. In *Zemletryaseniya v SSSR v 1968 g.* (Earthquakes in the USSR in 1968), pp. 59-91. Moscow: Nauka.
Zalesskii, B. V. and Belikov, B. P. (1948) *Tr. IGD AN Ser. Petrogr.*, vyp. 89, 28, 78-126.
Zapolskii, K. K. (1971) In *Eksperimental'naya seismologiya* (Experimental seismology), pp. 20-36. Moscow: Nauka.
Zhurkov, S. N. (1957) *Vestn. AN SSSR*, 11, 78-82.
Zhurkov, S. N. and Narzullaev, B. N. (1953) *Zh. Teor. Fiz.*, **23**, 10, 119.
Zisman, W. A. (1933) *Proe. Nat. Acad. Sci. US*, **19**, 666.
Zorin, Yu. A. and Novoselova, M. R. (1974) In *Voprosy kolichestvennoi otsenki seismicheskoi opasnosti* (Questions of a quantitative appraisal of seismic danger), pp. 82-85. Moscow: Nauka.
Zverev, S. M. and Galperin, E. I. (eds.) (1962) *Glubinnoe seismicheskoe zondirovanie zemnoi kory v SSSR* (Deep seismic sounding of Earth's crust in the USSR). Leningrad: Gostoptekhizdat.
Zverev, S. M. and Kapustin, N. N. (1980) *Seismicheskoe issledovanie litosfery Tikhogo okeana* (Seismic studying of the Pacific Ocean lithosphere). Moscow: Nauka.
Zverev, S. M. and Kosminskaya, I. P. (eds.) (1980) *Seismicheskie modeli litosfery osnovnykh geostruktur territorii SSSR* (Seismic models of lithosphere of the basic geological structures on USSR territory). Moscow: Nauka.

Index